W9-ADA-460

Reactive Molecules

REACTIVE MOLECULES

The Neutral Reactive Intermediates in Organic Chemistry

CURT WENTRUP
Department of Chemistry
University of Marburg
West Germany

A Wiley-Interscience Publication
JOHN WILEY & SONS
New York • Chichester • Brisbane • Toronto • Singapore

Library of Congress Cataloging in Publication Data:

Wentrup, Curt.
Reactive molecules.

"A Wiley-Interscience publication."
Includes bibliographical references and index.
1. Chemistry, Physical organic. 2. Chemical reaction,
Conditions and laws of. 3. Spectrum analysis. I. Title.
QD476.W44 1984 547.1′39 83-16824
ISBN 0-471-87639-9

Printed in the United States of America

10 9 8 7 6 5 4 3 2 1

PREFACE

The purpose of this book is to provide a detailed overview of the chemistry of neutral reactive intermediates—radicals, carbenes, nitrenes, strained rings, and antiaromatics—with an emphasis on their intrinsic nature, the "things in themselves" in the words of Kant.† The reactions of reactive intermediates often appear bewildering at first sight; at other times, they are deceptively simple. The road to a complete comprehension is, therefore, long and arduous. Entirely erroneous conclusions have been drawn from apparently sound experimental evidence in the past, and this will certainly continue in the future. It would be foolish to believe that we now know the absolute truth. However, I thought it worthwhile to provide a textbook that not only describes the chemistry of each major class of reactive intermediates, from the basic principles to the present state of research, but also presents details of the ways in which the knowledge was obtained.

Therefore, experimental methods, for example, ESR spectroscopy, the CIDNP effect, and other spectroscopic methods, are described in considerable detail. This will enable the uninitiated reader to interpret and evaluate the information obtainable using such techniques. Likewise, basic principles of frontier molecular orbital theory and the estimation of thermochemical properties are set forth in the first chapter and used in the main text wherever appropriate.

The book addresses itself primarily to the advanced undergraduate and graduate student of organic chemistry. However, much material that is not normally required in examinations is included, and this together with fairly extensive references to the original literature and to review articles should make the book useful for anybody who wishes to embark on research on reactive intermediates.

Even so, a textbook of this format cannot possibly cover the entire existing literature. In order to include some material which, for reasons of space, could not

†"Dieses Experiment der reinen Vernunft hat mit dem der Chymiker, welches sie mannigmal den Versuch der Reduktion, im allgemeinen aber das synthetische Verfahren nennen, viel Ähnliches. Die Analysis des Metaphysikers schied die reine Erkenntnis a priori in zwei sehr ungleichartige Elemente, nämlich die der *Dinge als Erscheinung,* und dann der *Dinge an sich selbst.*"

(This experiment in pure reasoning has much in common with that of the chemist, often referred to as reduction, but more generally known as the synthetic method. The analysis of the metaphysicist separated pure cognition a priori into two very disparate elements, namely that of the *things as appearances,* and then the *things in themselves.*) Immanuel Kant, *Critik der reinen Vernunft,* J. F. Hartknoch, Riga, 1787 p. XXI.

be extensively described in the main text, each chapter except the first, introductory one, terminates with a Problems section. While some of the problems have a direct bearing on the main themes, others are intended to extend the range of reactions covered and therefore can be quite demanding. The answers to the problems may be found using the references to the original literature.

The present volume is based in an earlier paperback (C. Wentrup, *Reaktive Zwischenstufen,* Vol. I), published by Thieme Verlag, Stuttgart, in 1979. The book has been entirely rewritten, extended, and updated. The limitation to *neutral* reactive intermediates is not a strict one: Radical ions, for example, are included in the radical chapter. A sequel on *Charged Reactive Intermediates* may be published at a later date.

The use of units and symbols follows the recommendations of The Symbols Committee of The Royal Society, London, 1975, with addenda 1981, with the exception that the kcal/mol (1 cal = 4.184 J) is retained as the energy unit. There are many reasons for this violation of the International System of Units (SI). Prime among them are the facts that most of the energy data quoted here were reported in kcal/mol in the original literature, that most English language journals continue to use the kcal/mol, and that most readers will be familiar with this unit. In order to facilitate the transition from the ''old'' to the ''new'' literature, these pages are followed by a list of units and constants, giving numerical values in both SI and other commonly used units.

CURT WENTRUP

Marburg, West Germany
January 1984

CONTENTS

UNITS, CONSTANTS, AND ABBREVIATIONS

energy (E)	1 kcal/mol = 4.184 kJ/mol 1 eV = 23.069 kcal/mol = 96.521 kJ/mol
entropy (S)	1 cal K^{-1} mol^{-1} = 1 entropy unit (e.u.) = 4.184 J K^{-1} mol^{-1}
length (l)	1 angstrom (Å) = 10^{-1} nm = 10^{-10} m 1 micron (μm) = 10^{-6} m
magnetic field (**H**)	1 gauss (G) = 10^{-4} tesla (T) = 0.1 mT
pressure (p)	1 Torr = 101325/760 Pa ≈ 133.322368 Pa 1 mm Hg = 13.5951 × 980.665 × 10^{-2} Pa ≈ 133.322387 Pa 1 atm = 101325 Pa 1 bar = 10^5 Pa
temperature (T)	0°C = 273.15 K
volume (V)	1 liter (l *or* L) = 10^{-3} m^3
Avogadro's number	$N_A = 6.022045 \times 10^{23}$ mol^{-1}
Boltzmann's constant	$\mathrm{k} = R/N_A = 3.29986 \times 10^{-24}$ cal/K

(Boltzmann's constant is in roman type to avoid confusion with rate constants k.)

gas constant	$R = 1.98719$ cal K^{-1} mol^{-1} = 8.31441 J K^{-1} mol^{-1}
Bohr magneton	$\mu_B = eh/4\pi m_e = 2.216558 \times 10^{-24}$ cal/T $= 9.274078 \times 10^{-24}$ J/T
magnetic moment of proton	$\mu_p = 3.371456 \times 10^{-27}$ cal/T $= 1.4106171 \times 10^{-26}$ J/T

Planck's constant

$$h = 1.583694 \times 10^{-34} \text{ cal} \cdot \text{s}$$
$$= 6.626176 \times 10^{-34} \text{ J} \cdot \text{s}$$
$$\hbar = h/2\pi = 2.520527 \times 10^{-35} \text{ cal} \cdot \text{s}$$
$$= 1.0545887 \times 10^{-34} \text{ J} \cdot \text{s}$$

elementary charge $e = 1.6021892 \times 10^{-19}$ coulomb (C)

g-factor for free electron $g = 2.0023$

rest mass of electron $m_e = 9.109534 \times 10^{-31}$ kg

velocity of light in vacuum $c = 2.997924580 \times 10^{8}$ m/s

Conversion factor for second-order rate constant:

$$1 \text{ liter mol}^{-1} \text{ s}^{-1} (= 1 \text{ M}^{-1} \text{ s}^{-1}) = 6.022045 \times 10^{20} \text{ cm}^3 \text{ molecule}^{-1} \text{ s}^{-1}$$

Base of natural logarithms $e = 2.71828182846$

(In the following, expressions in quotation marks are less correct terms often used in the literature.)

ΔH_f^0 enthalpy (or ''heat'') of formation in the gas-phase at 1 atm and 298.15 K

ΔH_r^0 enthalpy of reaction

DH^0 bond dissociation enthalpy (''bond dissociation energy'') = BDE

E_S strain enthalpy (''strain energy'')

E_{ST} singlet–triplet splitting (enthalpy (''energy'') difference between singlet and triplet states, positive when triplet is ground state)

ISC intersystem crossing

OF FIRE AND WATER

The symbol △ used to denote heat in this book is one of the old alchemical symbols for the four Aristotelian elements,

△	▽	🜁	🜃
Fire	Water	Air	Earth

These symbols, used extensively by alchemists since the 13th century, appear to be derived in part from Plato's regular polyhedral symbols for the elements, the tetrahedron for Fire, the icosahedron for Water, the octahedron for Air, and the cube for Earth.

Empedocles of Agrigent (~440 BC) is usually credited with the conception of these four elements. Thales of Miletus (~600 BC) believed in Water, Anaximenes (546 BC) in Air, and Heraclitus (~500 BC) in Fire (πῦρ) as the basic element.† In view of this long tradition, I shall use the term "pyrolysis" as synonymous with "thermolysis," even though the heat (θερμόν) in chemistry is rarely produced by direct fire.

Regrettably, certain books and journals have adopted an inverted delta, ▽, to denote heat in chemical reactions. This is not only a misconception, but a sin against the history of chemistry. The Water symbol, ▽, was still in common use with its proper meaning about a hundred years ago.§ The Fire symbol, △, is the only one still with us, and it would be a shame to alter it.

†J. R. Partington, *A History of Chemistry,* Macmillan, London. 1970, Part 1.
§H. Kopp, *Geschichte der Chemie,* Zweiter Theil, Vieweg und Sohn, Braunschweig, 1844, p. 423.

Reactive Molecules

Chapter 1

GENERAL PRINCIPLES

1.1. DEFINITION OF AN INTERMEDIATE

A chemical reaction A → B may be either concerted or nonconcerted. If it is concerted, the free energy and the enthalpy of the system rise continuously until a maximum, the transition state (TS), is reached, and then decrease again until a stable configuration of the product B is obtained. In the nonconcerted case an intermediate associated with an energy minimum is formed (Figure 1.1).

A complete description of a chemical reaction requires an energy surface as in Figure 1.1, together with a knowledge of the atomic and molecular movements that take place along the reaction coordinate. Enthalpies of formation and entropies of stable molecules (e.g., A and B) can be measured or estimated with considerable precision as will be shown later. Free energies are then derived from the equation

$$\Delta G^0 = \Delta H^0 - T\Delta S^0$$

Activation energies are obtained from kinetic measurements with the aid of the Arrhenius equation,

$$k = Ae^{-E_a/RT}$$

The Arrhenius parameters A and E_a are related to activation enthalpies and entropies through the equations,

$$E_a = \Delta H^{\ddagger} + RT$$

and

$$k = \frac{\mathrm{k}T}{h} \mathrm{e}^{\Delta S^{\ddagger}/R} \mathrm{e}^{-\Delta H^{\ddagger}/RT}$$

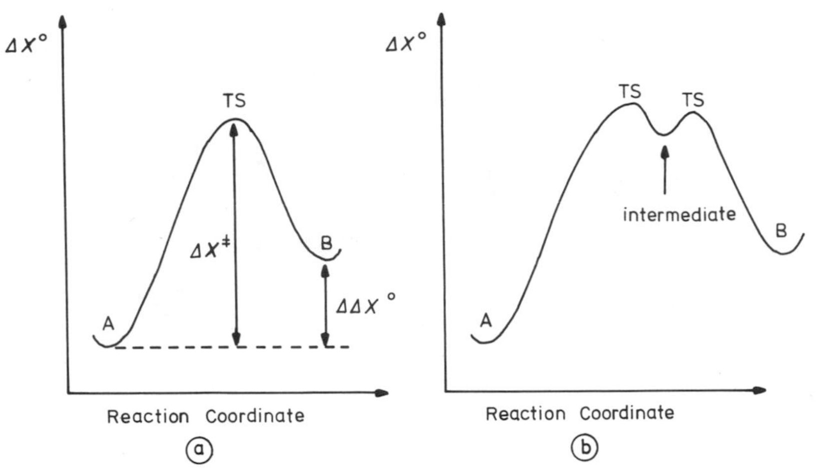

Figure 1.1. (a) Concerted reaction; (b) reaction with reactive intermediate. ΔX is a thermodynamic quantity, usually ΔG or ΔH. The reactions A → B are endothermic as shown.

for *unimolecular* reactions in the gas-phase and for all reactions in solution. For *bimolecular* reactions in the gas-phase, the following relationships apply:[1]

$$E_a = \Delta H^{\ddagger} + 2RT$$

and

$$k = e^2 \frac{kT}{h} e^{\Delta S^{\ddagger}/R} e^{-E_a/RT}$$

In the case of the reaction in Figure 1.1(b), a measurement of the activation energy will yield only the energy of the *first* transition state (or, in general, of the *highest* one). It does not tell anything about the concertedness of the reaction. The real difficulty is, therefore, to find out whether or not an *intermediate* is formed. Sometimes the intermediate lies in a sufficiently deep energy well so that it can either be isolated or detected by physical means (e.g., IR, UV, ESR, microwave, or photoelectron spectroscopy) using low-temperature matrix isolation or gas-phase techniques. More often the existence of the intermediate is inferred only from its peculiar reactions, as will be discussed in the main body of this book.

One may ask, Where is the limit between a transition state and an intermediate? By definition, a transition state represents an energy maximum, a state of continuous change. An intermediate is a stable molecule inasmuch as small energy barriers to further reaction exist. The vibration frequencies of organic molecules are of the order of 10^{13} s^{-1}, that is, a vibration takes place in 10^{-13} s. If the molecule does not survive for this time interval, it will have reacted further within the duration of the one vibration, and it can hardly be said to be a true intermediate. Therefore, we must require that the molecules undergo at least *one* normal vibration before rearranging or decomposing into other molecules. Experimentally, it is not possible to distinguish such a short-lived intermediate from a transition state. The practicable limit is about 10^{-12} s. Thus our definition of a reactive intermediate is as follows: It is a molecular species which undergoes at least 10 normal vibrations before reacting further. Even this broader definition presents an extremely tough challenge to the experimentalist.

1.2. KINETICS OF REACTIONS INVOLVING INTERMEDIATES

If a reactive intermediate cannot be observed directly, it may often be detected by kinetic methods. Consider the reaction

$$\mathrm{A} \underset{k_{-1}}{\overset{k_1}{\rightleftharpoons}} \mathrm{B} + \mathrm{C} \xrightarrow[k_2]{\mathrm{D}} \mathrm{E} + \mathrm{C}$$

$$\text{rate} = -\frac{dA}{dt} = k_1\,A - k_{-1}\,B\,C \tag{1}$$

where in the rate equation (1) A, B, and C denote concentrations.

Examples of such a reaction would be a homolytic or heterolytic dissociation giving a radical or carbocation intermediate B, which is then trapped in a second-order reaction with the reagent D to give the observed product E. For example,

$$(\mathrm{CH_3})_3\mathrm{C{-}Cl} \underset{k_{-1}}{\overset{k_1}{\rightleftharpoons}} \left[(\mathrm{H_3C})_2\overset{\oplus}{\mathrm{C}}{-}\mathrm{CH_3} + \mathrm{Cl}^{\ominus} \right] \xrightarrow[k_2]{\mathrm{H_2O}} (\mathrm{CH_3})_2(\mathrm{H_3C})\mathrm{C{-}OH} + \mathrm{Cl}^{\ominus} + \mathrm{H}^{\oplus}$$

Applying the steady-state approximation,

$$\frac{dB}{dt} = k_1\,A - k_1\,B\,C - k_2\,B\,D = 0$$

or

$$B = \frac{k_1 A}{k_{-1}\,C + k_2\,D} \tag{2}$$

and inserting the expression for B (2) into (1), we get

$$\text{rate} = \frac{k_1\, k_2\, A\, D}{k_{-1}\, C + k_2\, D} \tag{3}$$

From Eq. (3), it can be seen that if the intermediate B reacts on to product much faster than it returns to starting material (A), that is,

$$k_{-1}\, C \ll k_2\, D \tag{4}$$

we then obtain

$$\text{rate} \cong k_1\, A$$

In other words, we have the ordinary first-order rate expression with the observed rate constant

$$k_{\text{obs}} = k_1$$

Thus the observation of first-order kinetics will be a proof for the existence of *an* intermediate (B) in the reaction. Chemical intuition decides the nature of B. The condition (4) can generally be achieved by using a very efficient trapping agent D and maintaining the concentration of D higher than that of C.

If, on the other hand, the reverse reaction is not negligible, the observed rate will be retarded by the term $k_{-1}\, C$ in Eq. (3). Since the concentration of C increases as the reaction proceeds, the rate will be seen to decrease with time. The rate depression will be augmented if the species C is deliberately added to the reacting system. In the preceding example of the S_N1 hydrolysis of *tert*-butyl chloride, the added reagent could be NaCl. In carbocation chemistry, this phenomenon is known as the *common ion rate depression*. It applies equally well to reversible fragmentations forming neutral molecules and radicals.

1.3. ESTIMATION OF THERMOCHEMICAL PROPERTIES

The heats of formation for various molecules can be measured by combustion calorimetry. For example, for an alkane,

$$C_nH_{2n+2} + \frac{3n+1}{2}\, O_2 \rightarrow n\, CO_2 + (n+1)H_2O + \Delta H_c^0$$

$$\Delta H_c^0 = \text{n}\, \Delta H_f^0(CO_2) + (\text{n} + 1)\, \Delta H_f^0(H_2O) - \Delta H_f^0(C_nH_{2n+2})$$

Since the heats of formation of carbon dioxide and water are known, the heat of formation of the alkane can be calculated if the heat of combustion, ΔH_c^0 is measured.

The term "heat of formation" signifies the enthalpy change when forming the compound from the elements in *their standard reference states*. The standard states of the elements are their stable forms at 1 atm at the desired temperature, where by definition $\Delta H_f^0 = 0$. For example, $\Delta H_f^0(H_2)$ (g) = 0, $\Delta H_f^0(O_2)$ (g) = 0, and ΔH_f^0(graphite) (s) = 0 at 298 K.

The superscript ^{0} indicates that the element is in its standard state. When nothing else is specified, the temperature is 298.15 K.

In order to compare ΔH_f^0 values of organic compounds, it is necessary that they all be referred to a common state—the ideal gas state at 1 atm and 298 K. Similarly, to derive meaningful heats of reaction, the heats of formation of all compounds involved should be gas-phase values. If only the heat of formation of a solid (s) or a liquid (l) is available, the heat of sublimation or vaporization must be added before comparison with other gas-phase values can be made:

$$\Delta H^{\circ}(g) = \Delta H^{\circ}(s) + \Delta H^{\circ}(\text{subl.})$$

$$= \Delta H^{\circ}(l) + \Delta H^{\circ}(\text{vap.})$$

$$= \Delta H^{\circ}(s) + \Delta H^{\circ}(m) + \Delta H^{\circ}(\text{vap.})$$

where m stands for melting.

If the required heats of sublimation or vaporization are unknown, the values for closely related compounds may sometimes be used as an approximation.

For example, the heats of sublimation of stilbene and azobenzene are quite similar.

C_6H_5–CH=CH–C_6H_5 C_6H_5–N=N–C_6H_5

From a knowledge of ΔH_f^0(s) for azobenzene and ΔH^0(subl) for styrene one can thus derive a reasonable value for ΔH_f^0(g) for azobenzene.†

Heats of atomization are the enthalpy changes associated with the decomposition of molecules into constituent atoms.

Extensive tables of heats of formation of organic and inorganic compounds exist.[2,3] A few values are found in Table 1.1, and several others can be derived from the data given later in Table 2.2. The reader is advised to acquire a copy of a set of tables (e.g., the ones in Ref. 2a) containing most of the heats of formation that will be useful in this book.

Inspection of the list of heats of formation of alkanes (Table 1.1) reveals that each extra CH_2 group causes a decrease of approximately 5 kcal/mol in ΔH_f^0. The average value for a large number of alkanes is 4.95. Thus we can assign to the CH_2 group a *group increment*[2] of $\Delta H_f^0 = -4.95$ kcal/mol. Note that ΔH_f^0 is in considerable error only for the first member of the series. This is often the case.

†*trans*-Azobenzene: ΔH_f^0(s) = 77.40; ΔH^0(subl) = 22.43; ΔH_f^0(g) = 96.83 kcal/mol.[4a] *trans*-Stilbene: ΔH_f^0(s) = 31.8 ± 0.2; ΔH^0(subl) = 23.71 ± 0.2; ΔH_f^0(g) = 55.5 ± 0.4 kcal/mol.[4b]

TABLE 1.1. Heats of Formation of Some Hydrocarbons[a]

Compound	ΔH_f^0 (kcal/mol)	$\Delta\Delta H_f^0$ (kcal/mol)
CH_4	−17.9	
		2.34
C_2H_6	−20.24	
		4.58
C_3H_8	−24.82	
		5.33
C_4H_{10}	−30.15	
		4.85
C_5H_{12}	−35.00	
		4.96
C_6H_{14}	−39.96	

[a]From Refs. 2 and 3.

From the heat of formation of butane, say, we can now calculate the group increment for the CH_3 group:

$$CH_3{-}CH_2{-}CH_2{-}CH_3$$

$$\Delta H_f^0 = -30.15 = 2\,[CH_3] + 2\,[CH_2] = 2\,[CH_3] - 2 \cdot 4.95$$

or

$$[CH_3] = -10.12 \text{ kcal/mol}$$

The average value from a larger set of data is

$$[CH_3] = -10.08 \text{ kcal/mol}$$

Likewise, we can dissect isopentane in order to obtain the group increment for CH:

$$H_3C{-}\underset{\displaystyle CH_3}{\overset{\displaystyle H}{C}}{-}CH_2{-}CH_3$$

$$\Delta H_f^0 = -36.92 = 3[CH_3] + [CH_2] + [CH]$$

$$[CH] = -36.92 + 3 \cdot 10.08 + 4.95 = -1.73$$

the average value is $[CH] = -1.90$ kcal/mol.

The group increments for all other kinds of groups are derived in the same manner.[2] The final values for hydrocarbons are reported in Table 1.2. From such

TABLE 1.2. Thermochemical Group Increments for Hydrocarbons and Some C—N Groups[a,b]

Group	$\Delta H_f^0(g)$ (kcal/mol)	Group	$\Delta H_f^0(g)$ (kcal/mol)
$C(H)_3(C)$	−10.08	$C_d(H)_2$	6.26
$C(H)_2(C)_2$	−4.95	$C_d(H)(C)$	8.59
$C(H)(C)_3$	−1.90	$C_d(C)_2$	10.34
$C(C)_4$	0.50	$C_d(C_d)(H)$	6.78
		$C_d(C_d)(C)$	8.88
$C(C_d)(C)(H)_2$	−4.76	$C_d(C_B)(H)$	6.78
$C(C_d)_2(H)_2$	−4.29	$C_d(C_B)(C)$	8.64
$C(C_d)(C_B)(H)_2$	−4.29	$C_d(C_t)(H)$	6.78
$C(C_t)(C)(H)_2$	−4.73		
$C(C_B)(C)(H)_2$	−4.86	$C_B(H)$	3.30
$C(C_d)(C)_2(H)$	−1.48	$C_B(C)$	5.51
$C(C_t)(C)_2(H)$	−1.72	$C_B(C_d)$	5.68
$C(C_B)(C)_2(H)$	−0.98	$C_B(C_t)$	5.7
$C(C_d)(C)_3$	1.68	$C_B(C_B)$	4.96
$C(C_B)(C)_3$	2.81		
		C_a	34.20
$C(H)_x(C)_y(N_A) = C(H)_x(C)_{y+1}$			

$2N_A(C)$ (*trans*) = (C)—N=N—(C) (trans) = 52.3 kcal/mol

$2N_A(C)$ (*cis*) = (C)—N=N—(C) (cis) ≧ 54.2 kcal/mol

$N_A(\dot{N}_A)(C)$ = (C)—N=$\dot{N}$ = 60.2 kcal/mol

[a]Data for hydrocarbons are from Ref. 2; data for nitrogen groups are from P. S. Engel, J. L. Wood, J. A. Sweet, and J. L. Margrave, *J. Am. Chem. Soc.*, **96,** 2381 (1974).
[b]A group is identified by its central atom and its ligands (in parentheses). $C(H)_3(C)$ is the CH_3 group bonded to C. C_d is a double-bond C; the other C_d of the double bond is not specified as ligand. C_B is a benzenoid C. C_t is a triple bond C. C_a is the central C in allene; the other C's in allenes are treated as C_d. N_A is an N in the azo group. $\dot{N}_A$ is the N in the diazenyl radical, which is used in Chapters 2 and 3.

increments, one can derive the heat of formation of any desired molecule, known or unknown, by simple addition. The error is almost as small as in combustion calorimetry, approximately ±1 kcal/mol. However, life would be too easy if this were all that is required. The increments are based on essentially strain-free molecules. If the desired molecule is strained, the strain energy must be added. We can calculate the heat of formation of cyclohexane precisely as 6 $[CH_2]$ = −29.7 kcal/mol (experimental value: −29.43 kcal/mol) because cyclohexane is unstrained. An analogous calculation for cyclopropane is in error by 27.5 kcal/mol, which is the strain energy of cyclopropane. Strain and strain energies are discussed

TABLE 1.3. Ring Corrections (Strain Energies) to be Applied in Estimates for Cyclic Hydrocarbons (kcal/mol)

Cyclopropane	28	Cycloheptane	6.4
Cyclopropene	54	Cycloheptene	7
Methylenecyclopropane	40.9	1,3-Cycloheptadiene	6.6
Cyclobutane	27	1,3,5-Cycloheptatriene	4.7
Cyclobutene	30	Cyclooctane	9.9
Cyclopentane	6.3	*cis*-Cyclooctene	7.7
Cyclopentene	5.9	*trans*-Cyclooctene	16.7
Cyclopentadiene	6.0	1,3,5-Cyclooctatriene	8.9
Cyclohexane	0	Cyclooctatetraene	17.1
Cyclohexene	2		
1,3-Cyclohexadiene	4.8		
1,4-Cyclohexadiene	0.5		

in detail in Chapter 5. It suffices here to give a list of strain corrections that should be added to estimated heats of formation of cyclic compounds (Table 1.3).

A correction should also be added for each *gauche* interaction in an alkane. Isopentane has at least one *gauche* interaction, and this raises the energy of the molecule by 0.80 kcal/mol.

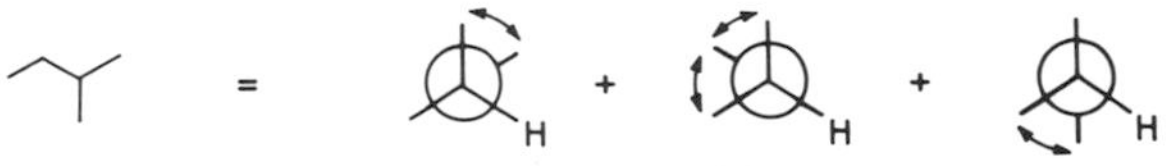

A *cis*-disubstituted alkene has a higher energy than the *trans*-alkene because of steric interaction between the groups. For simple groups like methyl, a *cis*-correction

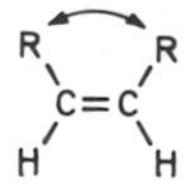

of 1 kcal/mol applies. If one R is *tert*-butyl, the correction is 4, and when both are *tert*-butyl, 10 kcal/mol must be added. *Ortho*-disubstitution on an aromatic ring requires a *cis*-correction of approximately 0.6 kcal/mol. Again, bulky groups increase the steric interaction and necessitate larger corrections.

Entropies can be estimated from group increments in the same way as ΔH_f^0, and tables are available.[2] We will not normally need entropies here, and it suffices to know that they are composed of translational, rotational, vibrational, and symmetry contributions. If these factors do not change very much overall in the course of a chemical reaction, we may be justified in setting $\Delta S^0 \cong 0$, and hence $\Delta H^0 \cong \Delta G^0$ for the reaction. The entropies of different molecules that have nearly the same mass and structure are almost identical. Thus, instead of estimating the entropy of a molecule, it may be quicker simply to use the known entropy of a structurally related compound. Compare, for example, the following molecules,

$S^0_{298} = 70.5$ cal K^{-1} mol^{-1} $\quad$ $S^0_{298} = 70.6$ cal K^{-1} mol^{-1}

The entropy unit (e.u.) is cal K^{-1} mol^{-1}. In the older German literature, the unit Clausius/mol = cal K^{-1} mol^{-1} is sometimes used.

The reaction

is rather complicated and passes through at least one and possibly three additional reactive intermediates. $\Delta S^{\ddagger}$ for the reaction in unknown and unmeasurable (at the moment), but probably it is large and negative. Nevertheless, we can be quite sure that ΔS^0 for the reaction is close to zero, for there is no change of mass, and the overall structural change is negligible.

1.4. RESONANCE

In 1880, Julius Thomsen[5] measured the heat of combustion of benzene and found that this value was 42.6 kcal/mol higher than expected for a compound containing three double bonds. In other words, the benzene was 42.6 kcal/mol too stable. Thomsen concluded as follows: *"Die starke Wiederstandsfähigkeit des Benzols deutet auf die Abwesenheit mehrfacher Bindungen."*† Consequently, he preferred the saturated structures of Ladenburg[6] (1869) and Claus[7] (1867) to Kekulé's[8] un-

saturated structures of 1865–1872.[9] In fact, Thomsen's heat of combustion was in error by +7.3 kcal/mol. Therefore, the benzene was really only 35.3 kcal/mol "too stable." This coincides with the value commonly known today as the resonance energy of benzene.

A word of explanation is in order since the value for the benzene resonance energy used in this book is only 20 kcal/mol. The commonly quoted value of 35 or 36 kcal/mol is based on heats of combustion or hydrogenation in comparison with the expectation values for the hypothetical molecule "cyclohexatriene." The

†"The strong resistance of benzene indicates an absence of multiple bonds."

TABLE 1.4. Resonance Energies of Aromatic Hydrocarbons (kcal/mol)[a]

Benzene	20.0	Triphenylene	57.0
Naphthalene	30.5	Perylene	60.2
Anthracene	37.8	Pyrene	54.2
Phenanthrene	44.0		
Chrysene	56.3		

[a]From Ref. 10.

comparison is of little practical value because even if a "cyclohexatriene" without benzene resonance existed, it would possess some conjugative interaction between the double bonds. The activation energy for rotation in butadiene is approximately 5 kcal/mol:

This value can be designated as the conjugative stabilization energy of planar butadiene.[10,11] In the transition state between the two planar forms, there is no interaction between the double bonds. It would be more reasonable, therefore, to compare benzene with a cyclohexatriene, which possesses this normal interaction between double bonds. This can be done using the thermochemical group increment for the fragment $C_d(C_d)(H)$, which is 6.78 kcal/mol (Table 1.2):

$$
\begin{aligned}
\Delta H_f^0 \text{ (calc)} &= 6.78 \cdot 6 = 40.68 \\
\Delta H_f^0 \text{ (expt)} &= \underline{19.82} \\
\text{resonance energy} & \quad 20.86 \text{ kcal/mol}
\end{aligned}
$$

This is the *thermochemical resonance energy of benzene*, and it makes it directly comparable with nonaromatic polyunsaturated molecules. Theoretical calculations[10,11] also yield a "low" value of about 20–25 kcal/mol for the resonance energy of benzene. Some further resonance energies are given in Table 1.4.

1.5. HAMMOND POSTULATE AND BELL–EVANS–POLANYI PRINCIPLE

The Hammond postulate[12] states that in a very exothermic reaction the transition state resembles the reactant(s), structurally and energetically. In an endothermic reaction, the transition state resembles the product(s). Consider, for example, the first part of the reaction in Figure 1.1(b) in which A is the reactant and the intermediate is the product. Since the transition state is only a little higher in energy than the intermediate, it is reasonable to assume that the two resemble each other. Therefore, the transition state is drawn close to the intermediate on the reaction coordinate. Structurally, the transition state is much farther away from A, a considerable energetic change having taken place. Likewise, in the exothermic reaction leading from the intermediate to the final product B, the transition state is close to the initial state.

This postulate has found general use in organic chemistry. An important consequence is that the measurement of activation energies tells us more about the intermediates—when they are involved—than it does about the reactants or products. If one compares a series of closely related reactions, a stabilizing effect of substituents on the intermediate is likely to be reflected in a lowering of the activation energy since the same stabilizing effect will be felt in the structurally related transition state.

Going a little further, one arrives at the Bell–Evans–Polanyi principle, which states that there is a proportionality between the activation energy and the heat of reaction,

$$E_a = \mathrm{A} + \mathrm{B}\ \Delta H_r^0$$

where A and B are constants and ΔH_r^0 is the heat of reaction referred to the standard state (ideal gas, 1 atm, 298 K). In other words, as a given reaction type becomes more exothermic, the activation energy decreases (Figure 1.2). The relationship

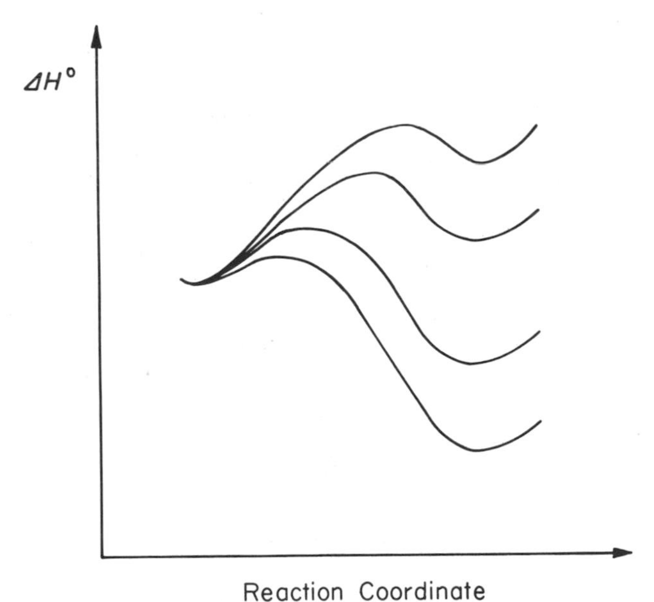

Figure 1.2. Bell–Evans–Polanyi principle.

has no foundation in thermodynamics, and it holds only for closely related reactions. Despite this limitation, it is, however, extremely useful. A more detailed description is given in Ref. 10.

1.6. THE REACTIVITY–SELECTIVITY PRINCIPLE

The reactivity–selectivity principle[13] stipulates that, in a competition system, the more reactive species should be the least selective. This is readily understood in terms of the Hammond postulate (Section 1.5): A reactive species leads to a low activation energy and an "early" transition state; little structural and energetic change will have taken place at the transition state, and the species will not, therefore, be in a position to choose among different reaction partners. An example of this is discussed in Figure 2.18, for the hydrogen abstraction by chlorine and bromine atoms. It may seem strange, then, that other systems are known in which the more reactive species is also the more selective one. Cases have been found[14] where selectivity is completely reversed merely by changing the temperature.

In the halogen-abstraction reaction of alkyl radicals with CCl_4 and $BrCCl_3$, bromine abstraction is faster than chlorine abstraction [$k_{Br} > k_{Cl}$; Eq. (5)].

$$R^{\bullet} \begin{cases} \xrightarrow[k_{Cl}]{CCl_4} R\text{—}Cl + \dot{C}Cl_3 \\ \xrightarrow[BrCCl_3,\ k_{Br}]{} R\text{—}Br + \dot{C}Cl_3 \end{cases} \qquad (5)$$

In agreement with the Hammond postulate and the Bell–Evans–Polanyi principle, we would expect more stable (i.e., low-energy) radicals, such as the tertiary ones, to react preferably with $BrCCl_3$, giving a large competition constant k_{Br}/k_{Cl}. A highly reactive (high-energy) radical such as $\cdot CH_3$ would react less selectively, giving a small competition constant. This is borne out in the experimental data measured at 0°C (see Figure 1.3). However, as the temperature increases, the selectivities decrease until at approximately 70°C all the radicals show nearly the *same* selectivity. This is the isoselective temperature T_{iso}. As the temperature increases still further, *opposite* selectivity is observed: The CH_3 radical is now the most selective, the tertiary radical the least selective one, although all selectivities decrease (Figure 1.3).

The reason[14] for this reversal is that we are not dealing with reactions where only the enthalpies ΔH need be considered (as is the case for the reactions shown in Figure 2.18). It is the differences in free energies of activation $\Delta\Delta G^{\ddagger}$ that really determine the relative rates of the various reactions (Figure 1.4). Since

$$\Delta\Delta G^{\ddagger} = \Delta\Delta H^{\ddagger} - T\,\Delta\Delta S^{\ddagger}$$

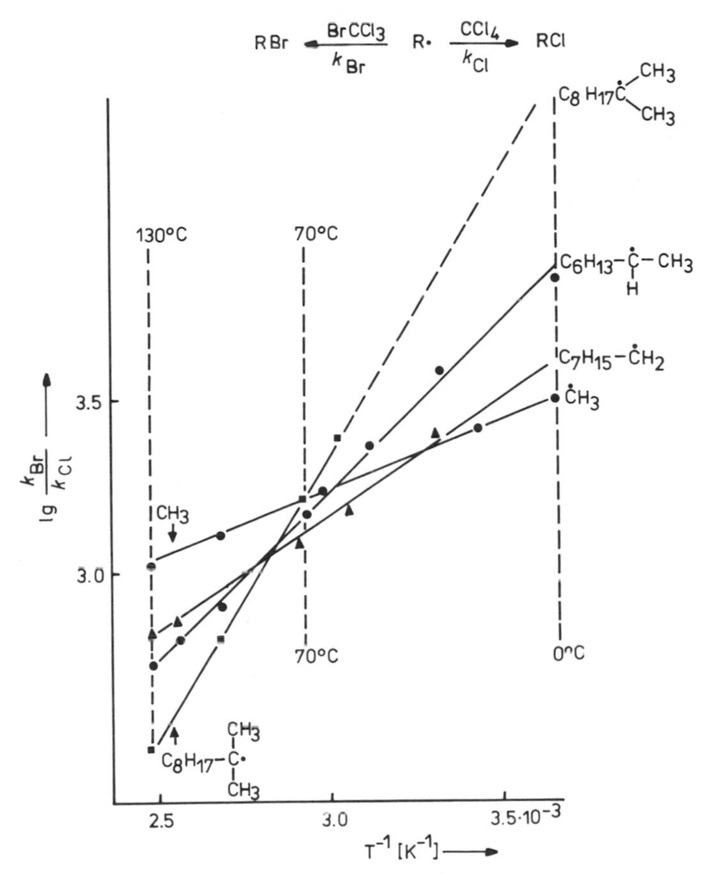

Figure 1.3. Temperature effect on the selectivities of alkyl radicals (Ref. 14). (Reproduced with permission of Verlag Chemie, Weinheim.)

it follows that if $\Delta\Delta H^{\ddagger}$ is not very large, the second term $T \cdot \Delta\Delta S^{\ddagger}$ may become important. In such cases, a temperature increase can lead to selectivity reversal. For the radicals shown in Figure 1.3, it was found[15] that, in fact, $\Delta\Delta H^{\ddagger}$ $(= \Delta H^{\ddagger}_{\mathrm{Cl}} - \Delta H^{\ddagger}_{\mathrm{Br}})$ increases from the primary to the tertiary radical. However, $\Delta\Delta S^{\ddagger}$ increases in the same series, so that above 70°C a reversal in $\Delta\Delta G^{\ddagger}$, and hence the selectivity, occurs.

The isoselective relationship[14] can be formulated:

$$\Delta\Delta H^{\ddagger} = T_{\mathrm{iso}} \cdot \Delta\Delta S^{\ddagger} + C$$

A consequence of this relationship is that, when relative reactivities are compared—and this is very often done in the chemistry of reactive intermediates—the effect of temperature should be studied in order to ascertain whether one is dealing with

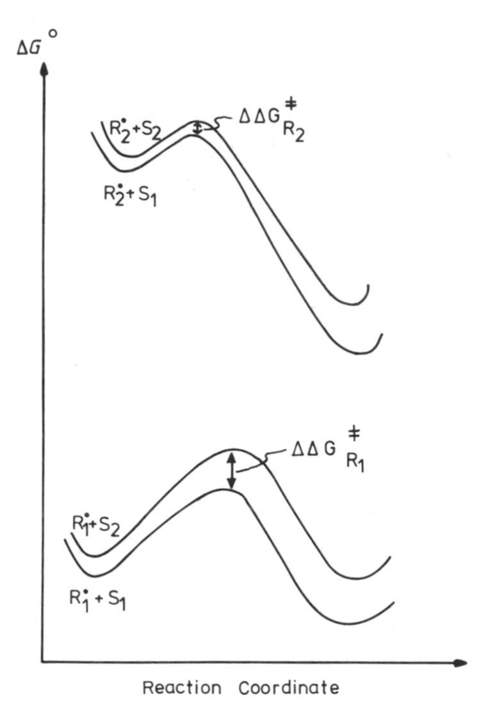

Figure 1.4. Bottom: competition of two substrates S_1 and S_2 for a relatively stable and unreactive radical $R_1^{\cdot}$. Top: competition for a less stable, more reactive radical $R_2^{\cdot}$.

enthalpy or entropy effects. The third factor that may be important is that of orbital interactions.

1.7. FRONTIER MOLECULAR ORBITAL THEORY

The interactions between two molecules in a chemical reaction can be described in terms of the corresponding interactions between molecular orbitals (MOs).[16] Usually, only the highest occupied and the lowest unoccupied molecular orbitals (HOMOs and LUMOs) need be considered. This approach is known as the *Frontier Orbital* or *Perturbation* MO method.[10,17,18] The interaction between two degenerate (i.e., of equal energy) orbitals Φ_a and Φ_b leads to two new orbitals, Ψ_1 and Ψ_2, described by linear combination of the two original ones. If both orbitals are occupied, no net energy gain results. However, if one of the orbitals were unoccupied, two electrons would experience a net energy lowering, $\delta\epsilon$ [Figure 1.5(a)]. Similarly, the interaction between two nondegenerate, localized MOs, usually a HOMO (Φ_a) and a LUMO (Φ_b) generates the two new MOs Ψ_1 and Ψ_2, but now each orbital retains some of its original character. The relative admixtures of Φ_a and Φ_b can be described by the parameters λ and μ (both < 1) [Figure 1.5(b)]. The stabilization energy $\delta\epsilon$ of the lower orbital is smaller than the destabilization energy $\delta\epsilon'$ of the higher one. Therefore, if the two orbitals Φ_a and Φ_b were both doubly occupied, a net destabilization of the two electrons residing in Ψ_2 would result; the interaction

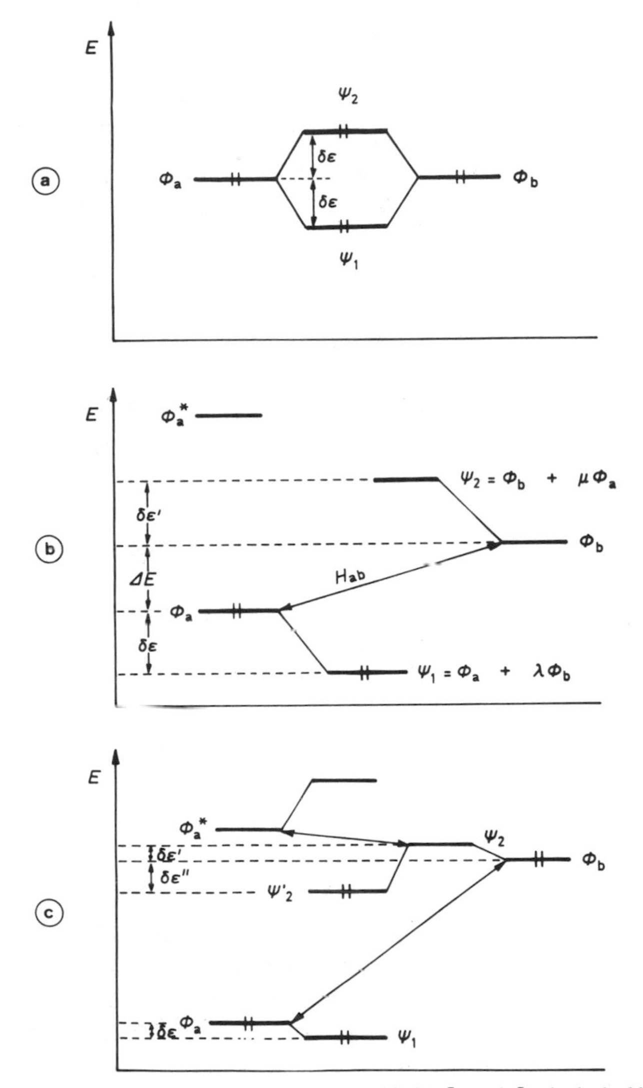

Figure 1.5. Orbital interactions; (a) two degenerate orbitals, Φ_a and Φ_b, both doubly occupied; (b) HOMO (Φ_a)–LUMO (Φ_b) interaction between nondegenerate orbitals; (c) LUMO (Φ_a^*) − HOMO (Φ_b) interaction dominates over (Φ_a) − (Φ_b) interaction; Φ_b is converted into Ψ_2^1.

would be repulsive. If one of the original MOs are unoccupied as in the case of our Φ_b, a net stabilization of the two electrons will occur.

The energy shifts $\delta\epsilon$ and $\delta\epsilon'$ depend on the strength of the interaction Hamiltonian H_{ab} and the energy gap ΔE between the original orbitals. H_{ab} is related to the overlap integral S_{ab}; no interaction will take place if Φ_a and Φ_b are of opposite symmetry, and it will be weak if the coefficients at the interacting atoms are small. The interaction will also diminish when ΔE increases.

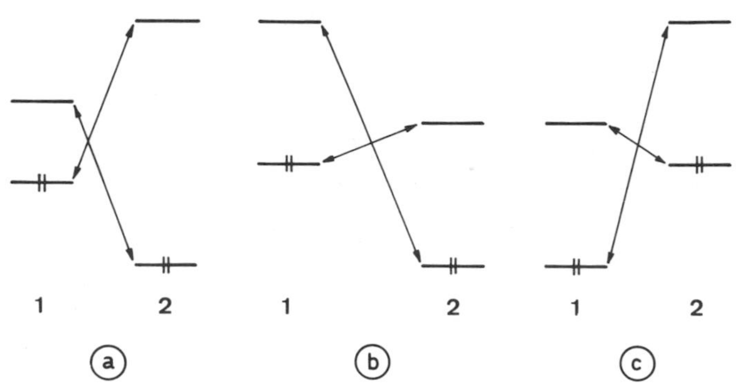

Figure 1.6. Three possibilities for HOMO–LUMO interaction between two molecules 1 and 2. (a) HOMO(1)–LUMO(2) and HOMO(2)–LUMO(1) equally strong; (b) HOMO(1)–LUMO(2) strongest; 1 is an electron donor; (c) HOMO(2)–LUMO(1) is strongest; 1 is an electron acceptor.

In Figure 1.5(b), we have created a new LUMO Ψ_2. This, of course, can interact in a second step with a high-lying vacant orbital Φ_a^* if such is available. This would not change the overall energy since the orbitals involved are unoccupied. If Ψ_2 is occupied, however, as in Figure 1.5(c), its two electrons would become stabilized by the amount $\delta\epsilon''$, and if the interaction is strong enough—that is, if the energy of Φ_a^* is low enough—net stabilization can again occur. This interaction will dominate if Φ_b is closer in energy to Φ_a^* than to Φ_a.

The high-lying molecular orbital Φ_b is typical of reactive intermediates. In a cation, it is vacant; in a radical, it is singly occupied; and in an anion, it is doubly occupied. The simple picture of Figure 1.5 shows that carbocations are stabilized by interaction with relatively high-lying HOMOs (Φ_a); carbanions by relatively low-lying LUMOs (Φ_a^*). Radicals can be stabilized by interactions in both directions (Φ_b and Ψ_2 singly occupied).

When two molecular systems react with each other, two sets of orbital interaction can take place, HOMO(1)–LUMO(2) and LUMO(1)–HOMO(2), as illustrated in Figure 1.6. In situation (a), the orbital energies are symmetrically arranged, and the two interactions are equally strong. In situation (b), the levels HOMO(1) and LUMO(2) are closer in energy; this is therefore the strongest interaction. The opposite holds for situation (c). Thus it is important to be able to judge the relative orbital energies. It may usually be assumed that electron-donating substituents (often designated $\ddot{X}$-substituents) increase the energies of both HOMO and LUMO. Electron-withdrawing substituents (so-called Z-substituents) have the opposite effect.[18,19]

So far, we have only considered variations in ΔE (cf. Figure 1.5). Also the coefficients of the interacting orbitals have to be considered since these determine the overlap integral S_{ab}. The frontier MOs will, therefore, combine preferentially in such a way that the atoms possessing the largest MO coefficients (of the right orbital symmetry) are connected. HOMO and LUMO energies can be obtained from photoelectron and UV spectroscopy, and the coefficients can be derived from ESR measurements on radical cations and radical anions that give information about the spin densities in the various atomic positions (cf. the McConnell equation, p. 38.).

A simple example will show how the frontier orbital analysis works. Consider the reactions of benzyl anion and benzyl cation with an olefin:

$$\text{C}_6\text{H}_5\text{–}\overset{\ominus}{\text{C}}\text{H}_2 + \text{CH}_2\text{=CH–X} \longrightarrow \textbf{1} \qquad (6)$$

$$\text{C}_6\text{H}_5\text{–}\overset{\oplus}{\text{C}}\text{H}_2 + \text{CH}_2\text{=CH–X} \longrightarrow \textbf{2} \qquad (7)$$

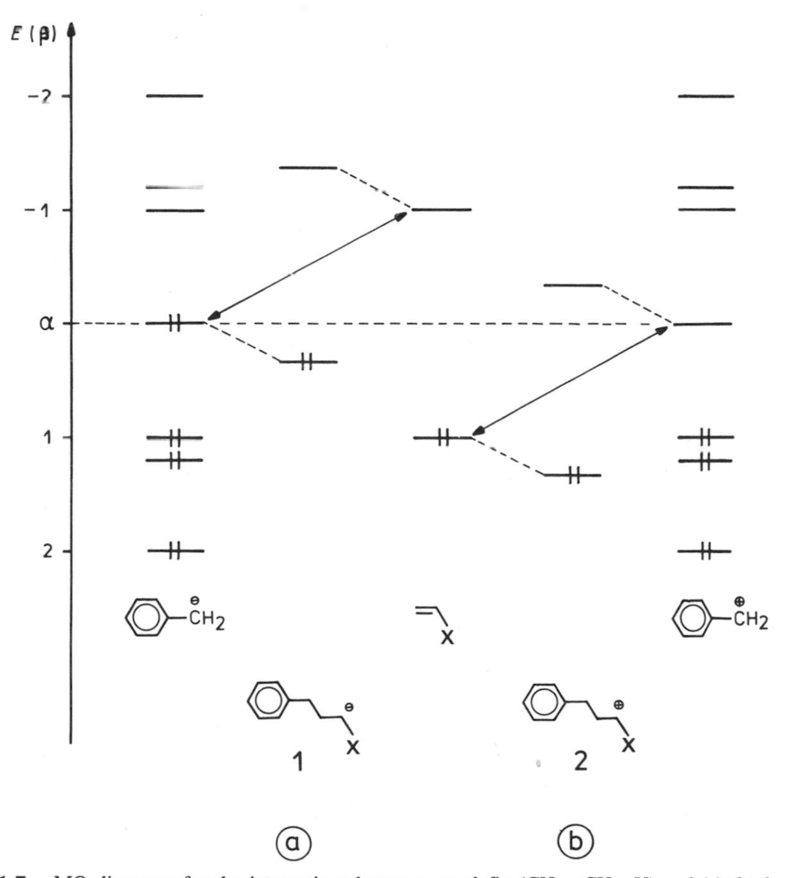

Figure 1.7. MO diagrams for the interactions between an olefin ($CH_2{=}CH{-}X$) and (a) the benzyl anion, (b) the benzyl cation, giving as products the ions **1** and **2**, respectively.

The orbitals of the benzyl system (Figure 1.7) are obtained from Hückel–MO calculations. In the anion, the essentially nonbonding HOMO is doubly occupied; in the cation, this orbital is vacant and has become the LUMO. The olefin we will regard as a derivative of ethylene where, as in all alternant hydrocarbons, the HOMO and LUMO levels are symmetrically arranged with respect to the α-level. In the anion, the HOMO(anion)–LUMO(olefin) interaction is the strongest; it is a situation like in Figure 1.6(b). The anion is a nucleophile, and the interaction will be stronger with a more electrophilic olefin because electron-withdrawing substituents will lower the LUMO energy. The cation, in contrast, is an electrophile; it interacts more strongly with the olefin HOMO, as shown in Figure 1.6(c).

Electron-donating substituents in the olefin will raise the energy of the HOMO and thus strengthen the interaction. Furthermore, electron-withdrawing substituents in the cation would have made this more electrophilic, lowered the cation LUMO, and again strengthened the interaction. All this can, of course, be understood simply by looking at the effects of X on the stabilities of the intermediates **1** and **2** in Eqs. (6) and (7). It may not be immediately obvious, however, that a more highly reactive benzyl anion, say, in which the HOMO is raised by introducing electron-donating substituents, might also be more *selective* than the less reactive parent benzyl anion. This means that in a competition system with two different olefins, the more reactive anion would react preferentially with the more reactive olefin (i.e., the highest HOMO would combine with the lowest LUMO). The reason for this is exactly that the orbital interaction increases as the energy difference ΔE decreases.[18,19]

The example with the benzyl anion was hypothetical. A real one is given in a study of the Wittig reaction.[20] There is much evidence that this reaction proceeds via an oxaphosphetane (**5**), formed from the phosphonium ylide **3** and the aldehyde **4**:

$$\underset{\mathbf{3}}{Ar_3\overset{\oplus}{P}\text{—}\overset{\ominus}{C}RR'} + \underset{\mathbf{4}}{Ar'\text{—}CH\text{=}O} \longrightarrow \underset{\mathbf{5}}{\begin{matrix} Ar_3P\text{—}CRR' \\ | \quad\;\; | \\ O\text{—}CHAr' \end{matrix}} \longrightarrow Ar_3P\text{=}O + RR'C\text{=}CHAr'$$

Depending on whether donor ($\ddot{X}$-) or acceptor (Z-) substituents are present in Ar and Ar′, the two transition states **6** and **7** may (in a simplified manner) be considered:

$$\underset{\mathbf{6}}{\begin{matrix} (\ddot{X}\text{—}C_6H_4)_3P\text{—}CRR' \\ \vdots \qquad \vdots \\ \ddot{O}\cdots CH\text{—}C_6H_4\text{—}Z \end{matrix}} \qquad\qquad \underset{\mathbf{7}}{\begin{matrix} (Z\text{—}C_6H_4)_3P\text{—}CRR' \\ \vdots \qquad \vdots \\ \ddot{O}\cdots CH\text{—}C_6H_4\text{—}\ddot{X} \end{matrix}}$$

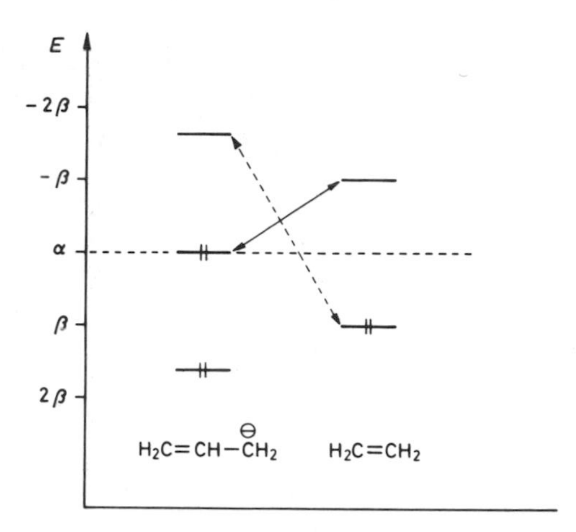

Figure 1.8. MO interactions between the allyl anion and ethylene.

Transition state **6** resembles the attack of a benzyl anion on an olefin and is accelerated by the $\ddot{X}$- and Z-substituents because they raise the ylide–HOMO and lower the aldehyde–LUMO. The opposite arrangement of substituents in **7** lowers the ylide–HOMO and raises the aldehyde–LUMO. It was found that the combination **6**, in which the relevant frontier orbital energies are closer, led not only to the highest reactivities, but also to higher selectivities.[20]

As we have just seen in the preceding section, the reactivity–selectivity parallelism could also have been due to entropy effects. However, it was established[20] that the ylide reactions took place well below the isokinetic temperature. Furthermore, all the reactions studied gave positive ρ values in the Hammett correlation,[13]

$$\log \frac{k}{k_0} = \sigma \rho$$

which implies that the phosphonium ylides are acting as nucleophiles, transferring charge to the aldehydes in the transition states. The frontier MO analysis appears, therefore, to be justified.

The frontier orbital treatment can be applied also to 1,3-dipolar and Diels–Alder cycloadditions:[19]

A 1,3-dipole can be likened to the allyl anion, with which it is isoelectronic. An interaction diagram is shown in Figure 1.8. In this case the ''dipole'' (allyl anion) will act as a nucleophile. However, if the MO levels of the alkene are raised, or

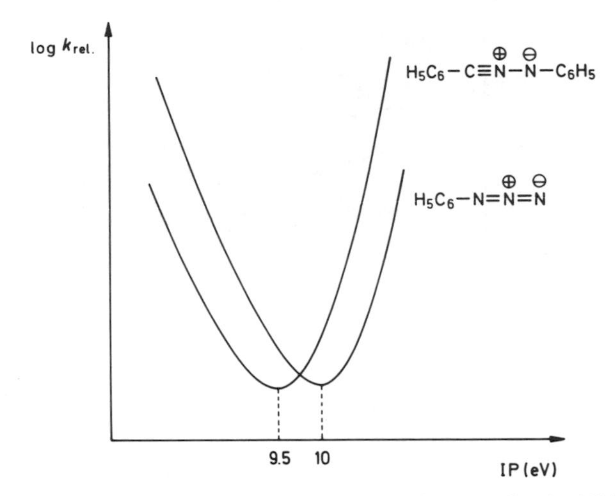

Figure 1.9. Schematic description of the variation of the relative rates of cycloaddition of 1,3-dipoles to substituted alkenes as a function of ionization potential (IP) of the alkene. Alkenes to the left of the figure are electron donors (e.g., enamines, enol ethers); to the right, they are attractors (e.g., fumarates, cyanoethylenes). For more details, see Ref. 19.

those of the dipole lowered, both interactions can become equally strong, and the dipole may eventually become an electrophile. This can be achieved by placing electron-donating substituents on the alkene or electron-attracting ones on the dipole. The ionization potentials of alkenes are measures of the relative HOMO energies; a high ionization potential corresponds to a low-energy HOMO. Provided that HOMO and LUMO energies are changed in the same manner on substitution, a high ionization potential also signifies a low-energy LUMO.

It has been found that several 1,3-dipoles add faster to both very electrophilic and very nucleophilic alkenes than to intermediate ones, thereby substantiating the idea that both dipole(HOMO)–alkene(LUMO) and dipole(LUMO)–alkene(HOMO) interactions can be dominant (Figure 1.9). Such dipoles may be said to be *ambiphilic:* They can be either nucleophilic or electrophilic, depending on the reaction partner.

The Diels–Alder reaction is isoelectronic with the 1,3-dipolar cycloaddition and subject to the same effects.[21,22] In all such cases of "hyperbolic reactivity" (cf. Fig. 1.9), the more reactive dipoles or dienes—be they electrophiles or nucleophiles—should also be the most selective when the reactions are frontier orbital controlled.[14,19]

It should be noted that a frontier orbital treatment considers only the *onset* of a reaction, and its success therefore suggests that the initial slope of the potential curve is often a fair indication of the height of the transition state. It has been pointed out, however, that the orbital energies are not additive to the same extent as the total energies, and the influence of substituents on orbital energies can therefore be more complicated than that on the total energies of starting materials, transition states, and products.[23]

REFERENCES AND NOTES

1. For more details on reaction kinetics, see, for example, K. J. Laidler, *Chemical Kinetics*, 2nd ed., McGraw-Hill, London, 1965; S. W. Benson, *The Foundation of Chemical Kinetics*, McGraw-Hill, New York, 1960; P. J. Robinson and K. A. Holbrook, *Unimolecular Reactions*, Wiley-Interscience, New York, 1972.
2. (a) S. W. Benson et al., *Chem. Rev.*, **69,** 279 (1969). (b) S. W. Benson, *Thermochemical Kinetics*, 2nd ed., Wiley, New York, 1976.
3. J. D. Cox and G. Pilcher, *Thermochemistry of Organic and Organometallic Compounds*, Academic Press, London, 1970; D. R. Stull, E. F. Westrum, and G. C. Sinke, *The Chemical Thermodynamics of Organic Compounds*, Wiley, New York, 1969; *JANAF Thermochemical Tables*, National Standards Reference Data System, National Bureau of Standards, U.S. Department of Commerce, Washington, D.C., 1971.
4. (a) F. W. Schulze, H.-J. Petrick, H. K. Cammenga, and H. Klinge, *Z. Phys. Chem.*, **107,** 1 (1977); E. Wolf and H. K. Cammenga, *ibid.* **107,** 21, (1977). (b) R. Shaw, *J. Phys. Chem.*, **75,** 4047 (1971); G. A. Lobanov and L. P. Karmanova, *Izv. Vyssh. Ucheb. Zaved., Khim. Khim. Tekhnol.*, **14,** 865 (1971) [*Chem. Abstr.*, **76,** 28438s (1972)]; E. Morawetz, *J. Chem. Thermodyn.*, **4,** 455 (1972).
5. J. Thomsen, *Ber. Dtsch. Chem. Ges.*, **13,** 1808 (1880); J. Thomsen, *Thermochemische Untersuchungen*, Verlag J. A. Barth, Leipzig, 1886, Vol. 4, p. 270.
6. A. Ladenburg, *Ber. Dtsch. Chem. Ges.*, **2,** 140 (1869).
7. A. Claus, *Theoretische Betrachtungen und deren Anwendung zur Systematik der Organischen Chemie*, O. Hollander, Freiburg, 1867; *Ber. Dtsch. Chem. Ges.*, **15,** 1405 (1882).
8. A. Kekulé, *Bull. Soc. Chim. Fr.*, **3,** 98 (1865); *Bull. Acad. R. Belg.*, **19,** 551 (1865); *Liebigs Ann. Chem.*, **137,** 129 (1866); *ibid.* **162,** 77 (1872).
9. For a detailed presentation of the developments leading to Kekulé's formula for benzene, see J. P. Snyder, *Nonbenzenoid Aromatics*, Academic Press, New York, 1969, Vol. 1, pp. 1–31. For discussions of the concept of aromaticity, see D. Lewis and D. Peters, *Facts and Theories of Aromaticity*, Macmillan, London, 1975; P. J. Garratt, *Aromaticity*, McGraw-Hill, Maidenhead, 1971; P. Garratt and P. Vollhardt, *Aromatizität*, Georg Thieme Verlag, Stuttgart, 1973.
10. M. J. S. Dewar, *The Molecular Orbital Theory of Organic Chemistry*, McGraw-Hill, New York, 1969.
11. H. Kollmar, *J. Am. Chem. Soc.*, **101,** 4832 (1979) and references therein.
12. G. S. Hammond, *J. Am. Chem. Soc.*, **77,** 334 (1955).
13. See, for example, J. E. Leffler and E. Grunwald, *Rates and Equilibria of Organic Reactions*, Wiley, New York, 1963.
14. B. Giese, *Angew. Chem.*, **89,** 162 (1977); *Angew. Chem. Int. Ed. Engl.*, **16,** 125 (1977).
15. B. Giese and K. Keller, *Chem. Ber.*, **112,** 1743 (1979).
16. For the "Woodward–Hoffman Rules," see R. B. Woodward and R. Hoffmann, *The Conservation of Orbital Symmetry*, Verlag Chemie, Weinheim, 1970.
17. K. Fukui, *Fortschr. Chem. Forsch.*, **15,** 1 (1970); *Acc. Chem. Res.*, **4,** 57 (1971); L. Salem, *Chem. Brit.*, **5,** 449 (1969); M. J. S. Dewar and R. C. Dougherty, *The PMO Theory of Organic Chemistry*, Plenum Press, New York, 1975.

18. I. Fleming, *Frontier Orbitals and Organic Chemical Reactions*, Wiley, London, 1976.

19. R. Sustmann, *Pure Appl. Chem.*, **40,** 569 (1974); K. N. Houk, *Acc. Chem. Res.*, **8,** 361 (1975).

20. B. Giese, J. Schoch, and C. Rüchardt, *Chem. Ber.*, **111,** 1395 (1978).

21. J. Sauer, *Angew. Chem.*, **79,** 76 (1967); *Angew. Chem. Int. Ed. Engl.*, **6,** 16 (1967).

22. A. I. Konovalov and B. N. Solomonov, *Dokl. Akad. Nauk. SSSR*, **211,** 1115 (1973); A. I. Konovalov, B. N. Solomonov, and O. Y. Chertov, *J. Org. Chem. USSR*, **11,** 103, 107 (1975).

23. J. R. Murdoch, *J. Am. Chem. Soc.*, **104,** 588 (1982).

Chapter 2

RADICALS

. . . les groupes hydrocarbonés d'atomicité impaire ne sauraient exister à l'état de liberté; car dans un tel groupe 1 atome de carbone manifesterait un nombre impair d'atomicités, comme le montre la formule suivante:

```
     H  H
     |  |
  H—C—C
     |  |
     H  H
```

Radical éthyle

—A. Wurtz, *Dictionnaire de Chimie*, L. Hachette et cie, Paris, 1869, Vol. 1, p. 456.

2.1. REACTIVITY AND STABILITY

A radical is an atom, molecule, or ion carrying an unpaired electron. If it has more than one such electron, it is a diradical, triradical, or so forth. Oxygen is a naturally occurring triplet diradical. Many other native free radicals have been detected in interstellar space, primarily using millimeter-wave spectroscopy. These include CH, CN, C_2, OH, NO, SO, HCO, C_2H, C_3N, and C_4H.[1] The near-perfect vacuum and low temperature (3 K microwave background radiation; about 20 K in medium density clouds due to gas kinetic motion) makes the interstellar medium an excellent laboratory for the synthesis and preservation of otherwise highly unstable entities. Over 50 radicals, molecules, and ions have now been identified, and several unstable compounds were found in space before they were terestrially known.[1] Of course, the hydrogen atom, which we may also regard as a free radical, is of widespread cosmic significance.

NO and NO_2 are more familar stable free radicals, and ClO_2 is an explosive, but nonetheless isolable free radical that can be prepared in the laboratory. Diphenylpicrylhydrazyl (DPPH) is a commercially available stable free radical that is used as a standard in ESR spectroscopy, and the nitroxides† form another class of extraordinarily stable radicals, a fact taken advantage of in the spin trapping technique (see Section 2.2.4). In fact, the first man-made free radical, Fremy's salt, $(KSO_3)_2NO$, isolated in 1845, was a nitroxide.

$\cdot N{=}O$ $\quad$ $O{=}N{-}O\cdot$ $\quad$ $O{-}Cl{-}O\cdot$

DPPH $\qquad$ Nitroxide ($R_2N{-}O\cdot$)

The reason why radicals like DPPH and the nitroxides are stable can readily be seen in the MO diagram shown in Figure 2.1(a).

The interaction between the singly occupied free radical *p* orbital and a filled lone-pair orbital (*n*) localized on nitrogen leads to a net energy gain for one electron. In a similar manner, delocalization through interaction with a C=C double bond (allyl and benzyl radicals) leads to resonance stabilization (discussed later).

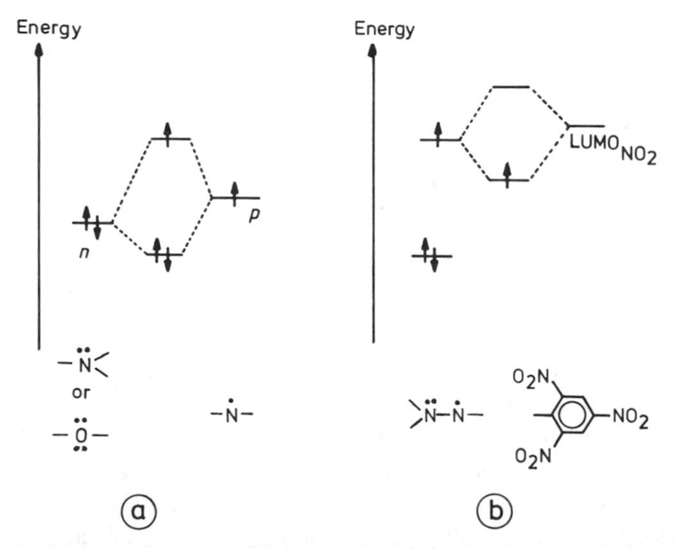

Figure 2.1. (a) Interaction between a filled lone-pair orbital (*n*) and a free radical *p* orbital (*p*); (b) further interaction with a low-lying LUMO.

†The IUPAC name is "aminooxy radical"; also, "aminyloxide" is used as synonym for "nitroxide" in the literature.

If now the stabilized radical (Figure 2.1(a)) possesses also a low-lying LUMO with which the *p* orbital can interact, a further stabilization is possible, as shown in Figure 2.1(b), for the LUMO originating from the trinitrophenyl residue. In other words, we have a situation similar to that described in Figure 1.5(c). This explains qualitatively the extraordinary stability of DPPH and many other radicals substituted by both *donor and accepter substituents*. Such radicals have been described as merostabilized, push-pull stabilized, or stabilized by the capto-dative effect.[2] An interesting case is the morpholinyl radical **1**, which exists in equilibrium with its dimer in solution at room temperature.[2b] The synergistic effects of donor and acceptor substituents can be described as in Figure 2.1(b), or, more simply, in terms of resonance structures (see also p. 106):

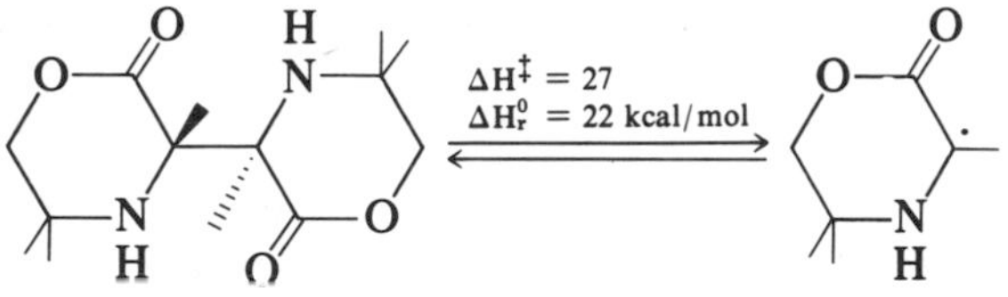

$\Delta H^{\ddagger} = 27$
$\Delta H_r^0 = 22$ kcal/mol

1

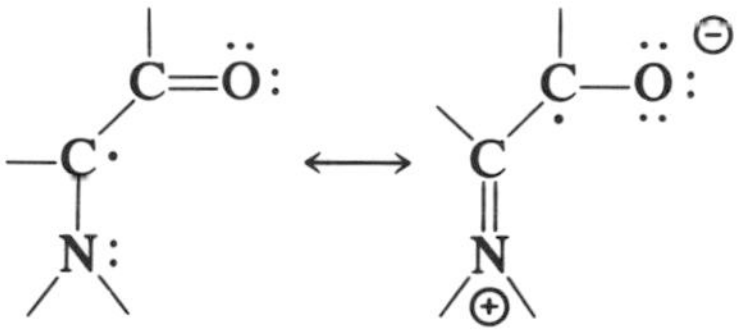

Most of the radicals we will encounter are not "stable"; that is, they are not persistent. (A persistent radical is one that is kinetically stable under the conditions of study.[3a]) It may take drastic conditions to generate them, and once formed, they may react extremely rapidly, with half-lives of the order of 10^{-10} s. For example, the homolytic cleavage of ethane into two methyl radicals [Eq. (1)] requires an enthalpy of activation ($\Delta H^{\ddagger}$) of 88 kcal/mol, which is equal to the "bond dissociation enthalpy" of the C—C bond, DH^0 (C—C).

$$CH_3{-}CH_3 \rightleftharpoons CH_3\cdot + CH_3\cdot \qquad \Delta H^{\ddagger} = 88 \text{ kcal/mol} \qquad (1)$$

The same enthalpy is released when two methyl radicals recombine to give ethane; the rate constant of the latter reaction in the gas phase has a value of the order

$$k \sim 10^{10} \text{ liter mol}^{-1} \text{ s}^{-1}$$

The upper limit for the rate of a bimolecular reaction is given by collision theory as

$$k = 10^{11} \text{ liter mol}^{-1} \text{ s}^{-1}$$

TABLE 2.1. Absolute Rate Constants for Radical Recombination ($2R\cdot \rightarrow R{-}R$)[a]

Radical	k (cm^3 molecule^{-1} s^{-1})	k (liter mol^{-1} s^{-1})
$\cdot CH_3$	$(4.0 \pm 1.0) \times 10^{-11}$	$(2.4 \pm 0.6) \times 10^{10}$
$\cdot C_2H_5$	$(1.3 \pm 0.3) \times 10^{-11}$	$(7.8 \pm 1.8) \times 10^{9}$
$(CH_3)_2\dot{C}H$	$(8.3 \pm 2.0) \times 10^{-12}$	$(5.0 \pm 1.2) \times 10^{9}$
$(CH_3)_3\dot{C}$	$(4.0 \pm 1.0) \times 10^{-12}$	$(2.4 \pm 0.6) \times 10^{9}$

[a]D. A. Parkes and C. P. Quinn, *J. Chem. Soc. Faraday Trans. 1*, **1976**, 1952. The exact values of recombination rate constants are under considerable debate. The present data are representative and were obtained by direct UV spectroscopic monitoring of the radicals in the gas phase.

Rate constants for some radical recombinations are given in Table 2.1. Since simple and sterically unhindered radicals recombine with virtually zero activation energy, the following relationship exists between the enthalpy of the reaction, $AB \rightarrow A\cdot + B\cdot$, and the bond dissociation enthalpy, $DH^0(AB)$:

$$\Delta H_r^0 \equiv \Delta H_f^0(A) + \Delta H_f^0(B) - \Delta H_f^0(AB) = DH^0(AB) \qquad (2)$$

This is also expressed in the enthalpy–reaction coordinate diagram shown in Figure 2.2.

The lack of an activation energy also implies that the two radicals will recombine on virtually every encounter. The only rate-limiting factor is, therefore, the rate

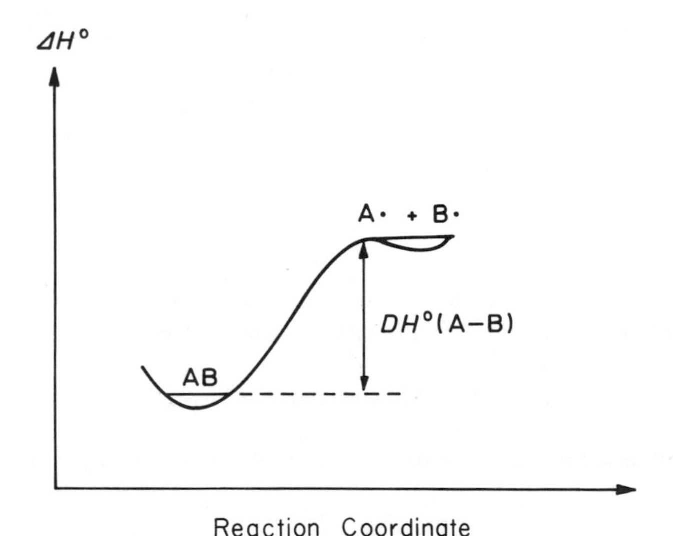

Figure 2.2. Enthalpy diagram for homolytic dissociation/recombination.

with which they approach each other, that is, the rate of diffusion. Such reactions are said to be diffusion controlled. From Eq. (2), we are now able to calculate the heats of radical reactions with the aid of tables of bond dissociation enthalpies and heats of formation. Some selected values are given in Table 2.2.

Consider the two possible thermal dissociations of ethane shown in Eqs. (1) and (3).

$$C_2H_5\text{—}H \longrightarrow C_2H_5\cdot + H\cdot \qquad (3)$$

The C—H bond dissociation enthalpy is 98 kcal/mol. Hence the reaction in Eq. (3) is 10 kcal/mol more endothermic than the cleavage in Eq. (1), and the high-temperature pyrolysis of ethane may be expected to produce methyl rather than ethyl radicals.

The list of bond strengths (Table 2.2) is also very practical for predicting which molecules will be useful radical sources and which will not. The C—H bond strengths in Table 2.2 indicate that it is easier to generate a tertiary alkyl radical than less highly substituted ones:

$$R_3C\text{—}H \qquad R_2HC\text{—}H \qquad RH_2C\text{—}H \qquad H_3C\text{—}H$$

$$\xrightarrow{\hspace{6em}}$$

C—H bond strength

Conversely, the tertiary radicals are less reactive (more stable) than secondary, primary, or unsubstituted ones:

$$R_3C\cdot \qquad R_2HC\cdot \qquad RH_2C\cdot$$

$$\xrightarrow{\hspace{6em}}$$

increasing reactivity

Thus the experimental determination of C—H bond strengths gives information on the intrinsic thermodynamic stability of radicals. For the radicals formed by C—H bond cleavage, the conditions laid down in Eq. (2) and Figure 2.2 will still apply; that is, the recombination with a hydrogen atom will take place with zero activation energy. However, when two tertiary butyl radicals recombine to hexamethylethane, an activation energy has to be overcome and the rate of recombination is, therefore, slower than that for the methyl radical (see Table 2.1). Therefore, Eq. (2) and Figure 2.2 do *not* strictly apply in this case. The activation energy has its origin in *steric strain* in the dimer, hexamethylethane.†

When bulky groups are attached to two neighboring carbon atoms, there will be

†So-called F-strain = front strain.

TABLE 2.2 Bond Dissociation Enthalpies and Heats of Formation (kcal/mol)[a]

$\Delta H_f^\circ(\dot{R})$ \ R—X \ $\Delta H_f^\circ(\dot{X})$ $\Delta H_f^\circ(R^+)$ \ $\Delta H_f^\circ(X^-)$		52.1 H 34.7	34.0 CH_3 (~34)	25.7 C_2H_5	20.8 n-C_3H_7	17.8 i-C_3H_7	7.5 t-C_4H_9	18.9 F −60	29.0 Cl −54.3	26.7 Br −50.9	25.5 I −45.1	9.0 OH −33.2	3.4 OCH_3 (≳ −30)	47.2 NH_2 30
H—	**52.1** 366	**104.2** 401	**104** (418)	**98**	**98**	**95**	**92**	**135.8** 370	**103.1** 334	**87.4** 324	**71.3** 315	**119** 390	**105** (386)	**110** 407
CH_3—	**34** 260	**104** 313	**88.2** (315)	**84.8**	**85.2**	**84.2**	**81.8**	**108.8** 256	**83.6** 227	**69.8** 219	**56.1** 212	**91.1** 274	**81.4** (274)	**86.7** 296
C_2H_5—	**25.7** 219	**98** 274	**84.8** (278)	**81.8**	**81.6**	**80.4**	**77.9**	**107.1** 220	**80.8** 191	**67.6** 184	**53.2** 176	**90.9** 242	**80.8** (241)	**84.3** 261
n-C_3H_7—	**20.8** 208	**98** 268	**85.2** (273)	**81.6**	**81.5**	**80.4**	**77.5**	**107.3** 214	**80.1** 185	**68.0** 178	**53.4** 171	**91.0** 235	**81.0** (236)	**84.8** 255
i-C_3H_7—	**17.8** 191	**95** 251	**84.2** (258)	**80.4**	**80.4**	**78.2**	**74.2**	**106.1** 200	**80.4** 170	**68** 164	**53.1** 156	**91.9** 222	**81.4** (221)	**85** 241
t-C_4H_9—	**7.5** 167	**92** 234	**81.8** (242)	**77.7**	**77.5**	**74.2**	**68.8**	**108.4** 189	**80.2** 157	**66.1** 149	**50.4** 140	**91.2** 208	**80.6** 207	**83.6** 226
[cyclopropyl]—	**61.3**	**100.7**	**89.6**	**86.2**	**86.3**	**85.4**	**82.7**	**110.9**	**85.4**	**73.7**	**58.6**	**97.5**	**87.8**	**91.2**
[cyclobutyl]—	**51.2**	**96.5**	**85.8**	**82.5**	**82.5**	**81.6**	**79**	**107.2**	**81.7**	**69.9**	**54.9**	**93.7**	**84**	**87.4**
[cyclopentyl]—	**24.3**	**94.8**	**83.6**	**80.3**	**80.5**	**79.6**	**77**	**105.2**	**79.7**	**67.9**	**52.9**	**91.3**	**82**	**85.4**
[cyclohexyl]—	**13.9**	**95.5**	**84.9**	**80.6**	**80.9**	**80.4**	**77.8**	**106**	**82**	**68.7**	**51.6**	**91.3**	**82.8**	**86.2**
[cycloheptyl]—	**12.2**	**92.5**	**81.5**	**78.2**	**78.2**	**77.3**	**74.7**	**102.9**	**77.4**	**65.6**	**50.6**	**89.4**	**79.7**	**83.1**
$H_2C{=}CH$—	**68.4**	**108**	**97.5**	**94.3**	**94.5**	**92.8**	**90.4**	**118.7**	**88.8**	**76.4**	**63**	**100.4**	**88.2**	
[phenyl]—	**77.7** 285	**110** 300	**99.7** (308)	**96.3**	**96.6**	**94.5**	**90.6**	**123.9** 253	**94.5** 219	**79.2** 210	**64.4** 202	**109.7** 275	**98.4** 272	**104.1** 294

$H_2C=CH-CH_2-$	**41.4**	**88.6**	**75.6**	**72.4**	**72.2**	**71.5**	**68.3**	**97.7**	**71.2**	**57.2**	**44.1**	**80**	**70.3**	**75.2**
$H_2C=CH-CH(CH_3)-$	**30.4**	**82.5**	**71.0**	**67.9**	**67.2**	**65.8**	**64.4**	**93.5**	**67.8**	**55.4**	**40.5**	**79.3**	**68.8**	**72.4**
[cyclopentenyl ring structure]	**38.4**	**82.3**	**71.6**	**67.8**	**67.2**	**65.6**	**61.6**	**93.5**	**67.8**	**55.4**	**40.5**	**79.3**	**68.8**	**72.4**
$HC\equiv C-CH_2-$	**86.2**	**93.9**	**81.1**	**77.5**	**77.4**	**76.3**	**73.4**	**103.2**	**76.7**	**63.9**	**49.3**	**86.9**	**76.9**	**80.7**
$(H_2C=CH)_2CH-$	**48.3**	**75.2**	**64.2**	**60.4**	**60.4**	**58.2**	**54.2**	**86.1**	**60.4**	**48**	**33.1**	**71.9**	**61.4**	**65**
$H_2C=CH-CH=CH-CH_2-$ *trans*	**48.3**	**82.3**	**69.5**	**65.9**	**65.8**	**64.7**	**61.8**	**91.6**	**65.1**	**52.3**	**37.7**	**75.3**	**65.3**	**69.1**
[cyclopentadienyl ring structure]	**61**[d]	**81.2**												
[cyclohexadienyl ring structure]	**44**	**69.8**	**59**	**55.2**	**55.2**	**53**	**49**	**80.9**	**55.2**	**42.8**	**27.9**	**66.7**	**56.2**	**59.8**
[cyclohexadienyl ring structure]	**44**	**70.1**	**59.3**	**55.5**	**55.5**	**53.3**	**49.3**	**81.2**	**55.5**	**43.1**	**28.2**	**67**	**56.5**	**60.1**
[cycloheptatrienyl ring structure]	**65** 212[e]	**73.2** 200	**61.2** 205	**57.9**	**57.9**	**57**	**54.4**	**81.9** 147	**56.4** 121	**44.6** 111	**29.6** 102	**68.4** 170	**58.7** 169	**62.1** 189
$C_6H_5-CH_2-$	**44.9** 213[e]	**85** 247	**71.8** (251)	**68.7**	**69**	**67.8**	**65**	**96.3** 196	**69.4** 166	**54.7** 157	**40** 149	**77.9** 215	**66.9** 212	**71.9** 236

[a]Data reproduced, with permission, from a more extensive compilation by K. W. Egger and A. T. Cocks, *Helv. Chim. Acta.*, **56,** 1516 (1973), unless otherwise indicated. Boldface numbers are homolytic bond dissociation enthalpies or heats of formation of radicals; numbers in regular typeface are for the corresponding heterolytic (gas-phase) processes. Some of the homolytic values may be in error by 2–3 kcal/mol, some of the heterolytic ones by as much as 10 kcal/mol. Newer determinations suggest a value of $\Delta H_f^\circ(t\text{-}C_4H_9\cdot)$ of 9.4 kcal/mol (see footnote *b*) but the old value of 7.5 kcal/mol has also been confirmed (see footnote *c*). Recent heats of formation of ions are given in *Gas-Phase Ion Chemistry*, M. T. Bowers, Ed., Academic Press, New York, 1979; H. M. Rosenstock, K. Draxl, B. Steiner, and J. T. Herron, *J. Phys. Chem. Ref. Data* (Suppl.) **1,** 1 (1977).

[b]A. L. Castelhano and D. Griller, *J. Am. Chem. Soc.*, **104,** 3655 (1982) and references therein.

[c]R. G. McLoughlin and J. C. Traeger, *J. Am. Chem. Soc.*, **101,** 5791 (1979).

[d]F. P. Lossing and J. C. Traeger, *J. Am. Chem. Soc.*, **97,** 1579 (1975).

[e]F. P. Lossing, *Can. J. Chem.*, **49,** 357 (1971).

serious van der Waals repulsion between the groups, and the central C—C bond will be weakened:[3]

$$(CH_3)_3C{-}C(CH_3)_3 \longrightarrow 2\ (CH_3)_2\dot{C}{-}CH_3 \qquad \Delta H^{\ddagger} = 68\ \text{kcal/mol} \qquad (4)$$

$$(\times)_3C{-}C(\times)_3{-}H \longrightarrow 2\ (\times)_2\dot{C}{-}H \qquad \Delta H^{\ddagger} = 36\ \text{kcal/mol} \qquad (5)$$

The preceding activation enthalpies may be compared with that given in Eq. (1) where $\Delta H^0 = \Delta H^{\ddagger} = DH^0$.

Obviously, we are dealing with two different effects:

1. Alkyl groups stabilize radicals inductively, making tertiary alkyl radicals more stable than secondary and secondary more stable than primary ones.
2. The more highly substituted alkanes decompose more easily, and the radicals formed experience an activation barrier toward recombination. The enthalpy profiles for the reactions given in Eqs. (4) and (5) have the form shown in Figure 2.3.

Consequently, a particular radical is "stable," long-lived, or persistent due to either (1) its own intrinsic *thermodynamic* stability or (2) the existence of activation barriers to further reaction in the system studied (i.e., *kinetic* stability). Kinetic stability always depends on the system under investigation. For example, the tertiary radical formed in Eq. (4) experiences an activation barrier, the exact value of which is a matter of dispute, toward dimerization to hexamethylethane. As we have seen, it would, however, immediately react with $H^{\bullet}$ atoms if such were available. The methyl radical $^{\bullet}CH_3$ is a highly reactive, kinetically unstable radical which is, therefore, difficult to observe and not normally persistent. If, however, the methyl radical existed in interstellar space (where it has not yet been detected), it would certainly be long-lived or "persistent" because reaction partners are far apart and no unimolecular decomposition pathways are available.

The free radical ClO_2 is a nice example of an intrinsically (thermodynamically) unstable compound that is nonetheless isolable in substance because no low-energy reaction channels exist under the conditions of the preparation. This compound can be converted photochemically to the isomeric peroxy radical $Cl{-}O{-}O^{\bullet}$ in a low-temperature matrix:[4]

$$O{\cdots}\dot{Cl}{\cdots}O \xrightarrow{h\nu} Cl{-}O{-}O^{\bullet}$$

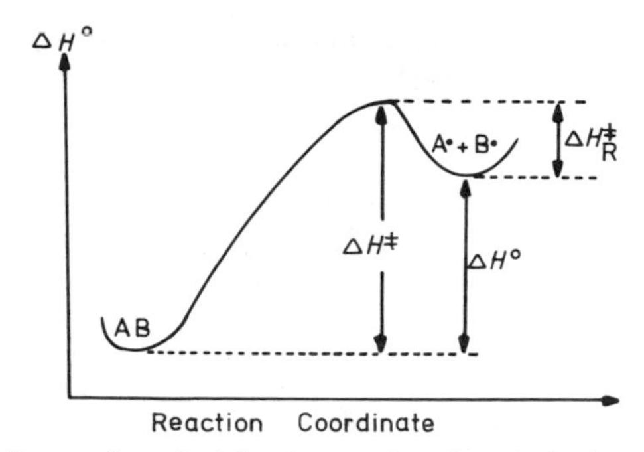

Figure 2.3. Enthalpy diagram for radical-forming reaction where the back reaction, A· + B· → AB has the activation enthalpy $\Delta H^{\ddagger}_{R}$.

It appears that Cl—O—O$^{\bullet}$ has a lower heat of formation than OClO; that is, it is thermodynamically more stable. Yet Cl—O—O$^{\bullet}$ is highly reactive and cannot be isolated at ordinary temperatures because of rapid secondary reactions.

The first stable hydrocarbon radical to be prepared, triphenylmethyl (**2**), is stabilized by a combination of steric and electronic effects. This radical was obtained by Gomberg in the year 1900 from the reaction between triphenylmethyl (trityl) bromide and silver:

$$Ph_3C\text{—}Br \xrightarrow[-AgBr]{Ag} Ph_3C^{\bullet} \rightleftharpoons (Ph)_2C{=}C_6H_5(H)(CPh_3)$$

2 **3**

$Ph_3C^{\bullet}$ (**2**) $\xrightarrow{NO}$ $Ph_3C\text{—}NO$

$Ph_3C^{\bullet}$ (**2**) $\not\rightarrow$ $Ph_3C\text{—}CPh_3$ (**4**)

The radical exists in equilibrium with its dimer (**3**) (12% dissociated at 25°C in a 0.01 M solution), but the dimer is not the expected hexaphenylethane (**4**) because steric hindrance due to the large phenyl groups in **2** makes a direct approach very difficult (vide infra). Instead, the dimer has the structure **3** in which one radical has attacked a benzene ring of another radical in the *para*-position.[5] Although the steric hindrance in **3** is less than that in **4**, the aromaticity of the central ring is sacrificed in the process. The aromaticity of benzene amounts to 20 kcal/mol (see Table 1.4) and this is a major reason why **3** dissociates back to **2** so easily. If a sterically demanding t-butyl group is introduced into the p-position, the dimer **(5)** will experience severe steric strain,

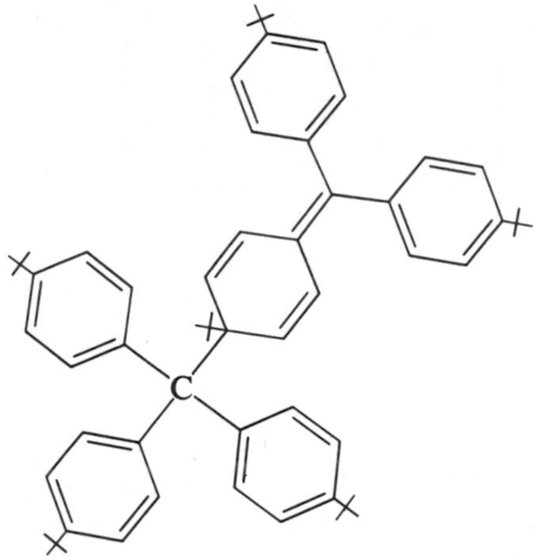

5

thereby facilitating dissociation.[6]

Likewise, steric hindrance toward radical dimerization can be increased by substituents in the o-position, as in **6**:

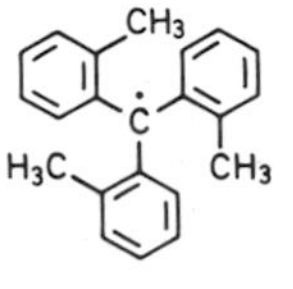

6

87% dissociated
(0.1 M in benzene at 20°C)

TABLE 2.3. Dissociation of Tetraphenylethanes (7)

7	R_1	R_2	R_3	ΔH^0 (kcal/mol)	ΔS^0 (cal K^{-1} mol^{-1})	Percent Dissociation[a]
a[b]	H	H	H	—	—	0
b[b]	CH_3	CH_3	H	22	45	0.35
c[b]	C_2H_5	C_2H_5	H	12	40	0.61
d[b]	t-Bu	H	H	19.7	18	1.97
e[b]	i-Pr	i-Pr	H	—	—	100
f[c]	t-Bu	H	t-Bu	—	—	100

[a]0.01 M solutions of **7** in m-xylene at 25°C.
[b]K. H. Fleurke, J. de Jong, and W. T. Nauta, *Rec. Trav. Chim.*, **84,** 1380 (1965).
[c]W. Theilacker and F. Koch, *Chem. Ber.*, **102,** 2020 (1969).

Even tetraphenylethanes (**7**) can be caused to dissociate when bulky groups are introduced in the *o*-positions (see Table 2.3).

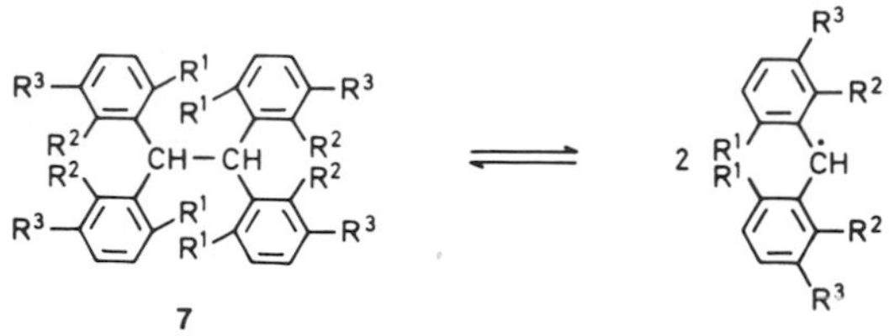

7

The barrier to dimerization is not the only reason why these radicals are stable: They are also intrinsically stabilized due to resonance with the benzene rings. If triphenylmethyl were planar,[7] the free electron would be equally distributed over all three rings:

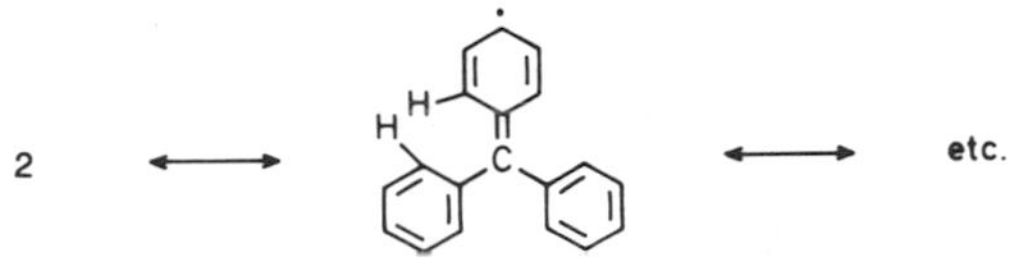

2 ⟷ ⟷ etc.

Steric hindrance makes such coplanarity impossible, so that resonance is diminished, but still significant. According to Hückel MO theory, the free electron density at the benzylic carbon atom is decreased to 4/7 in the benzyl radical.

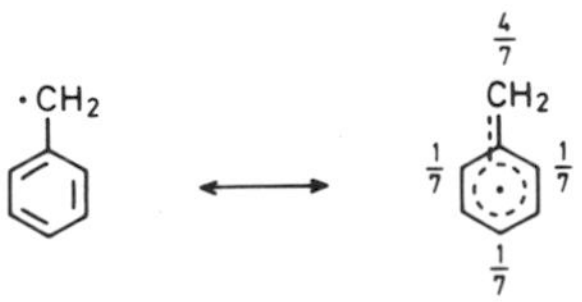

Benzyl radicals are, therefore, much more stable than, for example, $CH_3^{\bullet}$, which has a full electron localized on carbon. The benzyl resonance lowers the heat of formation of the radical by approximately 12 kcal/mol (see Table 2.4). These radicals are not unreactive, however. Triphenylmethyl reacts rapidly with NO to give nitrosotriphenylmethane, and with oxygen it gives the peroxide,

$$Ph_3C\text{-}O\text{-}O\text{-}CPh_3$$

The conjugated molecule **8** (Čičibabin's hydrocarbon[8]) is an example of "radicalization" due to aromatization: The resonance energy gained (20 kcal/mol per benzene ring) is sufficient to allow an equilibrium with the singlet diradical **9** in solution. In the singlet state, the two "free" electrons are still coupled; that is, they have opposite spins. However, heating the molecule is enough to uncouple the electrons so that a triplet diradical **10** is formed (we shall discuss singlets and triplets in Section 2.2.2 and in Chapter 3). The most general reaction of radicals is hydrogen abstraction from the solvent (Section 2.5.1), which, in the present case, gives rise to the monoradical **11**:

9 ⇌ **8**

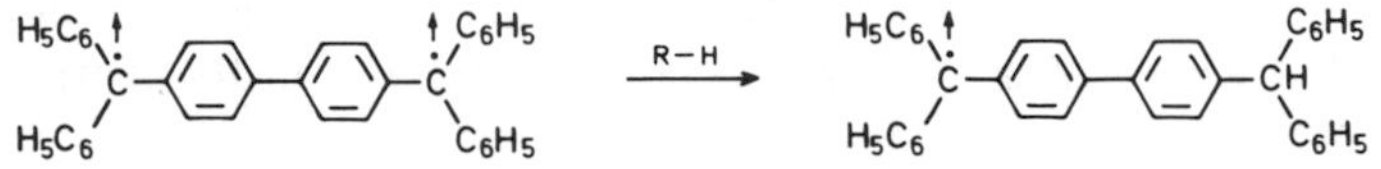

9 ⇌ (Δ) **10**

10 —(R–H)→ **11**

In contrast, it is not possible to write a nonradical singlet ground-state structure for the cross-conjugated diradical **12** (Schlenk's hydrocarbon[9]). As a consequence, **12** exists in the ground state as a triplet diradical, and it reacts in the same way as

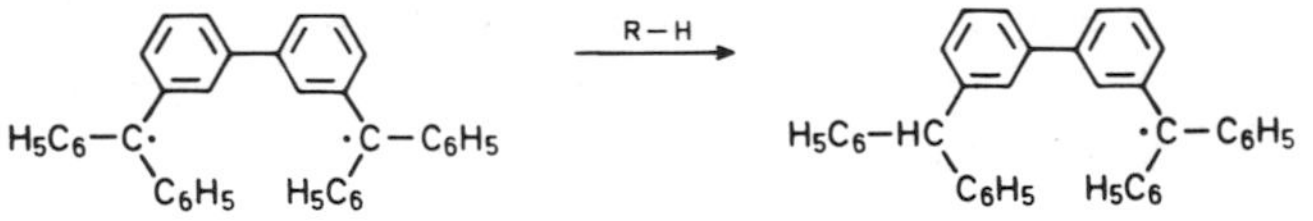

12 —(R–H)→ monoradical —(R–H)→ **13**

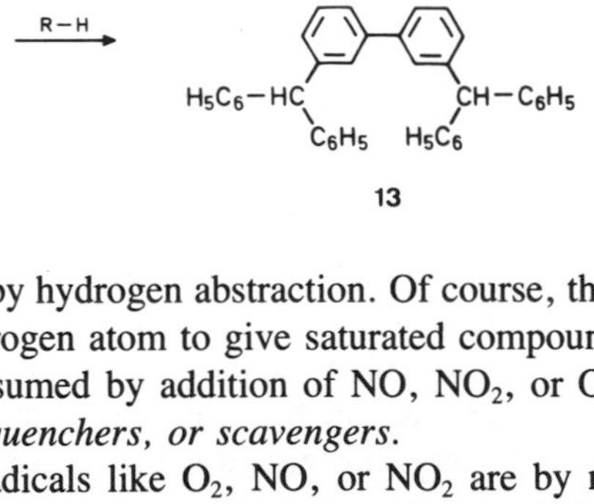

the triplet **10**, giving monoradicals by hydrogen abstraction. Of course, these monoradicals may abstract another hydrogen atom to give saturated compounds (e.g., **13**). All these radicals will be consumed by addition of NO, NO_2, or O_2, which therefore are called *radical traps, quenchers, or scavengers.*

Thus it is seen that "stable" radicals like O_2, NO, or NO_2 are by no means unreactive. When brown NO_2 gas is condensed, the colorless dimer N_2O_4 is formed,

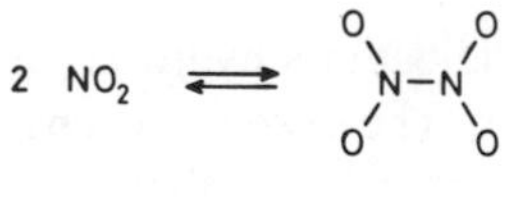

$$2\ NO_2 \rightleftharpoons O_2N{-}NO_2$$

and even NO dimerizes:

$$2\ NO \rightleftharpoons \begin{matrix} N{=}O \\ | \\ N{=}O \end{matrix} \qquad \Delta H^0 = -2.45\ \text{kcal/mol}$$

TABLE 2.4. Extra Resonance Energies (RE) in Radicals and Ions[a]

Radical or Ion	RE(kcal/mol)	Radical or Ion	RE(kcal/mol)
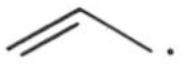	9.6 ± 1.5	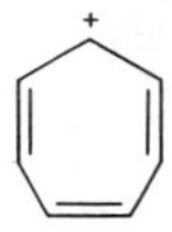	(49)
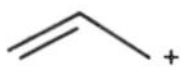	(12)		
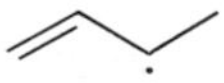	12.6 ± 1.5		
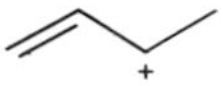	(13)	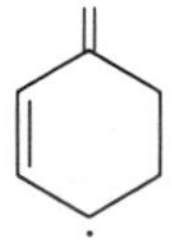	23.4 ± 2.5
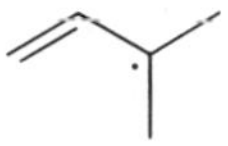	13.1 ± 1.5 (7.8 ± 1.5)		
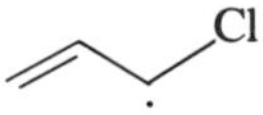	9.4 ± 2	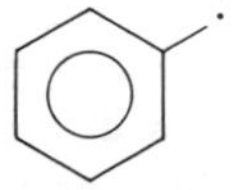	12.5 ± 1.5
	11.4 ± 2.2 (10.8 ± 2.5)		
	12.6 ± 1.5	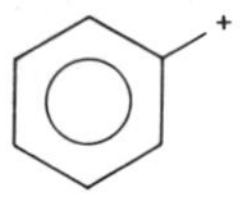	(21)
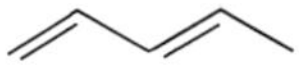	20 ± 2		4–9
		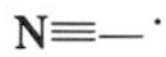	5–7[b]
		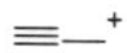	24
	15–20	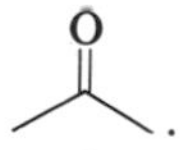	(0)
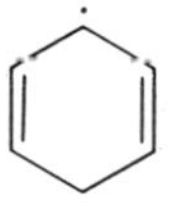	24.6 ± 2.5	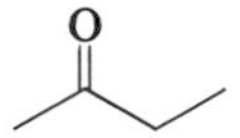	(2.7) (7)
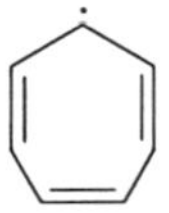	21 ± 2	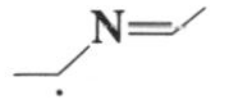	14.5 ± 2
		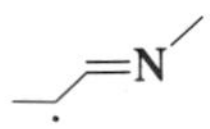	~7 ±3

[a]Data compiled by K. W. Egger and A. T. Cocks, *Helv. Chim. Acta,* **56,** 1537 (1973) and reproduced with permission.
[b]D. Belluš and G. Rist, *Helv. Chim. Acta,* **57,** 194 (1974); K. D. King and R. D. Goddard, *Int. J. Chem. Kinet.,* **10,** 453 (1978).

with an energy gain of 2.45 kcal/mol.†

When these molecules exist as radicals in the gas phase, it is because the small dissociation enthalpy is outweighed by the *entropy* gained when two molecules are formed from one:

$$\Delta G^0 = \Delta H^0 - T\,\Delta S^0$$

If ΔH^0 is small and the temperature is high, $T\,\Delta S^0$ will be the main term, lowering the free energy ΔG^0. This does not change the fact that ΔH^0 for dimerization of

†The van der Waals bond strength of $(NO)_2$ is only 260 cal/mol. $(O_2)_2$ exists as a van der Waals dimer with an energy gain of 233.5 cal/mol. Many other weakly bound molecules exist, such as $(He)_2$ (20 cal/mol), $(Ar)_2$ (238 cal/mol), and $(Xe)_2$ (439 cal/mol) (Ref. 10).

NO is slightly negative; that is, the dimerization is exothermic. The important fact is that the *formation of chemical bonds almost always results in a lowering of the energy*. This is the main reason for the reactivity of "reactive intermediates" and for the existence of molecules.

2.2. DETECTION OF RADICALS AND STRUCTURE DETERMINATION

The distinguishing feature of a free radical is its unpaired electron. Any molecule or atom that has one or more unpaired electrons is paramagnetic; it acts as a small magnet and is drawn into a magnetic field. This magnetic property can be detected in various ways, for example by magnetic susceptibility measurements. We shall be concerned here mainly with ESR spectroscopy, which is a direct observation, and the CIDNP effect, which provides an indirect method for probing the radical origin of reaction products. Results obtained from low-temperature infrared spectroscopy, theoretical calculations, and chemical studies will also be mentioned in the appropriate places.

2.2.1. ESR Spectroscopy and the Structure of Radicals

Let us first review briefly the NMR phenomenon (Figure 2.4). In an external magnetic field, a nucleus possessing a nuclear spin I can exist in distinct nuclear magnetic energy states characterized by the magnetic quantum number m_I; for protons $m_I = \pm\frac{1}{2}$. The NMR phenomenon depends on a slight overpopulation of the lower energy state, $m_I = +\frac{1}{2}$, and an experimentally induced transition from the lower to the upper energy state, which requires an energy $\Delta E = \mu \mathbf{H}_0/\mathrm{I}$, where μ is the nuclear magnetic moment.

The ESR (electron spin resonance) phenomenon[11] is completely analogous (Figure 2.5). The electron has the spin $S = \frac{1}{2}$, which in a magnetic field is allowed to

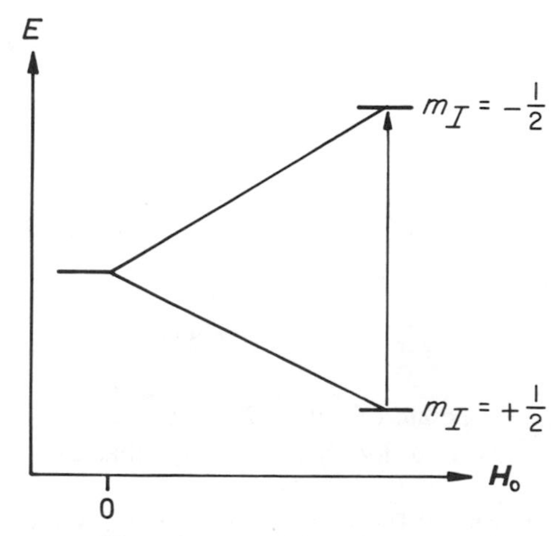

Figure 2.4. The NMR phenomenon.

take the values $m_S = \pm\frac{1}{2}$. For one or more unpaired electrons, one defines the total *spin multiplicity* as $(2S + 1)$. Thus one unpaired electron has the spin multiplicity $2 \cdot \frac{1}{2} + 1 = 2$, and the system is called a doublet, corresponding to the two values of m_S. Two unpaired electrons have the spin multiplicity $2 \cdot 1 + 1 = 3$, and the ensemble is said to be a triplet. It has three values of m_S: -1, 0, and $+1$, formed by vectorial addition of the m's for the individual electrons. A diamagnetic molecule possessing no unpaired electrons has the spin multiplicity $2 \cdot 0 + 1 = 1$ and is said to be a singlet. Paired electrons have opposite spin quantum numbers, one with $S = +\frac{1}{2}$, the other with $S = -\frac{1}{2}$, so the combined spin equals zero. There is only one value of m_S; that is, there is no energy level splitting in an applied magnetic field, and ESR cannot be observed.

Associated with the electron spin is a magnetic moment μ

$$\mu = -m_S g \mu_B = \mp\tfrac{1}{2}\, g\mu_B, \qquad E = -\mu \mathbf{H}$$

where g is the so-called g-factor, a dimensionless factor which for the free electron has the value 2.0023. It differs slightly, but measurably, for unpaired electrons in different molecules. μ_B is the Bohr magneton, which transforms the dimensionless quantity into units of magnetic moment (see Units and Constants). The two possible values of μ correspond to two states of equal energy. In the presence of an external magnetic field $\mathbf{H}$, the energies become different (Figure 2.5). The state with $m_S = -\frac{1}{2}$ will be stabilized; the one with $m_S = +\frac{1}{2}$ destabilized. The stronger the field, the greater the energy difference, $\Delta E = 2\mu\mathbf{H} = g\mu_B\mathbf{H}$.

The splitting of the energy levels is called the Zeeman effect, and the energy levels sometimes the Zeeman levels. Insertion of the values of g and μ_B for a field $\mathbf{H}$ of 0.3400 T will show that the value of ΔE is extremely small, ~1 cal/mol. The population of the two levels is given by the Boltzmann distribution, $\exp(-\Delta E/RT)$. Because the energy difference is so small, the overpopulation of the lower level is only of the order of one in a thousand, but this is sufficient that absorption of microwave radiation, $h\nu = \Delta E$, can be observed and recorded. (In NMR spec-

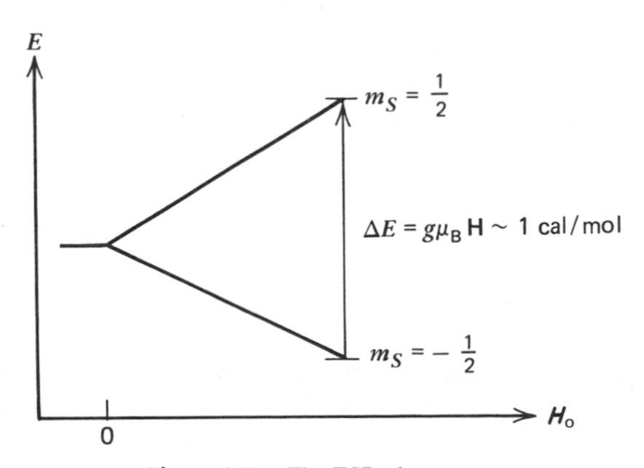

Figure 2.5. The ESR phenomenon.

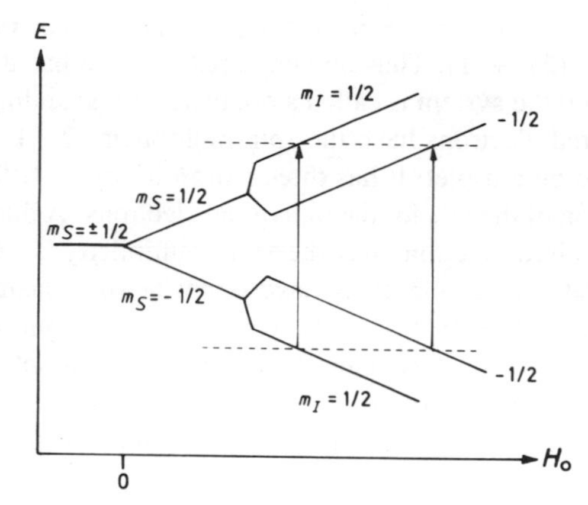

Figure 2.6. Coupling between electron and nuclear spins. Selection rules for the transitions: $\Delta m_S = \pm 1$; $\Delta m_I = 0$.

troscopy, the overpopulation of the lower level is only one in 10^5 at room temperature, and $\Delta E \sim 6 \cdot 10^{-3}$ cal/mol. Thus it is easier to observe ESR than NMR.) In order for the resonance absorption to continue, it is necessary that the excited spins ($m_S = +\frac{1}{2}$) relax again so that the lower level can be repopulated. The mechanism of relaxation is by spin–lattice and spin–spin interaction as in NMR spectroscopy.

Also in complete analogy with NMR spectroscopy, the unpaired electron can couple with nuclear spins, for example ^{13}C or ^{1}H, which both have a nuclear spin $m_I = \pm\frac{1}{2}$. This is illustrated in Figure 2.6. Designating the electron spins m_S and the nuclear spins m_I, the selection rule is

$$\Delta m_S = \pm 1 \qquad \Delta m_I = 0$$

Hence only two transitions can occur (Figure 2.6), but these are of equal energy and intensity. Only the magnetic fields are different, the difference a_H (for coupling with protons) being the *hyperfine splitting constant* (hfs). a_H may be likened to J_{HH} in proton NMR spectroscopy. An absorption spectrum such as that depicted in Figure 2.7(a) would result. In practice, however, derivative spectra [Figure 2.7(b)] are recorded. This is a technical stratagem that serves to increase the sensitivity and to narrow the linewidths.

The magnitude of the coupling with the ith nucleus N_i is given by the equation

$$a_{N_i} = Q_{N_i} \rho_i$$

Where Q is a factor that varies from nucleus to nucleus, and ρ_i is the electron spin density at the ith nucleus.† Spin densities can be calculated by MO methods;

†The equation for coupling with protons, $a_{H_i} = Q_{H_i}\rho_i$ is known as the *McConell equation*. Generally, one may speak of McConnell-type equations (Ref. 12).

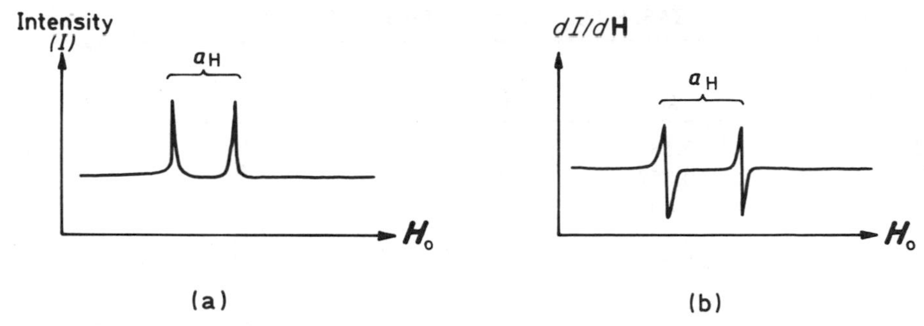

Figure 2.7. (a) "Normal" ESR spectrum, showing coupling, a_H; (b) derivative spectrum.

conversely, the experimental a_{N_i} give direct information about spin densities. This is especially important in extended conjugated systems such as aromatic hydrocarbons.

Consider the methyl radical **(14)**. This radical is planar, or very nearly so, and the evidence for this comes mainly from ESR spectroscopy. The radical is, therefore, sp^2 hybridized, and the free electron moves in a (pure) p orbital. This orbital has a node in the plane occupied by the protons. We might expect, therefore, that there would be no coupling; $\rho_i = 0$. However, a coupling $a_H \sim 2.3$ mT is observed. This is caused by spin polarization via the σ orbitals (see **14**). Being a small magnet, the radical electron will tend to polarize the nearest σ electron in the same direction. This follows from Hund's first rule, which states that two electrons in two different orbitals will tend to have the highest spin multiplicity (i.e., to be unpaired). The two σ electrons in the C–H bond have to be paired, however, due to the Pauli exclusion principle. The electron near the proton in **14** will then be spin-polarized in the direction opposite to the free electron, and a small *negative* spin density will be experienced by the proton. Hence a negative a_H is recorded.

Longer-range couplings exist as well; for protons on neighboring carbons (β-protons) the sign of a_H is usually positive. In the methyl radical, the electron will also couple with the carbon nucleus, as can be observed by ^{13}C-labeling. This coupling is positive and the magnitude is $a_{^{13}C} = 3.8$ mT.

The magnitude of the hyperfine splitting constant a is of great importance for the determination of radical structure. If a carbon radical could be induced to be unplanar **(15)** the electron would be in a sp^3-type orbital and accordingly have some s character. Since s orbitals are spherical, all the orbitals will now overlap to some extent, and neighboring electrons must be paired (see **16**). There will be a *positive* spin density at the positions of the protons.

Table 2.5 gives some constants for *free atoms*.

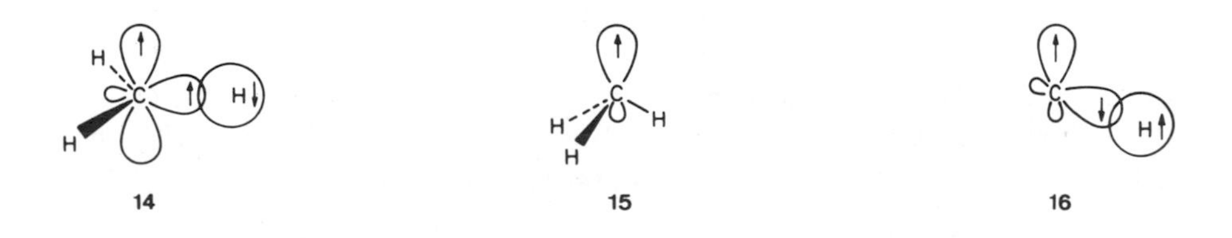

TABLE 2.5. Constants for Free Atoms[a]

Nucleus (N)	g_N	I_N	m_I	a_N(mT)
^{1}H	5.585	$\frac{1}{2}$	$\pm\frac{1}{2}$	51.0
D	0.857	1	0, ±1	7.8
^{13}C	1.404	$\frac{1}{2}$	$\pm\frac{1}{2}$	111.0
^{14}N	0.404	1	0, ±1	55
^{17}O	−0.757	$\frac{5}{2}$	$\pm\frac{1}{2}, \pm\frac{3}{2}, \pm\frac{5}{2}$	165
^{19}F	5.255	$\frac{1}{2}$	$\pm\frac{1}{2}$	1715

[a]J. R. Morton, *Chem. Rev.*, **64,** 453 (1964). The hyperfine splitting constants a_N are for electrons in pure *s* orbitals.

For the hydrogen atom, the hfs is ~51 mT. The C—H bonds attached to an sp^3-hybridized carbon atom have $\frac{1}{4}$ *s* character, so we may expect an $a_H^{sp^3} \sim 51/4 = 12.7$ mT. Unfortunately, the spin polarization mechanism (see **14**) has the opposite effect, so that the observed coupling may be much smaller. Thus the cyclopropyl radical (**17**) has a small coupling, $a_{H\alpha} = 0.65$ mT, but it is positive, indicating that the radical is *nonplanar:*

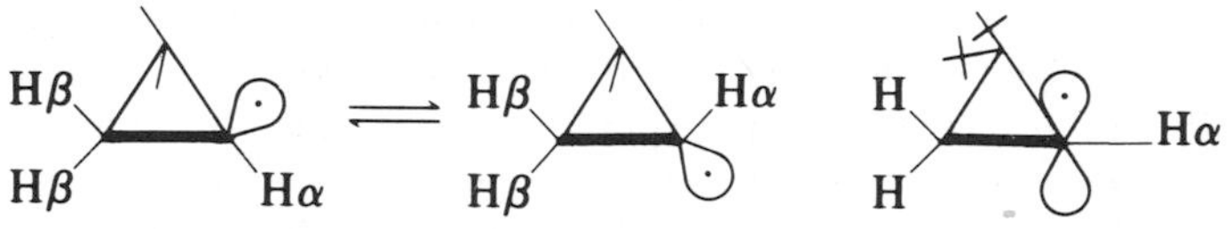

Hβ Hβ Hα ⇌ Hβ Hβ Hα   H H Hα

17a **17b** **18**

Chemical evidence for nonplanarity has been presented, but the barrier to inversion is very low (**17a** ⇄ **17b**).[13] Had the radical existed in one of the conformations **17a** or **17b** for a long time on the ESR time scale ($>10^{-10}$s), the two βH's would have been expected to give different coupling constants. They were, however, identical (2.37 mT).[13c] On the other hand, the ditertiary butyl substituted radical **18** was found by ESR spectroscopy to be *planar.*[13d] The reason for this may be sought in steric interactions between Hα and the *tert*-butyl substituents, which is minimized in the planar form.

Fluoroalkyl radicals are nonplanar. The ^{13}C splitting constants increase in the series:[14]

	$\cdot CH_3$	$\cdot CH_2F$	$\cdot CHF_2$	$\cdot CF_3$	
$a_{^{13}C}$	3.85	5.48	14.88	27.2	mT

For CF_3, the hfs of 27.2 mT is approximately one-fourth of the pure *s* value (Table 2.5), indicating the CF_3 is sp^3-hybridized. Thus, when going from CH_3 to CF_3, the angle of deviation from planarity should gradually increase from 0 to 17.8°, the regular tetrahedron value being 19.5°. Several oxygen containing radicals are also found to be nonplanar, for example,[15]

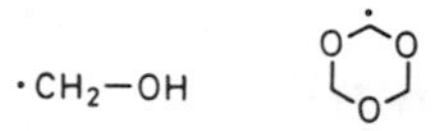

$\cdot CH_2{-}OH$ (trioxanyl radical)

$a_{^{13}C}$ 4.6 9.9 mT

A rationalization of why *electronegatively substituted radicals are nonplanar* has been given by Pauling.[16] An electronegative atom will attract electrons from the central carbon atom. This will be more efficient in a *p*-type orbital than in an *s*-type, or, conversely, if the C—F link were pure *s*, for instance, the F atom would distort the *s* symmetry so that a higher electron density would be created at fluorine. Hence the orbital would acquire *p* character and go from sp^2(planar) to sp^3(tetrahedral). This effect will be compensated to some extent by *p-p* back donation from the fluorine lone pairs (see **19**), which gives the C—F bond some double-bond character (**20**).

112°

19 **20**

For the same reason, alkoxy radicals possess some C=O bond character, which manifests itself in a reduced rate of rotation about the C—O bond (**21**). The

slow

21

nonplanarity of cyclopropyl (**17**) can be understood on the same basis. It is well known that the C—C bonds in cyclopropane have excessive *p* character because the 60° geometrical angles do not allow sp^3-hybridization (normal angle: 109°). The extra *p* character is taken from the C—H bonds which, of necessity, acquire more *s* character ($\sim sp^2$). Hence the electron in cyclopropyl will reside in an *s*-type orbital, and the radical will be a σ radical.

There is evidence from ESR spectroscopy that the *tert*-butyl radical may be nonplanar also. From the ^{13}C hfs of 4.62 mT measured on the perdeuterated radical in an adamantane matrix, a dihedral angle of ~26° was derived.[17] Although this conclusion has been criticized, recent ab initio calculations confirm a nonplanar geometry (**22**) with ~22° out-of-plane bending at the radical center.[18]

This bending leads to a better hyperconjugative interaction with the neighboring C—H bond, with which the radical *p* orbital is eclipsed. The calculations further indicate that all alkyl radicals other than methyl should show some (small) degree of pyramidalization at the radical center.[18]

22

Although these radicals are stabilized by eclipsing as illustrated in formula **22**, the calculated activation energies for rotation about the central C—C bonds are exceedingly small. This has been confirmed by low-temperature matrix IR studies of the ethyl radical[19] for which the eclipsed nonplanar structure **23** with a dihedral angle of 6.2° was calculated by ab initio methods.

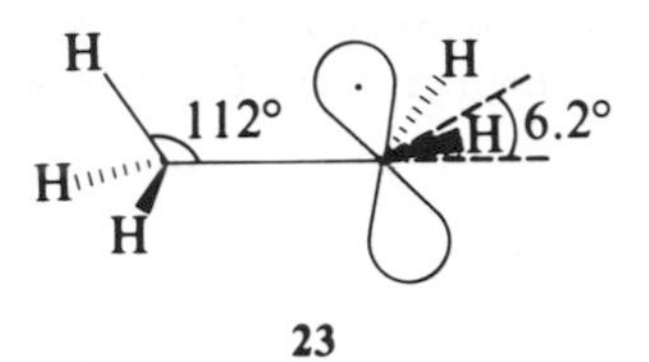

23

Using the deuterated radical HCD_2—$\dot{C}D_2$, only *one* methyl C—H stretching vibration was observed in the IR, thereby indicating that the three hydrogen atoms in the methyl group are identical; that is, the barrier to rotation is exceedingly small (~150 cal/mol). The infrared absorptions further demonstrate a weakening of the β-CH bonds and a strengthening (shortening) of the α-CH and α-CC bonds in alkyl radicals.[19] This is readily understood if the radical center in **23** is compared with the similarly hybridized methylene group in ethene. In other words, **23** may be regarded as halfway olefinic. The weakness of the eclipsed β-CH bond is reflected in the disproportionation of alkyl radicals, whereby an olefin and an alkane are formed (see Section 2.5.2).

ESR spectroscopy can give important information on the geometry at neighboring carbon atoms. In analogy with the Karplus rule in NMR spectroscopy, there exists a relationship between the βH coupling constant and the dihedral angle θ (cf. **24**):

$$a_{\beta H} = B_0 + B \cos^2 \theta$$

where B and B_0 are constants ($B_0 \lesssim 0.3$ mT and $B \sim 5.0$ mT per free electron).

When H_β is in the nodal plane in **24**, $\theta = 90°$ and $a \cong 0$. When H_β is eclipsed

24

with the *p* orbital, the hyperfine splitting constant will be at a maximum. The same type of relationship holds for other β-nuclei. For example, ^{19}F, ^{33}S, and ^{35}Cl couplings can be directly observed.

Some preferred radical conformations derived in this manner are shown in formulas **25–29**.[20] As indicated earlier, the radical centers may actually be slightly pyramidal in many cases.

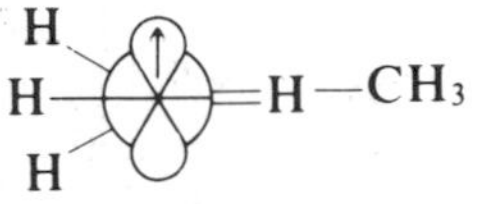

25 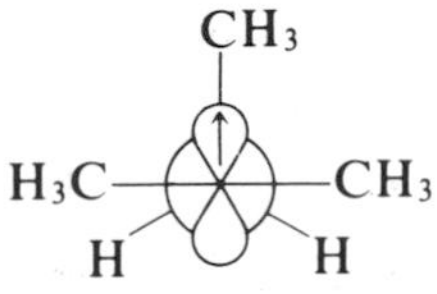 **26** 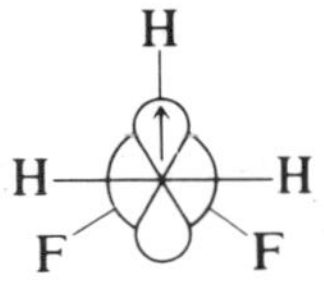 **27**

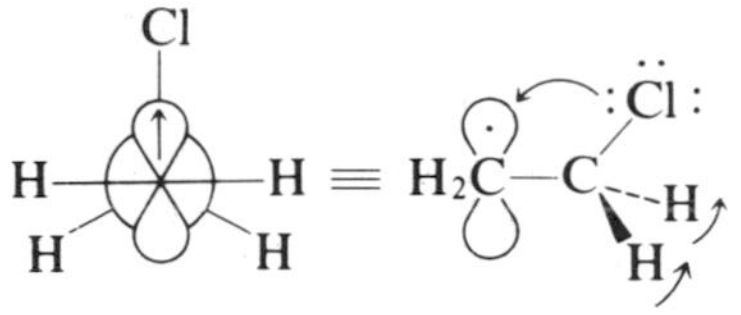

28 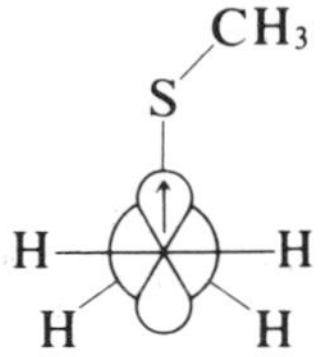 **29**

It is particularly noteworthy that, in most cases, the free radical *p* orbital is eclipsed with a neighboring group. The eclipsing of a β-chlorine or β-sulfur substituent can be ascribed to back donation from the substituent lone pairs as shown in formula **28**. Consequently, the β-substituent will tilt a little in the direction of the *p* orbital, and at the same time, the β-hydrogen atoms will tilt upward (see **28**). This tends to put the β-H's into the nodal plane and therefore reduces the coupling to ≈ 0. Such back donation *does not result in the formation of symmetrically bridged radicals,* however,

for the coupling constants with the α- and β-protons are found to be different. This observation does not exclude the possibility that unsymmetrically bridged radicals such as **28** can interconvert via a symmetrical transition state with a rate that is slow on the ESR time scale:

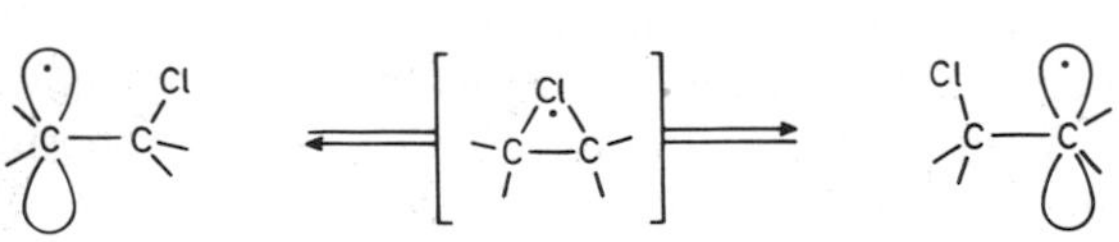

28a **28b**

Had the symmetrical species been the ground state, all protons would have been identical, and only *one* hfs would have been observed. Had the interchange between **28a** and **28b** been fast, the ESR instrument would have detected only an averaged hfs value.

There is strong chemical evidence that unsymmetrical chlorine- or bromine-bridged radicals do indeed interconvert; the β-bromine radical with a rate $>10^{11}$ s^{-1}; the β-chlorine radical with a rate $<10^{8}$ s^{-1}.[21] This was established in the following way. The β-bromo-alkyl radical **31**, labeled with ^{82}Br, is produced by abstraction of a secondary hydrogen in **30** by bromine atoms (for details of radical-forming reactions, see Section 2.5.1.). The outcome of the reaction is a bromination, yielding **33** and **34**. The observation of ~100% scrambling of radioactive bromine between the two positions in **33** and **34** proves the rapid interconversion of the bridged radicals **31** and **32**, *or* the exclusive presence of *one* symmetrically bridged intermediate, **35**.

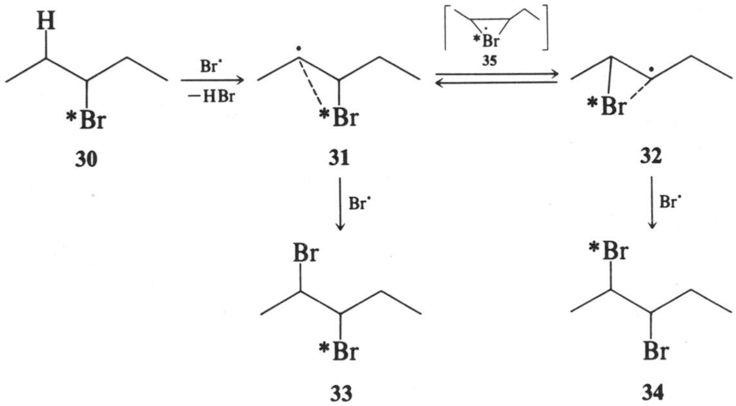

*Br = ^{82}Br

In Section 2.2.3, it will be shown that the unsymmetrically bridged radicals **31** and **32** represent the ground state.

Bromine bridging has the further effect of lowering the energy of activation for abstraction of the hydrogen atom indicated in **30**: The stabilization is felt already in the transition state. Furthermore, $\Delta S^{\ddagger}$ is lowered as well, indicating a "tight" transition state.[22] With the more electronegative (less electropositive) chlorine as bridging atom, the stabilization is weaker; the frequency of interchange is lower, and the weakness of the bridge bond allows rotation to occur. Accordingly, a partial loss of optical activity is observed when the enantiomerically pure substrate **36** is brominated:

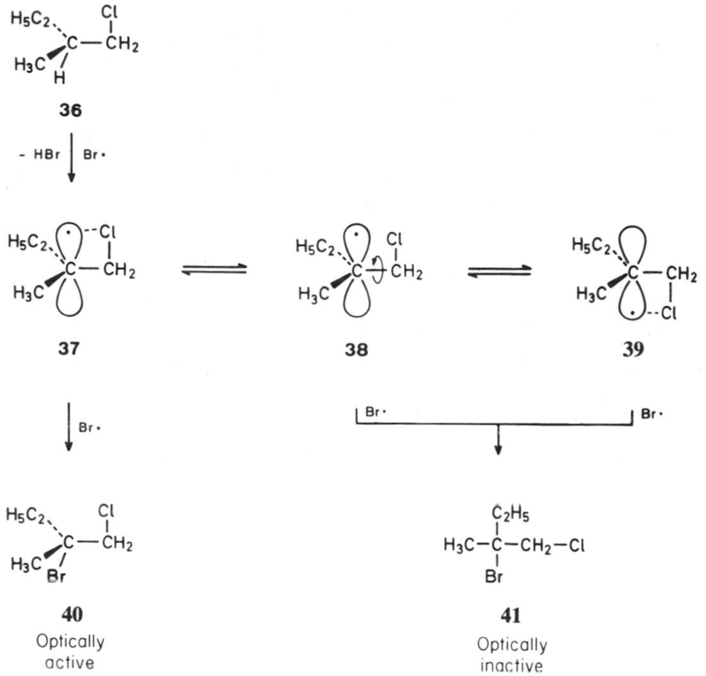

36

$-HBr$ | $Br\cdot$

37 ⇌ **38** ⇌ **39**

$Br\cdot$

40
Optically active

41
Optically inactive

If the product-forming step, **37** → **40**, is fast enough, **40** will be optically active due to backside attack of Br· on the bridged species **37**. If the reaction is slow—and it can be slowed down by decreasing the concentration of Br·—**37** will interconvert with **38** and **39** so that optical integrity is lost, giving **41** as the final product.

If two bridged radicals have different stabilities, the less stable one will rearrange to the more stable one; thus a primary radical can rearrange to a tertiary one within 10^{-10} s, which is too fast for an ESR observation of the primary radical.[23]

Semiempirical quantum-chemical calculations in the INDO approximation have confirmed the stability of the unsymmetrically bridged β-chloroethyl radical (**28**), yielding a barrier to rotation of 16 kcal/mol.[23b] This is sufficient to explain the partial retention of configuration in the reactions of **36**.

The unsymmetrical nature of the halo-radicals contrasts with the *halonium ions*, which are known to be cyclic and symmetrical.[24] For example,

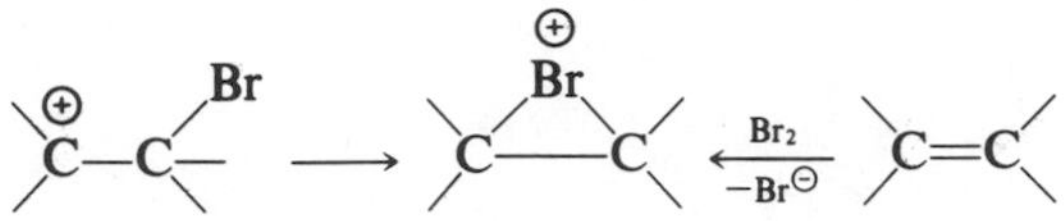

Here too, the bromonium ions are more strongly stabilized than the chloronium ions.[25] A cationic center is necessarily more electron deficient than a radical center, resulting in a stronger bonding with the halogen atoms.

2.2.2. The Triplet State

If a molecule contains two unpaired electrons, it will, according to Hund's first rule, be a triplet diradical. Examples are 1,*n*-diradicals (**42**), carbenes (**43**), and the excited triplet states of aromatic molecules such as naphthalene (**44**).

$\overset{\uparrow}{C}H_2-(CH_2)_{n-2}-\overset{\uparrow}{C}H_2$

42

43

S_0 S_1 T_1

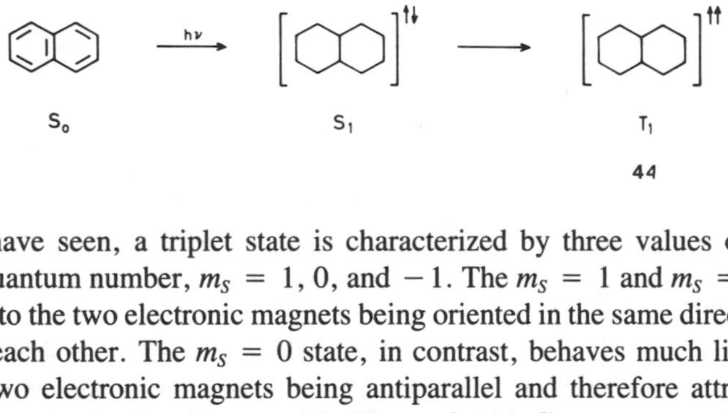

44

As we have seen, a triplet state is characterized by three values of the spin-magnetic quantum number, $m_S = 1, 0,$ and -1. The $m_S = 1$ and $m_S = -1$ states correspond to the two electronic magnets being oriented in the same direction; hence they repel each other. The $m_S = 0$ state, in contrast, behaves much like a singlet state, the two electronic magnets being antiparallel and therefore attracting each other. (This point is also illustrated in Figure 2.11.) Consequently, the three m_S levels do not possess the same energy, but the $m_S = 0$ level is stabilized and the $m_S = 1$ and -1 levels are destabilized, as shown in Figure 2.8. The energy separation is called D.[26]

Now in this triplet molecule, which could be triplet naphthalene (**44**), for example, each electron moves in the magnetic field of the other. If the molecule has less than cylindrical symmetry, there will be two different values of these internal fields (e.g., along the two principal axes of the molecular plane in naphthalene). These fields do not affect the $m_S = 0$ level, which has no overall spin and can be likened to a singlet state. They cause, however, a further splitting of the $m_S = 1$ and $m_S = -1$ levels. This splitting, called $2E$, is also shown in Figure 2.8. Since

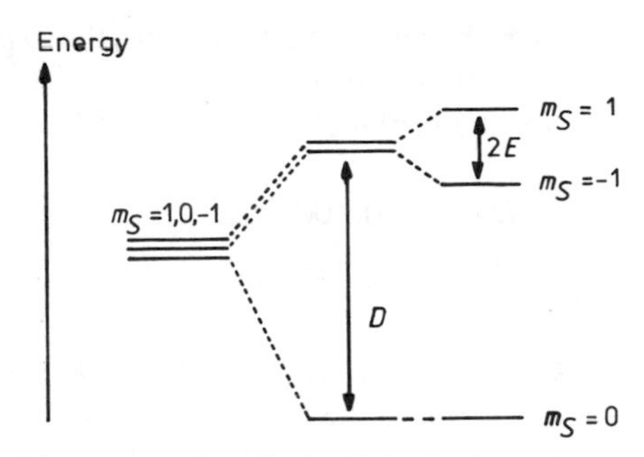

Figure 2.8. Zero-field splitting in the triplet state.

these splittings persist in the absence of an external magnetic field, the constants D and E are termed zero-field splitting parameters.

When an external field **H** is applied, the energy difference between the $m_S = 1$ and $m_S = -1$ levels will increase, just as is the case for a single electron in a magnetic field (Zeeman effect). The energy of the $m_S = 0$ level remains unchanged, however. Thus the three levels will split as illustrated in an idealized fashion in Figure 2.9.

The selection rule is $\Delta m_S = \pm 1$ (cf. also Figure 2.5). Although $\Delta m_S = 2$ is forbidden, this transition may be observed under certain conditions.

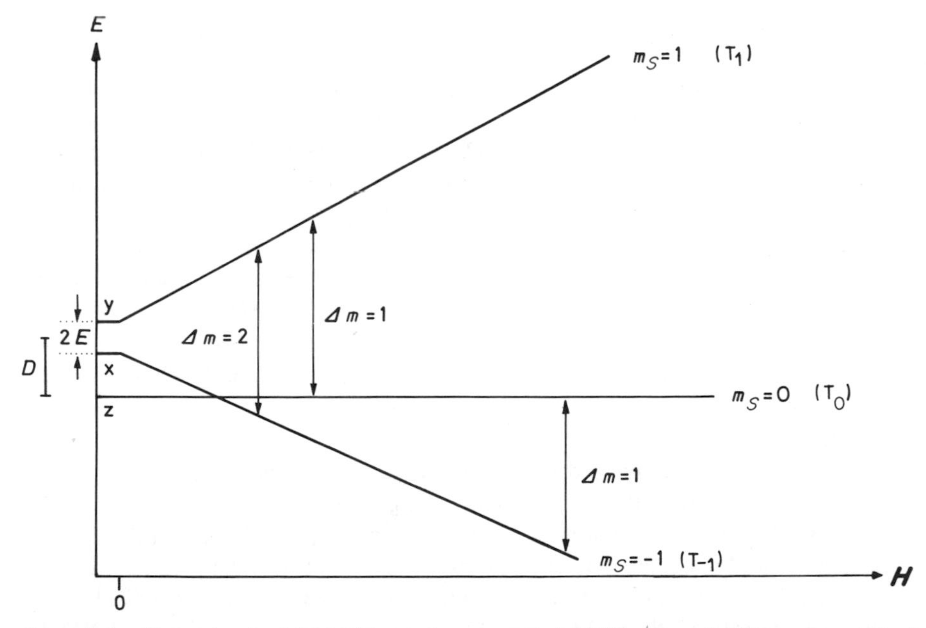

Figure 2.9. Energy levels of the triplet state in a magnetic field (**H**). The states (T) are characterized by the spin magnetic quantum number (m_S). D and E are the zero-field splitting parameters.

The parameters D and E, which can be extracted from the experimental spectra, are energy values. In practice, they are usually converted to wave number by dividing by hc. The values for the naphthalene triplet[27] (**44**) are

$$D/hc = 0.1003 \text{ cm}^{-1}$$

$$E/hc = 0.0137 \text{ cm}^{-1}$$

The information that can be derived from these parameters is one of electron distribution or delocalization. The more the two electrons are delocalized (i.e., the further they are apart, the smaller the value of the repulsive interaction D will be. The magnitude of E gives information on the deviation from cylindrical symmetry. Aromatic compounds with threefold or higher symmetry axes (e.g., benzene) have $E = 0$. A perfectly symmetrical species like linear methylene (a hypothetical molecule) would also have $E = 0$.

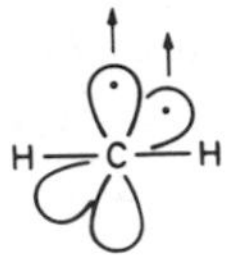

H—C—H

$$E = 0$$

$$D/hc \cong 1 \text{ cm}^{-1} \text{ (calc)}$$

The experimental observation that $E \neq 0$ therefore proves that methylene is nonlinear. The ESR spectra of carbenes and nitrenes will be discussed in Section 4.3.2.

2.2.3. CIDNP (Chemically Induced Dynamic Nuclear Polarization)

The CIDNP effect is the result of a reaction product being produced with an excess population of a particular nuclear spin. This can occur when the product results from an interaction between two radicals. The effect is observed by means of *nuclear* magnetic resonance spectroscopy. As we have seen (Figure 2.4), the nuclear magnetic energy state is split into two levels ($m_I = \pm\frac{1}{2}$) in a magnetic field, and the lower level ($m_I = +\frac{1}{2}$) is slightly overpopulated. If one could achieve a much higher degree of overpopulation of this level, much stronger absorption signals would be observed. On the other hand, if one could overpopulate the *upper* level ($m_I = -\frac{1}{2}$), this would undergo transitions to the lower energy level with *emission* of energy. As a result, a negative peak would be observed in the NMR spectrum. These situations will be referred to as *enhanced absorption* (A) and *emission* (E), respectively.

When a radical pair is formed in a solution reaction, it will exist initially as a singlet or triplet pair with the two partners in close proximity (Scheme 1).

The law of spin conservation[28] dictates that, in a thermal reaction, the pair $\overline{A\cdot\ B\cdot}$ must initially be a singlet. In a photosensitized reaction, the precursor could first be excited to the triplet state which then dissociates into a triplet radical pair, $^3\overline{A\cdot\ B\cdot}$. The history of the radical pair can be represented as in Figure 2.10. A triplet diradical has, as we have seen, three components in a magnetic field; T_{-1},

$$\text{Precursor} \longrightarrow \overline{\text{A}\cdot\ \text{B}\cdot} \longrightarrow \text{A—B (geminate product)}$$

$$\downarrow$$

$$\text{free radicals} \xrightarrow{\text{reencounter}} \text{A—B}$$
$$\xrightarrow{\text{solvent-H}} \text{A—H}$$
$$\longrightarrow \text{other reactions}$$

(A—B, A—H and other reactions: scavenging products)

Scheme 1

T_0, and T_1 (Figure 2.9). The same is true of a radical pair. When the *singlet* radical pair $^1\overline{\text{A}\cdot\ \text{B}\cdot}$ is just created, the singlet state will be of low energy, the triplets of high energy. As the two radicals move apart, the difference in energy between the states will decrease. At infinite separation, there is no difference between singlet and triplet pairs—there are only two independent free radicals (doublets). Thus the energies will converge on a certain value E_0.

When the singlet and triplet states have almost identical energies, transitions can take place between them without an energy input. This is referred to as mixing the singlet and triplet states. The mixing can take place where S and T_{-1} cross (Figure 2.10) or at the asymptotic level where T_0 and S acquire equal energies. A singlet–triplet conversion (intersystem crossing) is, however, spin-forbidden, and it takes a certain time. The pair will spend much more time in the region where S–T_0 mixing can occur than anywhere else, and this type of mixing is, therefore, most important.

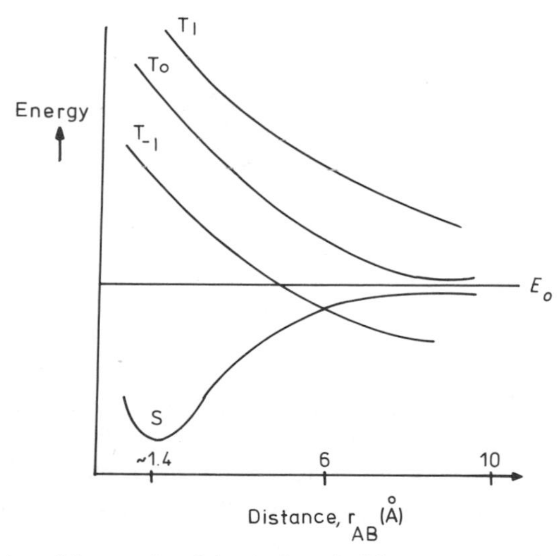

Figure 2.10. Variation of the energies of the singlet and triplet components of a radical pair with the distance r_{AB} between the two radicals. (Reprinted with permission from Ref. 29. Copyright 1972 American Chemical Society.)

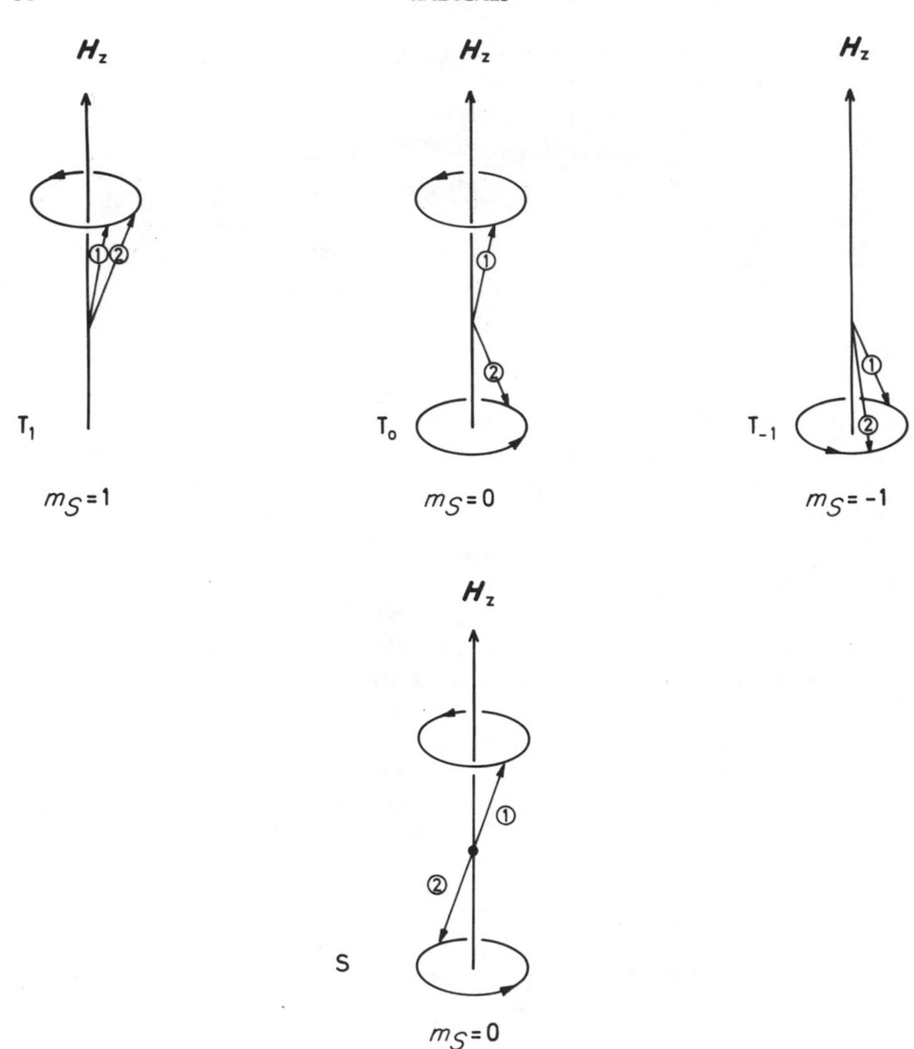

Figure 2.11. Precession of the spin magnetic vectors of two electrons (1) and (2) of a radical pair about the external magnetic field $\mathbf{H}_z$.

Even in this region, the mixing must take place rather rapidly before the radicals diffuse apart.

The distance between the radicals at which mixing can occur is 6–10 Å. The mechanism by which this spin-forbidden mixing can occur in a magnetic field **H** is illustrated in Figure 2.11. Each energy state of the pair corresponds to a set of directions of the electron spin magnetic vectors with respect to the field. Each spin magnetic field vector precesses around the external magnetic field vector $\mathbf{H}_z$ with a frequency

$$\omega = \frac{1}{\hbar} g\mu_B \mathbf{H}_0$$

Both in the singlet (S) and in the T_0 pair, the two electrons have opposite spin, $m_S = \pm\frac{1}{2}$. If the two electrons have different g-factors, the precession frequencies will also be different:

$$\omega_1 = \frac{1}{\hbar} g_1 \mu_B \mathbf{H}_0$$

$$\omega_2 = \frac{1}{\hbar} g_2 \mu_B \mathbf{H}_0$$

If we start with the S-pair, and if $g_1 > g_2$, electron 1 will precess faster than electron 2. After a certain time, the two electrons will be out of phase by 180°, and the system will be identical with the T_0 state. The precession will continue, and after another 180°, the S-state will again be reached. Likewise, if we start with a T_0-state, it will acquire singlet character. If the g-factors were equal, as they would be for a symmetrical pair, for example, $CH_3CH_2\cdot\ \cdot CH_2CH_3$, no such mixing would be possible. Whether there is mixing or not, this mechanism alone does not lead to the CIDNP effect.

We must now consider that the precessional frequencies are determined not only by the external field **H**, but also by *the internal hyperfine field* (cf. Section 2.2.1). Each electron moves in the field of the nuclei, and the interaction is described by the hyperfine splitting constant a. We must modify the precessional frequencies accordingly:

$$\omega_1 = \frac{1}{\hbar} g_1 \mu_B \mathbf{H}_0 + \frac{1}{\hbar} \mu_B \sum a_1 m_1 \qquad (6)$$

$$\omega_2 = \frac{1}{\hbar} g_2 \mu_B \mathbf{H}_0 + \frac{1}{\hbar} \mu_B \sum a_2 m_2 \qquad (7)$$

$$(\omega_1 - \omega_2) = \mu_B \hbar^{-1} [(g_1 - g_2)\mathbf{H}_0 + \sum a_1 m_1 - \sum a_2 m_2] \qquad (8)$$

where m_i is the spin magnetic quantum number (m_I) of the ith nuclus. It is now clear that not only Δg, but also $\Delta a \cdot m$ influence the rate of singlet–triplet mixing.

We continue the example used earlier with an S-pair where $g_1 > g_2$. If radical 1 is coupled to a proton that is in the $+\frac{1}{2}$ state ($m_I = +\frac{1}{2}$) and with a_1 positive, the frequency ω_1 will increase [cf. Eq. (6)], and the triplet state T_0 will be reached in a shorter time. If the proton is in the $-\frac{1}{2}$ state, this will decelerate the precession; the radical will exist for a longer time in the singlet state. *Since only the singlet states can recombine*, the $-\frac{1}{2}$ state will favor product formation, and the product will be formed with an enrichment of $m = -\frac{1}{2}$ protons. As we have seen, the $m = +\frac{1}{2}$ protons favor triplet formation, and since the triplets cannot recombine (the electron pair constituting a chemical bond has singlet spin multiplicity), they diffuse apart and undergo other reactions (''scavenging reactions,'' cf. Scheme 1). The scavenging product will be enriched in $m_I = +\frac{1}{2}$ protons.

It is clear from Eqs. (6)–(8) that the enrichment of proton spin states depends on the relative magnitudes of g_1 and g_2, a_1 and a_2, and their signs, as well as on whether the radical pair $\overline{A\cdot\ B\cdot}$ arose from a singlet or a triplet precursor. This is the mechanism by which specific nuclear spin energy levels are overpopulated. Overpopulation of the $m_I = \frac{1}{2}$ level causes enhanced absorption; of the $m_I = -\frac{1}{2}$ level, enhanced emission (cf. Figure 2.4).

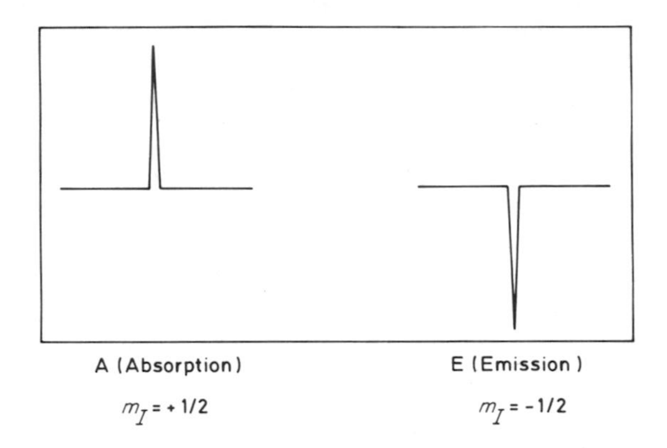

If the product contains several nuclei that are coupled to each other by the NMR-coupling constant J, the so-called multiplet effect results. Consider the example of two protons coupled by J. Combination of the two energy levels $m_I = \pm\frac{1}{2}$ for the two protons gives a total of four energy levels, resulting in the well-known splitting of each NMR peak into a doublet (Figure 2.12). The labeling of the levels corresponds to the values of m_I (α and β spin; $m_I = \pm\frac{1}{2}$). The ordering of the peaks depends on the sign of J.

It is seen that different spectra are obtained for different overpopulations, and for different signs of J. In total, one has to know g, a, and J. From this and the appearance of the spectrum, one can deduce the way in which the compound was formed (singlet or triplet; geminate or scavenging). The phasing of a multiplet is described as *AE* or *EA*, as the case may be, going from left to right (increasing field) (see Figure 2.12). Rather than going through the whole argument anew in each instance, one can apply the Kaptein rules[31] which, with a knowledge of the parameters involved, allow the immediate determination of polarization in a given case and vice versa. Kaptein defined two factors, Γ_n for the net effect (*A* or *E*) and Γ_m for the multiplet effect (*AE* or *EA*).

The signs of these factors are determined by multiplying the signs of the parameters, some of which we have seen earlier:

$$\Gamma_n \equiv \mu \cdot \epsilon \cdot \Delta g \cdot a_i = (+) \Rightarrow \text{net absorption } (A)$$
$$= (-) \Rightarrow \text{emission} \quad (E)$$
$$\Gamma_m \equiv \mu \cdot \epsilon \cdot a_i \cdot a_j \cdot J_{ij} \cdot \sigma_{ij} = (+) \Rightarrow EA \text{ multiplet}$$
$$= (-) \Rightarrow AE \text{ multiplet}$$

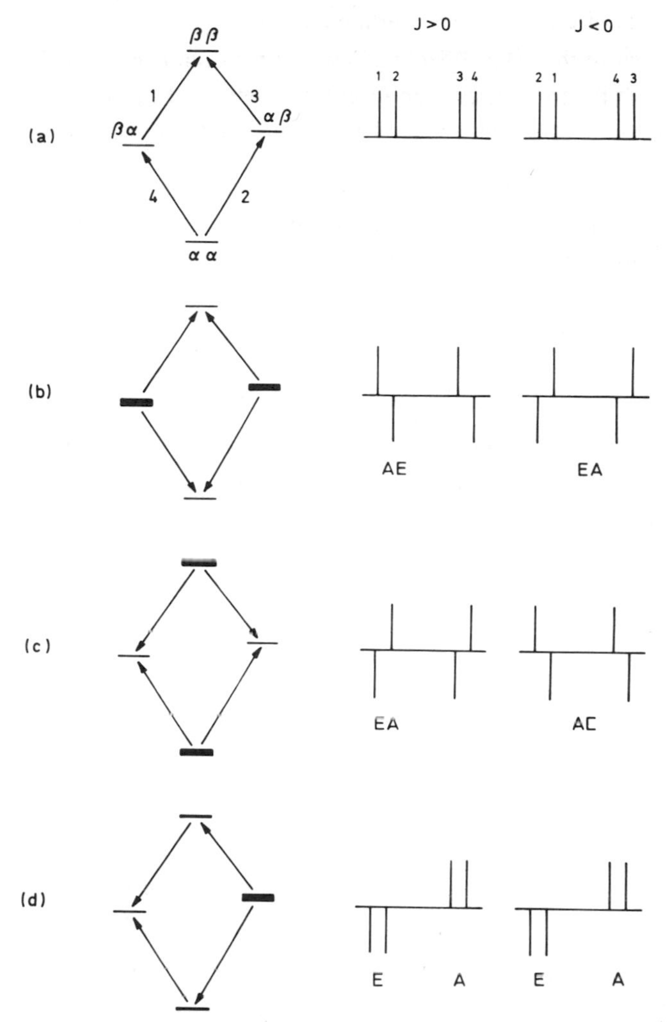

Figure 2.12. AB proton NMR spectra with $J_{AB} > 0$ and $J_{AB} < 0$. (a) Nuclear spin levels and line spectra for normal spin population; (b) βα- and αβ-levels overpopulated; (c) αα- and ββ-levels overpopulated; (d) αβ-level more highly overpopulated than the αα- and ββ-levels. (Adapted from Ref. 30a with the permission of The International Union of Pure and Applied Chemistry.)

The signs of the parameters are defined as follows:

μ	(+)	for triplet or free precursor (e.g., reencounter of radicals that had diffused apart)
	(−)	for singlet precursor
ϵ	(+)	for geminate products
	(−)	for scavenging products
Δg		the sign of $(g_i - g)$, where g_i is the g-factor for the radical containing nucleus i, and g is the g-factor for the other radical in the pair

a	the sign of the hyperfine splitting constant.
J_{ij}	the sign of the NMR coupling constant between nuclei i and j
σ_{ij}	(+) if the two nuclei i and j were originally in the same radical fragment
	(−) if nuclei i and j were originally in different radicals

EXAMPLE: In Section 2.2.1, we saw that the 2-chloroethyl radical existed as an unsymmetrically bridged species. No conclusion was reached as to whether the 2-bromoethyl radical was symmetrical (**45**) or existed as rapidly interconverting unsymmetrical structures (**46**).

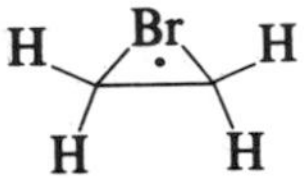

45 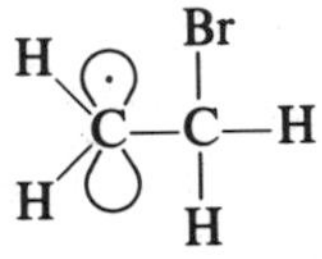 **46**

The radical has been produced in the following reaction:[32]

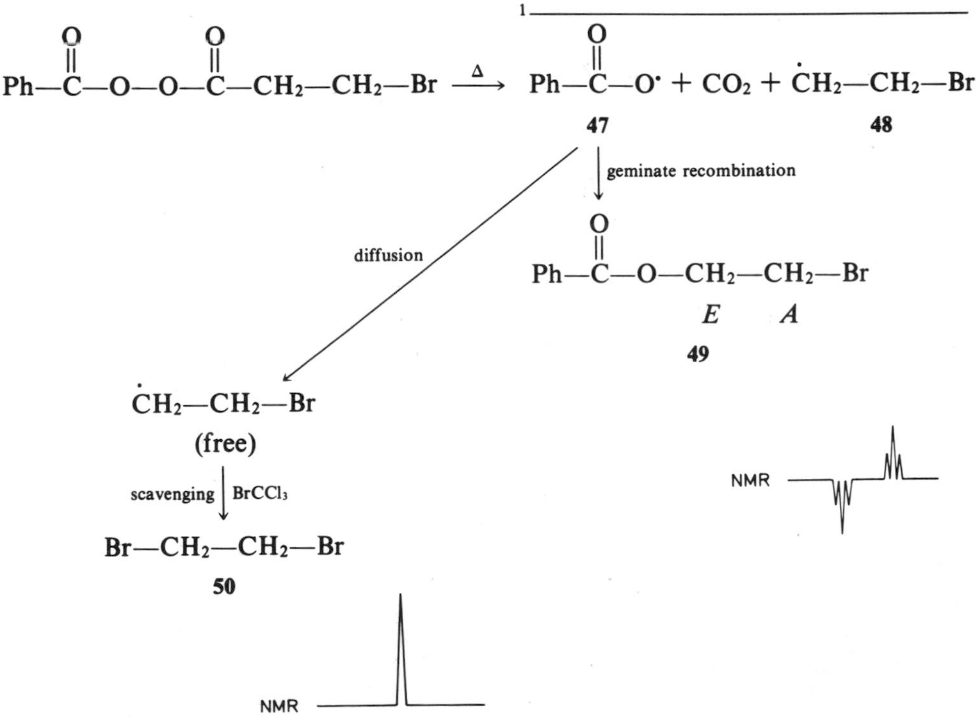

$$\text{Ph—C(=O)—O—O—C(=O)—CH}_2\text{—CH}_2\text{—Br} \xrightarrow{\Delta} \text{Ph—C(=O)—O}^\bullet + \text{CO}_2 + \dot{\text{C}}\text{H}_2\text{—CH}_2\text{—Br}$$

47 **48**

geminate recombination

$\text{Ph—C(=O)—O—CH}_2\text{—CH}_2\text{—Br}$

E *A*

49

NMR

diffusion

$\dot{\text{C}}\text{H}_2\text{—CH}_2\text{—Br}$ (free)

scavenging BrCCl$_3$

$\text{Br—CH}_2\text{—CH}_2\text{—Br}$

50

NMR

Scheme 2

From the different polarizations of the two methylene groups in **49**, it is immediately clear that the radical structure was **46**. In **45**, the four hydrogen atoms would be identical, and the four protons in the product would have been identically polarized.

Application of the rule for Γ_n gives for **49**:

$$\Gamma_n = \mu \cdot \epsilon \cdot \Delta g \cdot a_i = (-)(+)(-)(-) = (-) \qquad \text{for } \alpha\text{-radical proton}$$

$$= (-)(+)(-)(+) = (+) \qquad \text{for } \beta\text{-radical proton}$$

remembering that the hyperfine splitting constant a is negative for α-H's and positive for β-H's (Section 2.2.1). The g-factor of **47** was taken as being larger than that of **48** (Δg negative) by comparison with known g-factors of related radicals. The E and A peaks for the two sets of protons in **49** are in accord with the rules. The scavenging product **50** should show the opposite polarization. But in **50** the two CH_2-groups are identical, and therefore, only a singlet is observed. Since one-half of the singlet should be polarized A, the other E, there is no net effect: A normal, unpolarized peak results.

Note that for predicting the multiplet effect, it is not necessary to know Δg. Thus ethyl bromide formed in a similar reaction (Scheme 3) was polarized as shown.

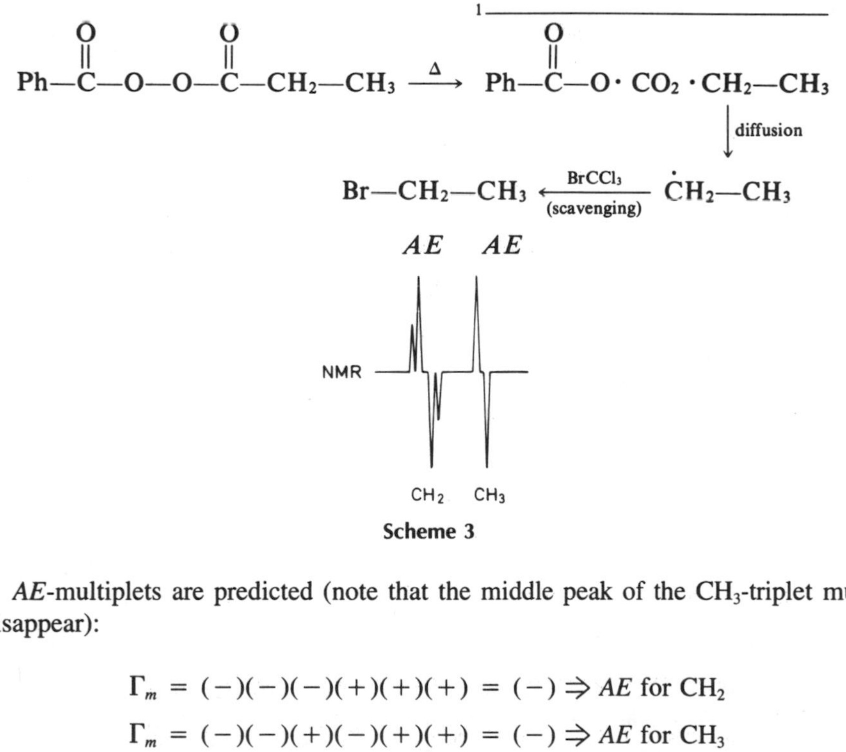

$$Ph{-}\overset{\overset{\displaystyle O}{\|}}{C}{-}O{-}O{-}\overset{\overset{\displaystyle O}{\|}}{C}{-}CH_2{-}CH_3 \xrightarrow{\Delta} \overline{Ph{-}\overset{\overset{\displaystyle O}{\|}}{C}{-}O\cdot\ CO_2\ \cdot CH_2{-}CH_3}^{1}$$

$$\downarrow \text{diffusion}$$

$$\underset{AE\quad AE}{Br{-}CH_2{-}CH_3} \xleftarrow[\text{(scavenging)}]{BrCCl_3} \dot{C}H_2{-}CH_3$$

NMR

CH_2 CH_3

Scheme 3

AE-multiplets are predicted (note that the middle peak of the CH_3-triplet must disappear):

$$\Gamma_m = (-)(-)(-)(+)(+)(+) = (-) \Rightarrow AE \text{ for } CH_2$$

$$\Gamma_m = (-)(-)(+)(-)(+)(+) = (-) \Rightarrow AE \text{ for } CH_3$$

(vicinal couplings J_{ij} are positive for protons).

Some values of a and g for various radicals, useful for the application of the Kaptein rules, are given in Table 2.6. For values of J, see Ref. 33.

TABLE 2.6. ESR Data for Selected Radicals (Hyperfine Splitting Constants in mT)[a]

Radical	a_H	$a_{^{13}C}$	g
$\cdot CH_3$	−2.304	+3.834	2.00255
$\cdot CH_2CH_3$	−2.238 (α) +2.687 (β)	+3.907 (α) −1.357 (β)	2.00260
$\cdot CH_2OH$	−1.768	+4.589	2.00334
$\cdot CCl_3$	—	9.19	2.0091
C_6H_5—$\dot{C}H_2$	+1.63 (α) +0.515 (*ortho*) −0.177 (*meta*) +0.618 (*para*)	2.445 (α) 1.445 (C1)	2.00260
$H_2C{=}\dot{C}H$ (σ)	+1.339	+10.757	2.00220
$C_6H_5\cdot$ (σ)	+1.74 (*ortho*) +0.59 (*meta*) +0.19 (*para*)	+12.9 (C1)	2.0024
H—$\dot{C}$=O (σ)	+13.59	+13.48	2.0009
C_6H_5—$\dot{C}$=O (σ)	0.021 (*ortho*) 0.118 (*meta*) $<$0.007 (*para*)	+14.96 (Cα)	2.0008
R—$\dot{N}$—R′ (aminyls, π-radicals)			2.0030–2.0046
Diphenylpicrylhydrazyl (DPPH)			2.0036
RR′C = $\dot{N}$ (iminyls, σ-radicals)			2.0025–2.0033
R—C(=O)—$\dot{N}$—R′ (π-radicals)[b]			2.0044–2.0053
R—SO_2—$\dot{N}$—R′ (π-radicals)[c]			2.0041–2.0042
R—N(O·)—R′ (nitroxides, π-radicals)			2.0055–2.0065
RR′C=N—O· (iminoxyls, σ-radicals)			2.000 –2.0065
R—O· (alkoxyls)			2.004 –2.03
R—O—O· (alkylperoxyls)			~2.01
CH_3—C(=O)—O· (acetyloxy)[d]			2.0058

[a]Data from H. Fischer, in *Free Radicals*, J. K. Kochi, Ed., Wiley, New York, 1973, Vol. II, p. 435, and *Landolt–Börnstein Numerical Data and Functional Relationships in Science and Technology*, New Series, Springer Verlag, Berlin, Heidelberg, New York, 1977, Vol. II/9b; 1979, Vol. II/9c1 and II/9c2, unless indicated otherwise.
[b]J. Lessard, D. Griller, and K. U. Ingold, *J. Am. Chem. Soc.*, **102,** 3262 (1980);
[c]W. C. Danen and R. W. Gellert, *J. Am. Chem. Soc.*, **102,** 3264 (1980).
[d]R. Kaptein, J. Brokken-Zijp, and F. J. J. de Kanter, *J. Am. Chem. Soc.*, **94,** 6280 (1972).

It is a useful rule of thumb that, within related series of radicals, the value of g increases with the atomic weight of the atom(s) that can carry the radical electron.

Useful though the CIDNP technique is, it should be borne in mind that the detection of CIDNP does not prove that all or even a major part of a particular reaction proceeds via radical pairs. Since strongly amplified NMR signals may be recorded, minor reaction pathways that might otherwise escape detection could be mistaken for the principal routes. Ideally, therefore, CIDNP should be used in conjunction with independent evidence.

2.2.4. Spin Trapping

The direct observation of free radicals by ESR spectroscopy requires, of course, a certain concentration of the radicals. In reacting systems, the lifetimes of the radicals are often so short that a sufficient concentration may not be attained. An elegant technique, spin trapping,[34,35] takes advantage of the fact that very stable nitroxide radicals[36] are formed on addition of (transient) free radicals to nitroso compounds or nitrones. For example,

$$R\cdot + C_6H_5{-}N{=}O \longrightarrow C_6H_5{-}N(R){-}O\cdot$$

$$R\cdot + C_6H_5{-}CH{=}\overset{\oplus}{N}(O^{\ominus}){-}C(CH_3)_3 \longrightarrow C_6H_5{-}CH(R){-}N(\dot{O}){-}C(CH_3)_3$$

$$R\cdot + \text{(cyclic nitrone, } \overset{\oplus}{N}{-}O^{\ominus}\text{)} \longrightarrow \text{(R-substituted cyclic nitroxide, } N{-}O\cdot\text{)}$$

Such reactions are extremely fast, with rate constants of $10^6 - 5 \cdot 10^8$ liter mol^{-1} s^{-1} at room temperature and activation energies in the range of 1–5 kcal/mol.[37] The elucidation of the structure of the reactive radical R· follows from the analysis of the ESR spectrum of the nitroxide spin adduct: From the hyperfine couplings between the unpaired electron and the nuclei in R, the structure of R may be inferred. The spin adducts of nitroso compounds are more useful for this purpose since the residue R is directly attached to the nitroxide moiety. Care should be exercised when applying this technique, because nitroso compounds in particular may undergo radical-forming reactions under a variety of conditions. Therefore, the stable nitroxide should be independently prepared by another route in order to establish its identity with the spin adduct observed in the trapping reaction.

As an example, *tert*-butylperoxy radicals are trapped by *t*-butyl phenyl nitrone (**51**):[38]

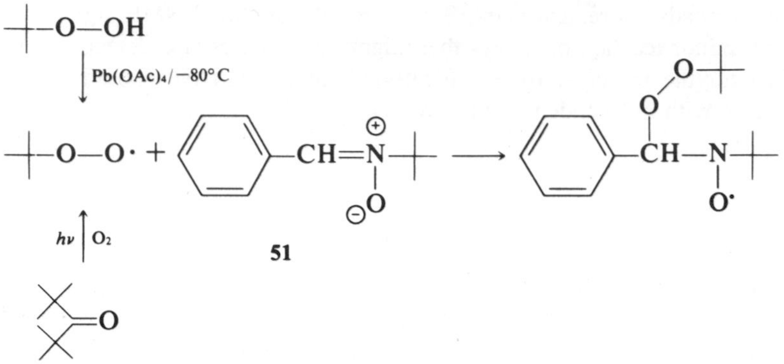

Benzoyloxy radicals (**53**) are formed in the thermal decomposition of dibenzoyl peroxide (**52**) (see also Section 2.4.4.). The benzoyloxy radicals normally undergo reaction with the solvent or decarboxylation to phenyl radicals (**54**) which, in turn, dimerize or abstract a hydrogen atom to yield benzene.

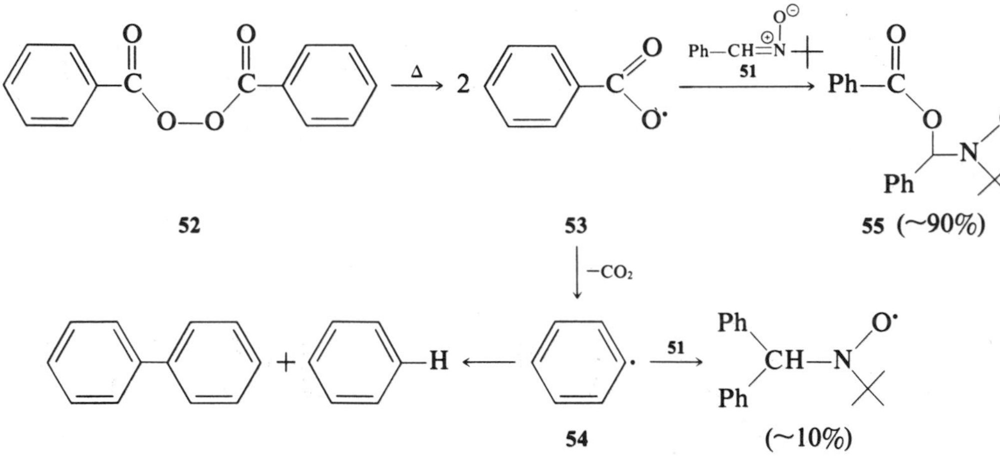

In the presence of the nitrone **51**, however, about 90% of the benzoyloxy radicals are trapped, and the decomposition of the peroxide **52** may be conveniently monitored by ESR spectroscopy of the spin adduct **55**.[39]

In Section 2.2.1. it was shown that α-chloroalkyl radicals (e.g., **28**) can undergo rapid rearrangement by chlorine atom migration. The secondary alkyl radical **56** rearranges to the more stable dichloromethyl radical **57** before being trapped by 2-methyl-2-nitrosopropane to give the spin adduct **58**.

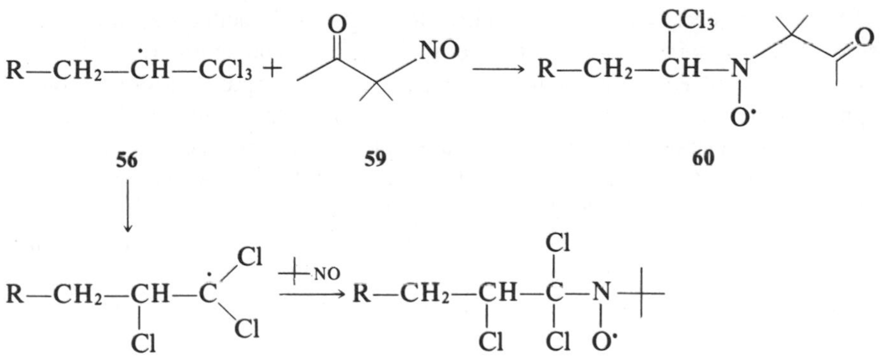

$$R{-}CH_2{-}\dot{C}H{-}CCl_3 + \text{59} \longrightarrow R{-}CH_2{-}CH(CCl_3){-}N(O^{\cdot}){-}C(CH_3)_2{-}C(O)CH_3$$

56 59 60

$$\text{56} \downarrow$$

$$R{-}CH_2{-}CH(Cl){-}\dot{C}Cl_2 \xrightarrow{tBuNO} R{-}CH_2{-}CH(Cl){-}CCl_2{-}N(O^{\cdot}){-}C(CH_3)_3$$

57 58

R = *t*BuO or CCl_3

However, the unrearranged radical **56** can be trapped by the more electrophilic nitroso compound **59** (due to the presence of the electron withdrawing acetyl group), thus leading to the product **60**.[40]

Spin trapping using nitrones and nitroso compounds is also becoming increasingly important in the study of free radical reactions in biological systems.[35] For example, nitrones have been used to trap radicals produced from potent carcinogens such as dimethylnitrosamine **(61)** and a variety of other compounds under the action of the lipid peroxidation system in rat liver microsomes.[41]

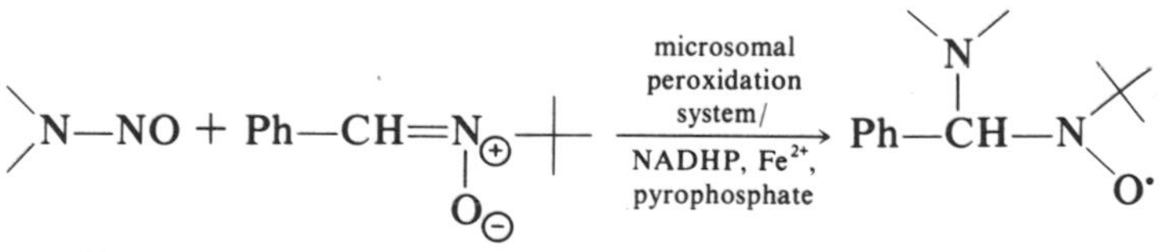

$$(CH_3)_2N{-}NO + Ph{-}CH{=}\overset{\oplus}{N}(O^{\ominus}){-}C(CH_3)_3 \xrightarrow[\text{NADHP, } Fe^{2+}\text{, pyrophosphate}]{\text{microsomal peroxidation system/}} Ph{-}CH(N(CH_3)_2){-}N(O^{\cdot}){-}C(CH_3)_3$$

61 62

2.2.5. Mass Spectrometry Methods

An entirely different kind of radical trapping is carried out in the source of a mass spectrometer using tetracyanoquinodimethane **(63)** or tetracyanoethylene **(64)** as electron acceptors.[42]

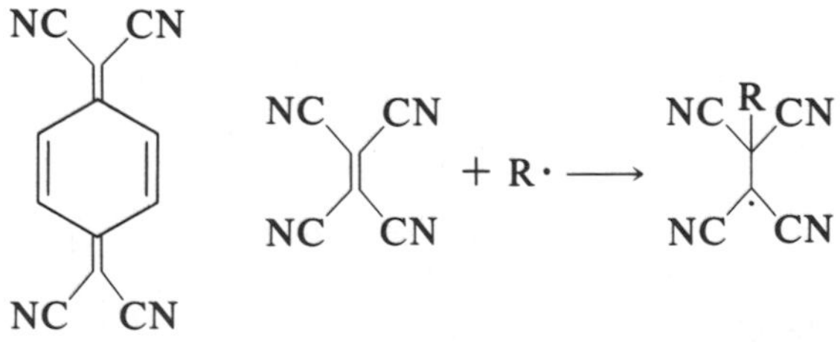

$$(NC)_2C{=}C_6H_4{=}C(CN)_2 \qquad (NC)_2C{=}C(CN)_2 + R\cdot \longrightarrow (NC)_2C(R){-}\dot{C}(CN)_2$$

63 64 65

The acceptor reacts with the free radical R· to give spin adducts such as **65**. The subsequently recorded mass spectrum of the long-lived adduct **65** gives information on the nature of R· at the moment of trapping. This trapping reaction is found to approach the rate of diffusion and to be even faster than methyl–methyl recombination. To illustrate the method, consider the reaction between an alkyl radical $\cdot CR_1R_2R_3$ and **63**. First, the spin adduct **66** is formed. Electron capture in the source of the mass spectrometer leads to the anion **67**, which now loses a CN radical to give the radical anion **68**. The chemical activation (CA) mass spectrum (induced by collision with an added gas and therefore also termed collisionally induced dissociation, or CID) of **68** then results in fragmentation, primarily involving the CR_3 substituent group.

NC CN / NC CN **63** + $\cdot CR_1R_2R_3$ ⟶ **66** $\xrightarrow{e^-}$ **67** $\xrightarrow{-CN\cdot}$

68 $\xrightarrow{CA}$ **69** + $R_1\cdot$

This may give, for example, the anion **69** by elimination of the radical $R_1^{\cdot}$. In a similar way, R_2 or R_3 might also have been eliminated. Thus the CA mass spectrum will allow the deduction of the structure of $CR_1R_2R_3$. The rearrangement of the 1-pentyl to the secondary 2-pentyl radical was studied in this way.[43]

70 ⟶ **71**

At high concentrations of **63**, 58% of the unrearranged primary radical **70** was trapped, the remainder rearranging to **71**. At lower concentrations of the trapping agent, correspondingly more rearrangement took place. From these data, a rate constant for the rearrangement of $\sim 10^2\ s^{-1}$ was derived.

The reaction **70** → **71** is a 1,4-hydrogen shift. The corresponding 1,5-shift in 1-hexyl was found to be faster still,

whereas 1,2- and 1,3-hydrogen shifts did not occur.[43] These results complement other studies of the hydrogen shift in alkyl radicals, which will be presented in Section 2.5.3.

Radicals that have a sufficiently long lifetime in the gas-phase without rearranging or decomposing may be examined directly by mass spectrometry without resorting to spin trapping methods. For such purposes a "radical gun," which is essentially a flash vacuum pyrolysis apparatus, attached to the source of a mass spectrometer has been described by Lossing.[44] For example, the mass spectrum of the benzyl radical, produced in the reaction

$$C_6H_5-CH_2-CH_2-ONO \xrightarrow{\Delta} Ph-CH_2-CH_2-O\cdot + \dot{N}O$$

$$\downarrow$$

$$Ph-\dot{C}H_2 + CH_2{=}O$$

can be measured directly.[45] Such studies are of interest not only from the point of view of radical chemistry, but also for the measurement of heats of formation of cations in the gas-phase. Thus the heat of formation of the benzyl cation[45] is derived from the measured ionization potential of the radical and its known heat of formation, the latter being obtained from kinetic studies.

A quantitative elaboration of the pyrolysis mass spectrometry technique, known as very low pressure pyrolysis (VLPP) allows the direct kinetic investigation of unimolecular radical-forming reactions. Many heats of formation of free radicals have been obtained from the evaluation of the kinetic data collected in this way.[46] As an example, the kinetics of formation of the allyl radical **(73)** by pyrolysis of diallyloxalate **(72)**, and its dimerization to 1,5-hexadiene **(74)**, led to a ΔH_f^0 (allyl·) = 39.1 ± 1.5 kcal/mol.[47]

844–1061 K

625–950 K

72 **73** **74**

Heats of formation and resonance energies of radicals such as benzyl, 1-naphthylmethyl, and 9-anthrylmethyl have been evaluated in a similar manner.[48]

2.3. GEOMETRY AND RATE OF FORMATION

As shown in Section 2.2.1, simple alkyl radicals are quasi-planar and sp^2-hybridized in the ground state, although small deviations from planarity may occur, as, for example, in the ethyl and *tert*-butyl radicals. Electronegative substituents may cause more marked pyramidalization, so that $\cdot CF_3$ is essentially sp^3-hybridized. The cyclopropyl radical **(17)** is a pyramidal σ-radical, but even here chemical and spectroscopic evidence as well as semiempirical calculations indicate a very low barrier, approximately 0.8 kcal/mol, for the interconversion of the two nonplanar forms. Furthermore, a steric perturbation (as in **18**) suffices to make such radicals planar.[13] All in all, it appears that the energy difference between planar and nonplanar radicals is very small. In sharp contrast, carbocations decidedly prefer a planar sp^2-hybridized configuration, whereas carbanions, in the absence of conjugating substituents, are pyramidal and sp^3-hybridized. Thus radicals are electronically and geometrically positioned between the carbocations and the carbanions.

It follows that it should be easier to generate bridgehead radicals—which are forced to be pyramidal—than bridgehead carbocations. This is indeed the case. Table 2.7 gives the relative rates of formation of radicals in the reaction

$$R{-}C(=O){-}O{-}O{-}tBu \xrightarrow{\Delta} R\cdot + CO_2 + \dot{O}tBu$$

and of ions in the solvolysis of the corresponding alkyl bromides:

$$R{-}Br \longrightarrow R^{\oplus} + Br^{\ominus}$$

Data for other radical-forming reactions and the nitrene insertion reaction with which we shall be concerned later are also reported in this table.

Although the rates of the radical-forming reactions decrease with increasing deviation from planarity of the radicals formed, it is seen that the effect on carbocation formation is larger by a factor of 10^8–10^{11}.

2.4. FORMATION OF RADICALS

The three major methods by which radicals can be produced are:

1. Homolysis: $A{-}B \rightarrow A\cdot + B\cdot$
2. Redox reaction, e.g.: $RO^- + Fe^{3+} \rightarrow RO\cdot + Fe^{2+}$
3. From other radicals, e.g.: $Br\cdot + C_2H_6 \rightarrow HBr + C_2H_5\cdot$

Method 3 will be discussed in Section 2.5.1. Method 2 is part of the general phenomenon of electron transfer, which is described in Section 2.6. Here we shall

TABLE 2.7. Relative Rates of Reaction at Bridgehead Hydrocarbon Centers (* = • or +)

Radical					Reference
Perester thermolysis	1	1.48	10^{-1}	10^{-3}	*a*
Azo-compound thermolysis	1	$4 \cdot 10^{-4}$	$5 \cdot 10^{-5}$	$5 \cdot 10^{-6}$	*a*
I-atom abstraction from R—I	1	0.7	0.57	0.18	*b*
Carbocation solvolysis of R—Br	1	10^{-3}	10^{-6}	10^{-14}	*c*
Nitrene insertion into R—H		1	0.3	0.07	*d*

[a] C. Rüchardt et al., Twenty-third IUPAC Congress, Special Lectures, Vol. 4, 1971, pp. 248 ff.; V. Golzke, F. Groeger, A. Oberliner, and C. Rüchardt, *Nouv. J. Chim.*, **2,** 169 (1977).

[b] W.C. Danen, T. J. Tipton, and D. G. Saunders, *J. Am. Chem. Soc.*, **93,** 5186 (1971).

[c] R. C. Fort and P. v. R. Schleyer, *Adv. Alicyclic Chem.*, 1, 284 (1966); G. J. Gleicher and P. v. R. Schleyer, *J. Am. Chem. Soc.*, **89,** 582 (1967).

[d] D. S. Breslow, E. I. Edwards, R. Leone, and P. v. R. Schleyer, *J. Am. Chem. Soc.*, **90,** 7097 (1968).

only mention as an example an application of *Fenton's reagent*,[49] by which hydrogen peroxide is reduced to ·OH radicals:

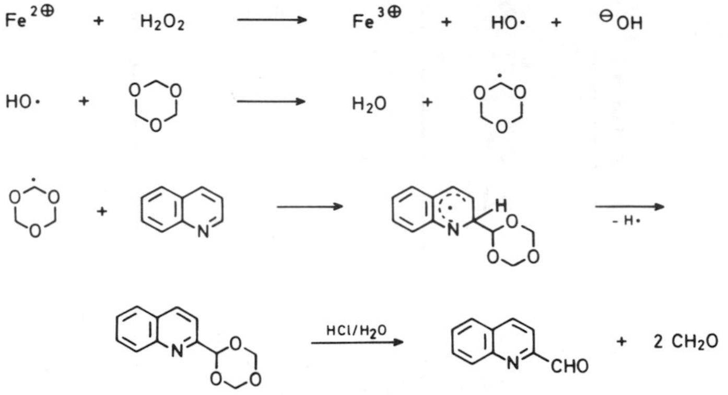

$$Fe^{2\oplus} + H_2O_2 \longrightarrow Fe^{3\oplus} + HO\cdot + {}^{\ominus}OH$$

$$HO\cdot + \ldots \longrightarrow H_2O + \ldots$$

$$\xrightarrow{-H\cdot}$$

$$\xrightarrow{HCl/H_2O} \ldots\text{CHO} + 2\ CH_2O$$

This example constitutes a useful method of preparation of heterocyclic aldehydes via free radical substitution.[50]

2.4.1. Homolysis and Heterolysis

A bond can be broken in two ways: homolytically or heterolytically.

$$A\text{—}B \longrightarrow A\cdot + B\cdot \qquad \text{homolysis}$$

$$A\text{—}B \longrightarrow A^{+} + B^{-} \qquad \text{heterolysis}$$

It is always easier to break a bond homolytically, at least in the gas phase. This can be seen by considering the total energy of a charged system $A^{+}B^{-}$, for example the hydrogen atom. The total energy is the sum of the kinetic and the potential energies:

$$E = T + V = \frac{1}{2}mv^2 - \frac{e^2}{r}$$

Thus the smaller the separation r, the lower the potential energy. If the particles are forced apart, the electrostatic attraction $-e^2/r$ has to be overcome. The thermodynamic result is that ions have much higher heats of formation than radicals. For example,

$$CH_3\text{—}CH_3 \longrightarrow \dot{C}H_3 + \dot{C}H_3 \qquad \Delta H_r = +88\ \text{kcal/mol}$$

$$\Delta H_f^0: \quad -20 \qquad 34 \qquad 34$$

$$CH_3\text{—}CH_3 \longrightarrow CH_3^{\oplus} + CH_3^{\ominus} \qquad \Delta H_r \cong +314\ \text{kcal/mol}$$

$$\Delta H_f^0: \quad -20 \qquad 260 \qquad \sim 34$$

(cf. Table 2.2). Consequently, *heterolysis in the gas-phase is practically unknown.* In *solution*, heterolysis is a well-known reaction (e.g., the S_N1 reaction). The reason is that *ions in solution are solvated.* The heats of solvation are very large and negative, so that overall energy is gained. This can be regarded as a gain in electrostatic energy between the ions and the coordinating solvent molecules.

Although heterolysis does not take place in the gas-phase, *transition states* may have polar character. Figure 2.13 shows a relationship between the energies of activation for the process

$$\underset{\mathrm{H}\quad\mathrm{X}}{-\overset{|}{\underset{|}{\mathrm{C}}}-\overset{|}{\underset{|}{\mathrm{C}}}-} \xrightarrow[\text{gas-phase}]{\Delta} \mathrm{>C{=}C<} + \mathrm{HX}$$

and the heterolytic bond dissociation enthalpies $DH^0(R^+X^-)$ for various halides. Although the data correlate, indicating a polar transition state, the activation energies are much smaller than $DH^0(R^+X^-)$. No ions are formed; the HX molecules are covalent in the gas-phase.

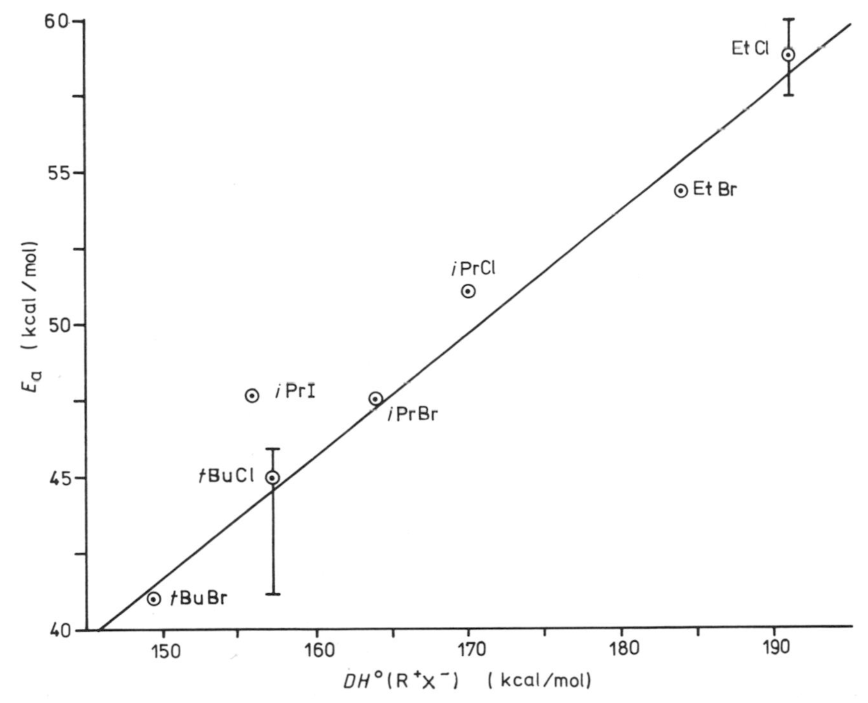

Figure 2.13. Correlation between the energy of activation (E_a) for gas-phase elimination and the heterolytic bond dissociation enthalpy [DH^0 (R^+X^-)]. Data from A. Maccoll [in *The Transition State*, Special Publication No. 16, The Chemical Society, London, 1962, p. 167; *Adv. Phys. Org. Chem.*, **3**, 91 (1965)] and Table 2.2. [Adapted with the permission of Professor Maccoll, The Royal Society of Chemistry, and *Advances in Physical Organic Chemistry* (Copyright: Academic Press Inc. (London) Ltd.]

The rates of homolysis reactions may be estimated from the Arrhenius equation

$$k = Ae^{-E_a/RT}$$

with a knowledge of bond dissociation enthalpies (see Tables 2.2 and 2.8) and using an average A-factor of 10^{15} s^{-1} for simple bond-fission reactions.

The energy required to break the bond may be provided by photolysis or thermolysis. From the shortest wavelength absorption band in the UV spectrum of the compound, one can calculate the maximum energy input in a photolysis reaction:

$$E = h\nu = \frac{hc}{\lambda}$$

Chlorine, for example, absorbs UV light down to 200 nm, corresponding to an energy of 143 kcal/mol. The bond dissociation energy is only 58 kcal/mol.

$$Cl_2 \xrightarrow{h\nu} Cl\cdot + Cl\cdot$$

The chlorine atoms formed will therefore contain a large excess of energy—up to 85 kcal/mol between them. This energy will be dissipated through collisions and reactions.

The most frequent reaction of a free radical is a *chain reaction*, as illustrated in the chlorination of an alkane:

$$\begin{array}{lll} Cl_2 \longrightarrow 2\,Cl\cdot & & \text{initiation} \\ Cl\cdot + R{-}H \longrightarrow HCl + R\cdot & \rbrace & \text{chain} \\ R\cdot + Cl_2 \longrightarrow R{-}Cl + Cl\cdot & \rbrace & \\ \hline Cl_2 + R{-}H \longrightarrow R{-}Cl + HCl & & \end{array}$$

Thousands of such reaction cycles in which R—Cl and HCl are formed with regeneration of the chain-carrying radical Cl· may take place before the chain is terminated by recombination of the radicals:

$$\left.\begin{array}{r} 2\,Cl\cdot \longrightarrow Cl_2 \\ 2\,R\cdot \longrightarrow R{-}R \\ R\cdot + Cl\cdot \longrightarrow R{-}Cl \end{array}\right\} \text{termination}$$

(Another termination reaction, i.e., disproportionation, will be described later.) The cleavage of the chlorine molecules might also be achieved thermally. From the Arrhenius equation, one calculates (with $A = 10^{15}$)

$$k\ (25°C) \cong 10^{-27}\ s^{-1}$$

TABLE 2.8. Some Average Bond Dissociation Enthalpies

Bond	DH^0 (kcal/mol)
C–H	98.7
C–C	82.6
N–N	39
O–O	35–37
C–N	73
N–O	53
Cl–O	52
C–I	51
Cl–Cl	58

which is an extremely slow reaction. In order for the reaction to proceed at a convenient rate (e.g., 10^0 s^{-1}) a temperature above 500°C is required.

The two classes of organic compounds most often used to start radical chain reactions (and hence called initiators) are peroxides and azo-compounds. Of these, the most frequently used are

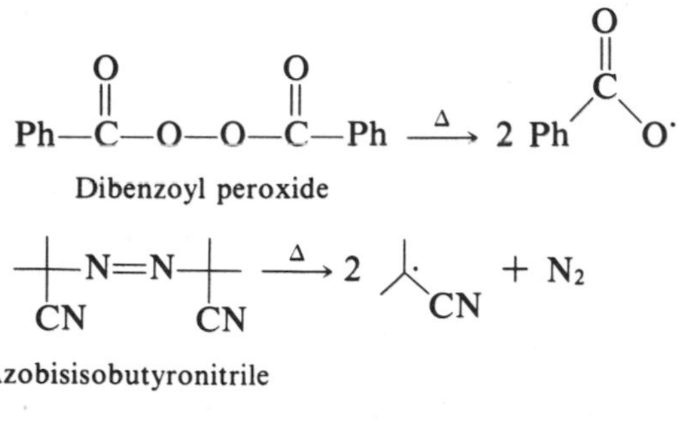

$$\text{Ph—C(=O)—O—O—C(=O)—Ph} \xrightarrow{\Delta} 2\ \text{Ph—C(=O)—O}^{\cdot}$$

Dibenzoyl peroxide

$$(CH_3)_2C(CN)\text{—N=N—}C(CN)(CH_3)_2 \xrightarrow{\Delta} 2\ (CH_3)_2\dot{C}\text{—CN} + N_2$$

Azobisisobutyronitrile

The O—O bond cleavage requires an activation energy of approximately 35 kcal/mol (Table 2.8). Therefore, a peroxide will decompose with a measurable rate ($\sim 10^{-4}$ s^{-1}) at 100°C. This is slow, but sufficient. The normal C—N bond dissociation energy is ~73 kcal/mol, however. Obviously, there must be something special about the azo-compounds; otherwise, they would decompose extremely slowly. This will be discussed in Section 2.4.3. For the special case of anchimerically assisted homolysis, see Ref. 91.

2.4.2. Peroxy Compounds

The initial bond-breaking steps in peroxy-compound thermolyses may be presented as follows:

$$\underset{\text{Dialkyl peroxides}}{R{-}O{-}O{-}R'} \longrightarrow \underset{\text{Alkoxy radicals}}{RO\cdot + \cdot OR'} \tag{9}$$

$$\underset{\text{Diacyl peroxides}}{R{-}\overset{O}{\overset{\|}{C}}{-}O{-}O{-}\overset{O}{\overset{\|}{C}}{-}R'} \longrightarrow \underset{\text{Acyloxy radicals}}{R{-}C({=}O)O\cdot + \cdot OC({=}O){-}R'} \tag{10}$$

$$\underset{\text{Peresters}}{R{-}C({=}O){-}O{-}O{-}R'} \longrightarrow R{-}C({=}O)O\cdot + \cdot O{-}R' \tag{11}$$

$$\longrightarrow R\cdot + CO_2 + \cdot OR' \tag{12}$$

The acyloxy radicals are of particular interest because they decompose further to carbon radicals:

$$R{-}C({=}O)O\cdot \longrightarrow R\cdot + CO_2 \tag{13}$$

The distinction between stepwise and concerted decomposition [Eqs. (11) and (12)] is discussed later.

Table 2.9 gives the activation parameters for some selected peroxy compounds. The important trend to be noted is that the more stable the radicals formed, the lower the activation energy (i.e., the easier the decomposition). *Tert*-butyl radicals form easier than methyl (compare entries 4 with 5 and 9 with 10). When only alkoxy radicals are formed, the alkyl substituent does not show such an effect (compare entries 2 and 3). Phenyl radicals, which are unstable σ-radicals, are particularly difficult to produce (entry 6), but the resonance-stabilized benzyl radicals form easily (entries 7, 11, and 15–17). Alkoxymethyl radicals (entry 18) are also resonance stabilized (see Section 2.2.1). When the radical stability manifests itself in the activation energy, *there must be partial radical character already in the transition state*. That is,

$$R{-}C({=}O){-}O{-}OR' \longrightarrow \underset{TS}{\left[R\cdots C({=}O)\cdots O\cdots O{-}R'\right]^{\ddagger}} \longrightarrow R\cdot \quad CO_2 \quad \cdot OR' \tag{14}$$

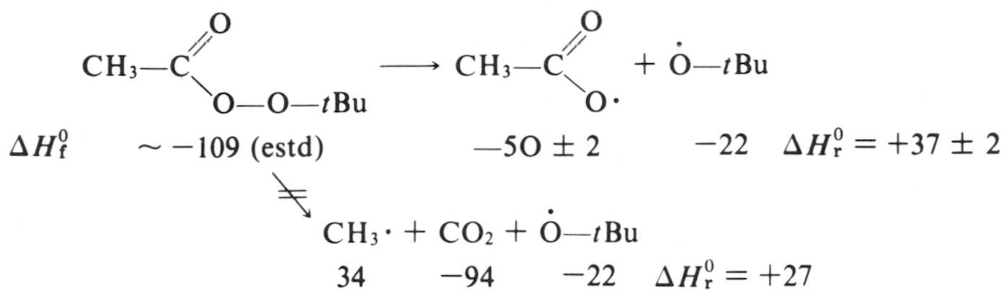

$$\mathrm{CH_3{-}C({=}O){-}O{-}O{-}\mathit{t}Bu} \longrightarrow \mathrm{CH_3{-}C({=}O){-}O\cdot} + \dot{\mathrm{O}}\mathrm{{-}\mathit{t}Bu}$$

ΔH_f^0 ~ −109 (estd) | −5O ± 2 | −22 | $\Delta H_r^0 = +37 \pm 2$

$$\not\searrow \quad \mathrm{CH_3\cdot} + \mathrm{CO_2} + \dot{\mathrm{O}}\mathrm{{-}\mathit{t}Bu}$$

34 | −94 | −22 | $\Delta H_r^0 = +27$

Scheme 4. (Heats of formation in kcal/mol).

When stabilized radicals R· are formed, the reaction should be concerted [see Eqs. (12) and (14)]. Such a reaction has a thermochemical advantage in producing CO_2 ($\Delta H_f^0 = -94$ kcal/mol) and the reaction will therefore be less endothermic by this amount. On the other hand, the transition state in Eq. (14) is entropically unfavorable since it represents a synchronization of the bond vibrations. This reveals itself in small or negative entropies of activation (negative $\Delta S^\ddagger \mathrel{\widehat{=}}$ ordered or rigid transition state; compare entries 11–19 in Table 2.9).

When less stable radicals such as $CH_3\cdot$ or Ph· are formed, the nonconcerted pathway [Eq. (11)] may be more favorable. In that case, the initial O—O bond cleavage would not be influenced much by the substituents, and all reactions that have $E_a > \sim 35$ kcal/mol may be expected to proceed essentially by the nonconcerted mechanism [Eq. (11)].

Consider the two specific reactions shown in Schemes 4 and 5. The thermochemistry derives from known or estimated enthalpies of formation. With the aid of the activation energies in Table 2.9, the energy–reaction coordinate diagrams in Figure 2.14 can be constructed. The peracetate in Scheme 4 has virtually identical enthalpies of activation ($\Delta H^\ddagger$) and reaction (ΔH_r^0), indicating one-bond homolysis leading to two radicals that have almost zero activation energy for recombination.

The phenylperacetate in Scheme 5, in contrast, has a $\Delta H^\ddagger$ that is *lower* than the endothermicity of the one-bond homolysis ($\Delta H_r^0 = 37 \pm 2$). This indicates that the reaction *must* be concerted. The lowering of the transition state is approximately

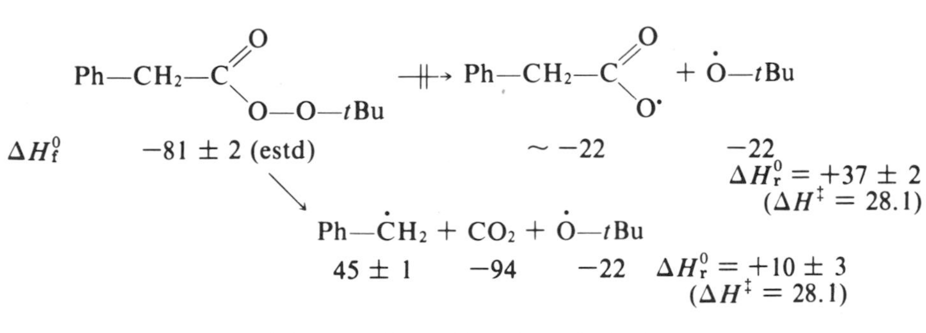

$$\mathrm{Ph{-}CH_2{-}C({=}O){-}O{-}O{-}\mathit{t}Bu} \not\longrightarrow \mathrm{Ph{-}CH_2{-}C({=}O){-}O\cdot} + \dot{\mathrm{O}}\mathrm{{-}\mathit{t}Bu}$$

ΔH_f^0 −81 ± 2 (estd) | ~ −22 | −22 | $\Delta H_r^0 = +37 \pm 2$ ($\Delta H^\ddagger = 28.1$)

$$\searrow \quad \mathrm{Ph{-}\dot{C}H_2} + \mathrm{CO_2} + \dot{\mathrm{O}}\mathrm{{-}\mathit{t}Bu}$$

45 ± 1 | −94 | −22 | $\Delta H_r^0 = +10 \pm 3$ ($\Delta H^\ddagger = 28.1$)

Scheme 5. (Heats of formation in kcal/mol)

TABLE 2.9. Activation Parameters for Peroxy Compounds

Compound	Phase	log [A (s^{-1})]	E_a (kcal/mol)	Reference
1. tBu-O—O—H	Gas	15.6	37.4	*a*
2. CH_3—O—O—CH_3	Gas	15.6	36.9 ± 1.1	*b*
3. tBu-O—O—tBu	Gas	15–16	36–39	*b*
4. H_3C—C(=O)—O—O—C(=O)—CH_3	Gas, solution	14–16	31 ± 2	*b*
	Gas	14.25	29.5	*a*
5. iPr—C(=O)—O—O—C(=O)—tBu	Solution	—	27.3	*b*
6. Ph—C(=O)—O—O—C(=O)—Ph	Solution	15.9	~30.2	*b*
7. $PhCH_2$—C(=O)—O—O—C(=O)—CH_2Ph	Solution	—	22.5 ± 0.5	*b*

		$\Delta S^{\ddagger}$ (cal mol^{-1} k^{-1})	E_a (kcal/mol)	
8. $Ph-C(=O)-O-O-tBu$	Solution	10	34.1	*c*
9. $CH_3-C(=O)-O-O-tBu$	Solution	17	38	*b*
10. $tBu-C(=O)-O-O-tBu$	Solution	13	30.6	*b*
11. $Ph-CH_2-CO_2OtBu$	Solution	3.9	28.7	*b*
12. $p-NO_2-C_6H_4CH_2-CO_2OtBu$	Solution	4.4	29.8	*b*
13. $p-CH_3-C_6H_4CH_2-CO_2OtBu$	Solution	−0.3	26.3	*b*
14. $p-CH_3O-C_6H_4-CH_2CO_2OtBu$	Solution	0.0	25.2	*b*
15. $Ph-C(CH_3)_2CO_2OtBu$	Solution	5.8	26.1	*b*
16. Ph_2CH-CO_2OtBu	Solution	−1.0	24.3	*b*
17. Ph_3CCO_2OtBu	Solution	5	24.1	*d*
18. $CH_3-O-CH_2-CO_2OtBu$	Solution	4	25	*b*
19. $tBuO-C(=O)-C(=O)-O-O-tBu$	Solution	6	26.9	*b*

[a]Data from S. W. Benson and H. E. O'Neal, *Kinetic Data on Gas-Phase Unimolecular Reactions*, U.S. Department of Commerce, Washington, 1970.

[b]Data from R. C. P. Cubbon, *Prog. React. Kinet.*, **5,** 29 (1970).

[c]D. L. Tuleen, W. G. Bentrude, and J. C. Martin, *J. Am. Chem. Soc.*, **85,** 1938 (1963).

[d]J. P. Lorand and P. D. Bartlett, *J. Am. Chem. Soc.*, **88,** 3294 (1966).

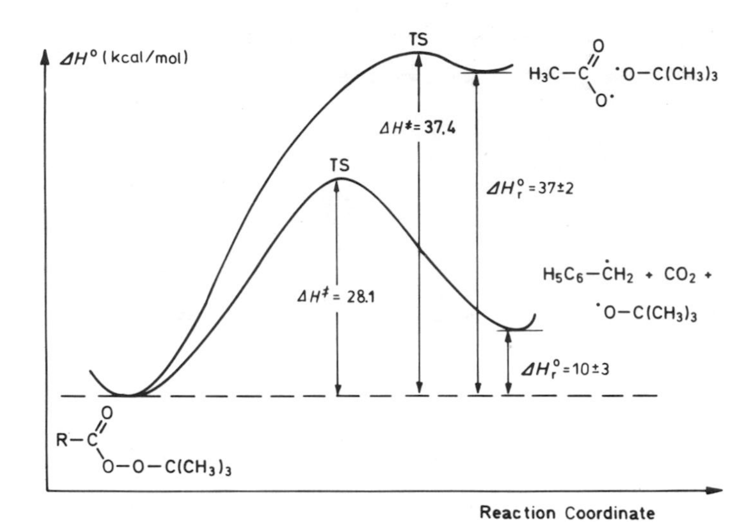

Figure 2.14. Enthalpy–reaction coordinate diagram for decomposition of peresters. R—C(=O)—OO-*t*-Bu; R = CH_3 or Ph—CH_2. $\Delta H^{\ddagger}$ is derived from $\Delta H^{\ddagger} = E_a - RT$, where RT at room temperature ≈ 0.6 kcal/mol.

9 kcal/mol. This is at least 75% of the extra benzyl radical resonance energy (see Table 2.4). It is clear that the full resonance energy need not be ''felt'' in the transition state, for the Hammond postulate dictates that the transition states will become less and less product-like as the reactions become less endothermic. Thus, *at the transition state, there will be unsymmetrical cleavage.* The weakest O—O bond is split first:

R---C(=O)---O-------O—*t*—Bu

We may now understand why the bridgehead radicals shown in Table 2.7 do not show very great variations in their rates of formation by perester homolysis: If the radicals are destabilized because of increasing nonplanarity, they will not be formed in a concerted way. What one measures will be mainly the rates of one-bond homolysis, as, for example, in the thermolysis of the perester **75**:

O=C(OO*t*Bu) **75** —(Δ, rate-determining step)→ O=C(O·) —(- CO_2)→ **76**

Hence the stabilities of the bridgehead radicals themselves will not be correctly reflected in the rates of their formation. A truer picture is obtained using the azo-compound **77** as the radical precursor:

N=N —Δ→ (·N=N) +

77 76

Whether or not this azo-compound decomposes concertedly, one radical **(76)** *must* be formed in the rate-determining step. The rate of decomposition will therefore correctly reflect the stability of the bridgehead radical. As Table 2.7 shows, it is indeed more difficult to generate highly strained radicals from azo-compounds than from peresters.

Now if it is difficult to produce the 1-norbornyl radical **(76),** one might also expect the bond dissociation enthalpy in norbornane **(78)** to be higher than usual. This is indeed the case.[51] The measured bond dissociation enthalpies indicate an

H ⟶ + H· $DH^0 = 96.4 \pm 2.5$ kcal/mol

78 **76**

CH_3, H_3C, CH_3, H ⟶ CH_3, H_3C, CH_3 + H· $DH^0 = 92$ kcal/mol

extra radical strain energy of ~4 kcal/mol in **76.** Even this is not very much and again shows that, energetically, a deviation from planarity is not too costly for a radical. One can compute the rate equivalence of these 4 kcal/mol of extra strain energy. If k_1 is the rate constant for homolysis of a *t*-butyl compound (e.g., perester) and k_2 the rate constant for the analogous norbornyl compound (e.g., **75**), then

$$\frac{k_1}{k_2} = \frac{A_1}{A_2} \cdot e^{4000/RT} = 10^3 \qquad \text{for } A_1 = A_2$$

Thus, if the two compounds have equal *A*-factors, **75** will decompose a thousand times slower than the *t*-butyl compound. This, indeed, is the value reported in Table 2.7.

2.4.3. Azo-compounds

Azo-compounds are well suited for the study of substituent effects on radical formation,

$$\underset{\mathbf{79}}{R{-}N{=}N{-}R'} \longrightarrow [R\cdots\cdots N \overset{\cdot\cdot}{\equiv} N\cdots\cdots R']^{\ddagger} \longrightarrow R\cdot \quad N{\equiv}N \quad \cdot R' \qquad (15)$$

since the complication just described for the peresters is avoided. The "driving force" for the reaction is the formation of the very stable nitrogen molecule $[\Delta H_f^\circ(N_2) \equiv 0]$. The reaction shown in Eq. (15) may be exothermic (*vide infra*), and according to the Hammond postulate the radical character in the transition state need not, therefore, be strongly developed. Nevertheless, the R—N bonds will be partly stretched in the transition state, whereas in the peresters (Section 2.4.2), only the weakest O—O bond need be stretched. In an *unsymmetrical* azo-compound **(79)**, the weakening of the bond to the more stable radical R will be most advanced in the transition state.

From the measured enthalpies of activation for some azo-compounds, the heats of formation of the corresponding radicals, and the measured or estimated heats of formation of the azo-compounds (Table 2.10), one may calculate the enthalpies of the radical-forming reactions:

$$\Delta H_r^0 = \sum \Delta H_f^0(\text{products}) - \Delta H_f^0(\text{precursor})$$

TABLE 2.10. Thermochemical and Activation Parameters for Symmetrical Azo-Compounds R—N=N—R

R	ΔH_f^0(R—N=N—R)	$\Delta H_f^0(R\cdot)$	$\Delta H^\ddagger$	ΔH_r^0	$\Delta S^\ddagger$	Reference
$CH_3\cdot$	35.5	34	51.5	32.5	16.9	*a, b*
Et·	22.4	25.7	49	29	16.0–16.5	*a*
i-Pr·	8.6	17.8	46.5	27	16.5–17.4	*a, b*
t-Bu·	−8.7	7.5	42.5	24	16.0–17.4	*a, b*
Ph-$CH_2\cdot$	86	45	36.6	4	11.5	*a*
Ph·	96.8	80	53	63	—	*a–c*
$CH_2{=}CHCH_2\cdot$	73	41.4	35	9.8	~11	*b, d*

ΔH values in kcal/mol; ΔS values in cal mol^{-1} K^{-1}. Enthalpies of reaction ΔH_r^0 are estimated on the basis of data reported by:

[a] S. W. Benson and H. E. O'Neal, *Kinetic Data on Gas-Phase Unimolecular Reactions,* U.S. Department of Commerce, Washington, 1970.

[b] P. S. Engel, J. L. Wood, J. A. Sweet, and J. L. Margrave, *J. Am. Chem. Soc.*, **96,** 2381 (1974); P. S. Engel, *Chem. Rev.*, **80,** 99 (1980).

[c] A. Leiba and I. Oref, *J. Chem. Soc. Faraday Trans. 1,* **75,** 2694 (1979); E. Wolf and H. K. Cammenga, *Z. Phys. Chem.*, **107,** 21 (1977).

[d] R. J. Crawford and K. Takagi, *J. Am. Chem. Soc.*, **94,** 7406 (1972); B. H. Al-Sader and R. J. Crawford, *Can. J. Chem.*, **48,** 2745 (1970).

The data in Table 2.10 are translated into energy profiles in Figure 2.15. It can be seen that the more stable the radicals formed (the more exothermic the reaction), the lower the transition state, and the lowering equals approximately the increase in the enthalpy of reaction ΔH_r^0. Therefore, these reactions illustrate the Bell–Evans–Polanyi principle:

$$\Delta E_a \sim \alpha \Delta \Delta H_r^0$$

In Figure 2.15, we have also taken the Hammond postulate into account, drawing the transition states closer to the starting materials for the more exothermic reactions. Figure 2.15 indicates that one can directly use $\Delta H^\ddagger$ as a measure of radical stability (assuming that $\Delta S^\ddagger$ is unchanged; to be precise, we ought to calculate $\Delta G^\ddagger$ as $\Delta H^\ddagger - T \cdot \Delta S^\ddagger$). For example, $\Delta H^\ddagger$ for azotoluene (**80**) is approximately 15 kcal/mol lower than that for azomethane (**81**), that is, more than the value of the resonance energy gained when forming *one* benzyl radical (12.5 ± 1.5 kcal/mol). More data are collected in Table 2.11.

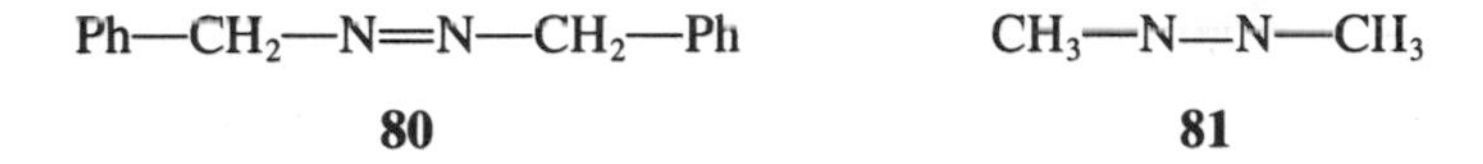

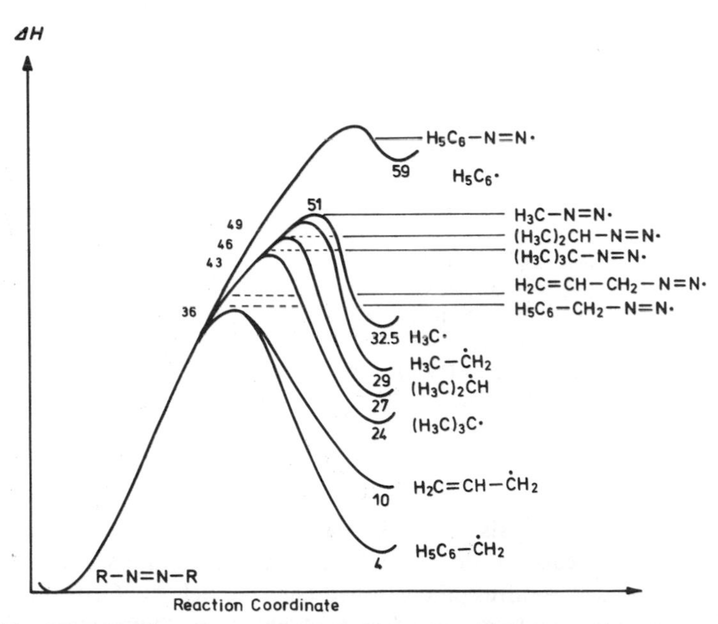

Figure 2.15. Decomposition of azo-compounds. Enthalpies of activation and reaction in kcal/mol. Horizontal lines correspond to $\Delta H^\ddagger$ for the reactions R—N=N—R→ R· + Ṅ=N—R. If these are higher than the experimental values of $\Delta H^\ddagger$, the reactions must be concerted, R—N=N—R → 2 R· + N_2. The activation energy reported for azobenzene [53.4 ± 2.9 kcal/mol: A. Leiba and I. Oref, *J. Chem. Soc. Faraday Trans. 1* **75**, 2694 (1979)] is suspect since it is lower than the calculated enthalpy of reaction.

TABLE 2.11. Activation Parameters for (Un)symmetrical Azo-Compounds R—N=N—R′

R	R′	$\Delta H^{\ddagger}$ (kcal/mol)	$\Delta S^{\ddagger}$ (cal mol⁻ K⁻¹)	Medium	Reference
1. $CH_3\cdot$	$\cdot CH_2CH=CH_2$	34.4	5	gas	[a]
2. n—$C_3H_7\cdot$	$\cdot CH_2CH=CH_2$	34.6	7	gas	[a]
3. *t*—Bu·	$\cdot CH_2CH=CH_2$	28.8	−3	gas	[a]
4. $CH_2=CHCH_2\cdot$	$\cdot CH_2CH=CH_2$	35.1	10–11	gas, liquid	[a,b]
5. $CH_2=CHC(CH_3)_2\cdot$	$\cdot C(CH_3)_2CH=CH_2$	25.3	5.5	liquid	[b]
6. $HC\equiv C-C(CH_3)_2\cdot$	$\cdot C(CH_3)_2-C\equiv CH$	26.2	8.2	liquid	[b]
7. $Ph\dot{C}H(CH_3)$	$\dot{C}H_3$	37.6	14	liquid	[c]
8. $Ph\dot{C}H(CH_3)$	$Ph\dot{C}H(CH_3)$	32.2	7	liquid	[d]
9. $Ph\dot{C}(CH_3)_2$	$Ph\dot{C}(CH_3)_2$	29	11	liquid	[d]
10. $Ph_2\dot{C}H$	$Ph_2\dot{C}H$	26	2.3	liquid	[e]
11. Ph·	$Ph_3C\cdot$	27	—	liquid	[f]
12. $CH_3-\dot{C}(CN)(CH_3)$	$CH_3-\dot{C}(CN)(CH_3)$	31	10	liquid	[g]
13. $Ph-\dot{C}(CF_3)_2$	$Ph-\dot{C}(CF_3)_2$	32.8	20	liquid	[h]

[a]R. J. Crawford and K. Takagi, *J. Am. Chem. Soc.*, **94,** 7406 (1972).
[b]P. S. Engel and D. J. Bishop, *ibid.*, **94,** 2148 (1972).
[c]S. Seltzer and F. T. Dunne, *ibid.*, **87,** 2628 (1965).
[d]S. F. Nelson and P. D. Bartlett, *ibid.*, **88,** 137 (1966).
[e]S. G. Cohen and C. H. Wang, *ibid.*, **77,** 2457 (1955).
[f]M. G. Alder and J. E. Leffler, *ibid.*, **76,** 1425 (1954).
[g]J. P. van Hook and A. V. Tobolsky, *ibid.*, **80,** 779 (1958).
[h]J. B. Levy and E. J. Lehmann, *ibid.*, **93,** 5790 (1971).

Cage Effect. The thermolysis of azoalkanes in solution shows a pronounced *cage effect;* that is, the radicals and the N_2 molecule are formed in a solvent cage from which they escape with a diffusion controlled rate (see Scheme 6). If the viscosity of the solvent is very high (e.g., in a glassy solvent at low temperature), diffusion will be slow, and it may take a considerable time before the free radicals recombine. During this time, they may be observed by ESR spectroscopy. A striking example of this cage effect is seen when a mixture of $CH_3—N=N—CH_3$ and $CD_3—N=N—CD_3$

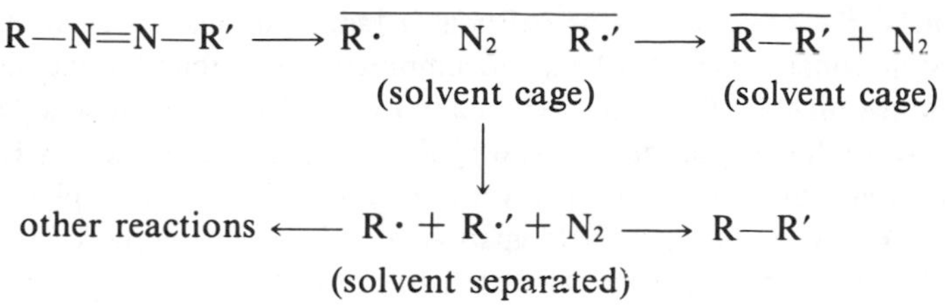

$$R—N{=}N—R' \longrightarrow \overline{R\cdot \quad N_2 \quad R\cdot'} \longrightarrow \overline{R—R' + N_2}$$

(solvent cage) (solvent cage)

$$\downarrow$$

$$\text{other reactions} \longleftarrow R\cdot + R\cdot' + N_2 \longrightarrow R—R'$$

(solvent separated)

Scheme 6

is decomposed in solution: Only C_2H_6 and C_2D_6 are formed, proving exclusive recombination within the cage (*cage return*):

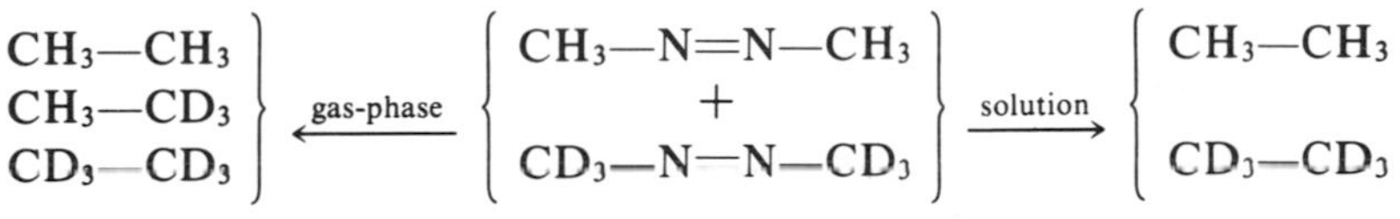

$$\left.\begin{matrix} CH_3—CH_3 \\ CH_3—CD_3 \\ CD_3—CD_3 \end{matrix}\right\} \xleftarrow{\text{gas-phase}} \left\{\begin{matrix} CH_3—N{=}N—CH_3 \\ + \\ CD_3—N{=}N—CD_3 \end{matrix}\right\} \xrightarrow{\text{solution}} \left\{\begin{matrix} CH_3—CH_3 \\ \\ CD_3—CD_3 \end{matrix}\right.$$

In the gas-phase, no such solvent effect is possible, and the three recombination products C_2H_6, C_2D_6, and $C_2H_3D_3$ are formed in the statistical ratio of 1:1:2.

Only when the radicals escape from the solvent cage (Scheme 6) can they undergo other reactions. The reaction with a thiol, for example,

$$R\cdot + R''SH \longrightarrow RH + R''S\cdot$$

is a very efficient *radical scavenging* process, which can be used to measure the concentration of free radicals. Azomethane gives no methane in the presence of a thiol, again indicating complete cage return. The frequently used radical chain initiator, azoisobutyronitrile (AIBN) **(82)**, however, gives only 20% of the cage product **(83)** in the presence of high concentrations of butanethiol:

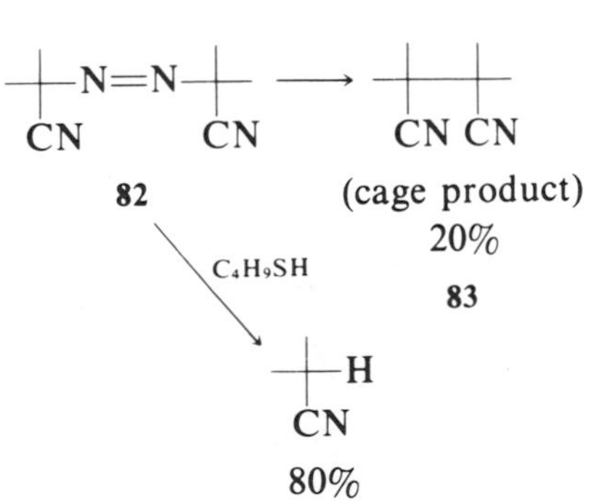

$$(CH_3)_2C(CN)—N{=}N—C(CN)(CH_3)_2 \longrightarrow (CH_3)_2C(CN)—C(CN)(CH_3)_2$$

82 → (cage product) 20% **83**

82 $\xrightarrow{C_4H_9SH}$ $(CH_3)_2C(CN)—H$ 80%

Accordingly, the effectiveness of AIBN as a radical initiator is at most 80%.

An extraordinary example of cage recombination was found in the photolysis of pure *meso*-azobis-3-methyl-2-phenyl butane **(84)** in frozen benzene solution at the temperature of liquid nitrogen (−196°C).[52] At this temperature, an ESR signal, stable for nine days, could be observed, corresponding to the triplet radical pair **85,** where the two radicals are held apart at 6–7 Å by the interposed N_2 molecule. (The radical separation can be calculated from the ESR splitting constant D, cf. Section 2.2.2.) Such a long separation rules out the possibility of having formed only the triplet excited azo-compound **87,** in which the N—N bond length would be approximately 1.4 Å. When the radical pair **85** was warmed, pure *meso*-**86** was

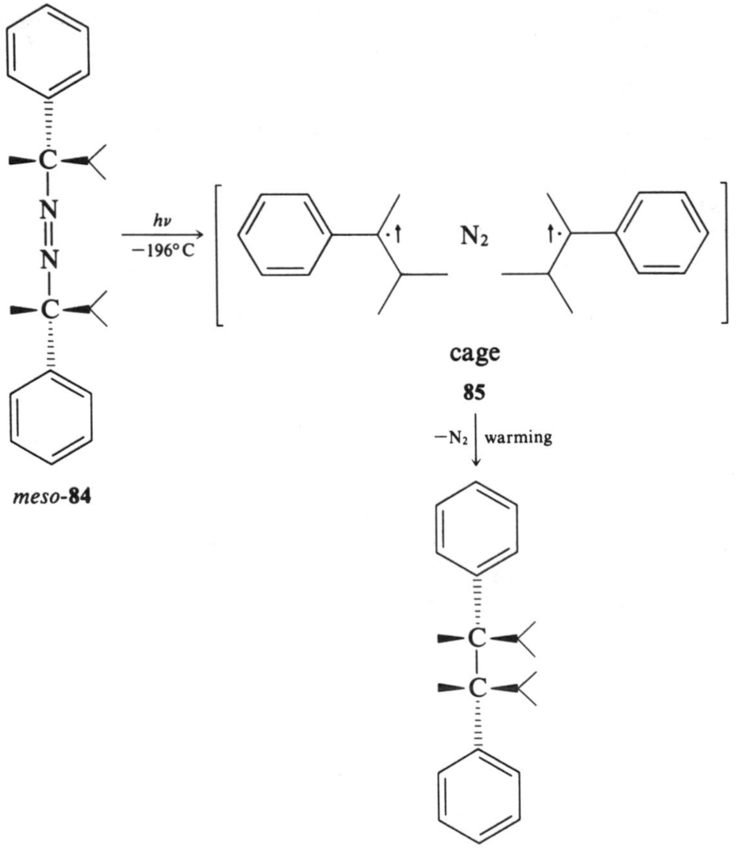

hν, −196° C

N_2

cage

85

$-N_2$ warming

meso-**84**

meso-**86**

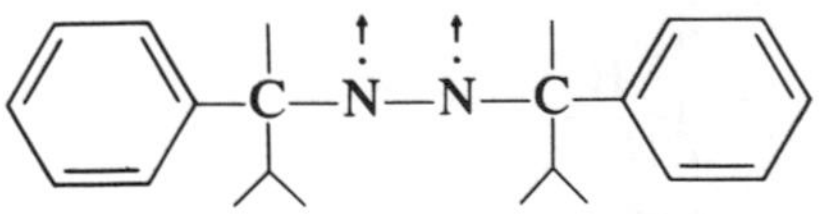

87

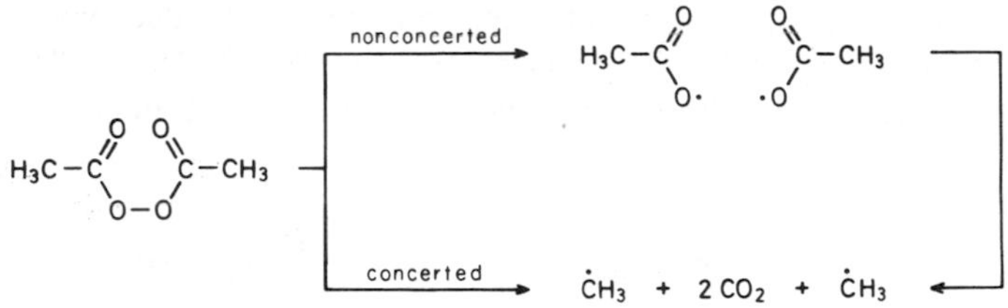

Scheme 7

obtained, indicating that the solvent cage at $-196°C$ is so rigid that no independent rotation of the two radical fragments can take place. When the photolysis was carried out at higher temperatures, stereochemical specificity was progressively lost, producing *d,l*-**86.**

2.4.4. Concerted or Nonconcerted Decomposition

Peroxides. Azo-compounds and peroxides may decompose by either a concerted or a nonconcerted pathway, as shown for diacetyl peroxide in Scheme 7. With the aid of ^{18}O-labeling it was shown[53] that the nonconcerted pathway is followed in solution, resulting in scrambling of the label between the four oxygen atoms (Scheme 8). This pathway appears to be general for all di-*n*-acyl-peroxides.[54] The same conclusion may be obtained by observing the viscosity dependence of the rate of formation of methyl radicals from diacetyl peroxide. If a radical pair is formed in a solvent cage, as in Scheme 8, a viscous solvent will make it more difficult for the radicals to diffuse apart. Conversely, it will be easier for the initial radicals to recombine, and the rate of fragmentation to $CH_3\cdot$ and CO_2 will be lower. In agreement with this, the decomposition of diacetyl peroxide shows a strong viscosity dependence.[55]

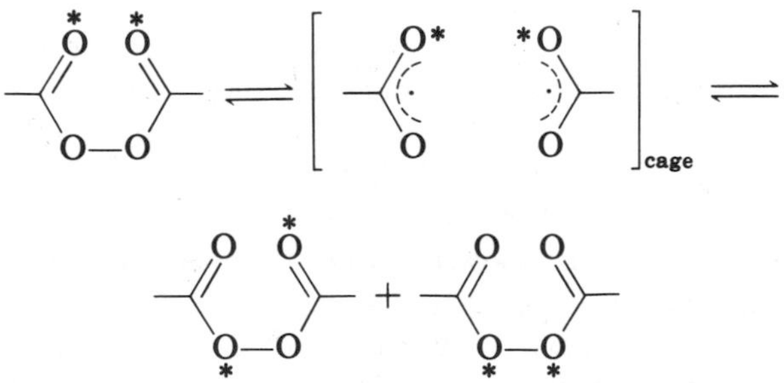

Scheme 8

By the same criterion, the *tert*-butyl arylperacetates (**88**) decompose via the *concerted* mechanism (no viscosity dependence) when R = CH_3 or CH_3O. However, when R = H or NO_2, viscosity dependence was observed, indicating (at least

$$R{-}C_6H_4{-}CH_2{-}\overset{O}{\overset{\|}{C}}{-}O{-}O{-}tBu \longrightarrow \left[R{-}C_6H_4{-}\dot{C}H_2\ CO_2\ \dot{O}{-}tBu \right]$$

88

$$O_2N{-}C_6H_4{-}CH_2{-}\overset{O}{\overset{\|}{C}}{-}O{-}O{-}tBu \longrightarrow$$

89

$$\left[O_2N{-}C_6H_4{-}CH_2{-}C(=O)O^{\bullet} + \dot{O}{-}tBu \right]$$

90

partial) one-bond scission. This was particularly pronounced for R = NO_2 (**89** → **90**).[55] It was found in another study[56] that the rate of decomposition of the arylperacetates **88** followed the Hammett relationship

$$\log \frac{k}{k_0} = \rho\sigma$$

or

$$\log \frac{k}{k_0} = \rho\sigma^+$$

where the correlation with σ^+ has often been taken as evidence for charge separation in the transition state: Substituents that can stabilize a positive charge (CH_3, CH_3O) accelerate the reaction; electron-withdrawing substituents (NO_2) retard it.

However, the observed correlation with σ^+ does not necessarily mean that there is partial carbenium ion character in the transition state. It need only mean that the substituents can conjugate with—and stabilize—the *radical* center in the transition state for the decomposition of **88** (see Ref. 57).

The correlation of rates wth σ^+ is shown in Figure 2.16. It is seen that the "*fast*" peresters are those that show no viscosity dependence and decompose by the concerted mechanism (R = CH_3 and CH_3—O). The "slowest" perester (R = NO_2) was found by the viscosity test to undergo nonconcerted fragmentation (**89** → **90**). A closer inspection of the Hammett correlation (Figure 2.16) now reveals that the point for R = NO_2 falls, in fact, outside the straight line given by the other points.

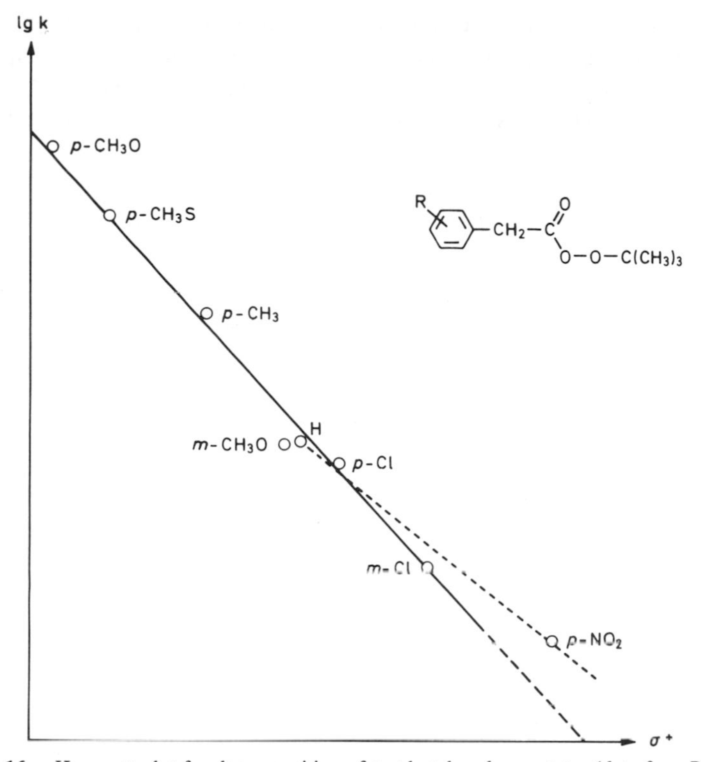

Figure 2.16. Hammett plot for decomposition of *tert*-butyl arylperacetates (data from Ref. 56).

The break is even more pronounced if the correlation is made with σ rather than σ^+. A break in a Hammett plot signifies a change in mechanism. Since the point for R = NO_2 is *above* the curve, the perester **89** decomposes *faster* than would be expected if the destabilizing conjugation with the NO_2 group were of importance in the transition state. The reason is, of course, that **89** does not give a benzyl radical, but rather the carboxyl radical **90,** in which the nitro-group has little influence on the stability of the radical center.

The mesomeric stabilization in $Ph_2\dot{C}H$ is larger than in $CH_3O—C_6H_4—\dot{C}H_2$, and in agreement herewith, the compound **91** decomposes faster than the arylper-

$$Ph_2CH—\overset{\overset{\displaystyle O}{\|}}{C}—O—O—\mathit{t}Bu$$

91

acetates (it has also a lower activation energy, see Table 2.9). Furthermore, **91** shows no viscosity dependence, so that the evidence is overwhelmingly in favor of two-bond (concerted) decomposition.

In the diacetyl peroxide decomposition (Scheme 8), about 34% of the acetoxy radicals recombine within the solvent cage. The remaining radicals diffuse apart to yield, in toluene solution, methyl acetate (15%), ethane (5%), and methane (48%).[58] With the aid of CIDNP, it was shown[54b,59] that both methyl acetate and ethane derive from secondary geminate radical pairs (Scheme 9):

$$(CH_3CO_2)_2 \overset{\Delta}{\rightleftharpoons} [CH_3COO\cdot \quad \cdot OOCCH_3]_{cage\ 1} \longrightarrow \text{free radicals}$$

$$\downarrow -CO_2$$

$$\underset{E}{CH_3COO{-}CH_3} \longleftarrow [CH_3\cdot \quad CH_3COO\cdot]_{cage\ 2} \longrightarrow CH_3\cdot + CH_3COO\cdot$$

$$\downarrow -CO_2 \qquad\qquad CH_3\cdot \xrightarrow{CCl_3COCCl_3} \underset{A}{CH_3{-}Cl}$$

$$\underset{E\quad E}{CH_3{-}CH_3} \longleftarrow [CH_3\cdot \quad \cdot CH_3]_{cage\ 3} \longrightarrow 2\ \dot{C}H_3 \xrightarrow{CCl_3COCCl_3} \underset{A}{CH_3{-}Cl}$$

Scheme 9

The observation of emission (E) for the ester protons in the NMR spectrum of methyl acetate during the reaction demonstrates that this product is formed by recombination within cage 2. Application of Kaptein's rules (Section 2.2.3) gives:

$\mu(-)$	thermal, singlet reaction
$\epsilon(+)$	cage product
$\Delta g(-)$	oxy-radicals have larger g-factors than carbon radicals
$a_i(-)$	hfs in $CH_3\cdot$; negative for α-protons

$$\Gamma_n = \mu\epsilon\Delta g\ a_i = (-) \Rightarrow \textit{net emission}$$

Methyl chloride is formed by reaction of hexachloroacetone with methyl radicals that have escaped from the cage (Scheme 9). Therefore, ϵ will be negative, and the opposite polarization, that is, enhanced absorption (A) is observed. Since ethane is also polarized (E), and the two methyl radicals in cage 3 cannot give rise to a net effect ($\Delta g = 0$), the polarization must originate in cage 2.

92

The phenyl radical **(92)** is an unstable σ-radical ($\Delta H_f^0 \cong 80$ kcal/mol), which, evidently, cannot adopt the more favorable sp^2-configuration (cf. Section 2.3).

Conversely, the C—H bond in benzene is very strong (110 kcal/mol). Having already established that diacetyl peroxide decomposes by the nonconcerted pathway, we may predict (correctly) that dibenzoyl peroxide will do the same:

$$Ph{-}C(=O){-}O{-}O{-}C(=O){-}Ph \underset{}{\overset{\Delta}{\rightleftharpoons}} 2\ Ph{-}C(=O){-}O^{\bullet}$$

Subsequently, the benzoyloxy radicals can eliminate CO_2 to yield phenyl radicals that either dimerize or react with the solvent:

$$Ph{-}C(=O){-}O^{\bullet} \longrightarrow Ph\cdot \begin{cases} \nearrow Ph{-}Ph \\ \overset{SH}{\searrow} Ph{-}H + S\cdot \end{cases}$$

The solvent radicals S· so formed can start a chain reaction by attacking the peroxide itself in a *bimolecular homolytic substitution* (S_H2) reaction:

$$S\cdot + Ph{-}C(=O){-}O{-}O{-}C(=O){-}Ph \longrightarrow Ph{-}C(=O){-}O{-}S + {}^{\bullet}O{-}C(=O){-}Ph$$

As a result, the decomposition of dibenzoyl peroxide proceeds by a mixture of pure homolysis and *induced homolysis,* the latter being initiated by the benzoyloxy radicals formed in the homolysis reaction. For a very readable and instructive CIDNP study of dibenzoyl peroxide, see Ref. 60.

Azo-compounds. Like peroxides, azo-compounds may decompose by concerted or nonconcerted mechanisms (Scheme 10). Referring to Figure 2.15, it is apparent that azobenzene and simple azoalkanes decompose by the nonconcerted mechanism. However, when *two* resonance-stabilized radicals of the benzyl type are formed, the reaction can become concerted.

Since the diazenyl radicals **93** are highly unstable, the transition states for their formation will have a very high degree of radical character, and we can draw an energy profile such as that shown in Figure 2.17.

Let us now examine how one can determine the mechanism of azoalkane decomposition. Because the reaction in Figure 2.17 is subject to cage return, it can be retarded by an increase in pressure (or viscosity). The effect of pressure is

$$R{-}N{=}N{-}R' \xrightarrow[\text{or } h\nu]{\Delta} \begin{cases} \nearrow R{-}N{=}N^{\bullet} + \cdot R' \ (\mathbf{93}) \xrightarrow{-N_2} \\ \searrow R\cdot \quad N{\equiv}N \quad \cdot R' \end{cases}$$

Scheme 10

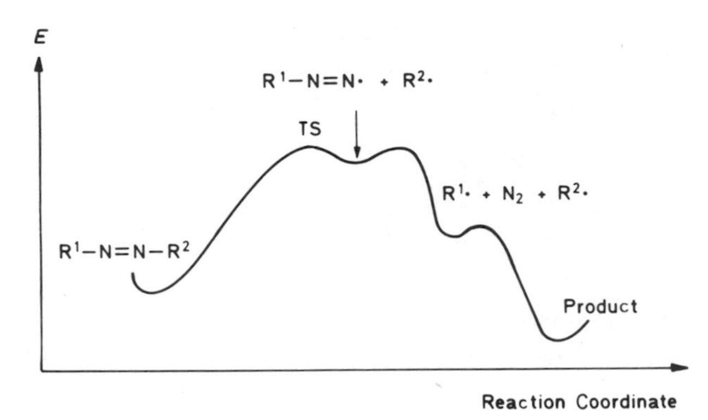

Figure 2.17. Energy profile for nonconcerted azoalkane homolysis.

measured in the *volumes of activation* $\Delta V^{\ddagger}$, which, therefore, are more informative than the energies of activation. A large volume of activation implies a nonconcerted fragmentation.[61] Thus, from the data in Table 2.12, it is again apparent that the decomposition is concerted when the two radicals formed are resonance stabilized, but nonconcerted when the highly unstable phenyl radical is one of the ultimate products. Porter et al.[62] photolyzed an optically active *trans*-azo-compound **(94)** and found that if the reaction was stopped before complete decomposition, both racemized *trans*-**94** and *cis*-**94** could be isolated. This is best explained in terms of the diazenyl radical pair intermediate **95:**

$H_5C_6-C(CH_3)(C_2H_5)-N=N-C_6H_5$ (*trans* 94) $\underset{}{\overset{h\nu}{\rightleftharpoons}}$ $[H_5C_6-\dot{C}(CH_3)(C_2H_5) \quad \cdot N=N-C_6H_5]$ (95) $\rightleftharpoons$ $H_5C_6-C(CH_3)(C_2H_5)-N=N-C_6H_5$ (*cis* 94)

The existence of the diazenyl radical was finally proven by CIDNP.[63] The thermal decomposition of *cis*-**96** in the probe of an NMR spectrometer resulted in polarization of several reaction products, as shown in Scheme 11. Most importantly, the *trans*-azo-compound (*trans*-**96**) was formed during the reaction, and the protons of the methyl groups were found to be in emission. *Trans*-**96** must, therefore, be formed at least in part by recombination of the primary radical pair **97.**

TABLE 2.12. Activation Volumes for Azoalkane Decomposition in Solution[a]

R—N=N—R′ R	R′	$\Delta V^{\ddagger}$(cm³/mol)
$Me_2C(CN)$		+4
$Me_2C(Ph)$		+5
$Me_2C(p\text{-}MeC_6H_4)$		+4
$Me_3C—O$		+4.3
Ph_3C	$p\text{-}NO_2—C_6H_4$	+20

[a]Data from Ref. 61.

Application of Kaptein's rules gives

$$\Gamma_n = \mu\epsilon\Delta g a_i = (-)(+)(+)(+) = (-) \Rightarrow E \qquad \text{for } \Delta g = (+)$$

Note that a_i is positive because the methyl-protons are β to the radical. The only parameter that cannot be predicted a priori is Δg. If the radical Ph—N=N· were an ordinary radical with the free electron in a p orbital, one would expect $g(\text{Ph—N=N}\cdot) > g(\text{Ph}\dot{\text{C}}\text{Me}_2)$ (see, e.g., the π-amino radicals, Table 2.6). How-

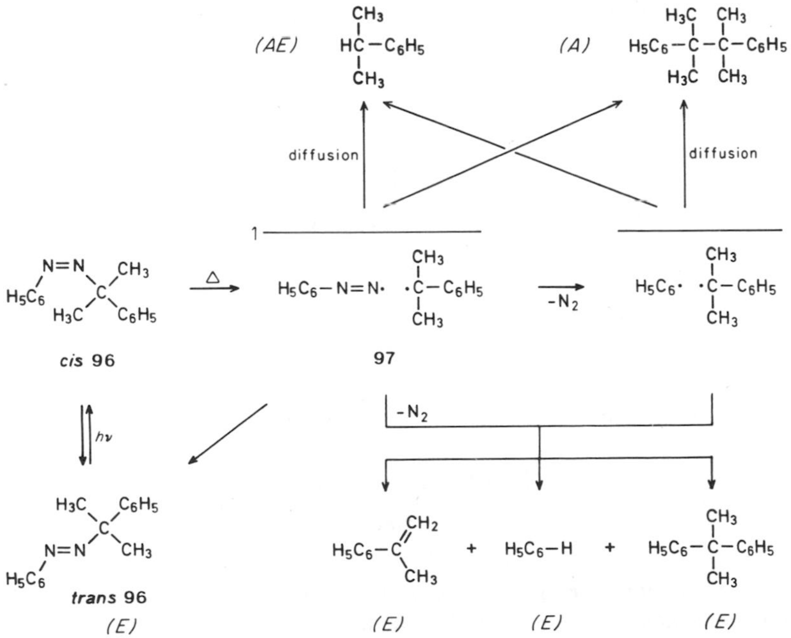

(AE) $HC(CH_3)_2-C_6H_5$

(A) $H_5C_6-C(CH_3)_2-C(CH_3)_2-C_6H_5$

diffusion

diffusion

1

$H_5C_6-N=N-C(CH_3)_2-C_6H_5$ (cis 96) $\xrightarrow{\Delta}$ $H_5C_6-N=N\cdot \; \cdot C(CH_3)_2-C_6H_5$ (97) $\xrightarrow{-N_2}$ $H_5C_6\cdot \; \cdot C(CH_3)_2-C_6H_5$

cis 96

97

$h\nu$

$-N_2$

trans 96 (E)

$H_5C_6-C(=CH_2)CH_3$ (E) + H_5C_6-H (E) + $H_5C_6-C(CH_3)_2-C_6H_5$ (E)

Scheme 11

ever, Ph—N=N· is a σ-radical (**98**) (compare the iminyls, Table 2.6); hence it is reasonable that $g(\text{Ph—N=N}\cdot) < g(\text{Ph}\dot{\text{C}}\text{Me}_2)$, resulting in the observed emission.

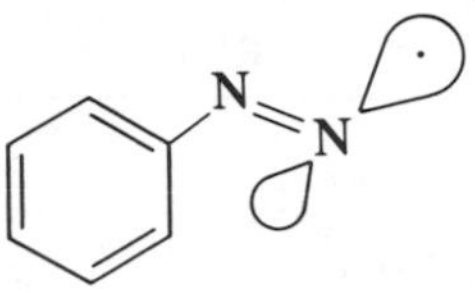

98

The other products shown in Scheme 11 can be assigned as cage or escape products with the knowledge that all cage products must show emission (E), and all escape or scavenging products must show either enhanced absorption (A) or AE multiplets.

The addition of $BrCCl_3$ (a radical scavenger) to the solution did not diminish the CIDNP intensity of *trans*-**96,** but the other radical products disappeared due to quenching reactions, for example,

$$\text{Ph}\dot{\text{C}}\text{Me}_2 + \text{BrCCl}_3 \longrightarrow \text{PhCMe}_2\text{Br} + \dot{\text{C}}\text{Cl}_3$$

From the known rates of reaction of $BrCCl_3$ with radicals, one can estimate $k_{\text{dec}}(\text{Ph—N=N}\cdot) = 10^7 - 10^9\ \text{s}^{-1}$ at 40°C. If the Arrhenius A-factor for decomposition of Ph—N=N· is $10^{16}\ \text{s}^{-1}$ and $k \geqslant 10^7$, E_a for the process will be $\leqslant 1.3$ kcal/mol. If k were $>10^{10}\ \text{s}^{-1}$, E_a would be <0.86 kcal/mol, and the phenyldiazenyl radical would no longer be detectable by CIDNP.

Application of the viscosity test to compounds similar to **96** also shows that these decompose by the stepwise mechanism (viscosity dependence). On the other hand, azocumene (**99**) and related compounds showed no viscosity dependence, indicating concerted fragmentation:[55]

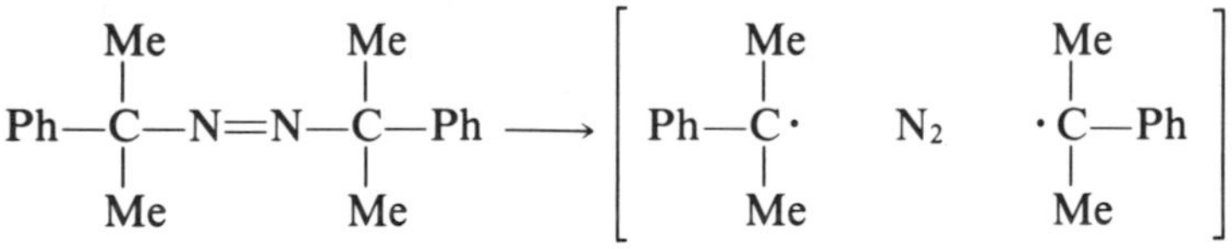

$$\text{Ph—C(Me)}_2\text{—N=N—C(Me)}_2\text{—Ph} \longrightarrow \left[\ \text{Ph—}\dot{\text{C}}\text{(Me)}_2 \quad \text{N}_2 \quad \cdot\text{C(Me)}_2\text{—Ph}\ \right]$$

99

The latter reaction is consistent with the concerted fragmentation of **84** (vide supra), and with the failure to observe CIDNP during the decomposition of **100** (Ref. 63).

$$\text{Ph}_2\text{CH—N=N—CHPh}_2$$

100

This means that either **100** fragments in a concerted manner, or else the intermediate diazenyl radical has a lifetime of less than $\sim 10^{-10}$ s (the rate of singlet–triplet mixing in radical pairs). It is quite possible that a diazenyl radical could be so short-lived because of the exothermicity of its decomposition (see Figure 2.17).

An intermediate that exists for 10^{-10} s still has time for about 1000 molecular vibrations. If the lifetime approaches 10^{-13} s, it becomes meaningless to talk of an intermediate, for it will only exist for about one vibration, and it will be indistinguishable from a transition state.

2.5. GENERAL TYPES OF REACTION

The chain reaction (see p. 66) is a normal occurrence in free radical chemistry, and it is of extreme industrial importance, being almost the only method available for functionalizing saturated hydrocarbons. The free radicals formed from an initiator or during a chain process may react in several general ways:

1. Atom abstraction: $R\cdot + X—Y \rightarrow R—X + Y\cdot$
2. Recombination: $R\cdot + R\cdot' \rightarrow R—R'$
3. Disproportionation: $R—CH_2—CH_2\cdot \rightarrow R—CH_2CH_3 + R—CH{=}CH_2$
4. Addition: $R\cdot + \rangle{=}\langle \rightarrow R\rangle\!—\!\langle\cdot$
5. Electron transfer: $R\cdot + M^{n\oplus} \rightleftharpoons R^{\oplus} + M^{(n-1)\oplus}$

The electron transfer reaction is a redox reaction. The involvement of electron transfer in electrophilic and nucleophilic substitution reactions will be discussed in Section 2.6.

2.5.1. Atom Abstraction

An atom abstraction process $R\cdot + X—Y \rightarrow R—X + Y\cdot$ will be faster the lower the bond dissociation energy in the molecule X—Y, the more stable the newly formed radical $Y\cdot$, and the stronger the new bond formed in R—X.

Compare the H-abstractions[64] by $Cl\cdot$ and $Br\cdot$:

$$\begin{array}{lcccl} Cl\cdot & + \; H—C_2H_5 \longrightarrow & HCl & + \; \dot{C}_2H_5 & k = 4 \cdot 10^9 \text{ liter mol}^{-1}\text{ s}^{-1} \\ \Delta H_f^0\text{: } 29 & -20 & -22 & 26 & DH^0(H—Cl) = 103 \text{ kcal/mol} \\ & & & & \Delta H_r^0 = -5 \text{ kcal/mol} \end{array}$$

$$\begin{array}{lcccl} Br\cdot & + \; H—C_2H_5 \longrightarrow & H—Br & + \; \dot{C}_2H_5 & k = 2 \cdot 10^2 \text{ liter mol}^{-1}\text{ s}^{-1} \\ \Delta H_f^0\text{: } 27 & -20 & -8.7 & 26 & DH^0(H—Br) = 87 \text{ kcal/mol} \\ & & & & \Delta H_r^0 = +10 \text{ kcal/mol} \end{array}$$

Since the reaction of the chlorine atom is exothermic and that of the bromine atom endothermic, the latter is decidedly slower. The difference is mainly due to the different bond strengths of HCl and HBr. Conversely, it is easier to abstract a hydrogen atom from HBr than from HCl (the reverse processes). This simple example demonstrates the usefulness of thermochemistry in predicting the outcome of reactions. The more exothermic a reaction, the nearer the transition state is to the initial state (Hammond postulate), and the lower the activation energy (Bell–Evans–Polanyi principle) is. Therefore, in strongly exothermic reactions, it is immaterial which kind of bond is broken in the transition state (e.g., primary, secondary, or tertiary C—H bonds). The attacking radical shows little *selectivity*. By this token, Br· is more selective than Cl· and preferably attacks tertiary C—H bonds since this makes the reaction more exothermic. These principles are illustrated in Figure 2.18.

The varying reactivities of different radicals are reported in Table 2.13. The reactivity order is always $CH_4 \leqslant$ primary < secondary < tertiary C—H, reflecting the order of bond dissociation enthalpies.

2.5.2. Recombination and Disproportionation

An encounter between two radicals can lead to either recombination or disproportionation.

$$\underset{\Delta H_f^0:\ 26}{CH_3{-}CH_2\cdot} + \underset{26}{CH_3{-}CH_2\cdot} \longrightarrow \underset{-30}{CH_3{-}CH_2{-}CH_2{-}CH_3} \qquad \Delta H_r^0 = -82 \tag{16}$$

$$\underset{\Delta H_f^0:\ 26}{CH_3{-}CH_2\cdot} + \underset{26}{CH_3{-}CH_2\cdot} \longrightarrow \underset{-20}{CH_3{-}CH_3} + \underset{12.5}{CH_2{=}CH_2} \qquad \Delta H_r^0 = -60 \tag{17}$$

Both reactions are strongly exothermic. The recombinations have virtually zero activation energies and high *A*-factors, resulting in rate constants of the order of

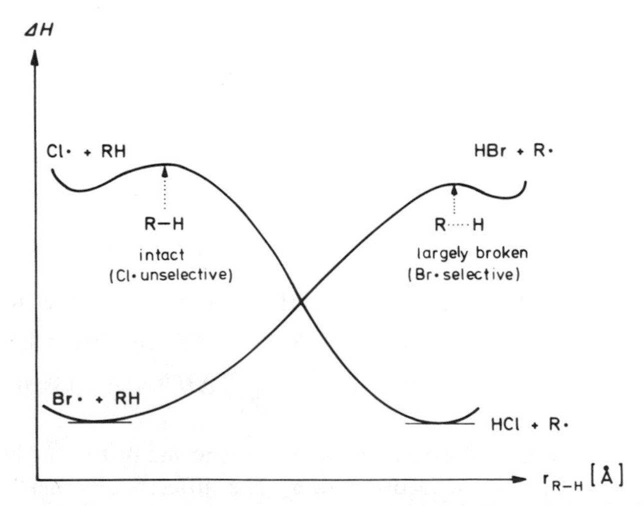

Figure 2.18. Application of the Hammond postulate and the Bell–Evans–Polanyi principle to atom abstraction reactions. Note: The two reaction profiles are not drawn to scale.

TABLE 2.13. Kinetic Data on Hydrogen Abstraction Reactions X· + R—H → XH + R· in the Gas-Phase[a]

X·	Temperature (K)	$k(E_a$, log $A)$			
		CH_3—H	*Prim*-C—H	*Sec*-C—H	*Tert*-C—H
F	298	0.7(1.2,10.5)	1(0.3,10.0)	1(0,9.8)	1.3(0,98)
Cl	523	0.03(3.8,9.8)	1(0.9,10.2)	2(0.6,10.4)	3(0.1,10.3)
OH	500	0.04(5.0,8.9)	1(1.6,8.8)	5(0.9,9.2)	8(0.2,9.1)
CF_3	455	0.04(11.0,8.4)	1(7.5,8.1)	7(5.5,8.0)	22(4.7,8.1)
CH_3O	455	0.04(11.0,8.2)	1(7.1,7.7)	2(5.2,8.0)	5(4.1,8.0)
CH_3	455	0.03(14.5,8.2)	1(11.8,8.4)	10(10.2,8.6)	64(8.1,8.4)
CCl_3	455	—	1(13.4,7.9)	40(10.3,7.9)	250(7.9,7.6)
C_2H_5	455	0.006(17.7,8.1)	1(12.6,7.8)	8(10.3,7.7)	32(8.9,7.6)
Br	300	0.003(18.4,10.5)	1(13.4,10.1)	640(10.0,10.5)	25,000(7.5,10.3)
I	300	0.0004(35,12)	1(26.4,11.1)	16(24.8,11.2)	6,200(21.0,11.1)

[a]Data from Ref. 66. The rate constants k(liter mol^{-1} s^{-1}), activation energies E_a(kcal/mol), and values of log [A (liter mol^{-1} s^{-1})] are reported.

10^9–10^{10} liter mol^{-1} s^{-1} in the gas phase (see Table 2.1). The disproportionations have almost competitive rates (ca. one-tenth) and very low activation energies. Therefore, the *A*-factors must be of the order 10^8–10^9 liter mol^{-1} s^{-1}.[66] The disporportionation can be viewed as an atom abstraction,

$$CH_3{-}CH_2\cdot \quad H{-}CH_2{-}CH_2\cdot \longrightarrow CH_3{-}CH_3 + CH_2{=}CH_2$$

and we have seen earlier that intermolecular abstractions also have *A*-factors in this range. Only the activation energies are much smaller for disproportionations, making these reactions considerably faster than the intermolecular abstraction reactions. This, too, is readily understood in terms of the Bell–Evans–Polanyi principle (cf. Figure 2.18).

The following abstraction reaction

$$CH_3{-}CH_2\cdot + CH_3{-}CH_3 \longrightarrow CH_3{-}CH_3 + CH_3{-}CH_2\cdot$$

$$\Delta H_r^0 = 0$$
$$E_a \cong 13 \text{ kcal/mol}$$

is thermoneutral and thus has no driving force. The activation energy is about 13 kcal/mol.[66] The reaction in Eq. (17), in contrast, has a powerful driving force (−60 kcal/mol), resulting in a very small activation energy (ca. 0.3 kcal/mol).

We have seen that the recombination of alkyl radicals can be decelerated by steric hindrance (Section 2.1). As a consequence, disproportionation now becomes faster than recombination, $k_{disprop} \sim 2\, k_{recomb}$.[66]

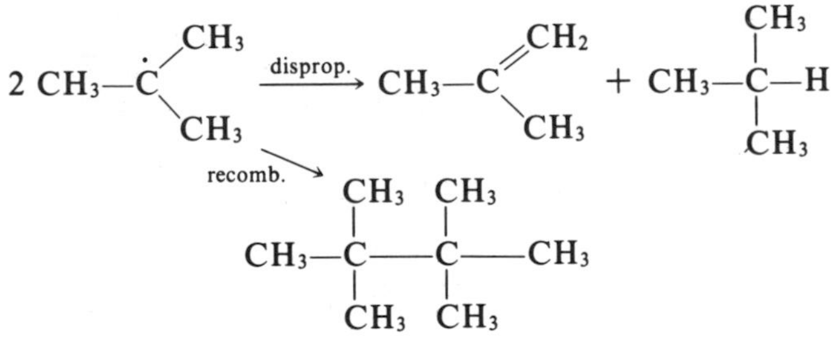

$$2\, CH_3{-}\dot{C}(CH_3)_2 \xrightarrow{\text{disprop.}} CH_3{-}C(=CH_2){-}CH_3 + CH_3{-}C(CH_3)_2{-}H$$

$$2\, CH_3{-}\dot{C}(CH_3)_2 \xrightarrow{\text{recomb.}} CH_3{-}C(CH_3)_2{-}C(CH_3)_2{-}CH_3$$

2.5.3. Intramolecular H-Abstraction and Rearrangement

1,2-migrations of hydrogen occur frequently and with very small activation energies in carbenium ions and in singlet carbenes (see Chapter 4). It might be expected, therefore, that such shifts would be observed in radicals also:

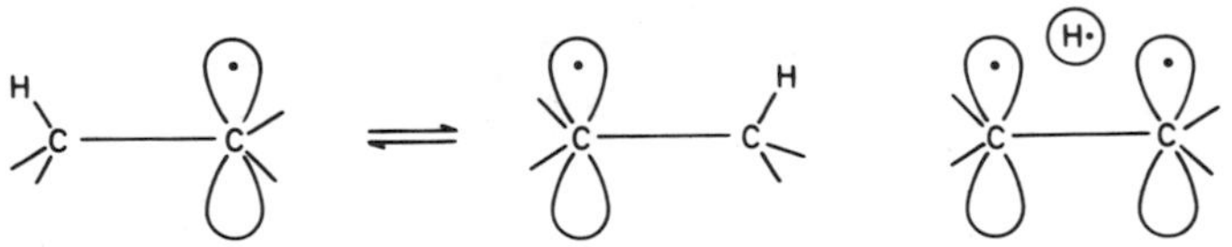

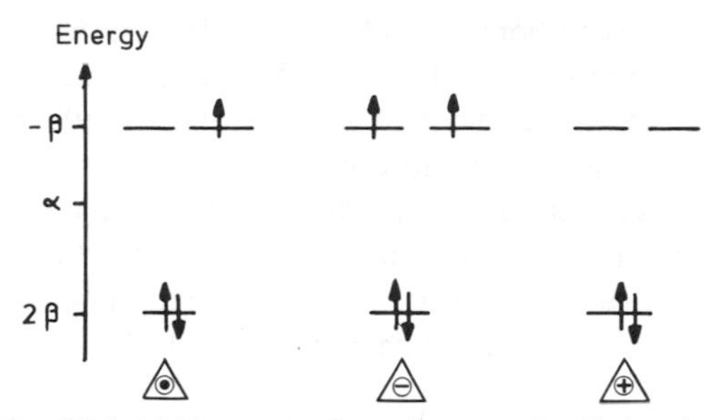

Figure 2.19. Hückel MO energies for cyclopropenyl radical, anion, and cation.

Such processes are very rare, however, occurring with activation energies of 30–35 kcal/mol.[65] In spite of a rather general opinion that 1,2-hydrogen shifts in radicals are *impossible*, one may observe that there is nothing wrong with such a rearrangement. The rarity of the reaction is due to the fact that, normally, other reactions with lower activation energies will be possible. The high activation energies for the 1,2-shift may be understood in terms of the transition state **101**, in which three electrons are cyclically arranged. Such a system is said to be *isoconjugate* with the cyclopropenyl radical, the MO levels of which are shown in Figure 2.19. Here, one electron must occupy an antibonding orbital, and the cyclopropenyl radical is, therefore, antiaromatic. This is true also for the carbanion (two electrons in antibonding orbitals), but in the carbenium ion, the antibonding orbitals will be vacant and therefore the system is aromatic, as can be seen from Hückel's $4n + 2$ rule. Consequently, the carbenium ion rearrangement corresponding to **101** ($H^{\oplus}$ in place of H·) has an "aromatic" transition state and the reaction is, in fact, very fast.

1,3-hydrogen shifts in radicals (e.g. **102**) are also rare processes requiring high activation energies:[65]

H $E_a \sim 30$–35 kcal/mol H

102

1,4- and 1,5-shifts, involving five- and six-membered transition states, respectively, are more frequent:[65,69]

H $E_a \sim 15$–21 kcal/mol H

H $E_a \sim 8$–11 kcal/mol H

The *intermolecular* H abstractions by carbon radicals have $E_a \sim 10$ kcal/mol, and A-factors $\sim 10^{8.5 \pm 1}$ liter mol^{-1} s^{-1} (from the data in Table 2.13).

An *intramolecular* reaction ought to have a higher probability of occurring because the reaction termini are in close proximity, corresponding to an approximately 10 molar local concentration of radicals. From transition state theory, one can then calculate that an intramolecular atom transfer reaction should have an A-factor of $\sim 10^{11}$ s^{-1}.† Although still under debate, experimental A-factors for the 1,4- and 1,5-hydrogen shifts are much too low [10^7–10^9 s^{-1}].[67] This has been attributed to:

(1) a loss of configurational entropy in the transition state[67] (the transition state is rigid);
(2) a lack cf suitable vibrational modes that would make the two ends of the radical collide;[68] and
(3) the postulate that the three atoms involved in hydrogen transfer must be collinear in the transition state.[69]

There are good reasons to believe that the latter explanation is the correct one. Viewed as an abstraction process, the hydrogen migration can take place via an interaction between the singly occupied radical p orbital (SOMO) and the vacant σ^* orbital of the C—H bond (LUMO) [Figure 2.20(a)]. This interaction is maximized in an end-on, ''linear'' approach. The ''perpendicular'' approach [Figure 2.20(b)], in contrast, would lead to a strong interaction between the radical p orbital and the filled σ orbital of the C—H bond (HOMO). This latter interaction is repulsive, again representing a three-electron system isoconjugate with the cyclopropenyl radical.

For geometric reasons, the arrangement shown in Figure 2.20(a) is best obtained for a five-carbon chain; hence the 1,5-H shift is the fastest. There is an analogy here with carbene reactions: As will be shown in Chapter 4, triplet carbenes *abstract* hydrogen via ''linear'' transition states, whereas singlet carbenes *insert* into C—H bonds via nonlinear transition states. The triplet carbene resembles a radical, but the singlet carbene possesses a vacant p orbital that can interact advantageously with the filled σ C—H orbital.

It should be noted, however, that different radicals may show different geometry preferences in the transition states. In a more electrophilic radical, the energy of the p orbital (SOMO) would be lowered, and a stronger and more favorable interaction with the σ C—H orbital (HOMO) would be possible; that is, the situation indicated in Figure 2.20(b) would obtain. Raising the HOMO energy would have the same effect. As an extreme, imagine that the C—H bond in Figure 2.20(b) be replaced by C—Cl ; the HOMO with which the radical can interact will now be a chlorine lone pair, and, as we have seen, a 1,2-shift of chlorine can be very rapid (see Section 2.2.1).

†$A = (e^2 kT/h)e^{\Delta S^{\ddagger}/R}$. A is obtained from an estimate of $\Delta S^{\ddagger}$.

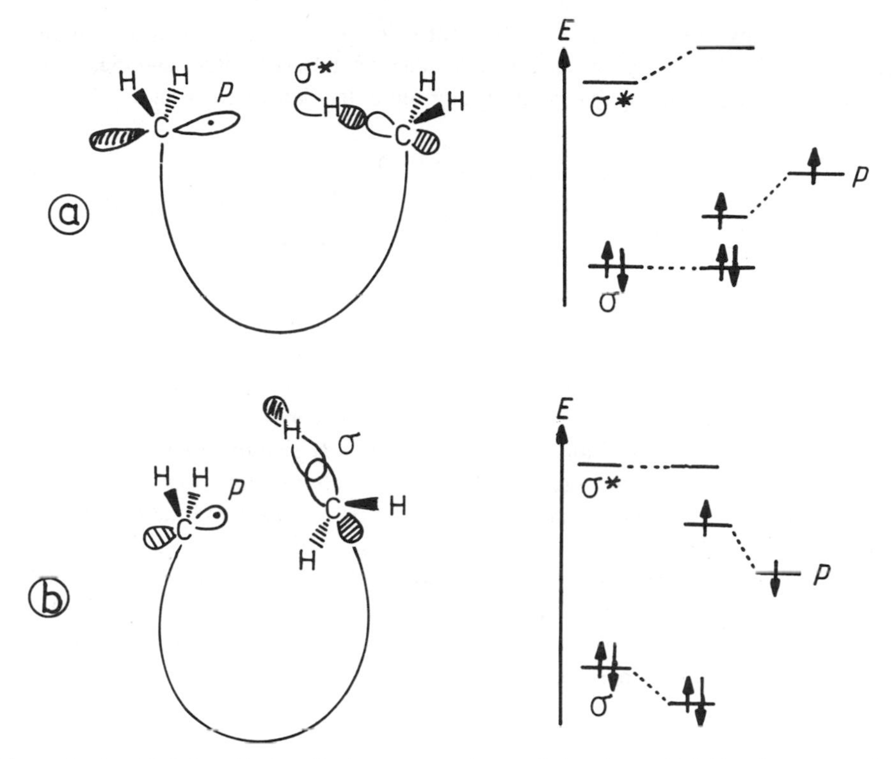

Figure 2.20. (a) "Linear" interaction between singly occupied *p* orbital (SOMO) and vacant σ* C–H orbital (LUMO); (b) "perpendicular" interaction between singly occupied *p* orbital (SOMO) and doubly occupied σ C–H orbital (HOMO).

Contrasting the slow 1,2-shifts of hydrogen (and alkyl groups), aryl groups migrate relatively easily. This is known as the *neophyl rearrangement:*

$$\text{R-C}_6\text{H}_4\text{-}\overset{|}{\underset{|}{\text{C}}}\text{-}\dot{\text{C}}\text{H}_2 \longrightarrow \cdot\overset{|}{\underset{|}{\text{C}}}\text{-CH}_2\text{-C}_6\text{H}_4\text{-R}$$

103 **104**

If the rearranged radicals **(104)** are tertiary, there will be a thermodynamic driving force so that the reaction may go completely to the right. The radicals **104** can undergo several subsequent reactions:

$$\text{abstraction} \Longrightarrow \text{H}\text{—}\overset{|}{\underset{|}{\text{C}}}\text{—CH}_2\text{—Ar}$$

$$\text{elimination} \Longrightarrow \;\rangle\text{C=CH—Ar} \quad \text{or} \quad \rangle\!\!=\text{—CH}_2\text{—Ar}$$

The reason why the neophyl rearrangement takes place so easily can be sought in the availability of low-lying vacant molecular orbitals (LUMOs) associated with the benzene rings and capable of accepting the extra radical electron:

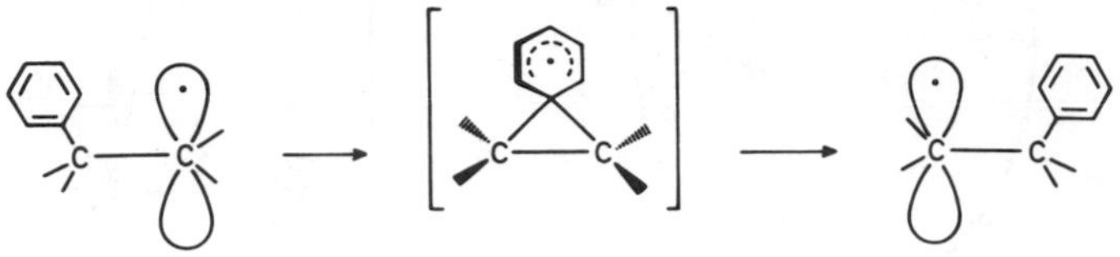

105 **106** **107**

This leads ultimately to the intermediate **106** in which the radical electron has been transferred completely into what was formerly the benzene ring. The reaction should, therefore, go even faster if the benzene ring contains electronegative substituents such as *ortho-* or *para*-nitro or -cyano groups, which have the double function of lowering the benzene LUMO and stabilizing the cyclohexadienyl radical **106**. The data given in Table 2.14 demonstrate the correctness of these ideas.

The intermediates **106** are extremely short-lived and cannot be detected by ESR spectroscopy even at 100 K.[70] However, the spiro[2.5]octadienyl radical **109** has been generated by hydrogen abstraction from **108** and directly observed by optical absorption and fluorescence spectroscopy:[70]

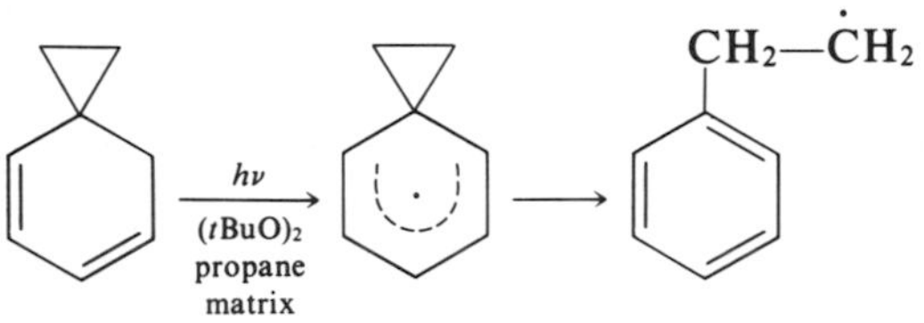

$h\nu$, $(tBuO)_2$, propane matrix

$CH_2—\dot{C}H_2$

108 **109** **110**

TABLE 2.14. Relative Rates and Activation Parameters for the Neophyl Rearrangement[a]

Ar	k_{rel}	Ar	log A	E_a (kcal/mol)
p-$CH_3O—C_6H_4—$	0.36	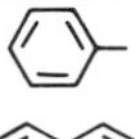	11.75	57
p-$F—C_6H_4—$	0.40			
p-$CH_3—C_6H_4—$	0.72	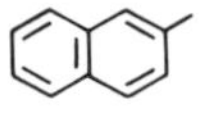	11.75	47.3
$C_6H_5—$	≡1			
p-$Cl—C_6H_4—$	1.76	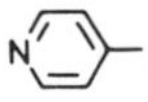 N	11.8	49.4
p-$Br—C_6H_4—$	1.91			
p-$NO_2—C_6H_4—$	31			
p-$NC—C_6H_4—$	35			

[a]C. Rüchardt and R. Hecht, *Chem. Ber.*, **98,** 2471 (1965); B. Maillard and K. U. Ingold, *J. Am. Chem. Soc.*, **98,** 1224, 4692 (1976).

The half-life of **109** at 243 K was approximately 170 ns.

The rearranged radical **110** can be observed by ESR.[70] A CIDNP study of the neophyl rearrangement failed to give any evidence for spin polarization arising from **109**, but this experiment was carried out at 105°C, where the half-life of the radical may well be below the detectable limit (ca. 10^{-10} s).[71]

Like phenyl, vinyl groups also migrate; thus homoallyl radicals **(111)** interconvert via cyclopropyl-methyl radicals **(112)**. This reaction leads to deuterium scrambling using properly labeled substrates [eq. (19)]:

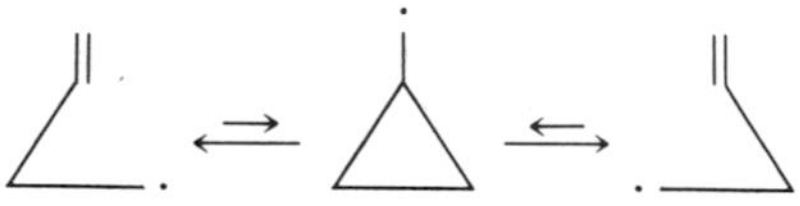

111 ⇌ 112 ⇌ 111 (18)

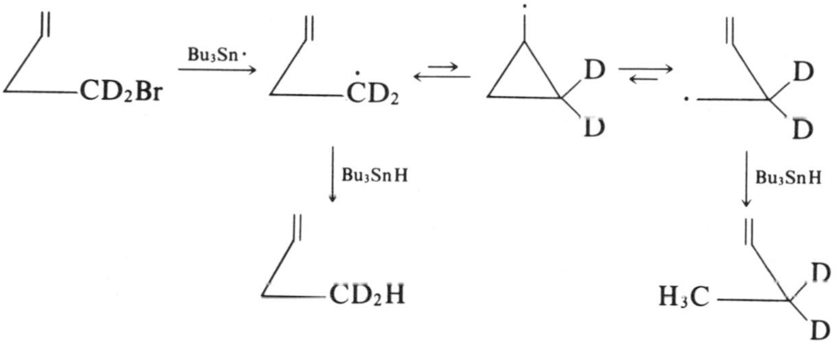

CD_2Br —$Bu_3Sn\cdot$→ $\dot{C}D_2$ ⇌ D D ⇌ D D

Bu_3SnH Bu_3SnH (19)

CD_2H H_3C D D

This has been followed by NMR and ESR spectroscopy.[72] The equilibrium is strongly in favor of the homoallyl radical **(111)**, and although **112** can be independently generated, such radicals undergo ring opening even at −100°C (see also p. 99). The equilibrium constant for the reaction **112** ⇌ **111** at 25°C was determined as approximately 10^4 in favor of **111**.[72]

A somewhat related rearrangement takes place in the bicyclo[3.1.0]hexenyl rad-

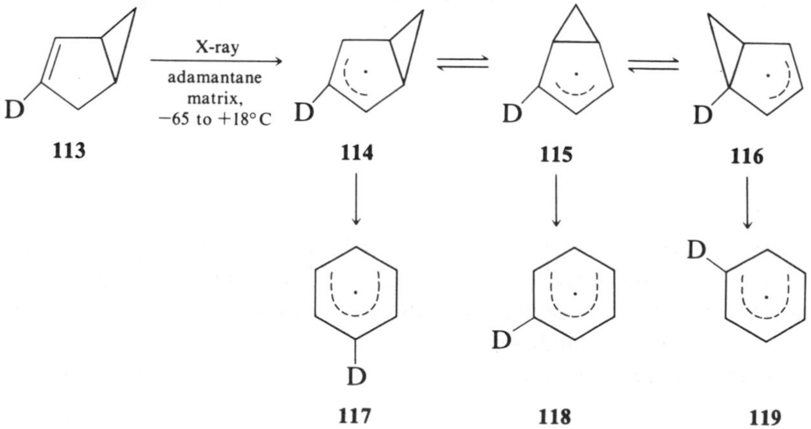

D —X-ray, adamantane matrix, −65 to +18°C→ D ⇌ D ⇌ D

113 114 115 116

(20)

D D D

117 118 119

icals (e.g., **114**). These radicals are generated by X-ray irradiation of the precursor **113** in an adamantane matrix and undergo ring opening to the corresponding cyclohexadienyl radicals (e.g., **117**) as observed by ESR spectroscopy. However, the ring opening is preceded by a rapid equilibration of the bicyclic radicals (**114** ⇌ **115** ⇌ **116**), resulting in deuterium scrambling.[73] The free energy of activation $\Delta G^{\ddagger}$ for the reaction **114** → **117** was measured as 16.4 ± 0.5 kcal/mol; accordingly, the reaction **114** ⇌ **115** has $\Delta G^{\ddagger} < 16.4$ kcal/mol. Using **120** as the starting material, it was shown that the methylene bridge substituents undergo *exo*/*endo* interchange during the reaction:

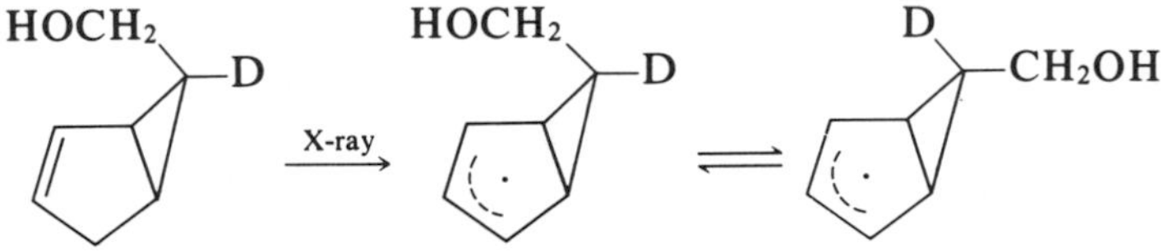

HOCH$_2$ D X-ray HOCH$_2$ D ⇌ D CH$_2$OH

120

Consequently, the methylene walk (**114** ⇌ **115** ⇌ **116**) probably takes place via transient cyclopentadienylmethyl radicals of the type **121**.[74] **121** is a homoallyl radical, and its cyclization gives a "cyclopropylmethyl" radical (e.g., **114**) which, however, is now stabilized by allylic resonance due to the presence of the extra double bond. Therefore, the stability order observed in Eq. (18) has been reversed.

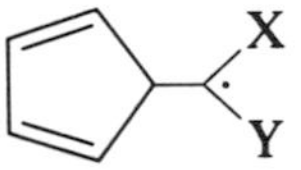

X Y

121

Interestingly, the analogous carbenium ion rearrangement [Eq. (21)] takes place in solution at about −90°C with an activation energy of 15 ± 1 kcal/mol.[75] Thus the radical and the carbenium ion rearrange with approximately equal energies of activation. However, there is a subtle difference in the mechanisms of the two

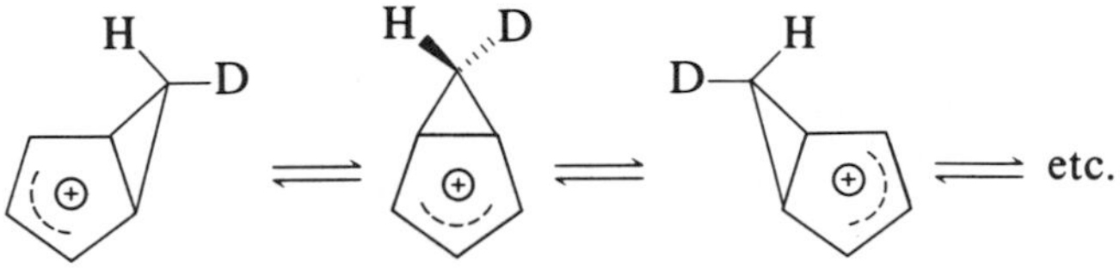

H D ⇌ H D ⇌ D H ⇌ etc. (21)

reactions: No *exo*/*endo* interchange was observed for the reaction given in Eq. (21). Therefore, the latter reaction is concerted, not involving discrete cyclopentadienylmethyl cations, in agreement with the rules of orbital symmetry and the high energy of primary carbenium ions.

When the radical reaction [Eq. (20)] is carried out in fluid solution, no rearrangement is observed even at temperatures as high as +70°C, the reason being that other, bimolecular, reactions (e.g., hydrogen abstraction and recombination) predominate. Thus, again the rarity of radical rearrangements is due not so much to any intrinsic impossibility as to the presence of other, faster, reaction channels.

2.5.4. Addition to Double Bonds

The interactions between a radical *p* orbital (SOMO) and the π and π^* orbitals of an olefin (HOMO and LUMO) are illustrated in Figure 2.21. The *p*-π interaction will be favored in the reaction between a relatively electrophilic radical with an electron-rich olefin (low-lying SOMO, high-lying HOMO); the *p*-π^* interaction, in a nucleophilic radical with an electron poor olefin (high-lying SOMO, low-lying LUMO). If the energy of the *p* orbital is roughly midway between the π and π^* orbitals, both interactions will be important; that is, the radical can show both electrophilic and nucleophilic properties.

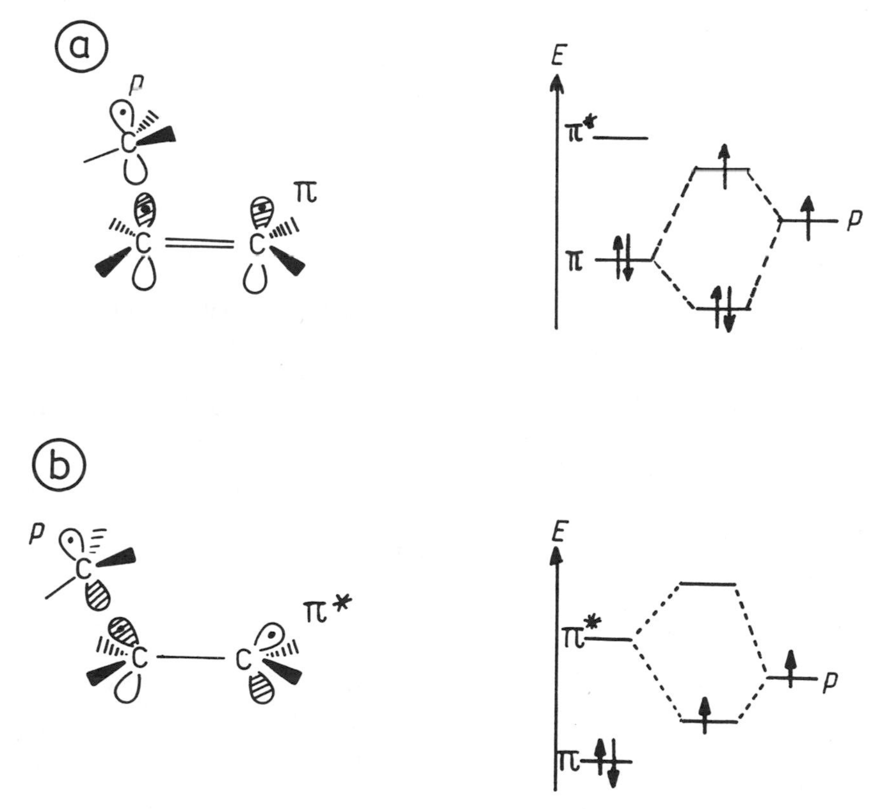

Figure 2.21. Addition of a radical to a double bond. (a) *p*-π (SOMO–HOMO) interaction; (b) *p*-π^* (SOMO–LUMO) interaction.

The "electrophilic" interaction [Figure 2.21(a)] is not likely to involve a symmetrical transition state **(122)** because this would be isoconjugate with the antiaromatic cyclopropenyl radical (see also Figure 2.19). We should expect, therefore, the addition to a double bond to proceed in an unsymmetrical manner [Eq. (22)].[77]

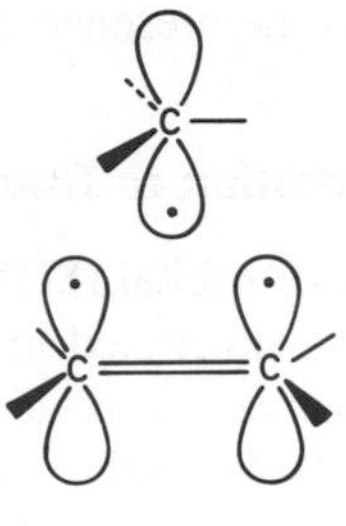

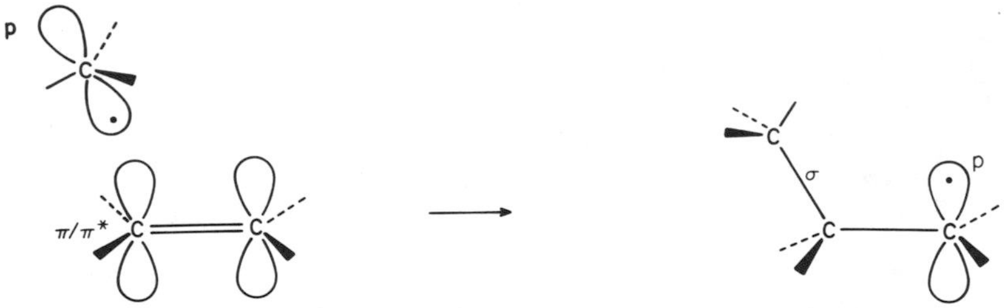

122

123 → **124** (22)

We shall see in Section 4.4.3 that the approach and the orbital interactions in the addition of carbenes to olefins are very similar to those of the radicals. The "electrophilic" and "nucleophilic" interactions illustrated in Figure 2.21 are nicely borne out in the radical-initiated copolymerization of vinyl acetate and dimethyl fumar-

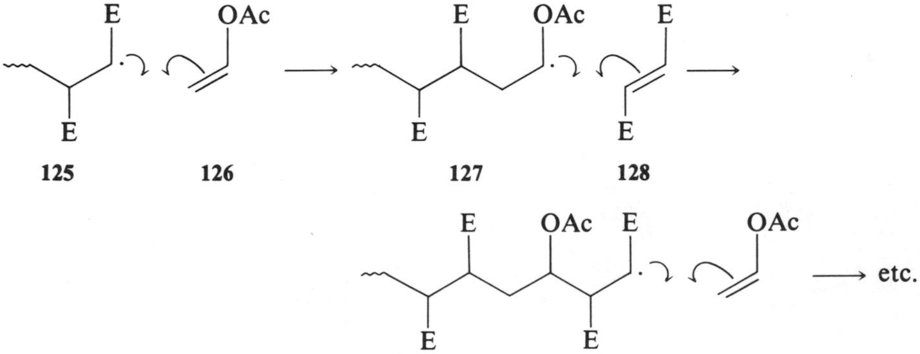

125 **126** → **127** **128** → ... → etc.

E = $COOCH_3$ Ac = $COCH_3$

ate.[76] The alternating incorporation of monomer units in the polymer chain can be understood in terms of the electrophilic (due to the α-ester group) radical **125** attacking the nucleophilic (due to the oxygen lone pair) olefin **126**. The newly formed radical **127** is now relatively nucleophilic (α-oxygen lone pair) and attacks the electron-poor fumarate **128** and so on.

The reverse of Eq. (22) is a homolytic fragmentation. Such reactions can take place when they are thermodynamically favorable (i.e., exothermic). The principle of microscopic reversibility demands that the same orbital interactions (Figure 2.21) be involved in the reverse of Eq. (22). That is, *the p and σ C—C orbitals must lie*

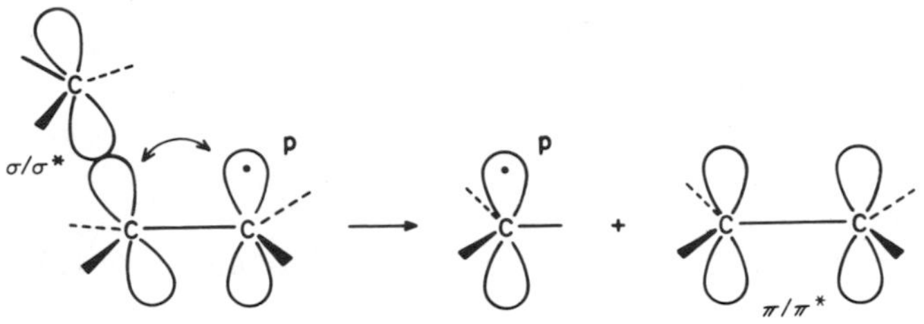

(23)

in one plane[77] [Eq. (23)]. This makes it understandable that the cyclopropyl radical (**129**) does not undergo ring opening (a "fragmentation") very easily[69,78] in spite of the fact that the reaction would be exothermic by about 30 kcal/mol:

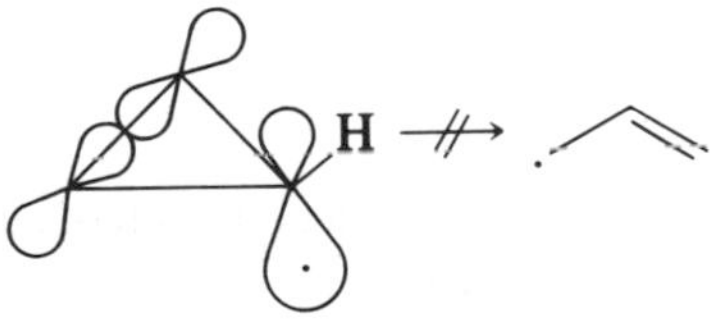

$$\Delta H_r^0 \sim -30 \text{ kcal/mol}$$

129

The orbitals involved are orthogonal in this case. In cyclopropylmethyl, in contrast, the orbitals can be properly aligned, and ring opening is extremely rapid:[72]

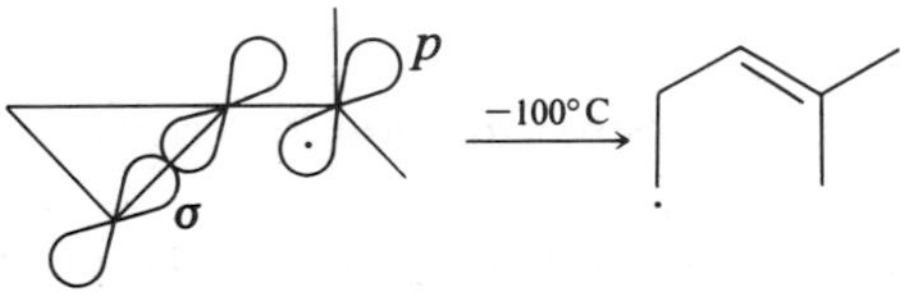

The *bimolecular* addition of a radical to a substituted olefin normally occurs at the least substituted end of the double bond, thereby producing the thermodynamically more stable product radical (e.g., **125** + **126** → **127**). This, as well as the exception with fluoroolefins, is also in agreement with frontier orbital theory.[76]

However, *intramolecular* radical additions often take a contrathermodynamic course:[69,79]

k_{131}

$k_{131}/k_{132} \sim \infty$

129 **130** **131** ΔH_f^0 **(132)** $<$ ΔH_f^0 **(131)**

132

k_{135}

$k_{135}/k_{136} \sim 76$ (25° C)

133 **134** **135** ΔH_f^0 **(136)** $<$ ΔH_f^0 **(135)**

k_{136}

136

This, too, may be understood in terms of the unsymmetrical transition states shown in Figure 2.21: had a symmetrical approach **(122)** been involved, we should have expected the formation of the thermodynamically favored products **132** and **136**. In contrast, in the nonsymmetrical approach (Figure 2.21), the radical *p* orbital will be able to overlap better with the *near* end of the C═C bond. Furthermore, the preferred conformations of the cyclizing radicals are not **129** and **133**, but, for steric reasons, **130** and **134**; these can only cyclize to the products **131** and **135**, respectively.

2.5.5. Structure and Reactivity

Philicity of Radicals. In many reactions, free radicals can be regarded as electrophilic. However, we have also seen that the radical electron can be donated into vacant orbitals in suitable substrates. This is a nucleophilic property. We shall examine first the electrophilic properties of free radicals in some detail. The more electrophilic the radical, the more reactive it is, the lower the transition state is, and the more exothermic the reaction. For example,

$$R\cdot + H{-}R' \longrightarrow RH + \cdot R' \qquad (R' = CH_3, C_2H_5, i\text{-}C_3H_7) \qquad (24)$$

There is a direct proportionality between log ($k_{C_2H_6}/k_{CH_4}$) and the dissociation enthalpy DH^0(R—H) of the bond formed when

$$R = F, Cl, Br, I, ROO, CD_3, CF_3, O, H, \text{ and } OH$$

This is a statement of the Hammond postulate, and we have already described one specific example (Figure 2.18). The relationship is expressed quantitatively by the Polanyi equation:

$$\Delta E_a = \Delta\Delta H_r^0 \cdot \alpha + \text{const}$$

where ΔE_a is the difference in activation energy for different radicals, and $\Delta\Delta H_r^0$ is the difference in the heats of reaction (remember that the Polanyi equation has no foundation in thermodynamics).

The data obtained from the reactions indicated in Eq. (24) can be fitted to the Polanyi equation to give[80]

$$\alpha = 0.015\ DH^0(\text{R—H}) + 2.03$$

When α varies from 0 to 1, we go from the initial state (reactants) to the final state (products) (cf. Figure 2.18). The transition state for the most reactive radical (F·) is almost indistinguishable from the initial state ($\alpha \sim 0$), while for the quite unreactive I·, the transition state is very close to the products ($\alpha = 0.89$).

In the reactions of $CH_3\cdot$, $CF_3\cdot$, Cl·, $C_2F_5\cdot$, $n\text{-}C_3H_7\cdot$, H·, and O·, the transition state is symmetrical, i.e., about midway between the initial and the final states. In these reactions, the bonds formed and broken have approximately equal dissociation enthalpies (for bond dissociation enthalpies, see Table 2.2). This simple relationship holds only for reactions with hydrocarbons. It will not be true if polar or resonance effects are involved. For example,

$$CH_3\cdot + CHCl_3 \longrightarrow CH_4 + \dot{C}Cl_3$$

$$CH_3\cdot + CH_3\text{—Ph} \longrightarrow CH_4 + \dot{C}H_2\text{—Ph}$$

In the latter case, however, another structure-reactivity relationship applies, namely, the Hammett equation,

$$\log \frac{k}{k_0} = \rho\sigma$$

where σ is the "substituent constant" for X in the substrate,

$$R\cdot + CH_3\text{—}C_6H_4\text{—}X \longrightarrow RH + \dot{C}H_2\text{—}C_6H_4\text{—}X$$

and ρ is the "reaction constant" or "sensitivity," which differs from reaction to reaction. The more negative ρ is, the more electrophilic the radical R·, and the aromatic substance will act as an electron donor in the transition state. Consequently, donor substituents X accelerate the reaction; acceptors slow it down.

Electronegative substituents in the radical R· itself make it more electrophilic, resulting in more negative ρ values. Hence the reactivity order:

$$t\text{Bu}\cdot < \dot{C}H_3 < \dot{C}Cl_3 < \dot{C}F_3$$

Some reactivity data for H abstraction from toluenes are given in Table 2.15.[81] Further data for different substrates can be found in Ref. 82.

The positive ρ value for the *tert*-butyl radical in Table 2.15 is particularly interesting since it implies that this radical is nucleophilic in character. This value has been cast in serious doubt, and it has been concluded that there are as yet no substantiated positive ρ values for hydrogen abstraction by alkyl radicals.[83] Nevertheless, a redetermination confirmed the positive ρ value for the *tert*-butyl radical.[84]

Be that as it may, there is no doubt that alkyl radicals, and in particular the *tert*-butyl radical, behave as *nucleophiles* in their *addition* reactions.[85] Thus the *tert*-butylation of 3-substituted pyridines takes place exclusively in the electron-deficient

R R
—H·
N⊕ N⊕
H H

TABLE 2.15. Hammett Correlation for Hydrogen Abstraction from Toluenes by Radicals[a]

Radical	Temperature (°C)	ρ
H·	40	−0.10
$CH_3\cdot$	100	−0.17
$(CH_3)_3C\cdot$	30	+0.99[b]
$C_6H_5\cdot$	60	−0.19
p-CH_3—$C_6H_4\cdot$	60	−0.12
p-O_2N—$C_6H_4\cdot$	60	−0.52
$\cdot CCl_3$	50	−1.59
$(CH_3)_3C$—O·	50	−0.34
Cl· (in CCl_4)	70	−0.84
Br· (in CCl_4)	80	−1.88

[a]Data from Ref. 81.

[b]A redetermination gave a value of +0.49 at 80°C (Ref. 84).

6-position with a ρ value of +5.5. The protonated pyridine is more reactive and more electrophilic than pyridine itself. The 3-cyano compound reacts 4380 times faster than the unsubstituted protonated pyridine, and the 3-methoxy compound, possessing an $\ddot{X}$-substituent, does not react at all.[85]

Likewise, the primary 5-hexenyl radical **(134)** reacts faster with 4-Z-substituted than with 4-$\ddot{X}$-substituted pyridines. This particular radical was chosen because the rate k_c of the competing cyclization to **135** was known from ESR measurements, thereby allowing an evaluation of the absolute value of k_a, the rate constant for the

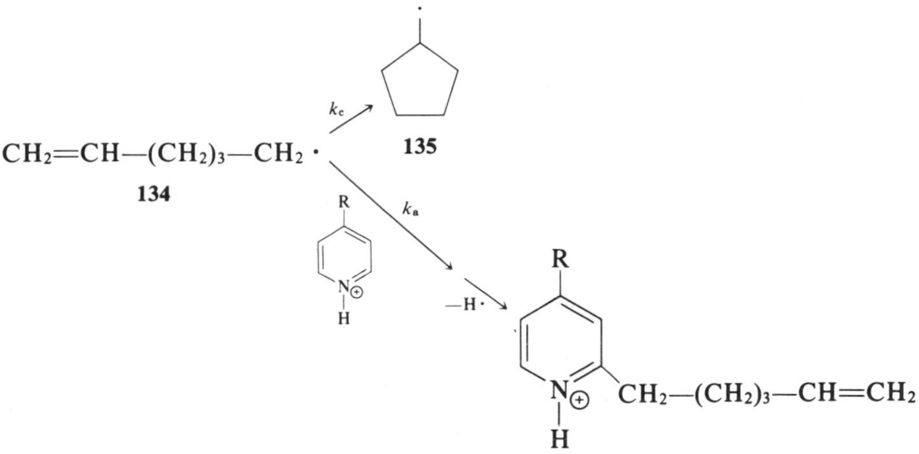

$CH_2{=}CH{-}(CH_2)_3{-}CH_2\cdot$ **134** $\xrightarrow{k_c}$ **135**; with 4-R-pyridinium (N⁺–H) $\xrightarrow{k_a}$ $\xrightarrow{-H\cdot}$ **137** (2-$CH_2{-}(CH_2)_3{-}CH{=}CH_2$, 4-R pyridinium, N⁺–H)

addition reaction. The activation energies (kcal/mol) for the addition reactions were 6.85(R = CH_3) and 4.9(R = CN), whereas the Arrhenius *A*-factors were nearly identical for these two reactions (i.e., entropy differences are insignificant). With a still more electrophilic heterocycle, protonated quinoxaline **(138)**, the activation energy dropped to 2.8 kcal/mol:

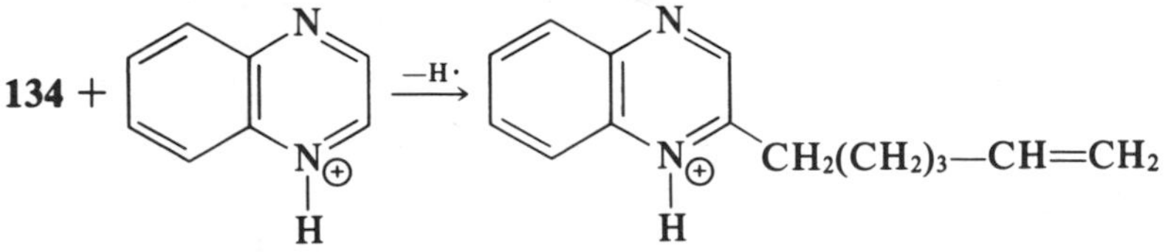

134 + **138** (protonated quinoxaline, N⁺–H) $\xrightarrow{-H\cdot}$ 2-$CH_2(CH_2)_3{-}CH{=}CH_2$ quinoxalinium (N⁺–H)

These heterocyclic bases are 10^2–10^5 times more reactive than benzene toward alkyl radicals. Furthermore, the opposite effect of substituents is seen in benzene derivates: anisole, possessing an $\ddot{X}$-substituent, reacts 3.4 times faster than benzene with the *n*-butyl radical. Since entropy differences are unimportant, and the thermochemistry (ΔH) should work in the same direction in the two series, there must be another explanation.

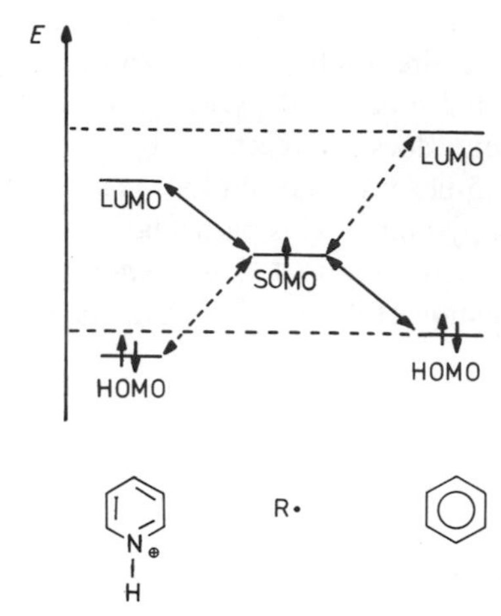

Figure 2.22. Frontier MO interactions between a free radical (*center*), protonated pyridine (*left*), and benzene (*right*).

The reversal is readily explained in terms of frontier MO theory.[85] Compared with benzene, the protonated heterocyclic bases have low-lying HOMOs and LUMOs; they are electrophilic. Z-substituents (CN) lower both levels, and $\ddot{X}$-substituents (OCH_3) raise them in both series. The radical SOMO can, therefore, interact strongly with the pyridine or quinoxaline LUMO; in the benzene series, the interaction will be primarily with the HOMO (Figure 2.22). A methoxy substituent then leads to an increased and *electrophilic* interaction with the benzene HOMO, whereas the same substituent in the pyridine series widens the SOMO–LUMO gap and decreases the reactivity. Since the interactions are stronger in the heterocyclic series, the radicals also show higher selectivities.

One extreme of a SOMO–LUMO interaction is the complete transfer of an electron from the radical to the heteroaromatic LUMO. This can be observed via the ESR spectrum of the radical **142** obtained by photolysis of benzophenone in the presence of the pyridinium salt **140.** The photolysis of benzophenone gives the

$$\underset{S_0}{Ph_2CO} \xrightarrow{h\nu} \underset{S_1}{Ph_2CO} \longrightarrow \underset{T_1}{Ph_2CO} \xrightarrow[-R\cdot]{RH} \underset{\mathbf{139}}{Ph\text{—}\dot{C}(OH)\text{—}Ph}$$

$$\underset{\mathbf{139}}{Ph\text{—}\dot{C}(OH)\text{—}Ph} + \underset{\mathbf{140}}{\text{4-CN-}C_5H_4N^{\oplus}\text{—}CH_3} \longrightarrow \underset{\mathbf{141}}{Ph\text{—}C^{\oplus}(OH)\text{—}Ph} + \underset{\mathbf{142}}{\text{4-CN-}C_5H_4N^{\cdot}\text{—}CH_3}$$

radical **139** by hydrogen abstraction from the triplet state. Electron transfer from **139** to **140** gives the cation **141** and the radical **142**.[86]

Positive ρ values were also determined for the addition of alkyl radicals to substituted styrenes:[87]

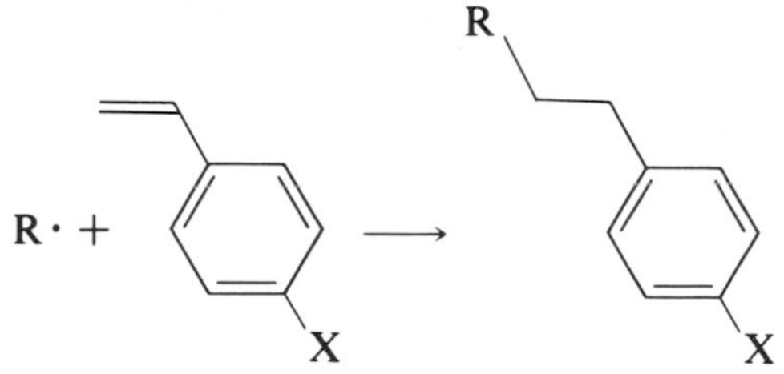

$$R\cdot + \text{(4-X-styrene)} \longrightarrow \text{(4-X-}C_6H_4CH_2CH_2R) \quad (25)$$

R = *t*Bu·, (cyclohexyl·), (n-pentyl·)

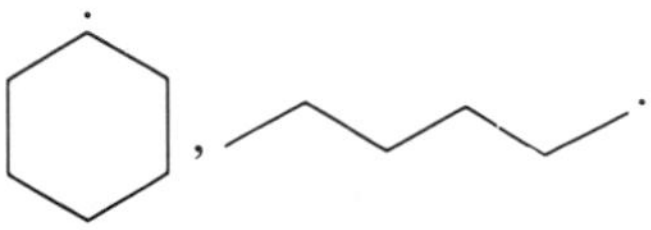

$\rho(-20°C) = 1.1$ 0.68   0.45

Here, too, the substituent effects in both radicals and olefins are readily explained by a frontier MO treatment [use Fig. 1.7(a), removing one electron from the SOMO of the benzyl anion to approximate the SOMO of an alkyl radical]. Alkyl substituents in the radical and Z-substituents in the olefin will decrease the SOMO–LUMO gap and thus lead to stronger interactions, faster and more selective reactions, and larger ρ values.

If a carbon radical is made more electrophilic by the presence of electron-withdrawing substituents, the SOMO energy will *decrease* and the interaction with an olefin HOMO can now dominate. Hence, in the series

$$\dot{C}H_3,\ \dot{C}H_2F,\ \dot{C}HF_2,\ \dot{C}F_3$$

the trifluoromethyl radical is both the most reactive and the most selective, reacting most readily with alkyl substituted (i.e., electron-rich) alkenes.[76b] Similarly, the radical **144** will have a lower energy SOMO than **143**, which explains its faster reaction with olefins.[88]

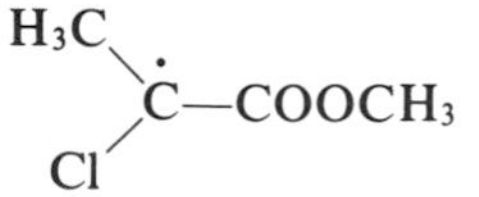

$$(H_3C)(Cl)\dot{C}{-}COOCH_3 \qquad (Cl)(Cl)\dot{C}{-}COOCH_3$$

143 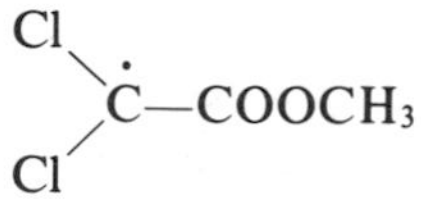   **144**

The donor–acceptor substituted or merostabilized radicals[2] are, as we have seen, a class apart. Such radicals do not add to normal olefins but rather undergo dimerization or disproportionation [see **1** and Figure 2.1 (b)].[2] They do, however, react with highly electrophilic double-bond systems by way of electron transfer.[2b] Thus, in all probability, **1** reacts with benzil to give the radical anion **145**:

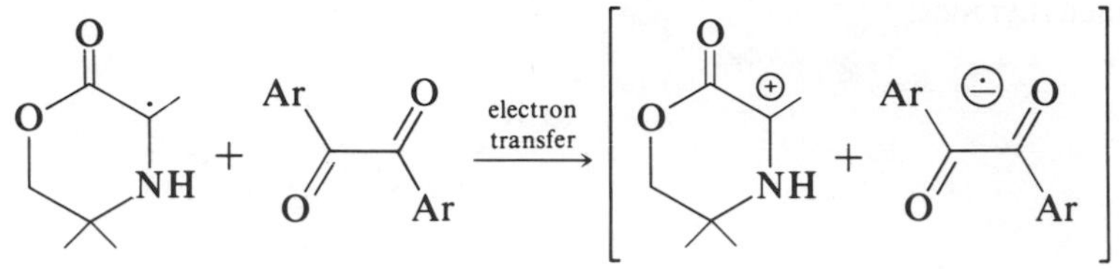

electron transfer

1 **145**

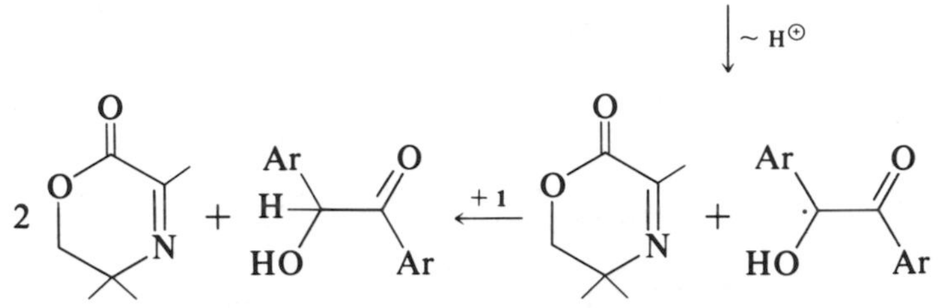

$\sim H^{\oplus}$

2 + ⟵ +1 +

Although the ESR spectrum of **145** could not be observed under these reaction conditions, that of the radical anions of tetracyanoethylene (TCNE) and other powerful electron acceptors were recorded during analogous reactions:[2b]

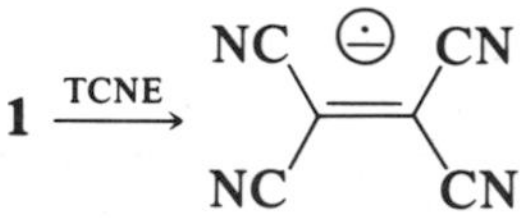

1 $\xrightarrow{\text{TCNE}}$

This implies that **1**, as well as other merostabilized radicals, have rather high-lying SOMOs. This is not so easily seen from a perturbation MO diagram as in Figure 2.1(b) since this considers only initial interactions between mentally separated orbitals. In merostabilized radicals, five π-electrons are distributed over four atomic centers as illustrated in the partial structure **146**.

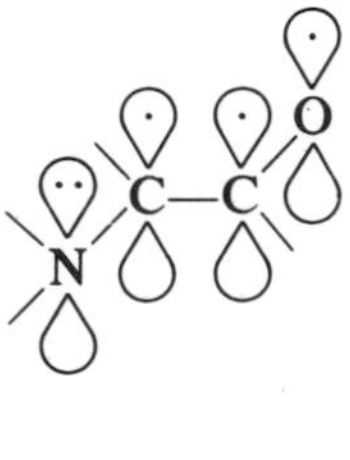

146 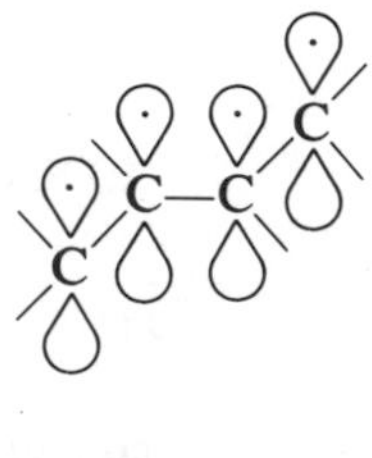 **147** 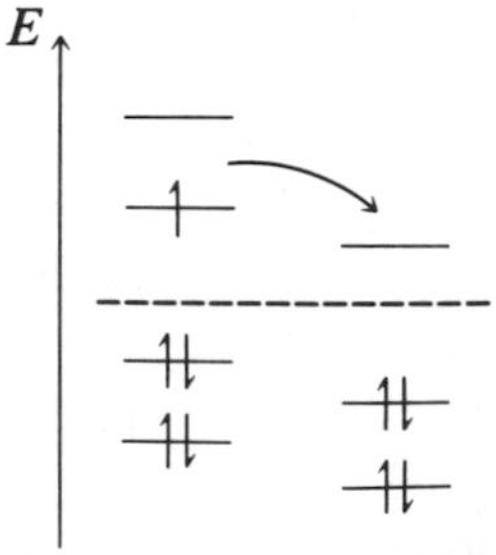 **146**

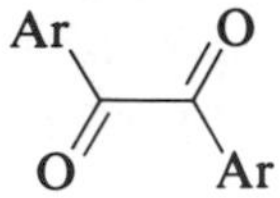

The four atomic *p* orbitals give rise to four molecular orbitals similar to those of butadiene (**147**); the fifth electron must, therefore, occupy an antibonding MO. Thus the merostabilization is due mainly to the lowering of the energy of two electron pairs (originally residing in the nitrogen lone pair and the C═O bond, respectively), whereas the delocalized "extra" electron maintains a high energy. Consequently, this "extra" electron can be transferred to another substance, such as benzil or TCNE, possessing a low-lying LUMO.

Another result of merostabilization is that donor–acceptor substituted olefins are excellent radical traps.[2a] For example, the olefin **148** traps the radical produced from azobisisobutyronitrile to furnish a merostabilized radical, **149**, which then dimerizes; the isobutyronitrile radical is otherwise unreactive toward olefins.

$$\underset{\text{AIBN}}{-\underset{CN}{\overset{|}{C}}-N{=}N-\underset{CN}{\overset{|}{C}}-} \xrightarrow{80^\circ C} \underset{\text{IBN}^\bullet}{(CH_3)_2\dot{C}CN} \xrightarrow[\mathbf{148}]{H_2C{=}C(SR)C{\equiv}N}$$

$$\underset{\mathbf{149}}{NC{-}C(CH_3)_2{-}CH_2{-}\dot{C}(SR)(C{\equiv}N)} \xrightarrow{\times 2} \underset{50\%\ (dl\ +\ meso)}{NC{-}C(CH_3)_2{-}CH_2{-}C(SR)(CN){-}C(SR)(CN){-}CH_2{-}C(CH_3)_2{-}CN}$$

The "radicophilic" olefins are such powerful radical traps that they inhibit polymerization of acrylonitrile and acrylic esters, trap phenyl radicals from the decomposition of dibenzoyl peroxide, and give spin adducts with both nucleophilic [$(CH_3)_2\dot{C}{-}OH$] and electrophilic radicals ($\dot{C}Cl_3$).[2a]

Steric Effects. Up until now, we have been concerned with enthalpic (thermodynamically stabilizing) and orbital effects. The latter are often described in an alternative manner as "polar" effects. There is a third "effect," the steric or entropic one, which can be of importance in any radical reaction, particularly when bulky groups are present in either reaction partner. Three relative rates out of many[89] illustrate the importance of steric inhibition of radical addition to the C═C bond:

$$\dot{C}H_3 + \text{>C=C<} \xrightarrow{k} CH_3{-}\overset{|}{\underset{|}{C}}{-}\dot{C}\text{<}$$

Alkene	k_{rel}
$H_2C{=}CH_2$	480
$H_2C{=}C(H)Ph$	32,000
$Ph_2C{=}CPh_2$	160

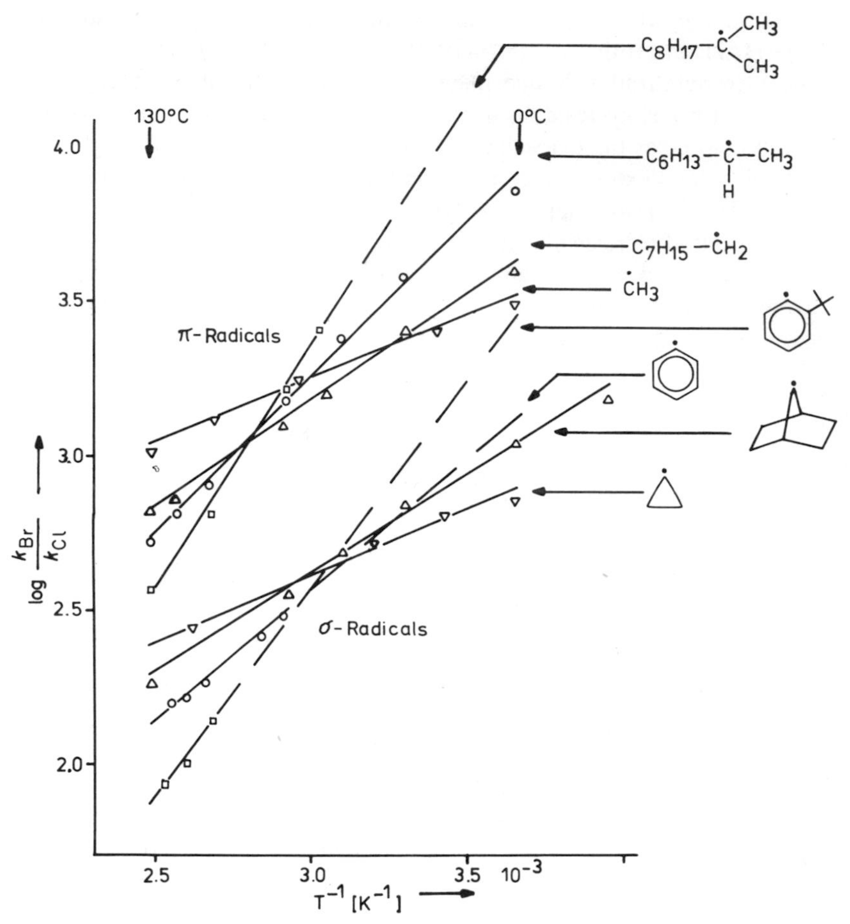

Figure 2.23. Radical selectivities as a function of temperature. (Adapted from Ref. 90 with permission of Verlag Chemie, Weinheim.)

Styrene reacts faster than ethylene, largely because the benzylic resonance energy is gained. This would also be gained in the reaction with tetraphenylethylene, and the slow rate must, therefore, be ascribed to a high entropy of activation.

Furthermore, the ρ values for the addition of alkyl radicals to styrenes [Eq. (25)] all decrease with increasing temperature, converging on a value of about +0.35 at 75°C.[87] Since the decrease is more dramatic for the bulky *tert*-butyl radical, the entropy term $T\Delta S^{\ddagger}$ in the equation

$$\Delta G^{\ddagger} = \Delta H^{\ddagger} - T \cdot \Delta S^{\ddagger}$$

is presumably overriding enthalpic and orbital effects at higher temperatures. It was shown in connection with the isoselective principle (Section 1.6) that $\Delta S^\ddagger$ and hence the temperature have a profound effect on the selectivities of alkyl radicals in halogen atom abstraction processes (Figure 1.3). The effect is different for

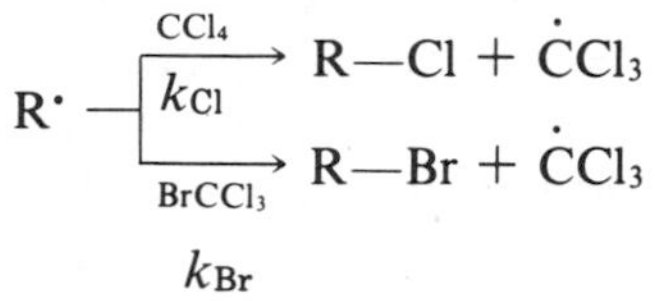

$$R^\cdot \begin{cases} \xrightarrow[k_{Cl}]{CCl_4} R{-}Cl + \dot{C}Cl_3 \\ \xrightarrow[k_{Br}]{BrCCl_3} R{-}Br + \dot{C}Cl_3 \end{cases}$$

different *classes* of radicals, and can be used, therefore, to distinguish π- and σ-radicals (Figure 2.23). The lower isoselective temperature and larger effect of $\Delta S^\ddagger$ for σ-radicals can be understood in terms of their higher reactivities, which lead to early transition states. According to the Hammond postulate, differences in $\Delta H^\ddagger$ are relatively unimportant for early transition states.[90]

2.6. ELECTRON TRANSFER AND RADICAL IONS

Ketyls **(150)** (Ref. 92) and the related semidiones **(151)** (Ref. 93) are examples of long-lived radical anions that are formed upon reduction of the corresponding carbonyl compounds with alkali metals. Such compounds have been extensively

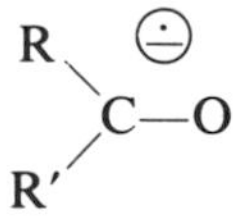

150  **151**

studied by ESR.[92,93] The formation of the radical anion of benzil **(145)** by electron transfer from a nucleophilic radical was mentioned in Section 2.5.5. In all such reactions the extra electron must enter a LUMO of the molecule being reduced. Naphthalene, when treated with sodium, readily forms the radical anion; benzene, because of the higher energy of its LUMO, is much more difficult to reduce in this way. Sodium naphthalenide **(152)** is characterized by an ESR spectrum in which

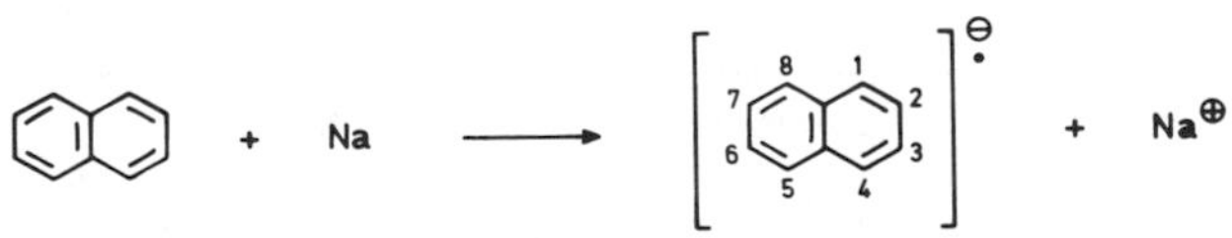

+ Na ⟶ [ ]$^{\cdot\ominus}$ + $Na^\oplus$

152

the extra electron couples with all eight hydrogen atoms. There are two sets of 4H each, $H_{1,4,5,8}$ and $H_{2,3,6,7}$. These couple differently with the electron, giving a pattern of five lines, each split into quintets.[94,95]

LUMO energies can be obtained by MO calculations or UV spectroscopy. A further experimental measure is available in the polarographic reduction potentials: The lower the potential, the lower the LUMO. Some reduction potentials for aromatic hydrocarbons together with the ESR splitting constants in the corresponding anions are presented in Table 2.16. From this, one can immediately predict which compounds will undergo *nucleophilic aromatic substitution* more readily and in which position the substitution is likely to take place. A large hyperfine splitting constant signifies a large electron density in the relevant position in the radical anion and consequently a large coefficient of the LUMO orbital in the parent

TABLE 2.16. Reduction Potentials ($E_{1/2}$) of Aromatic Compounds and Hyperfine Splitting Constants (a_H) in the Corresponding Radical Anions

Compound	$-E_{1/2}$[a] (Volts vs. Hg pool)	a_H(mT)[b] in position (1)	(2)	(3)	(4)	(5)
Benzene	(≥2.6)	0.375				
Pyridine	2.20		0.355	0.082	0.970	
Naphthalene	1.982	0.495	0.183			
Biphenyl	2.032		0.273	0.043	0.546	
Pyrimidine	1.822		0.072		0.978	0.131
Pyrazine	1.569		0.263			
Pyridazine	1.657			0	0.592	

[a]Data from K. B. Wiberg and T. P. Lewis, *J. Am. Chem. Soc.*, **92,** 7154 (1970).

[b]Data from F. Gerson: *Hochauflösende ESR-Spektroskopie,* Verlag Chemie, Weinheim, 1967, and B. C. Gilbert and M. Trenwith, *Physical Methods in Heterocyclic Chemistry,* A. R. Katritzky, Ed., Academic Press, London, 1974, Vol. 6, p. 96.

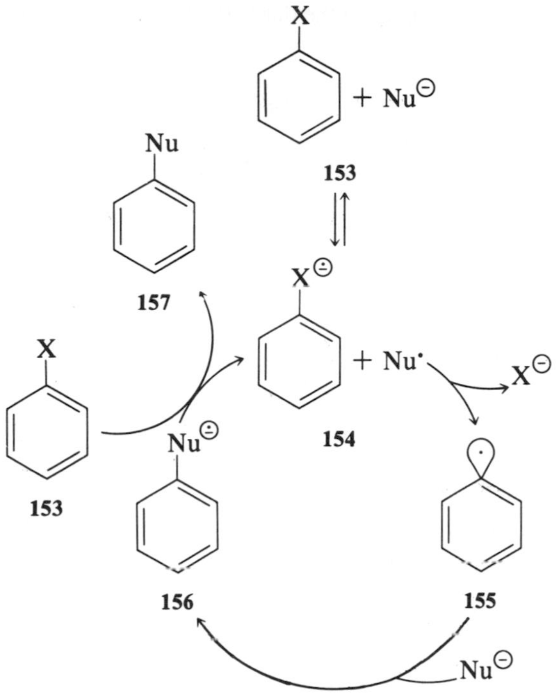

Scheme 12

aromatic compound. For example, halopyridines easily undergo nucleophilic substitution in the 2- and 4-positions, whereas reaction in the 3-position is extremely difficult. Pyrimidine and its benzo-derivative, quinazoline, react very fast (lower LUMO) and preferably in the 4-position (largest coefficient).[96] These are examples of the S_NAr reaction.[97] Simple halobenzenes do not undergo this reaction easily.

A newer method of nucleophilic aromatic substitution, however, involves activation of the substrate to give a radical ion.[98] The principle is illustrated in Scheme 12.[99] The starting material (e.g., chlorobenzene) takes up an electron coming from the nucleophile, often stimulated by irradiation, or from an alkali metal (often potassium in liquid ammonia, i.e., solvated electrons), deliberately added to the mixture. The radical anion **154** then expels $X^{\ominus}$, giving the radical **155.** This radical combines with the nucleophile to furnish a new radical anion **156,** which, by transferring the extra electron to a molecule of starting material, completes the cycle and gives the final product, a substituted benzene **157.** This is the $S_{RN}1$ mechanism (unimolecular nucleophilic substitution by way of radicals).[98] The evidence for this mechanism includes the inhibition by radical scavengers, the acceleration by solvated electrons or light, and, in a few cases, ESR spectroscopic detection of intermediates.

As an example,[98] bromobenzene arylates the potassium enolate of acetone under irradiation; no reaction takes place in the dark. Potassium metal can be used instead of irradiation:

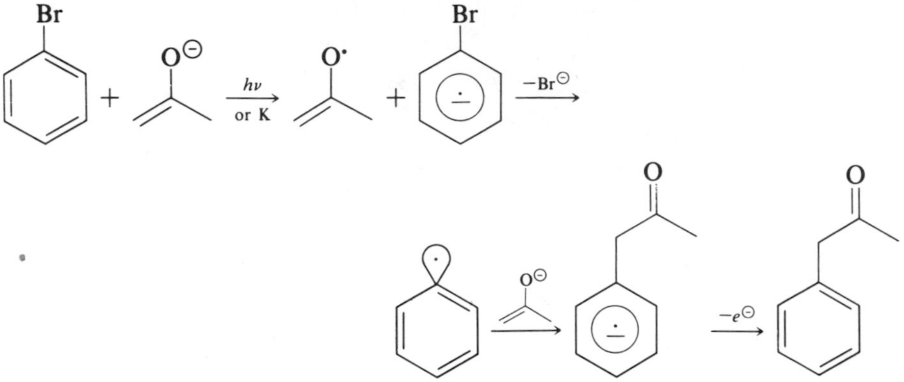

Various bromo-heterocycles, including the otherwise unreactive 3-bromopyridine, undergo the same type of reaction:[100]

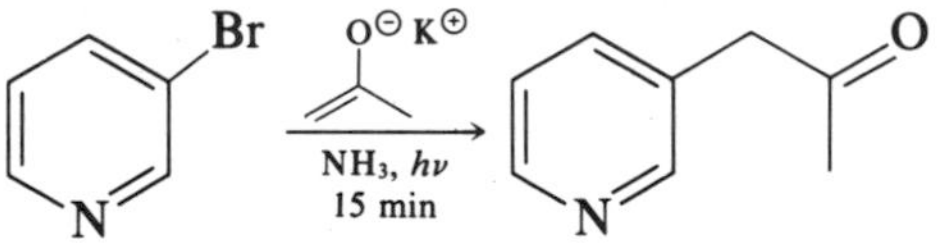

65%

$S_{RN}1$ reactions also take place in the benzylic position of substituted benzenes, particularly in *p*-nitro derivatives and other compounds possessing strongly electron-withdrawing groups and consequently low-lying LUMOs:[102,103]

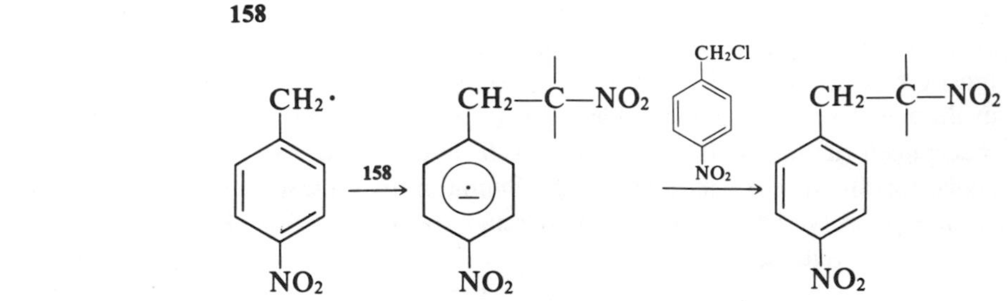

158

Scheme 13

Indeed, the fact that *C*-alkylation, not *O*-alkylation, took place with the salts of aliphatic nitro compounds (e.g., **158**) in these reactions was germane to the discovery of the $S_{RN}1$ mechanism. Benzyl chloride itself reacts with exclusive *O*-alkylation under these conditions, signifying that a normal S_N2 reaction takes place. In other cases, a competition between $S_{RN}1$ and S_N2 can be observed, indicating that it would be wrong to assume that all reactions hitherto regarded as S_N2-reactions should be reclassified as electron transfer reactions.[102,103]

Reactions of the type shown in Scheme 13 are insensitive to steric hindrance and are, therefore, of considerable synthetic potential. Thus C—C bonds joining two tertiary groups can be generated in this way, as illustrated by the folllowing example.[102]

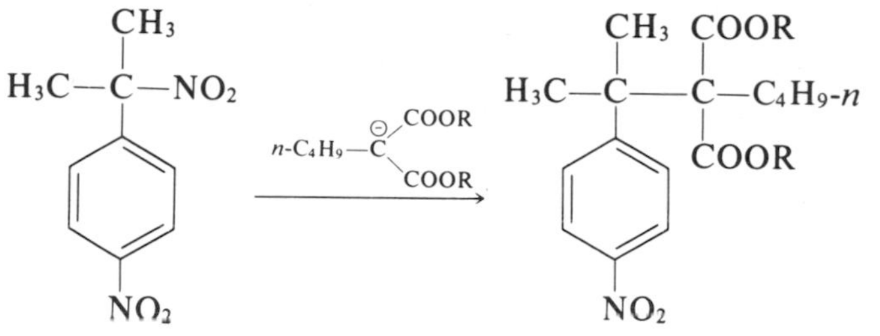

CH3 | H3C—C—NO2 (p-nitrophenyl) NO2 —[n-C4H9—C⊖(COOR)2]→ CH3 COOR | | H3C—C——C—C4H9-n | COOR (p-nitrophenyl) NO2

In all the preceding reactions, an electron was added to the LUMO of the (usually aromatic) substrate. One might also say that the nucleophile ($Nu^{\ominus}$ in Scheme 12) was oxidized or that one electron was removed from its HOMO. One might then expect that aromatic substrates possessing high-energy HOMOs could react with electrophiles (E) by way of electron transfer:

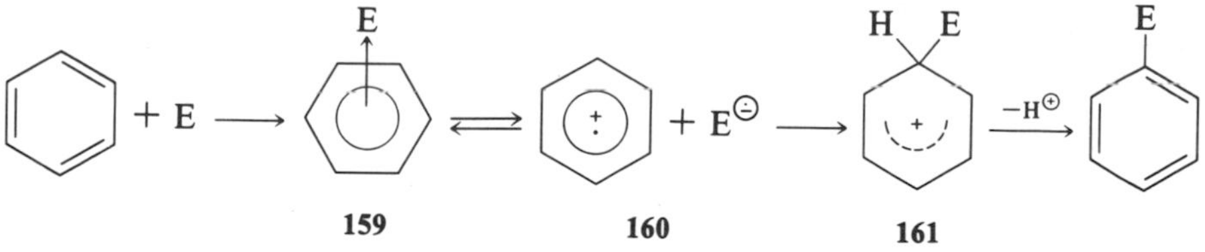

+ E ⟶ **159** ⇌ **160** + $E^{\ominus}$ ⟶ **161** —$-H^{\oplus}$→ (Ph–E)

Scheme 14

This is a question of whether the π-complex or donor-acceptor complex[104] **(159)** in a normal electrophilic aromatic substitution reaction would be able to undergo complete electron transfer to give the radical ion pair **160,** which then recombines to the σ-complex **161.** In principle, such a reaction is possible whenever the LUMO of the electrophile is of lower energy than the HOMO of the aromatic substrate. Some relevant HOMO and LUMO energies listed by Nagakura and Tanaka some time ago are given in Figure 2.24.[105] HOMO energies can be obtained experimentally by determination of the ionization potentials (IP) by mass spectrometry or, more

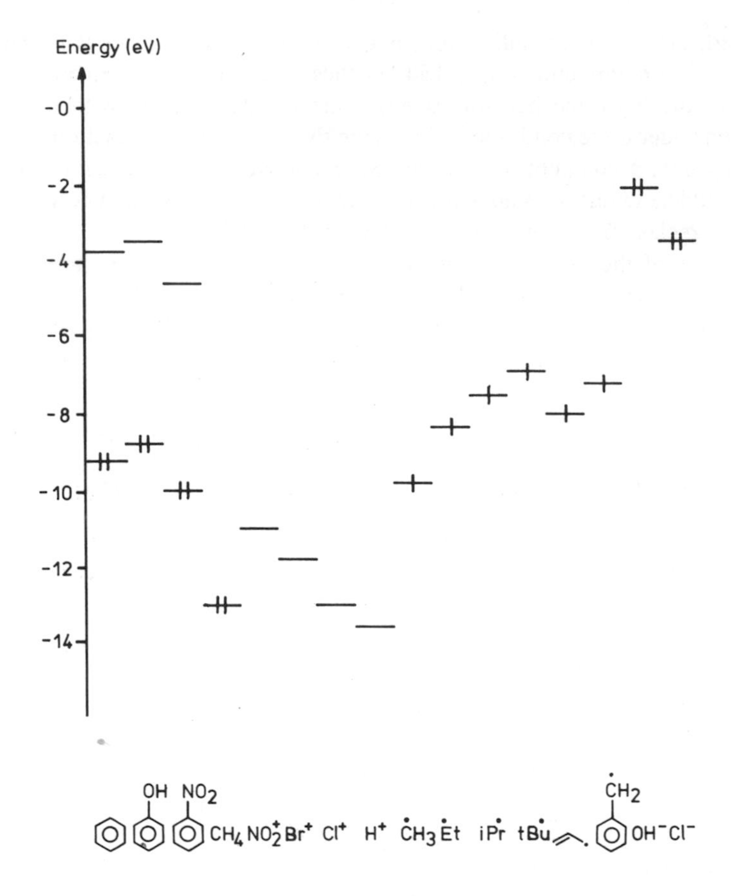

Figure 2.24. HOMO, LUMO, and SOMO energies for some molecules, radicals, and ions, based on data given by Nagakura and Tanaka (Ref. 105) and adjusted to newer values reported to D. W. Turner, *Molecular Photoelectron Spectroscopy,* Wiley, London, 1970, and F. P. Lossing, *Can. J. Chem.*, **49,** 357 (1971); F. P. Lossing and G. P. Semeluk, *ibid.*, **48,** 955 (1970). (Adapted with permission of the *Bulletin of the Chemical Society of Japan.*)

accurately, photoelectron spectroscopy. The ionization potential is linked to the orbital energy (ϵ) via Koopmans' theorem,

$$\mathrm{IP} \leqslant -\epsilon$$

Another measure of HOMO energies is provided by the polarographic oxidation potentials. Some data are given in Table 2.17. As it turns out, the rates of electrophilic substitution of various aromatics correlate very well with the ionization potentials.[106] Furthermore, radical cations of aromatic hydrocarbons are routinely

TABLE 2.17. Oxidation Potentials ($E_{1/2}$), Ionization Potentials (IP), and Hyperfine Splitting Constants in Radical Cations of Aromatic Hydrocarbons.

Compound	$E_{1/2}$ [a] (Volts vs. s.c.e.)	IP [b] (eV)	a_H (mT) [c] in position (1)	(2)	(3)	(4)	(5)	(6)	(9)
Benzene C_6H_6	2.30	9.24	—						
Naphthalene	1.54	8.15	0.277	0.103					
Phenanthrene	1.50	7.92	—						
Anthracene	1.09	7.41	0.306	0.138					0.653
Perylene	0.85	6.90	0.310	0.046	0.410				
Azulene	0.71	7.43	1.065	0.152	1.065	0.038	0.415	0.112	

[a] Data from E. S. Pysh and N. C. Yang, *J. Am. Chem. Soc.*, **85,** 2124 (1963); s.c.e. is the saturated calomel electrode.
[b] Data from D. W. Turner, *Molecular Photoelectron Spectroscopy*, Wiley, London, 1970; M. J. S. Dewar and S. D. Worley, *J. Chem. Phys.*, **50,** 654 (1969); M. J. S. Dewar and D. W. Goodman, *J. Chem. Soc. Faraday Trans. 2*, **68,** 1784 (1972); J. H. D. Eland, *Int. J. Mass. Spectrom. Ion Phys.*, **9,** 214 (1972).
[c] Data from F. Gerson, *Hochauflösende ESR-Spektroskopie*, Verlag Chemie, Weinheim, 1967; I. C. Lewis and L. S. Singer, *J. Chem. Phys.*, **43,** 2712 (1965); R. M. Dessau and S. Shih, *ibid.*, **53,** 3169 (1970).

produced for ESR spectroscopy by treatment with concentrated sulfuric acid. For example,

Here the electrophile is a proton. The positions in which electrophilic substitution occur are those that have the largest HOMO coefficients. These can be predicted from the hyperfine splitting constants in the radical cations (Table 2.17). There is a good correlation between the magnitudes of the hfs and the partial rate factors for electrophilic substitution.[106]

If both the radical cation and the radical anion formed in an electron transfer reaction (Scheme 14) are sufficiently long-lived, a stable salt may be formed (see also Problem 9):[107]

In view of the preceding considerations, it can be no surprise that free radicals may be formed by electron transfer under conditions resembling those of typical nucleophilic substitution reactions, for example, the following:[108]

$$Ph_3C{-}Cl + C_2H_5Li \longrightarrow \overline{Ph_3C\cdot + C_2H_5\cdot} + LiCl$$

diffusion → $Ph_3C\cdot$ (ESR spectrum); recombination → $Ph_3C{-}C_2H_5$ (CIDNP)

The triphenylmethyl radical was observed by ESR; the recombination product exhibited CIDNP.

When the reaction

$$Ph_3C^{\oplus}\ ClO_4^{\ominus} + tBuO^{\ominus}K^{\oplus} \longrightarrow \overline{Ph_3C\cdot\ tBuO\cdot} \longrightarrow Ph_3C{-}OtBu$$

was carried out in the probe of an ESR spectrometer, the $Ph_3C\cdot$ signal increased to maximum intensity about 3 minutes after mixing the reagents. When the base was changed to 2,4,6-tri-*t*-butylphenoxide, the phenoxy radical became observable by ESR, too:

$$(H_5C_6)_3C^{\oplus}\ ClO_4^{\ominus} + 2,4,6\text{-}[(H_3C)_3C]_3C_6H_2O^{\ominus}\ K^{\oplus} \xrightarrow{-KClO_4} (H_5C_6)_3C\cdot + 2,4,6\text{-}[(H_3C)_3C]_3C_6H_2O\cdot$$

$$\downarrow$$

$$(H_5C_6)_3C(H)C_6H_4{=}C(C_6H_5)_2$$

Steric hindrance now prevents recombination of the radicals, and the well-known dimer of triphenylmethyl is formed instead (cf. Section 2.1).[108]

Evidence for a free radical component of the Wurtz reaction [Eq. (26)] was given as early as 1956.[109] When the reaction between *n*-butyl iodide

$$R{-}Hal + 2\,M \longrightarrow R{-}R + 2\,M{-}Hal \tag{26}$$

and either lithium metal or *n*-butyl lithium was carried out in isopropylbenzene, up to 19% of 2,3-dimethyl-2,3-diphenylbutane was formed, indicating dimerization of benzylic radicals. A modern formulation of this reaction could be:

$$C_4H_9I + Li \xrightarrow{130^\circ C} C_4H_9I^{\ominus}\ Li^{\oplus} \xrightarrow{-I^{\ominus}} C_4H_9\cdot$$

$$\underset{24\%}{C_8H_{18}} \xleftarrow{\times 2} C_4H_9\cdot \xrightarrow{C_6H_5CH(CH_3)_2} \underset{20\%}{C_6H_5\dot{C}(CH_3)_2 + C_4H_{10}}$$

$$C_6H_5\dot{C}(CH_3)_2 \xrightarrow{\times 2} \underset{17\%}{C_6H_5C(CH_3)_2{-}C(CH_3)_2C_6H_5}$$

A radical component—the extent of which is unknown—is also supported by CIDNP in the related reaction between ethyl iodide and ethyl lithium. The development of an *AE* multiplet for ethyl iodide during this reaction is in agreement with the existence of a singlet pair of ethyl radicals from which the polarized C_2H_5I is formed after diffusive iodine abstraction from unpolarized C_2H_5I. CIDNP of radical recombination products has also been observed in related studies.[110]

$$C_2H_5I + C_2H_5Li \longrightarrow C_2H_5I^{\ominus} + C_2H_5Li^{\oplus}$$

$$\downarrow -I^{\ominus} - Li^{\oplus}$$

$$\underset{AE}{C_2H_5\cdot + C_2H_5I} \xleftarrow{C_2H_5I} \overline{\cdot C_2H_5 \quad \cdot C_2H_5}$$

$$\downarrow$$

$$C_4H_{10}$$

Much more experimental work aimed at a clear-cut demarcation of traditional S_N and electron transfer reactions[111] is highly desirable.

2.7. EXAMPLES AND PROBLEMS

The purpose of this section is to give practical examples and also to incorporate further material which, for reasons of space, has been treated only lightly in the main text. The material is presented in the form of problems.

1. Explain the following relative rate data:[112]

$$Cl\cdot + Ph{-}CH_2{-}CH_3 \xrightarrow{k_\alpha} Ph{-}CH(Cl){-}CH_3$$

$$Cl\cdot + Ph{-}CH_2{-}CH_3 \xrightarrow{k_\beta} Ph{-}CH_2{-}CH_2{-}Cl$$

$$k_\alpha / k_\beta\ (40°C) = 14.5$$

2. Explain why in bromination toluene is about 66 times more reactive than cyclohexane (per aliphatic hydrogen atom) and why in chlorination the ratio

is 0.5.[113] (For the purpose of argument, the chlorination ratio may be taken to be unity.)

$$C_6H_5CH_3 + C_6H_{12} \xrightarrow{X_2} C_6H_5CH_2X + C_6H_{11}X$$

X = Br, Cl

3. In the *Gomberg–Bachmann–Hey* reaction, the base is believed to extract the diazonium salt from the aqueous phase in the form of a diazo anhydride, Ar—N=N—O—N=N—Ar. Give a homolytic mechanism for the arylation.[114,116]

$$R\text{-}C_6H_4\text{-}N_2^{\oplus}Cl^{\ominus} + Ar'\text{—}H \xrightarrow[\text{base}]{\Delta} R\text{-}C_6H_4\text{-}Ar'$$

4. Complete the mechanism of the *Meerwein* reaction and explain the regioselectivities.[115,116]

$$Ar\text{—}N_2^{\oplus}Cl^{\ominus} \xrightarrow[-Cu^{II}Cl_2]{Cu^{I}Cl} [Ar\text{—}N{=}N\cdot] \xrightarrow{-N_2} Ar\cdot$$

$$Ar\cdot + \text{H—}C{=}C\text{<} \longrightarrow Ar - C{=}C\text{<}$$

$$C_6H_5\text{—}N_2^{\oplus}Cl^{\ominus} + CH_2{=}CH\text{—}COOH \longrightarrow C_6H_5\text{—}CH{=}CH\text{—}COOH$$

$$Cl\text{—}C_6H_4\text{—}N_2^{\oplus}Cl^{\ominus} + C_6H_5\text{—}CH{=}CH\text{—}CHO \longrightarrow C_6H_5\text{—}CH{=}C(CHO)\text{—}C_6H_4\text{—}Cl$$

5. Polyethylene formed by free-radical polymerization of ethylene is not completely linear but contains approximately 30 side chains per 1000 carbon atoms. Explain the formation of butyl side chains in a growing radical (use information from Section 2.5.3).[117]

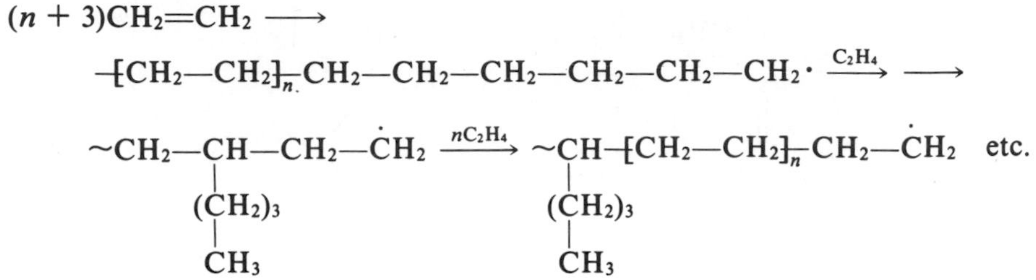

$$(n+3)CH_2{=}CH_2 \longrightarrow$$

$$-[CH_2-CH_2]_n-CH_2-CH_2-CH_2-CH_2-CH_2-CH_2\cdot \xrightarrow{C_2H_4} \longrightarrow$$

$$\sim CH_2-\underset{\underset{CH_3}{|}}{\underset{(CH_2)_3}{|}}{CH}-CH_2-\dot{C}H_2 \xrightarrow{nC_2H_4} \sim\underset{\underset{CH_3}{|}}{\underset{(CH_2)_3}{|}}{CH}-[CH_2-CH_2]_n-CH_2-\dot{C}H_2 \quad \text{etc.}$$

6. Rationalize the formation of the product of the following *Barton* reaction, using a hydrogen shift in a free radical.[118]

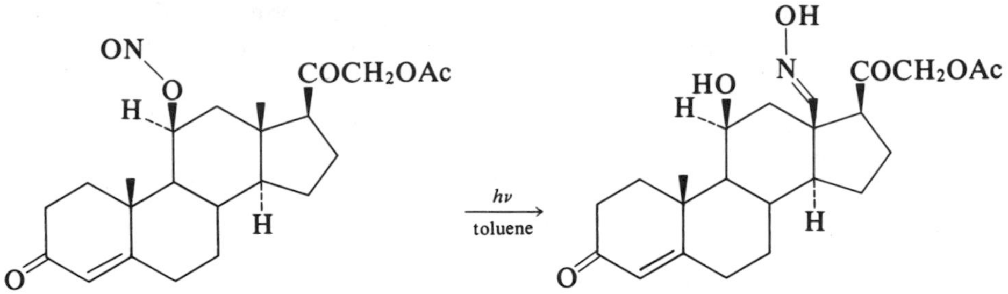

ON, O, H, COCH₂OAc, H, O $\xrightarrow[\text{toluene}]{h\nu}$ OH, N, HO, H, COCH₂OAc, H, O

7. Free radical polymerization of ethyl acrylate, often initiated by potassium persulfate, gives a viscous, cross-linked polymer. In spite of steric hindrance, cross-linking occurs most readily at the position indicated. Explain.[119]

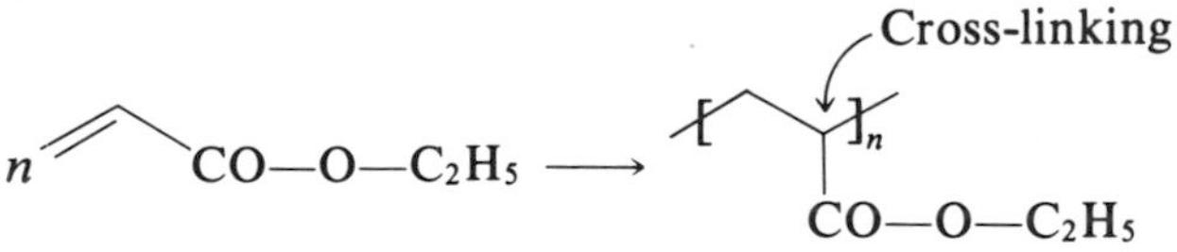

$$n\ CH_2{=}CH-CO-O-C_2H_5 \longrightarrow -[CH_2-CH(CO-O-C_2H_5)]_n-$$

Cross-linking

8. Explain the following homolysis rate data (cf. Section 2.2.1).[120]

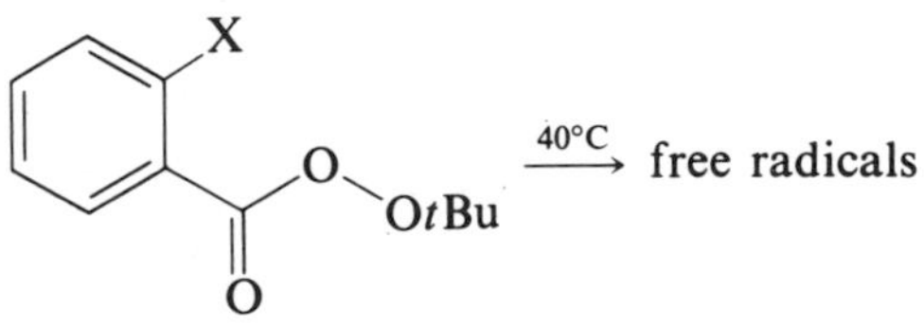

X, O, O*t*Bu, O $\xrightarrow{40°C}$ free radicals

X	k_{rel}	$\Delta H^{\ddagger}$(kcal/mol)	$\Delta S^{\ddagger}$(cal K^{-1} mol^{-1})
H	1	34	+10.0
C_6H_5—S	10^4	23	−3.4

9. Draw resonance structures for *Wurster*'s salts,

stable radical cations first prepared in 1879, to demonstrate that these are donor–acceptor stabilized (or merostabilized). See Ref. 2a.

10. Write an electron transfer mechanism for the *Sandmeyer* reaction with iodide ion.[121]

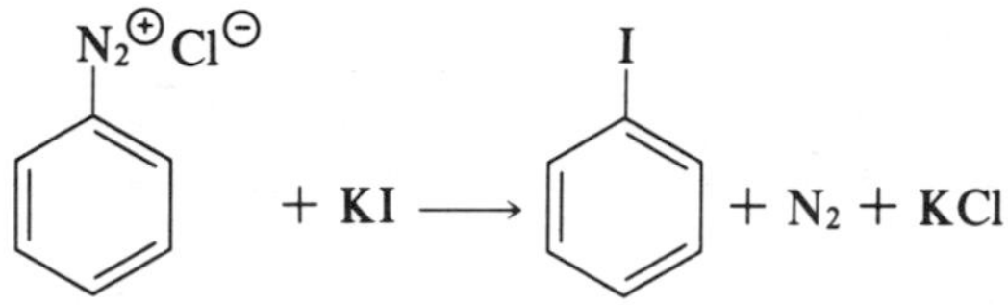

$N_2^{\oplus}Cl^{\ominus}$ + KI ⟶ I + N_2 + KCl

11. Write an electron transfer mechanism ($S_{RN}1$) for the following electrolysis.[101]

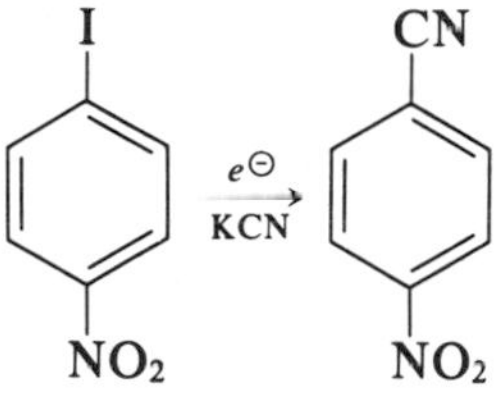

$e^{\ominus}$, KCN

12. Write electron transfer mechanisms for the following reactions in which cyclopentylmethyl radicals are formed (cf. **134** → **135**).

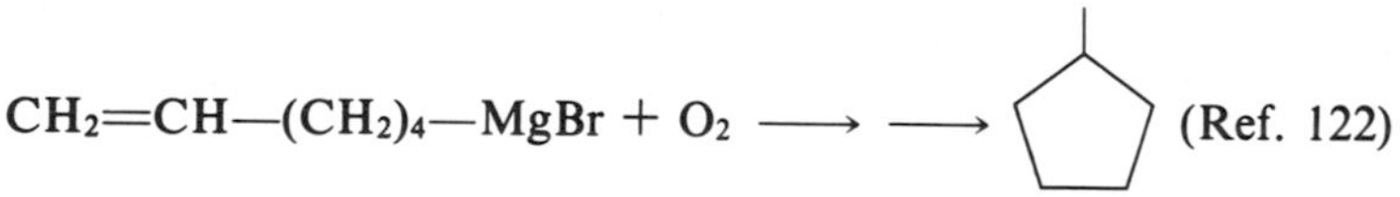

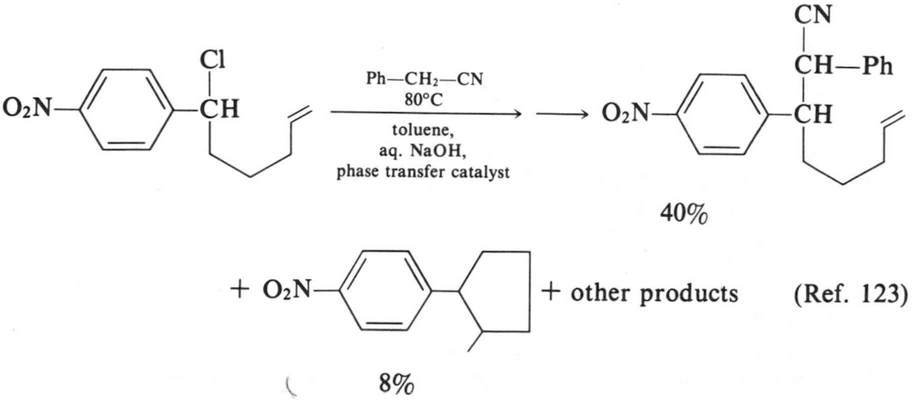

$CH_2{=}CH{-}(CH_2)_4{-}MgBr + O_2$ ⟶ ⟶ (Ref. 122)

$Ph{-}CH_2{-}CN$, 80°C, toluene, aq. NaOH, phase transfer catalyst

40%

+ O_2N– + other products (Ref. 123)

8%

REFERENCES AND NOTES

1. G. Winnewisser, *Top. Curr. Chem.*, **99**, 40 (1981); see also *Molecules in Interstellar Space*, A. Carrington and D. A. Ramsay, Eds., The Royal Society, London, 1982.
2. (a) Review: H. G. Viehe, R. Mérenyi, L. Stella, and Z. Janousek, *Angew. Chem.*, **91**, 982 (1979); *Angew. Chem. Int. Ed. Engl.*, **18**, 917 (1979). (b) J. M. Burns, D. L. Wharry, and T. H. Koch, *J. Am. Chem. Soc.*, **103**, 849 (1981).
3. (a) D. Griller and K. U. Ingold, *Acc. Chem. Res.*, **9**, 13 (1976). (b) C. Rüchardt and H.-D. Beckhaus, *Angew. Chem.*, **92**, 417 (1980); *Angew. Chem. Int. Ed. Engl.*, **19**, 429 (1980).
4. J. L. Gole, *J. Phys. Chem.*, **84**, 1333 (1980); R. F. Curl, *J. Chem. Phys.*, **37**, 779 (1962); M. G. K. Killai and R. F. Curl, *ibid.*, **37**, 2921 (1962).
5. See M. Gomberg, *Chem. Rev.*, **1**, 91 (1925); K. Ziegler, *Angew. Chem.*, **61**, 168 (1949); K. Ziegler and L. Ewald, *Justus Liebigs Ann. Chem.*, **473**, 163 (1929); H. Lankamp, W. T. Nauta, and C. MacLean, *Tetrahedron Lett.*, **1968**, 249; J. M. McBride, *Tetrahedron*, **30**, 2009 (1974); T. H. Colle, P. S. Glaspie, and E. S. Lewis, *J. Org. Chem.*, **43**, 2722 (1978).
6. See A. R. Forrester, J. M. Hay, and R. H. Thomson, *Organic Chemistry of Stable Free Radicals*, Academic Press, London, New York, 1968.
7. When steric hindrance between *ortho*-substituents is avoided, planar triarylmethyl radicals can be prepared:

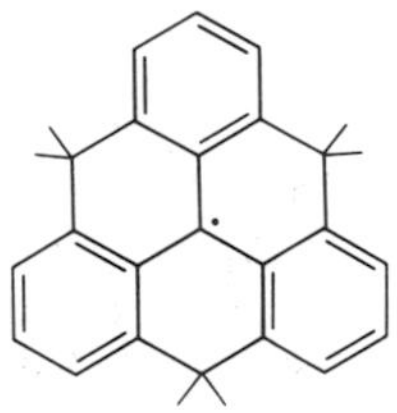

   F. A. Neugebauer, D. Hellwinkel, and G. Aulmich, *Tetrahedron Lett.*, **1978**, 4871. Steric hindrance can also be used to make benzyl radicals nonplanar:

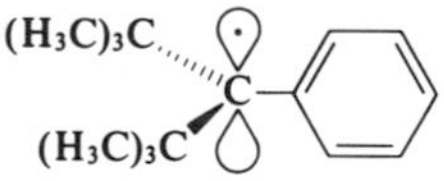

   In the preceding radical, the *p* orbital is nearly perpendicular to the benzene π orbitals, and the benzylic resonance is, therefore, virtually absent: K. Schreiner and A. Berndt, *Angew. Chem.*, **86**, 131 (1974); *Angew. Chem. Int. Ed. Engl.*, **13**, 144 (1974).
8. A. E. Tschitschibabin, *Ber. Dtsch. Chem. Ges.*, **40**, 1810 (1907); ESR spectrum: H.-D. Brauer, H. Stieger, and H. Hartmann, *Z. Phys. Chem. Neue Folge*, **63**, 50 (1969).
9. W. Schlenk and M. Brauns, *Ber. Dtsch. Chem. Ges.*, **48**, 716 (1915); ESR spectrum: G. Kothe, K.-H. Denkel, and W. Sümmermann, *Angew. Chem.*, **82**, 935 (1970); *Angew. Chem., Int. Ed. Engl.*, **9**, 906 (1970).
10. C. E. Dinerman and G. E. Ewing, *J. Chem. Phys.*, **53**, 626 (1970); *ibid.*, **54**, 3660 (1971); G. E. Ewing, *Angew. Chem.*, **84**, 570 (1972); *Angew. Chem. Int. Ed. Engl.*, **11**, 486 (1972).
11. For more detailed descriptions of ESR, see A. Carrington and A. D. McLachlan, *Introduction to Magnetic Resonance*, Harper International edition, New York, 1967; N. M.

Atherton, *Electron Spin Resonance,* Ellis Horwood, Chichester, 1973; J. E. Wertz and J. R. Bolton, *Electron Spin Resonance,* McGraw-Hill, New York, 1972.

12. H. M. McConnell, *J. Chem. Phys.*, **24,** 764 (1956); H. M. McConnell and D. B. Chesnut, *ibid.*, **28,** 107 (1958).

13. (a) H. M. Walborsky and J.-C. Chen, *J. Am. Chem. Soc.*, **93,** 671 (1971). (b) L. J. Altman and R. C. Baldwin, *Tetrahedron Lett.*, **1971,** 2531. (c) R. W. Fessenden, *J. Phys. Chem.*, **71,** 74 (1967). (d) V. Malatesta, D. Forrest, and K. U. Ingold, *J. Am. Chem. Soc.*, **100,** 7073 (1978).

14. R. W. Fessenden and R. H. Schuler, *J. Chem. Phys.*, **43,** 2704 (1965).

15. R. O. C. Norman, *Chem. Brit.*, **6,** 66 (1970); H. Fischer in *Free Radicals,* J. K. Kochi, Ed., Wiley, New York, 1973, Vol. II, p. 435.

16. L. Pauling, *J. Chem. Phys.*, **51,** 2767 (1969). Compare A. D. Walsh, *Discuss. Faraday Soc.*, **2,** 21 (1947); R. C. Bingham and M. J. S. Dewar, *J. Am. Chem. Soc.*, **95,** 7180, 7182 (1973).

17. D. E. Wood, L. F. Williams, R. F. Sprecher, and W. A. Lathan, *J. Am. Chem. Soc.*, **94,** 6241 (1972).

18. M. N. Paddon-Row and K. N. Houk, *J. Am. Chem. Soc.*, **103,** 5047 (1981); M. Yoshimine and J. Pacansky, *J. Chem. Phys.*, **74,** 5168 (1981).

19. J. Pacansky and M. Dupuis, *J. Chem. Phys.*, **68,** 4276 (1978); J. Pacansky and H. Coufal, *ibid.*, **72,** 5285 (1980); J. Pacansky and M. Dupuis, *J. Am. Chem. Soc,* **104,** 415 (1982).

20. P. J. Edge and J. K. Kochi, *J. Am. Chem. Soc.*, **94,** 6485 (1972); P. J. Krusic and J. K. Kochi, *ibid.*, **93,** 846 (1971); T. Kawamura and J. K. Kochi, *ibid.*, **94,** 648 (1972); A. J. Bowles, A. Hudson, and R. A. Jackson, *Chem. Phys. Lett.*, **5,** 552 (1970).

21. P. S. Skell, R. R. Pawlis, D. C. Lewis, and K. J. Shea, *J. Am. Chem. Soc,* **95,** 6735 (1973).

22. P. S. Skell and K. J. Shea, *J. Am. Chem. Soc.*, **94,** 6550 (1972).

23. (a) J. Cooper, A. Hudson, and R. A. Jackson, *Tetrahedron Lett.*, **1973,** 831. (b) I. Biddles and A. Hudson, *Chem. Phys. Lett.*, **18,** 45 (1973).

24. G. A. Olah and R. D. Porter, *J. Am. Chem. Soc.*, **93,** 6877 (1971).

25. J. W. Larsen and A. V. Metzner, *J. Am. Chem. Soc.*, **94,** 1614 (1972).

26. For more thorough mathematical descriptions of the triplet state and zero-field splittings, see Ref. 11 and also S. P. McGlynn, T. Azumi, and M. Kinoshita, *Molecular Spectroscopy of the Triplet State,* Prentice-Hall, New Jersey, 1969; E. Wasserman, L. C. Snyder, and W. A. Yager, *J. Chem. Phys.*, **41,** 1763 (1964). A readable, nonmathematical description is given by E. Wasserman, *Prog. Phys. Org. Chem.*, **8,** 319 (1971).

27. C. A. Hutchison and B. W. Mangum, *J. Chem. Phys.*, **34,** 908 (1961).

28. E. Wigner and E. Witmer, *Z. Phys.*, **51,** 859 (1928); F. A. Matsen and J. D. Klein, *Adv. Photochem.*, **7,** 1 (1969).

29. R. Kaptein, *J. Am. Chem. Soc.*, **94,** 6251 (1972).

30. (a) G. L. Closs, Twenty-third IUPAC Congress, Special Lectures, **4,** 19 (1971). (b) A. R. Lepley and G. L. Closs, Eds., *Chemically Induced Magnetic Polarization,* Wiley, New York, 1973. (c) H. R. Ward, *Acc. Chem. Res.*, **5,** 18 (1972); (d) R. G. Lawler, *ibid.*, **5,** 25 (1972).

31. R. Kaptein, *Chem. Commun.*, **1971,** 732.

32. J. P. Hargis and P. B. Shevlin, *Chem. Commun.*, **1973,** 179.

33. J. W. Emsley, J. Feeney, and L. H. Sutcliffe, *High Resolution Nuclear Magnetic Resonance,* Pergamon Press, London, 1966, Vol. 2, Chapter 10.

34. E. G. Janzen, *Acc. Chem. Res.*, **4,** 31 (1971); M. J. Perkins, in *Essays on Free Radical Chemistry,* R. O. C. Norman, Ed., Special Publication No. 24, The Chemical Society, London, 1970, p. 97; C. A. Evans, *Aldrichimica Acta,* **12,** 23 (1979).

35. E. G. Janzen in *Free Radicals in Biology,* W. A. Pryor, Ed., Academic Press, New York, 1980, Vol. 4, pp. 115–154.

36. H. G. Aurich, in *Chemistry of the nitro and nitroso groups,* S. Patai, Ed., Wiley, Chichester, 1982, Supplement F, pp. 565–622.

37. P. Schmid and K. U. Ingold, *J. Am. Chem. Soc.,* **100,** 2493 (1978); E. G. Janzen, D. E. Nutter, and C. A. Evans, *J. Phys. Chem.,* **79,** 1983 (1975).

38. J. A. Howard and J. C. Tait, *Can. J. Chem.,* **56,** 176 (1978).

39. E. G. Janzen, C. A. Evans, and Y. Nishi, *J. Am. Chem. Soc.,* **94,** 8236 (1972).

40. R. G. Gasanov, I. I. Kandror, and R. Kh. Freidlina, *Tetrahedron Lett.,* **1975,** 1485.

41. A. N. Saprin and L. H. Piette, *Arch. Biochem. Biophys.,* **180,** 480 (1977).

42. C. N. McEwen and M. A. Rudat, *J. Am. Chem. Soc.,* **103,** 4343 (1981).

43. M. A. Rudat and C. N. McEwen, *J. Am. Chem. Soc.,* **103,** 4349 (1981).

44. F. P. Lossing, *N.Y. Acad. Sci.,* **67,** 499 (1957).

45. F. P. Lossing, *Can. J. Chem.,* **49,** 357 (1971).

46. D. M. Golden, G. N. Spokes, and S. W. Benson, *Angew. Chem.,* **85,** 602 (1973); *Angew. Chem. Int. Ed. Engl.,* **12,** 534 (1973).

47. M. Rossi, K. D. King, and D. M. Golden, *J. Am. Chem. Soc.,* **101,** 1223 (1979).

48. D. F. McMillen, P. L. Trevor, and D. M. Golden, *J. Am. Chem. Soc.,* **102,** 7400 (1980).

49. H. J. H. Fenton, *J. Chem. Soc.,* **65,** 899 (1894); C. Walling, *Acc. Chem. Res.,* **8,** 125 (1975).

50. G. P. Gardini, *Tetrahedron Lett.,* **1972,** 4113.

51. H. E. O'Neal, J. W. Bagg, and W. H. Richardson, *Int. J. Chem. Kinet.,* **2,** 493 (1970).

52. P. D. Bartlett and J. M. McBride, *Pure Appl. Chem.,* **15,** 89 (1967).

53. J. W. Taylor and J. C. Martin, *J. Am. Chem. Soc.,* **87,** 3650 (1965); *ibid.,* **89,** 6904 (1967).

54. (a) R. C. P. Cubbon, *Prog. React. Kinet.,* **5,** 29 (1970). (b) R. A. Cooper, R. G. Lawler, and H. R. Ward, *J. Am. Chem. Soc.,* **94,** 545 (1972).

55. W. A. Pryor and K. Smith, *J. Am. Chem. Soc.,* **92,** 5403 (1970).

56. P. D. Bartlett and C. Rüchardt, *J. Am. Chem. Soc.,* **82,** 1756 (1960); C. Rüchardt and H. Böck, *Chem. Ber.,* **100,** 654 (1967).

57. A. A. Zavitsas and J. A. Pinto, *J. Am. Chem. Soc.,* **94,** 7390 (1972); see also D. D. Tanner, P. W. Samal, T. C.-S. Ruo, and R. Henriques, *ibid.,* **101,** 1168 (1979).

58. L. Herk, M. Feld, and M. Szwarc, *J. Am. Chem. Soc.,* **83,** 2998 (1961).

59. R. Kaptein, *Chem. Phys. Lett.,* **2,** 261 (1968).

60. B. Blank and H. Fischer, *Helv. Chim. Acta,* **54,** 905 (1971).

61. R. C. Neuman and E. W. Ertley, *J. Am. Chem. Soc.,* **97,** 3130 (1975); see also R. C. Neuman and M. J. Amrich, *J. Org. Chem.,* **45,** 4629 (1980).

62. N. A. Porter, M. E. Landis, and L. J. Marnett, *J. Am. Chem. Soc.,* **93,** 795 (1971); N. A. Porter and L. J. Marnett, *ibid.,* **95,** 4361 (1973); N. A. Porter, G. R. Dubay, and J. G. Green, *ibid.,* **100,** 920 (1978).

63. N. A. Porter, L. J. Marnett, C. H. Lochmüller, G. L. Closs, and M. Shobataki, *J. Am. Chem. Soc.,* **94,** 3664 (1972).

64. M. L. Poutsma, in *Methods in Free Radical Chemistry,* E. S. Huyser, Ed., Marcel Dekker, New York, 1969, Vol. 1, p. 79; W. A. Thaler, *ibid.,* Vol. 2, p. 121.

65. T. Ibuki, A. Tsuji, and Y. Takezaki, *J. Phys. Chem.,* **80,** 8 (1976).

66. J. A. Kerr, in *Free Radicals,* J. K. Kochi, Ed., Wiley, New York, 1973, Vol. I, p. 1.

67. K. J. Mintz and D. J. LeRoy, *Can. J. Chem.,* **51,** 3534 (1973).

68. L. S. Mayanz, *J. Phys. Chem. USSR,* **37,** 623 (1964); E. A. Hardwidge, C. W. Larsen, and B. S. Rabinovitch, *J. Am. Chem. Soc.,* **92,** 3278 (1970).

69. J. W. Wilt, in *Free Radicals,* J. K. Kochi, Ed., Wiley, New York, 1973, Vol I, p. 333.

70. A. Effio, D. Griller, K. U. Ingold, J. C. Scaiano, and S. J. Sheng, *J. Am. Chem. Soc.,* **102,** 6063 (1980).

71. P. B. Shevlin and H. J. Hansen, *J. Org. Chem.,* **42,** 3011 (1977).

72. A. Effio, D. Griller, K. U. Ingold, A. L. J. Beckwith, and A. K. Serelis, *J. Am. Chem. Soc.,* **102,** 1734 (1980) and references therein.

73. R. Sustmann and F. Lübbe, *Chem. Ber.,* **112,** 42 (1979).

74. F. Lübbe and R. Sustmann, *Chem. Ber.,* **112,** 57 (1979).

75. P. Vogel, M. Saunders, N. M. Hasty, and J. A. Berson, *J. Am. Chem. Soc.,* **93,** 1551 (1971).

76. (a) I. Fleming, *Frontier Orbitals and Organic Chemical Reactions,* Wiley, London, 1976, p. 183. (b) For an equivalent analysis of radical addition and substitution reactions in terms of polar effects, see J. M. Tedder, *Angew. Chem.,* **94,** 433 (1982); *Angew. Chem. Int. Ed. Engl.,* **21,** 401 (1982).

77. A. L. J. Beckwith, "Some Aspects of Free-Radical Rearrangement Reactions," Special Publication No. 24, The Chemical Society, London, 1970, p. 239. See also Ref. 79a. For theoretical calculations supporting linear transition states in C—H abstraction, and non-symmetrical ones in additions to double bonds, see H. Fujimoto, S. Yamabe, T. Minate, and K. Fukui, *J. Am. Chem. Soc.,* **94,** 9205 (1972), and references therein.

78. E. Haselbach, *Helv. Chim. Acta,* **54,** 2257 (1971); J. D. Roberts and D. Schuster, *J. Org. Chem.,* **27,** 51 (1962).

79. (a) A. L. J. Beckwith, *Tetrahedron,* **37,** 3073 (1981). (b) M. Julia, *Acc. Chem. Res.,* **4,** 386 (1971). (c) C. Walling and M. S. Pearson, *J. Am. Chem. Soc.,* **86,** 2262 (1964); C. Walling and A. Cioffari, *ibid.,* **94,** 6059, 6064 (1972).

80. I. B. Afanes'ev, *Russ. Chem. Rev.,* **40,** 216 (1971).

81. W. A. Pryor, T. H. Lin, J. P. Stanley, and R. W. Henderson, *J. Am. Chem. Soc.,* **95,** 6993 (1973).

82. Yu. L. Spirin, *Russ. Chem. Rev.,* **38,** 529 (1969); G. A. Russell, in *Free Radicals,* J. K. Kochi, Ed., Wiley, New York, 1973, Vol I, p. 275.

83. D. D. Tanner, P. W. Samal, T. C.-S. Ruo, and R. Henriques, *J. Am. Chem. Soc.,* **101,** 1168 (1979).

84. W. A. Pryor, F. Y. Tang, R. H. Tang, and D. F. Church, *J. Am. Chem. Soc.,* **104,** 2885 (1982).

85. F. Minisci, *Top. Curr. Chem.,* **62,** 1 (1976); A. Citterio, F. Minisci, O. Porta, and G. Sesano, *J. Am. Chem. Soc.,* **99,** 7960 (1977).

86. L. Grossi, F. Minisci, and G. F. Pedulli, *J. Chem. Soc., Perkin Trans. 2,* **1977,** 948.

87. B. Giese and J. Meister, *Angew. Chem.,* **89,** 178 (1977); *Angew. Chem. Int. Ed. Engl.,* **16,** 178 (1977). For radical addition to electron-deficient olefins, see B. Giese, G. Kretzschmar, and J. Meixner, *Chem. Ber.,* **113,** 2787 (1980).

88. J. Sorba, J. Fossey, and D. Lefort, *Bull. Soc. Chim. France,* **1977,** 967; see also A. Citterio, F. Minisci, A. Arnoldi, R. Pagano, A. Parravicini, and O. Porta, *J. Chem. Soc. Perkin Trans. 2,* **1978,** 519.

89. W. A. Pryor, D. L. Fuller, and J. P. Stanley, *J. Am. Chem. Soc.,* **94,** 1632 (1972). Further data are summarized by D. C. Nonhebel and J. C. Walton, *Free Radical Chemistry,* Cambridge University Press, Cambridge, 1974.

90. B. Giese and J. Stellmach, *Chem. Ber.,* **113,** 3294 (1980); see also B. Giese, *Angew. Chem.,* **89,** 162 (1977); *Angew. Chem. Int. Ed. Engl.,* **16,** 125 (1977).

91. Example of an anchimerically accelerated bond homolysis:

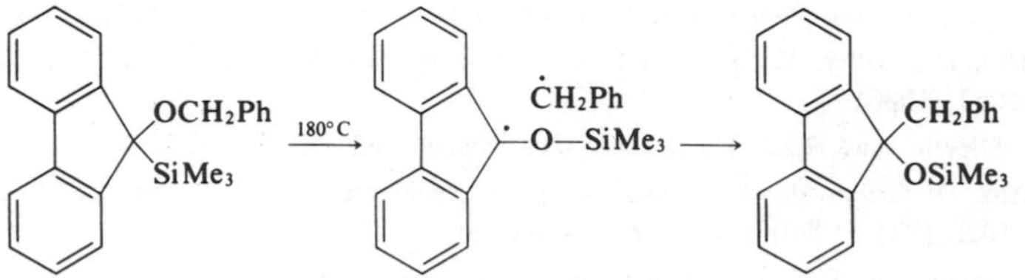

OCH_2Ph, $SiMe_3$ $\xrightarrow{180°C}$ $\dot{C}H_2Ph$, $\cdot$ $O{-}SiMe_3$ $\longrightarrow$ CH_2Ph, $OSiMe_3$

M. T. Reetz, *Angew. Chem.*, **91,** 185 (1979); *Angew. Chem. Int. Ed. Engl.*, **18,** 173 (1979).

92. N. Hirota, in *Radical Ions,* E. T. Kaiser and L. Kevan, Eds., Wiley, New York, 1968, p. 35.

93. G. A. Russell, in *Radical Ions,* E. T. Kaiser and L. Kevan, Eds., Wiley, New York, 1968, p. 87.

94. For a general review of radical anions of aromatic hydrocarbons, see N. L. Holy, *Chem. Rev.*, **74,** 243 (1974).

95. F. Gerson, *Hochauflösende ESR-Spektroskopie,* Verlag Chemie, Weinheim, 1967.

96. G. Illuminati, *Adv. Heterocycl. Chem.*, **3,** 285 (1964).

97. J. A. Zoltewicz, *Top. Curr. Chem.*, **59,** 35 (1975).

98. J. F. Bunnett, *Acc. Chem. Res.*, **11,** 413 (1978).

99. M. Chanon and M. L. Tobe, *Angew. Chem.*, **94,** 27 (1982); *Angew. Chem. Int. Ed. Engl.*, **21,** 1 (1982).

100. A. P. Komin and J. F. Wolfe, *J. Org. Chem.*, **42,** 2481 (1977).

101. D. E. Bartok, W. C. Danen, and M. D. Hawley, *J. Org. Chem.*, **35,** 1206 (1970).

102. N. Kornblum, *Angew. Chem.*, **87,** 797 (1975); *Angew. Chem. Int. Ed. Engl.*, **14,** 734 (1975).

103. For a general review, see I. P. Beletskaya and V. N. Drozd, *Russ. Chem. Rev.*, **1979,** 431.

104. Many neutral molecules form donor–acceptor complexes D,A $\longleftrightarrow$ $D^{+\cdot}A^{-\cdot}$ characterized by a long wavelength band in the UV spectrum [charge-transfer (CT) band]. The position of the CT band for a given acceptor correlates with the vertical ionization potential of the donor: $E_{CT} = h\nu_{CT} = \alpha \cdot IP_v + \text{const}$. Examples are the complex between iodine and benzene and that between mesitylene and trinitrobenzene. Data are given by E. M. Kosower, *An Introduction to Physical Organic Chemistry,* Wiley, New York, 1968.

105. S. Nagakura and J. Tanaka, *Bull. Chem. Soc. Jpn.*, **32,** 734 (1959).

106. E. B. Pedersen, T. E. Petersen, K. Torssell, and S.-O. Lawesson, *Tetrahedron,* **29,** 579 (1973). See also N. D. Epiotis, *J. Am. Chem. Soc.*, **95,** 3188 (1973).

107. L. R. Melby, R. L. Harder, W. R. Hertler, W. Mahler, R. E. Benson, and W. E. Hochel, *J. Am. Chem. Soc.*, **84,** 3374 (1962).

108. F. S. D'yachkovskii, N. N. Bubnov, and A. E. Shilov, *Dokl. Akad. Nauk SSSR,* **123,** 870 (1958); K. A. Bilevitch, N. N. Bubnov, and O. Yu. Okhlobystin, *Tetrahedron Lett.*, **1968,** 3465; K. A. Bilevitch, N. N. Bubnov, O. Yu. Okhlobystin, and N. G. Radzhabov, *Dokl. Akad. Nauk SSSR,* **191,** 119 (1970); K. A. Bilevitch and O. Yu. Okhlobystin, *Russ. Chem. Rev.*, **37**, 954 (1968); Z. V. Todres, *ibid.*, **47,** 148 (1978); **43,** 1099 (1974).

109. D. Bryce-Smith, *J. Chem. Soc.*, **1956,** 1603.

110. H. R. Ward, *Acc. Chem. Res.*, **5,** 18 (1972); H. R. Ward, R. G. Lawler, and R. A. Cooper, in *Chemically Induced Magnetic Polarization,* A. R. Lepley and G. L. Closs, Eds., Wiley, New York, 1973, p. 281; see also Ref. 99.

111. See S. Bank and D. A. Noyd, *J. Am. Chem. Soc.*, **95,** 8203 (1973); H. E. Ziegler, I. Angres, and D. Mathisen, *ibid.*, **98,** 2580 (1976).

112. G. A. Russell, A. Ito, and D. G. Hendry, *J. Am. Chem. Soc.*, **85,** 2976 (1963).

113. G. A. Russell and H. C. Brown, *J. Am. Chem. Soc.*, **77,** 4578 (1955); G. A. Russell, in *Free Radicals,* J. K. Kochi, Ed., Wiley, New York, 1973, Vol. I, p. 291.

114. W. E. Bachmann and R. A. Hoffman, *Org. React.*, **2,** 224 (1947); G. H. Williams, *Homolytic Aromatic Substitution,* Pergamon press, London, 1960; W. A. Pryor, *Free Radicals,* McGraw-Hill, New York, 1966, p. 251. See also J. I. G. Cadogan, *Acc. Chem. Res.*, **4,** 186 (1971).

115. C. S. Rondestvedt, *Org. React.*, **11,** 189 (1960).

116. For other diazonium salt decompositions via phenyl cations, see H. Zollinger, *Acc. Chem. Res.*, **6,** 338 (1973); H. Zollinger et al., *J. Am. Chem. Soc.*, **98,** 3301 (1976); *ibid.*, **100,** 2811, 2816 (1978); *Helv. Chim. Acta,* **60,** 1079 (1979); H. Zollinger, *Chem. Brit.*, **16,** 257 (1980); C. G. Swain et al., *J. Am. Chem. Soc.*, **97,** 783, 791, 796, 799 (1975); R. C. Horvat, S. D. Miller, and S. A. Safron, *ibid.*, **98,** 8274 (1976); H. B. Ambroz and T. J. Kemp, *Chem. Commun.*, **1982,** 172.

117. M. J. Roedel, *J. Am. Chem. Soc.*, **75,** 6110 (1953); A. H. Willbourn, *J. Polym. Sci.*, **34,** 569 (1959).

118. D. H. R. Barton and J. M. Beaton, *J. Am. Chem. Soc.*, **83,** 4083 (1961).

119. The radical formed by abstraction in the position indicated is not only tertiary, but also donor–acceptor stabilized. For the structures of these and related polymers, see, for example, K. J. Saunders, *Organic Polymer Chemistry,* Chapman and Hall, London, 1973, pp. 122 and 107; E. S. Huyser, *Free Radical Chain Reactions,* Wiley, New York, 1970, Chapter 12.

120. J. C. Martin, in *Free Radicals,* J. K. Kochi, Ed., Wiley, New York, 1973, Vol. II, p. 493; T. Koenig, *ibid.*, Vol. I, p. 113; P. Livant and J. C. Martin, *J. Am. Chem. Soc.*, **98,** 7851 (1976). A bridged radical is formed in an anchimerically assisted homolysis reaction:

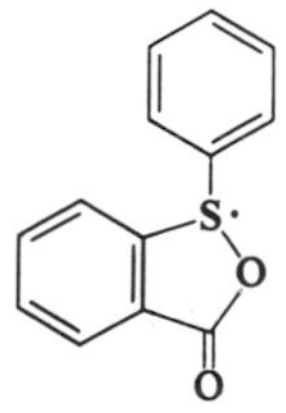

A *p*-methylthio-group has no comparable effect ($k_{rel} \sim 1$).

121. P. R. Singh and R. Kumar, *Aust. J. Chem.*, **25,** 2133 (1972).

122. C. Walling and A. Cioffari, *J. Am. Chem. Soc.*, **92,** 6609 (1970):

$$RMgBr + O_2 \longrightarrow R\cdot + O_2^{-\cdot} + \overset{+}{M}gBr$$

The free radical then cyclizes. Note, however, that, in the absence of oxygen, carbanions may also cyclize to a small extent: J. F. Garst, J. A. Pacifici, C. C. Felix, and A. Nigam, *J. Am. Chem. Soc.*, **100,** 5974 (1978) and references therein.

123. M. Julia, *Pure Appl. Chem.*, **40,** 553 (1974).

Chapter 3

DIRADICALS

In its present stage of evolution, chemistry compromises between the abstract principle and the naïve pictorial hypothesis.

—C. N. Hinshelwood, *The Structure of Physical Chemistry*, Clarendon Press, Oxford, 1958.

3.1. TRIMETHYLENES

The naïve pictorial representation of the singlet trimethylene diradical (**1**) suggests that this species should be a reactive intermediate residing in an energy well. However, the conclusion from highly sophisticated theoretical calculations[1] is that singlet trimethylene occupies an energy maximum; that is, it is a transition state. On the other hand, some *triplet* 1,3-diradicals (e.g., **2**) have been directly observed by ESR spectroscopy and are, therefore, true intermediates.[2]

CH_2

$\uparrow\dot{C}H_2$ $\dot{C}H_2\downarrow$

1 **2**

The first theoretical calculations on trimethylene, using the extended Hückel method,[3] predicted that the opening of a C—C bond in cyclopropane (**3**, only one of the Walsh orbitals[4] shown for clarity) could lead either to a "90,90"-intermediate (**4**) or a "0,0"-intermediate (**5**), these designations referring to the angles of twist of the terminal methylene groups with respect to the plane described by the three carbon atoms:

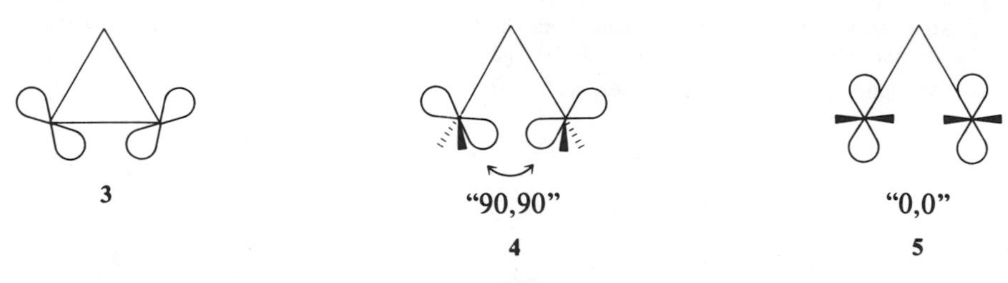

The "90,90"-intermediate corresponds merely to a cyclopropane with a stretched C—C bond, and it should collapse to cyclopropane within a molecular vibration. The "0,0"-intermediate may be described as a "π-cyclopropane," and the calculations[3] predicted this species to be a true intermediate. It was further found[3] that the HOMO of the "0,0"-intermediate corresponds to an antisymmetric arrangement of orbitals (A), whereas the LUMO is the corresponding symmetric one (S). Consequently, the reclosure of trimethylene to cyclopropane is expected to take place by a conrotatory motion of the terminal methylene groups:

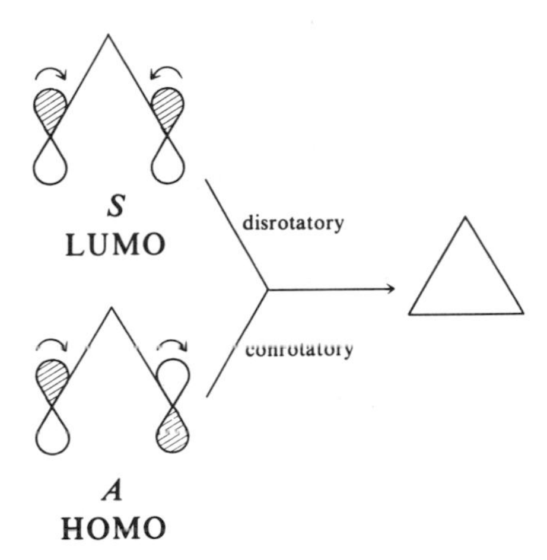

The principle of microscopic reversibility then demands that the ring opening of cyclopropane should also take place in a conrotatory manner and that the predominant thermal reaction of 1,2-disubstituted cyclopropanes should be double inversion:

X Y conrotatory Y X

Although more sophisticated theoretical treatments[1] indicate that the "0,0"-intermediate is an energy maximum (transition state), they do predict preferred double inversion by conrotatory motion. It should be noted that the preference for conrotation over disrotation is calculated to be very small (0.5–1 kcal/mol). Therefore, it is not surprising that other factors, such as steric ones, may outweigh the small preference for conrotation. Thus certain *cis*-disubstituted cyclopropanes undergo thermal double rotation in a disrotatory manner, presumably because this engenders less steric interaction between the groups:[5]

X Y disrotatory X Y

Experiments with deuterated cyclopropanes have revealed two types of thermal isomerization: *cis–trans* isomerization and structural rearrangement to propene:

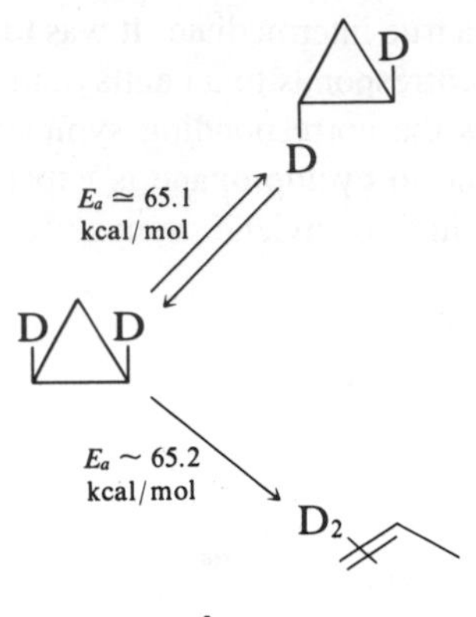

Scheme 1

Some kinetic data, useful in the subsequent discussion, are given in Table 3.1.

If the species **5** represents the transition state for *cis–trans* isomerization, then this must occur by a *concerted rotation* of two methylene groups, irrespective of the question of conrotatory or disrotatory motion. Consequently, the optically active cyclopropane **6** should undergo optical isomerization to **7**, resulting in racemization, by cleavage of the C1—C2 bond.

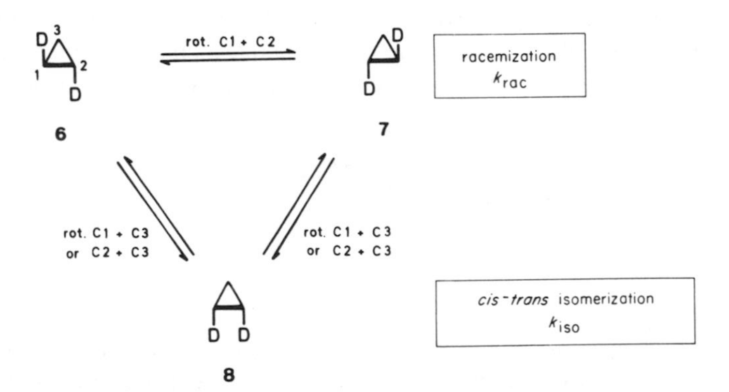

Since cleavage may also occur at the C1—C3 or C2—C3 bonds, *cis–trans* isomerization to **8** will also be observed, and the ratio of rate constants k_{rac}/k_{iso} for the two processes should equal 1. Indeed, this was found experimentally.[8] Had the trimethylene diradical **1** been an *intermediate* in these reactions, we should have expected the two terminal methylene groups to rotate independently, yielding $k_{iso}/k_{rac} = 1.5$. A *single* methylene rotation in **6** (with the rest of the molecule static) would have given $k_{iso}/k_{rac} = 2$. The experimental evidence[8] thus strongly supports the *concerted rotation* of two methylene groups in the opening and re-closure of cyclopropane. Unfortunately, the situation is not as clear-cut for other, substituted,

TABLE 3.1. Thermal Decomposition of Compounds Giving Formal 1,3-Diradical Intermediates or Transition States

Reaction	Reference	E_a (kcal/mol)	log A	$\Delta S^{\ddagger}_{473}$ (cal K^{-1} mol^{-1})
	[a]	65.1	16.11	12.3
	[a]	64.2	15.7	10.4
	[a]	65.2 ± 0.2	15.14 ± 0.03	7.8
	[b]	64.0 ± 2	14.52	5.0
 N=N	[c]	42.4 ± 0.3	15.93 ± 0.13	11.4
 N=N	[c]	41.0 ± 0.3	15.70 ± 0.15	10.4
 N=N	[c]	40.3 ± 0.3	15.54 ± 0.11	9.7
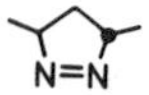 N=N	[c]	40.2 ± 0.3	15.67 ± 0.11	10.3
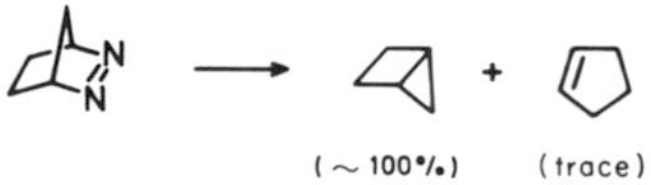 N N (~100%) (trace)	[c]	36.9 ± 0.2	14.74 ± 0.10	6.0
	[d]	37.3 ± 0.3	14.90 ± 0.10	6.7
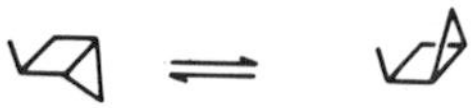	[e]	38.9 ± 0.8	14.45	4.7
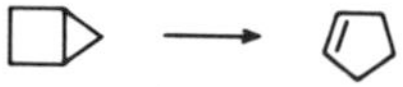	[f]	45.6	14.1	3.1

[a]Values quoted in Ref. 6

[b]D. M. Kulick, J. E. Taylor, and D. A. Hutchings, *Can. J. Chem.*, **52**, 216 (1974) (wall-less reactor).

[c]Reference 7.

[d]S. G. Cohen, R. Zand, and C. Steel, *J. Am. Chem. Soc.*, **83**, 2895 (1961).

[e]J. P. Chesick, *J. Am. Chem. Soc.*, **84**, 3250 (1962).

[f]C. Steel, R. Zand, P. Hurwitz, and S. G. Cohen, *J. Am. Chem. Soc.*, **86**, 679 (1964).

cyclopropanes; although double rotation is often reported to dominate, there are several examples of single rotation, mixtures of single and double rotation, and random intermediate [i.e., trimethylene (**1**)].[9]

If singlet trimethylene (**1**) is not an intermediate in the optical and *cis–trans* isomerization of cyclopropane, it is hardly so either in the structural isomerization to propene (Scheme 1), which has nearly the same activation energy but a lower *A*-factor, signifying geometrical constraints in the transition state, where a 1,2-hydrogen shift must occur concertedly with the ring opening. We may thus represent these two reactions as shown in Figure 3.1.

The "heat of formation" of the trimethylene *transition state* is obtained from the known heat of formation of cyclopropane plus the activation enthalpy for *cis–trans* isomerization:

$$\Delta H^{\ddagger}_{298} = E_a - RT = 65.1 - 0.6 = 64.5 \text{ kcal/mol}$$

In the next section, we shall see that thermochemical arguments for the existence of the *singlet trimethylene intermediate* in an energy well approximately 10 kcal/mol deep had been advanced[6] for the very reactions that were described as concerted in Figure 3.1. Analogously, the isomerization of methylcyclopropane to 1-butene was explained by the diradical mechanism:

Δ H

9 **10**

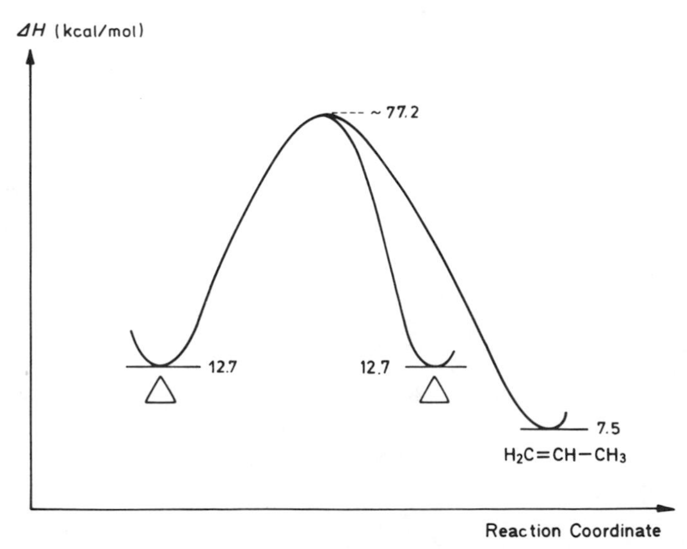

Figure 3.1. Energy profile for isomerization of cyclopropane.

However, there is experimental evidence that such hydrogen shifts can occur in a concerted manner, not involving discrete diradical intermediates. Thus the isomerization of 2,4-dehydrohomoadamantane (**11**) to homoadamantene (**12**) proceeded with undiminished rate (80 times faster than **9** → **10**):[10]

11 12

• = ^{13}C

In this case, structural constraints would force a hypothetical diradical (**13**) to adopt a "90,90"-conformation of the type shown in **4**:

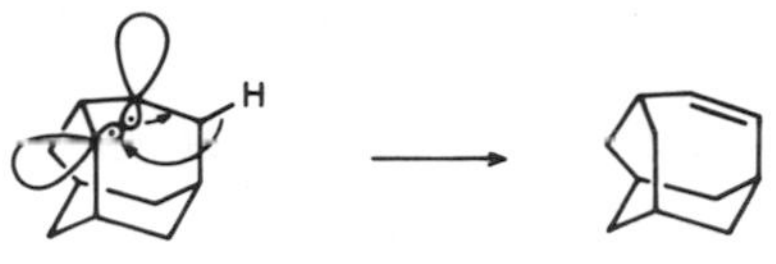

13 12

As mentioned previously, such a species is not expected to be an intermediate since it should collapse to cyclopropane without activation energy.

Whereas the weight of the evidence excludes the singlet-**1** intermediate, *there is evidence that the triplet trimethylene exists*. Irradiation of the azo-compound **14** at 5.5 K produced the triplet diradical **15** which could be directly observed by ESR spectroscopy. Since the signal persisted near the absolute zero (1.3 K), the triplet

hν, 5.5 K Δ ISC

14 15 16

is very likely the ground state. On warming, the diradical **15** cyclized to **16**, a process that requires spin inversion [intersystem crossing (ISC)]. From the temperature dependence of the disappearance of **15**, it could be determined that the barrier height is no more than 2 kcal/mol.[2] When the reaction was performed at room temperature using benzophenone as a triplet sensitizer, CIDNP was observed for **16** (all protons showing enhanced absorption), again indicating *triplet* **15** as the precursor.[2]

We can now draw an energy diagram for trimethylene (Figure 3.2) with the singlet as a transition state, the triplet as a true minimum. Mixing of the singlet and triplet states can, in principle, occur where the two curves cross. Therefore, the conversion of triplet **15** to singlet **16**, for example, does not require a singlet diradical intermediate.

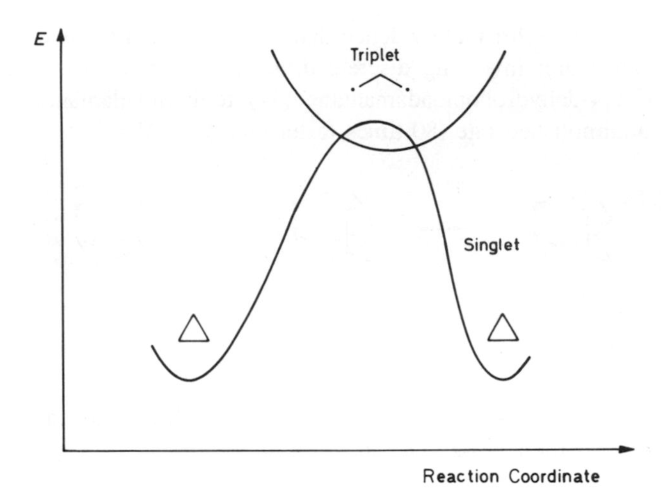

Figure 3.2. The trimethylene energy profile.

A trimethylene diradical stabilized by benzylic resonance **(18)** is obtained by photolysis of the cyclopropane **17**:[11]

$h\nu$, 77 K; 120 K

17 **18** **19**

The triplet **18**, which in all likelihood is the ground state,[12] has a relatively small value of the zero-field splitting constant, $D/hc = 0.026 \text{ cm}^{-1}$, due to delocalization into the naphthalene system (cf. **15**, 0.084 cm^{-1}). "Thermolysis" of triplet **18** at 120 K results in the "cyclopropane-propene" rearrangement, giving **19**, a reaction again requiring a crossing to the singlet energy surface.

Delocalization is also the reason for the relative ease of observation of the triplet trimethylenemethane diradical **(20)** and its derivatives:

N=N; $h\nu$, $-N_2$; C; $h\nu$, $-$CO; O

20

The low D/hc value (0.025 cm^{-1}) and the observation that the ESR spectrum consists of seven equidistant lines (coupling with all six protons) are in agreement with a delocalized structure:[13]

20

The disappearance of this species, as observed by ESR spectroscopy, takes place with an activation energy of 7 kcal/mol. However, recent theoretical calculations give values between 14.9 and 17.6 kcal/mol for the singlet–triplet splitting, the triplet being the ground state.[14] Although the discrepancy is disturbing, perhaps the experimental and theoretical values will converge on each other in future determinations, just as has been the case with the singlet–triplet splitting in methylene (see Section 4.1).

The cyclic trimethylenemethane **22** has been particularly well studied.[15] Photolysis of the azo-compound **21** gives first the *singlet* diradical **22**-S, which appears to be a discrete entity, decaying to the triplet **22**-T with an activation energy of less than 2.3 kcal/mol.[16] An excited state of this singlet has been directly observed by picosecond fluorescence spectroscopy.[17] Furthermore, **22**-S undergoes ring closure to the bicyclo[2.1.0]pentane **23**, an unstable compound which is the major product at $-78°C$ and ring-opens again above $-50°C$ to give, ultimately, the dimers of the ground-state triplet (**22**-T).[16] **23** can be trapped with acrylonitrile, and the kinetics of this system demonstrates that it is a two-step process; that is, **23** reverts to a diradical **22**-S before reacting with acrylonitrile to give the products **24** and **25** (Scheme 2).[18]

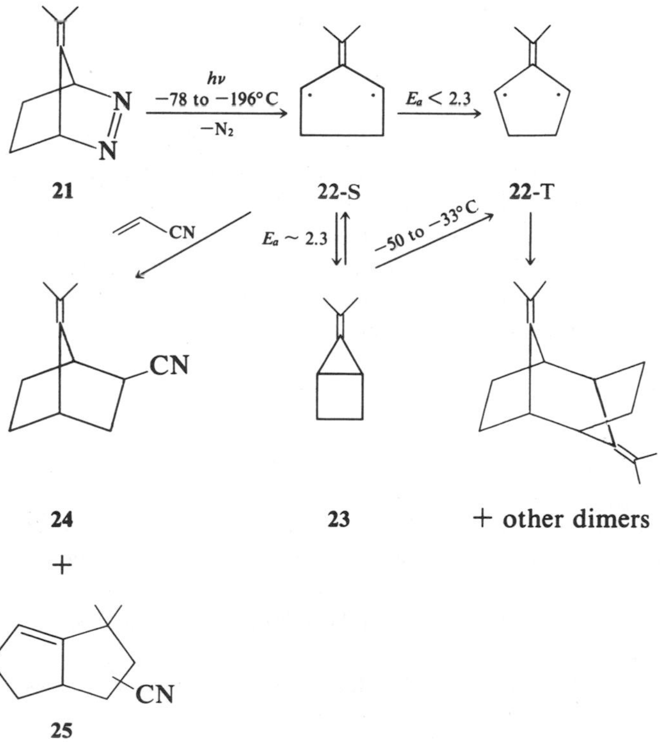

Scheme 2 (energy values in kcal/mol)

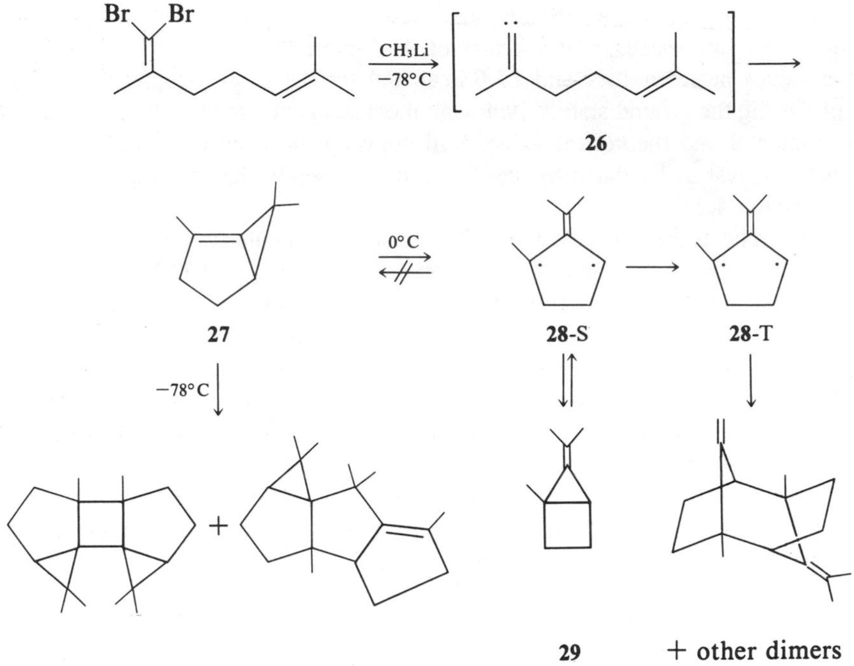

Br Br

CH_3Li, $-78°C$

26

0° C

27

28-S

28-T

$-78°C$

29 + other dimers

Scheme 3

Another interesting molecule, the bicyclo[3.1.0]hex-1-ene derivative **27** can be generated by cyclization of the carbene **26**. The intriguing chemistry summarized in Scheme 3 demonstrates that **27** enters the trimethylenemethane energy surface at about 0°C, giving **28**-S and then **28**-T. The bicyclopentane **29** can be observed by NMR spectroscopy when **27** is generated at $-30°C$.[19] Interestingly, the diradical **28**-T appears to be of lower energy than either of the valence-satisfied hydrocarbons **27** and **29**. In other words, excessive strain leads to formally negative bond dissociation enthalpies for the bridge bonds in **27** and **29**.

3.2. DIRADICALS AND THERMOCHEMISTRY

Thermochemical calculations[6] indicate that trimethylene (**1**) should be a stable species. Its heat of formation was estimated as outlined in Scheme 4:

$$\Delta H_f^0(\mathbf{1}) = 98 - 52.1 + 20.8 = 66.7 \text{ kcal/mol}$$

However, the transition state for the *cis–trans* isomerization of cyclopropane lies at

$$\Delta H^0 = \Delta H_f^0(\text{cyclopropane}) + \Delta H^{\ddagger}_{\text{isom}} \cong 77.2 \text{ kcal/mol}$$

$$\underset{\Delta H_f^0:\ -25.1}{\mathrm{H{-}CH_2{-}CH_2{-}CH_2{-}H}} \xrightarrow[98]{DH^0} \underset{20.8}{\mathrm{H{-}CH_2{-}CH_2{-}\dot{C}H_2}} + \underset{52.1}{\mathrm{H\cdot}}$$

$$\downarrow DH^0\ (98)$$

$$\underset{\mathbf{1}}{\mathrm{\dot{C}H_2{-}CH_2{-}\dot{C}H_2}} + \underset{52.1}{\mathrm{H\cdot}}$$

Scheme 4 (all enthalpy values in kcal/mol)

It thus appears that **1** is stable with respect to closure to cyclopropane, the barrier to this process being approximately 10 kcal/mol.

Why is this conclusion wrong? The procedure is Scheme 4 implies that the dissociation energies for the terminal CH_2—H bonds in propane and the propyl radical are identical. Only the first one is known (Table 2.2). It was assumed[6] that the dissociation energy for the second CH_2—H bond in the propyl radical, giving **1**, would be the same. However, the calculation ignores the existence of electronic states and produces an average of singlet and triplet states. Since there are three components of the triplet but only one singlet, what is calculated is a species lying a quarter of the S–T splitting above the triplet ground state.[20] As we have seen, the terminal methylene groups in the trimethylene singlet are not ignorant of each other, nor are they expected to be so in the triplet. This would almost certainly make the second terminal CH_2—H bond stronger than the first. Estimates of singlet–triplet splittings range from about 2 kcal/mol in **15** to 14–17 kcal/mol in **20**. This, together with a strengthening of the terminal CH_2—H bond in the propyl radical with respect to that in propane would bring the thermochemical estimate in line with the quantum mechanical and experimental results, according to which singlet trimethylene is not at an energy minimum.

In a similar manner we could calculate the heat of formation of a hypothetical 1,2-diradical (**30**):

$$\mathrm{CH_3{-}CH_3} \xrightarrow[98]{DH^0} \underset{25.7}{\mathrm{CH_3{-}\dot{C}H_2}} \xrightarrow[(98)]{DH^0} \underset{\mathbf{30}}{\mathrm{\dot{C}H_2{-}\dot{C}H_2}} + \underset{52.1}{\mathrm{H\cdot}}$$

$$98 = \Delta H_f^0\ (\mathbf{30}) + 52.1 - 25.7$$

$$\Delta H_f^0\ (\mathbf{30}) = 71.6\ \mathrm{kcal/mol}$$

It is known that geometrical isomerization of ethylene takes place thermally, with an activation energy of 64 kcal/mol:

$$\textit{trans}\text{-DHC=CHD} \underset{}{\overset{E_a = 64\ \mathrm{kcal/mol}}{\rightleftarrows}} \textit{cis}\text{-DHC=CHD}$$

The heat of formation of ethylene is 12.5 kcal/mol. The transition state for its isomerization thus lies at approximately 64 + 12.5 = 76.5 kcal/mol, and it would appear that **30** is about 5 kcal/mol more stable than this transition state.

Again, the two bond dissociation energies in ethane and ethyl are not likely to be identical, and the heat of formation calculated for **30** will not represent a real

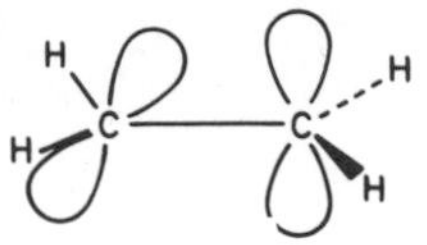

31

molecule. The transition state for geometrical isomerization of ethylene will resemble **31**, in which the two *p* orbitals are orthogonal. An infinitesimal rotation of one of the groups will lead to ethylene. The momentum necessary for such an infinitesimal rotation will be abundantly available since **31** itself is produced by just such a rotation in ground-state ethylene. It would be strange if the rotation were to stop at the crest of the energy barrier, 64 kcal/mol above ethylene.

One can also estimate[6] the heat of formation of a 1,4-diradical:

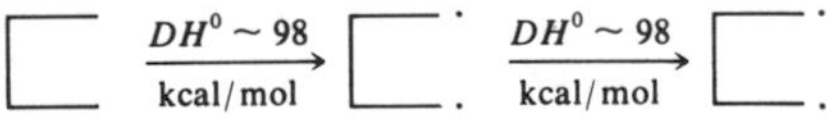

$DH^0 \sim 98$ kcal/mol; $DH^0 \sim 98$ kcal/mol

32

One obtains ΔH_f^0(**32**) ≅ 62 kcal/mol.

The cleavage of cyclobutane has $E_a \cong 62.5$ kcal/mol. Since ΔH_f^0(cyclobutane) = 6.3 kcal/mol, **32** should lie in an energy well about 7 kcal/mol deep. The assumption of identical bond dissociation energies is more justified in this example: The radical centers are further apart, and the stable conformation of the diradical is likely to be the *trans*-form **33** in which there is little interaction between the two

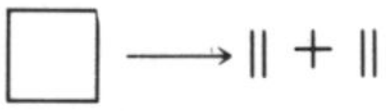

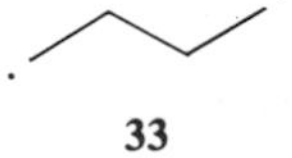

33

electrons. In the case of tetramethylene (**33**), it is possible that discrete singlet and triplet diradicals exist, although the singlet energy barrier is probably much less than 7 kcal/mol. Interestingly, quantum-chemical calculations agree in this case that the singlet represents a minimum[21] (Figure 3.3). However, more elaborate calculations[22] indicate that only **33**, not **32**, represents a true energy minimum, and that the barrier separating **33** from cyclobutane is about 4.7 kcal/mol.

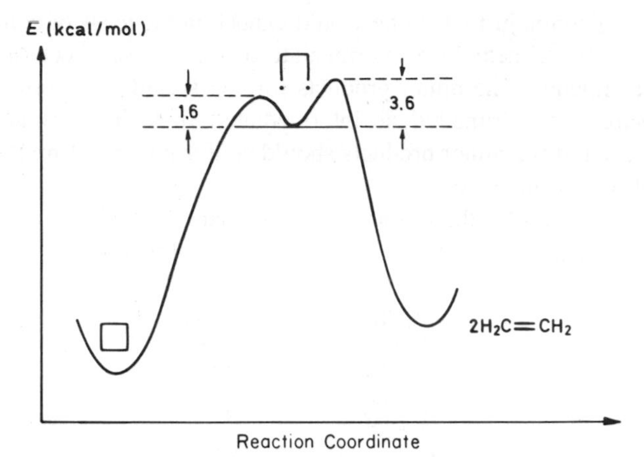

Figure 3.3. Energy surface for the cleavage of cyclobutane. (Adapted with permission from Ref. 21. Copyright 1974 American Chemical Society.)

3.3. PYRAZOLINES

1-Pyrazoline (**34**) and its substituted derivatives decompose thermally to give the same types of products obtained on pyrolysis of cyclopropane. One could reasonably expect, therefore, that a common trimethylene intermediate or transition state (**35**) would be involved:[23,24]

4
5 3
N=N
1 2 $\xrightarrow{\Delta}$ $[\cdot\wedge\cdot]^{\ddagger}$ $\longrightarrow$ $\triangle$ + propene

34 **35**

However, the results of stereochemical studies do not allow such a clear-cut conclusion. Many pyrazolines have been found to form cyclopropanes with predominant *single inversion* as illustrated for the following 3,5-dimethyl compounds:

H_3C CH_3 H_3C H
H N=N H H N=N CH_3

cis *trans*

minor major minor

H_3C CH_3 H_3C H
H H H CH_3

This single inversion is just what one would expect on the basis of conrotatory ring closure in a "0,0"-trimethylene intermediate of type **5** (see Section 3.1) as calculated by Hoffmann.[3] The minor products formed would then correspond to disrotatory closure in the trimethylene intermediate. Since **5** is planar, the stereochemistry of each of the minor products should conform to a 1:1 mixture of double retention and double inversion.

This is contradicted by the observed predominant ($\leq 87\%$) double retention in the thermolysis of the optically active diphenylpyrazoline **36**:[25]

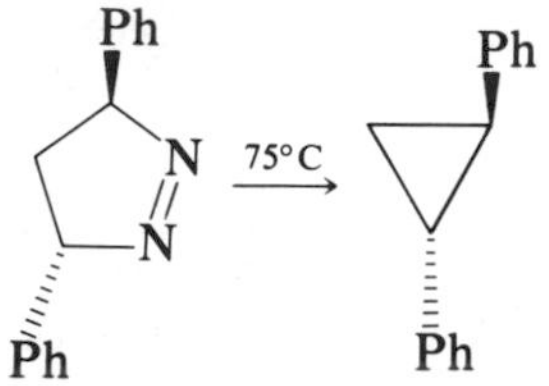

Ph N N 75° C Ph Ph Ph

36

Other studies using optically active *cis*- and *trans*-3-ethyl-5-methylpyrazolines also demonstrated retention of optical activity in the products, but both double retention, double inversion, and single inversion were occurring.[26] The conclusion from this study[26] was that none of the mechanisms put forward for pyrazoline decomposition can, per se, explain all the facts.

It is necessary to consider the possibility that pyrazolines undergo initial one-bond homolysis to give a diazenyl diradical, thereby bypassing the trimethylene species:

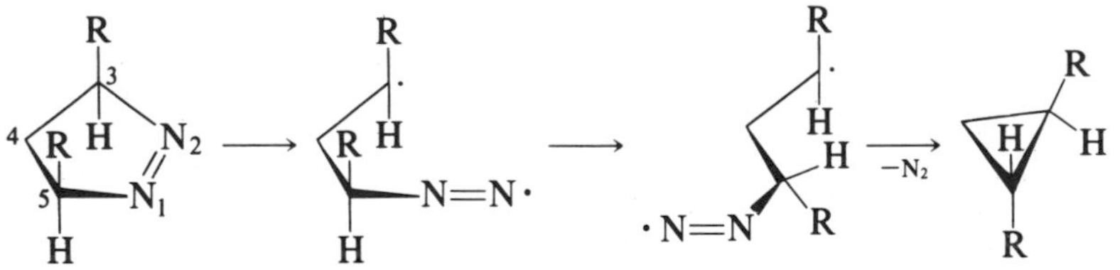

R 3 4 R H N_2 5 N_1 H → R R H N=N· H → R H H ·N=N R → $-N_2$ R H H R

37 **38** **39** **40**

Backside homolytic displacement of nitrogen in the diazenyl diradical **39** would lead to the single inversion observed in many systems (**37** → **40**). In fact, theoretical calculations (using STO-3G and 4-31G basis sets) predict this two-step path to be the lowest energy one.[27] Depending on the exact conformation of the diradical **39** and the degree of rotation at the carbon radical center C-3, varying degrees of single inversion, double inversion, and double retention are possible. The two-step mechanism could thus be flexible enough to accommodate the varied experimental observations.

There is no lack of evidence *against* the two-step mechanism just mentioned:

Each successive methyl group in the 3- and 5-positions in pyrazoline causes a decrease in the activation energy by approximately 1 kcal/mol (see Table 3.1). Furthermore, each successive vinyl substituent in compounds **41–43** causes a decrease of about 8 kcal/mol in the energy of activation.[28] These data are in agreement with rate-determining formation of a trimethylenelike species that can profit from double allylic resonance.

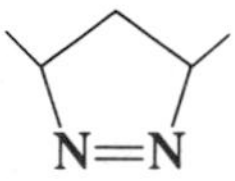

N=N N=N N=N
41 **42** **43**
E_a: 40 31 23 kcal/mol

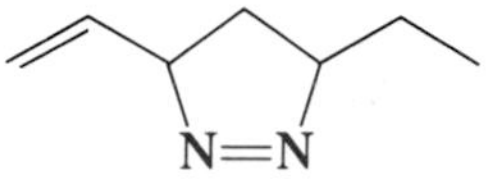

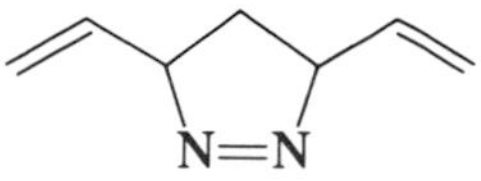

The most convincing evidence for a *two-step* mechanism via the diazenyl diradical comes from an examination of the secondary deuterium kinetic isotope effect on the decomposition of the 4-ethylidenepyrazolines **44** and **45**, which leads to the conclusion that in each case the C—N bond *anti* to the ethylidene group is broken preferentially.[28]

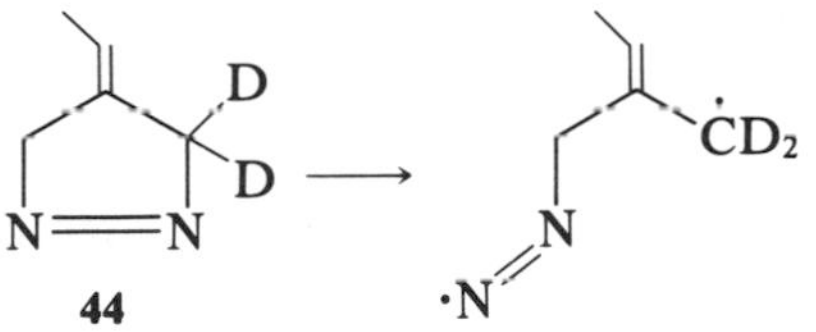

44 → $\dot{C}D_2$ / ·N=N

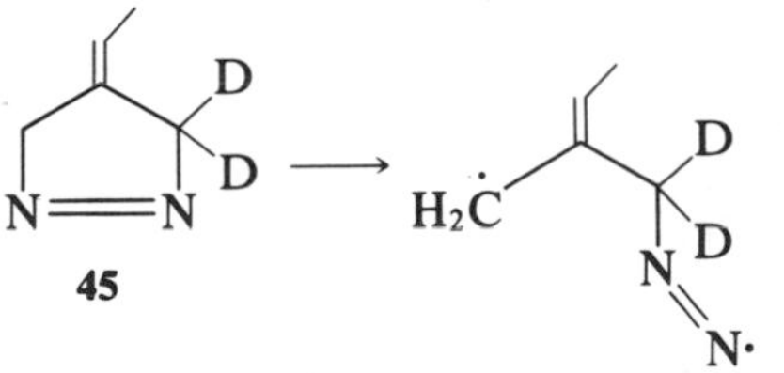

45 → $H_2\dot{C}$ / D D / N=N·

Evidence for a two-step mechanism has also been found in the oxadiazoline **46.**[29] Although this system does not necessarily have any bearing on the mechanism of pyrazoline decomposition, the connection is interesting. The slow step is assumed to be the ring-opening to the diradical **47** (rate constant k_1). When R = phenyl, the carbon radical in **47** will be stabilized, thereby leading to a low value of k_2, the rate constant for intramolecular hydrogen abstraction giving the product **48**. As a consequence, k_3 (which is little affected by the stability of the carbon radical) will dominate, and the major product is the oxirane **51**, presumably formed via diradical **49** and zwitterion **50**. When, however, R = CH_3, the carbon radical in **47** will be more reactive and hydrogen abstraction faster, and **48** is now the main

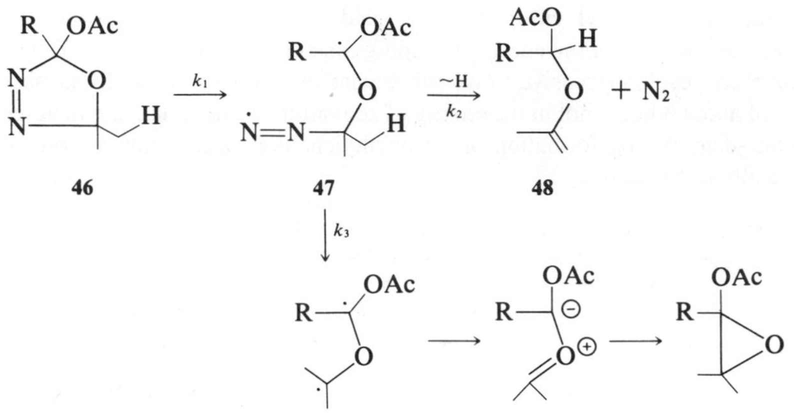

46 47 48

49 50 51

product found. A direct reaction **46** → **48** was excluded due to the lack of a deuterium isotope effect when the *gem*-dimethyl groups in **46** were replaced by CD_3.

In conclusion, the mechanism of pyrazoline decomposition is not fully understood, but it seems likely that, depending on substituents, both concerted and nonconcerted nitrogen elimination can occur.

3.4. 2,3-DIAZABICYCLO[2.2.1]HEPT-2-ENE

As mentioned earlier, low-temperature photolysis of 2,3-diazabicylo[2.2.1]hept-2-ene **(14)** gives the triplet diradical **15**. The reaction has been studied both thermally

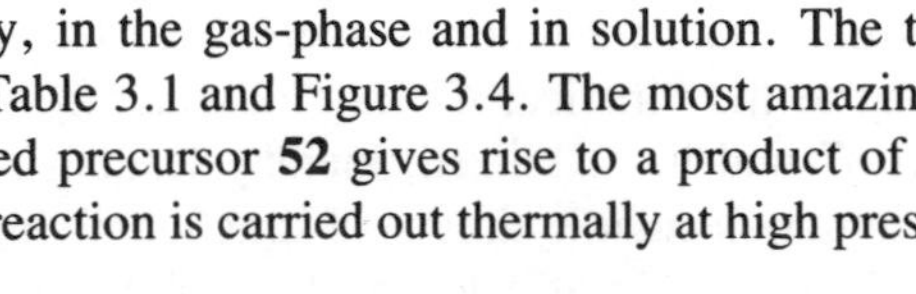

14 15 16

and photochemically, in the gas-phase and in solution. The thermochemical data are summarized in Table 3.1 and Figure 3.4. The most amazing observation is that the deuterium labeled precursor **52** gives rise to a product of predominant double inversion when the reaction is carried out thermally at high pressure (100–200 Torr) in the gas-phase:[30]

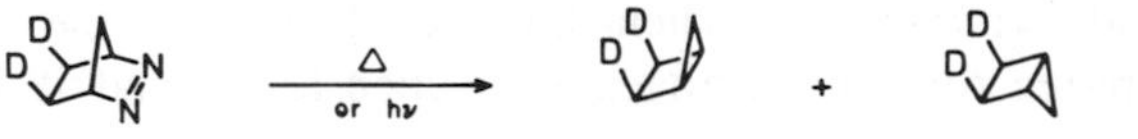

52 *syn* – 53 *anti* – 53

Syn/Anti Ratio for **53**	Conditions
1:0.5	$h\nu$, solid phase, −70°C
1:1.5	$h\nu$, solution, or gas-phase at 350–760 Torr
1:3	Δ, 180°C, 100–200 Torr
1:1	$h\nu$, gas-phase, 1–2 Torr

Although a complete explanation of these and other, similar reactions[31,32] is not in sight, some rationalizations can be made. First, the predominant double retention on photolysis of crystalline **52** at −70°C can perhaps be understood in terms of the geometric constraints imposed by the crystal lattice. Second, the predominant double inversion observed in fluid solution or in the gas-phase at relatively high pressures can be understood in terms of two-step cleavage via the diazenyl diradical **54.** Backside displacement of nitrogen gives rise to the required inversion. In

52 → **54** $\xrightarrow{-N_2}$ *anti*-**53**

contrast to the monocyclic pyrazolines, steric constraints make single inversion impossible.

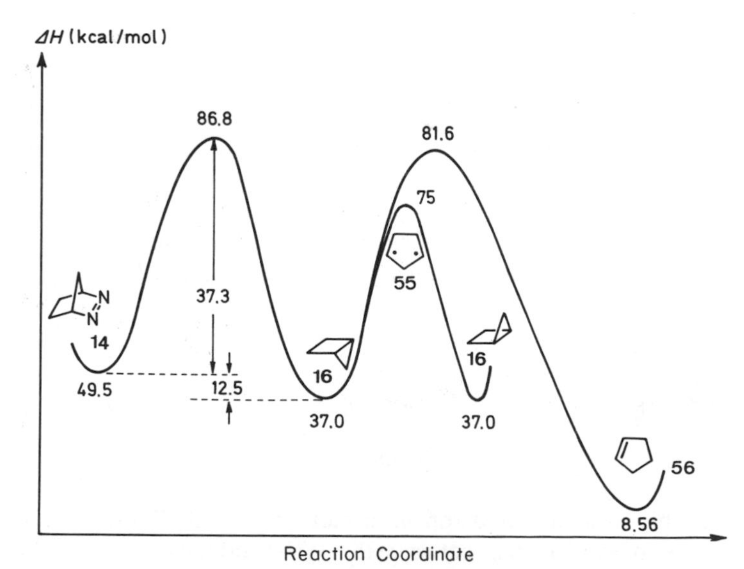

Figure 3.4. The $C_5H_8N_2$ energy profile.

If the gas-phase reaction is carried out photochemically, at lower pressure, (1–2 Torr), the initial product *anti*-**53** will carry excess vibrational energy, originating either from the irradiating photons or from the energy of activation for the process. As can be seen from Figure 3.4, the transition state for the decomposition of **14** lies at approximately 86.8 kcal/mol. The heat of formation of **16** is 37 kcal/mol, and the transition state for inversion in **16** lies at about 75 kcal/mol. Consequently, the initially formed **16** carries an excess energy of 86.8 − 37.0 = 49.8 kcal/mol, more than sufficient to cross the inversion barrier of about 38 kcal/mol. The excess energy will be dissipated rapidly by collisions at higher pressures, but not at lower ones. This is the phenomenon of chemical activation, which will be described more fully in Section 4.5. We may thus assume that, irrespective of the mechanism of decomposition of **14**, the low-pressure gas-phase photolysis will lead to rapid inversion in **16**. Evidence for the operation of the chemical activation effect is found in the formation of cyclopentene (**56**, see Figure 3.4) in the low-pressure reaction. This requires a transition-state energy of 81.6 kcal/mol, barely lower than the first activation barrier of 86.8 kcal/mol. Evidence for a CA effect in the thermolysis of a related bicyclic pyrazoline has also been reported.[32]

3.5. ALLOWED AND FORBIDDEN CYCLOREVERSIONS

A comparison of the decomposition rate constants for a series of polycyclic pyrazolines (Table 3.2) demonstrates a remarkable effect of structure. The very fast decompositions of **57, 60, 61,** and **62** indicate *concerted and orbitally allowed reations*. For example, the reaction of **62** is a $\sigma^2 + \sigma^2 + \pi^2$ cycloreversion, ''allowed'' by orbital symmetry, and occurring via a six-electron aromatic transition state:

6e ‡ + N_2

62

Likewise, thanks to the bulging Walsh orbitals in cyclopropane, the cyclopropane ring can act like a π-bond in the concerted fragmentation of **57**:

‡ + N_2

57 63

In order for this backside cyclopropane π-participation in **57** to be effective, there must be a reasonable overlap between the Walsh orbitals and the developing *p* orbitals at C-1 and C-4. In other words, the transition state **63** must be able to

TABLE 3.2. Polycyclic Pyrazolines

Compound	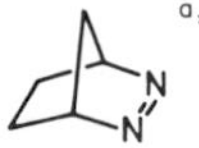 14 [a,b]	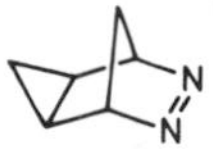 57 [b]	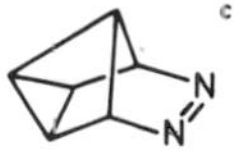 58 [c]	 59 [b]	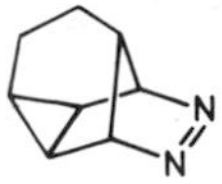 60 [b]	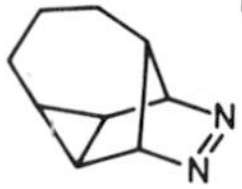 61 [b]	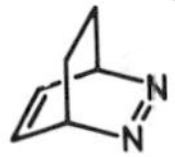 62 [d]
k_{rel}	1	2.2×10^{11}	7.1	9.2	5.2×10^{8}	1.9×10^{11}	Very fast
E_a(kcal/mol)	37.3	17.7	36.6	36.5	23.3	19.6	≤17
$\Delta S^{\ddagger}$ (cal K^{-1} mol^{-1})	6–7	−13	7.8	8.3	−5	−6	
Product(s)	 16		 (8%)  + others		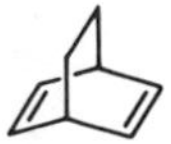		

[a]See Table 3.1.

[b]E. L. Allred and A. L. Johnson, *J. Am. Chem. Soc.*, **93**, 1300 (1971).

[c]T. J. Katz and N. Acton, *J. Am. Chem. Soc.*, **95**, 2738 (1973); N. J. Turro, C. A. Renner, W. H. Waddell, and T. J. Katz, *ibid.*, **98**, 4321 (1976).

[d]N. Rieber, J. Alberts, J. A. Lipsky, and D. M. Lemal, *J. Am. Chem. Soc.*, **91**, 5668 (1969).

flatten to some extent. This requirement is satisfied in **60** and **61,** where the polymethylene bridge pushes the two C_3-bridges apart. However, in **58** and **59** the bridge is too short for the transition state to flatten; hence the cyclopropane Walsh orbitals will be almost orthogonal to the developing *p* orbitals at C-1 and C-4. Compound **58** gives prismane only on photolysis at 30°C. Since the cyclopropane ring remains intact in the prismane, no participation can be involved in the transition state leading to this product. The products of ring opening, "Dewar-benzene" (bicyclo[2.2.0]hexa-2,5-diene) and benzene are also formed.

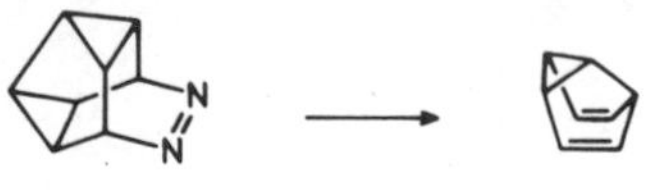

64 → 65

The azo-compound **64** has a geometry similar to that of **59,** but there are now two cyclopropane rings that can participate in the reaction, which occurs already during the preparation of **64,** giving semi-bullvalene **(65).**[33]

A cyclobutane ring can also be effective in π-participation, leading to rate enhancement, but the effect is less than for cyclopropane (Table 3.3). Note that all the *fast* reactions in Tables 3.2 and 3.3 have low values of $\Delta S^{\ddagger}$, implying "tight" transition states in which the electronic reorganizations occur in concert, in contrast to the nonsynchronous reaction of **14,** for example. Note also that an *endo*-cyclopropane ring has little or no effect on the rate (compound **66,** Table 3.3). Here, the Walsh orbitals project in the wrong direction so that no participation can take place.

Several compounds owe their existence and kinetic stability largely to the fact that their decomposition or rearrangement is "forbidden" by orbital symmetry. For example, 3,4-dihydro-1,2-diazete would be expected to fragment to ethylene and nitrogen in a strongly exothermic reaction:

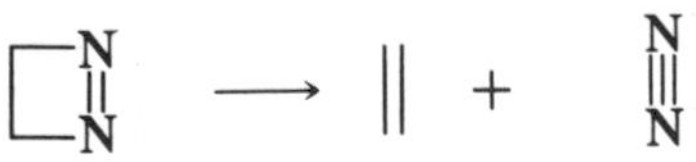

$\Delta H_f^0 \sim 65$ (estimated)   12.5   0 kcal/mol

However, if this reaction were concerted, it would constitute a thermally forbidden [2 + 2]-cycloreversion. The "forbiddenness" is reflected in the rather high activation energies for the decomposition of the following compounds (compare with the data in Table 3.2):[34]

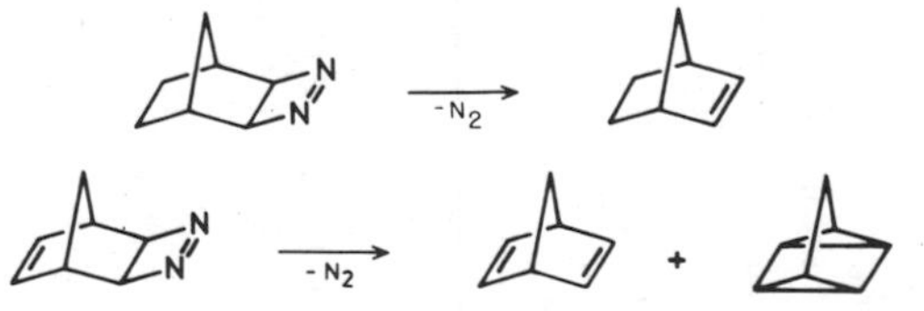

$-N_2$

$\Delta H^{\ddagger} = 33.7$ kcal/mol
$\Delta S^{\ddagger} = 3.0$ cal K^{-1} mol^{-1}

$-N_2$

$\Delta H^{\ddagger} = 33.1$ kcal/mol
$\Delta S^{\ddagger} = 5.4$ cal K^{-1} mol^{-1}

94 : 6

TABLE 3.3. Cyclobutane Participation[a]

Compound	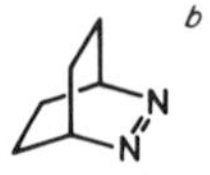 [b]	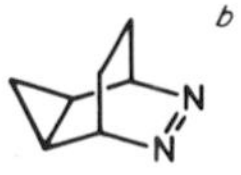 [b]	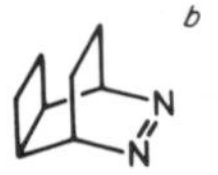 [b]	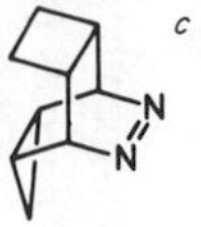 **66** [c]
k_{rel}	1	1.1×10^{17}	6.7×10^{4}	8.8×10^{2}
E_a (kcal/mol)	44.6	14.9[b] 2.3[d]	39.2	41.4
$\Delta S^{\ddagger}$ (cal K^{-1} mol^{-1})	+10.5	−21[b] +6.1[d]	+11	+10.3
Product	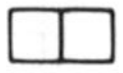		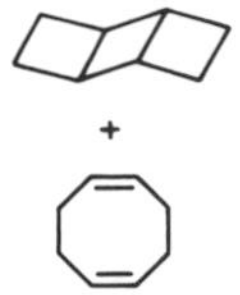 +	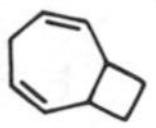

[a]For further examples, see K.-W. Shen, *J. Am. Chem. Soc.*, **93,** 3064 (1971); D. W. McNeil, M. E. Kent, E. Hedaya, P. F. D'Angelo, and P. O. Schissel, *ibid.*, **93,** 3816 (1971); L. A. Paquette and M. J. Epstein, *ibid.*, **95,** 6717 (1973); H. Schmidt, A. Schweig, B. M. Trost, A. B. Neubold, and P. H. Scudder, *ibid.*, **96,** 622 (1974); J. A. Berson, S. S. Olin, E. W. Petrillo, Jr., and P. Bickart, *Tetrahedron,* **30,** 1639 (1974); N. J. Turro, K. C. Liu, W. Cherry, J. M. Liu, and B. Jacobson, *Tetrahedron Lett.*, **1978,** 555; P. S. Engel, C. J. Nalepa, R. A. Leckonby, and W.-K. Chae, *J. Am. Chem. Soc.*, **101,** 6435 (1979); N. J. Turro, J.-M. Liu, H.-D. Martin, and M. Kunze, *Tetrahedron Lett.*, **1980,** 1299.

[b]E. L. Allred and J. C. Hinshaw, *Chem. Commun.*, **1969,** 1021.

[c]E. L. Allred and K. J. Voorhees, *J. Am. Chem. Soc.*, **95,** 620 (1973).

[d]J. P. Snyder and D. N. Harpp, *J. Am. Chem. Soc.*, **98,** 7821 (1976).

3.6. 1,4-DIRADICALS

A considerable body of experimental work demonstrates that six-membered cyclic azo-compounds such as **67,** 1,1-dialkyldiazenes **(68),** and cyclobutanes **(69)** afford the same thermal reaction products with the same stereochemistry.[35,36] Furthermore, the thermal cycloaddition of ethylene to *cis*- and *trans*-2-butene at 430°C and 12 atmospheres of pressure fits the same mechanistic scheme.[37] It is concluded that a

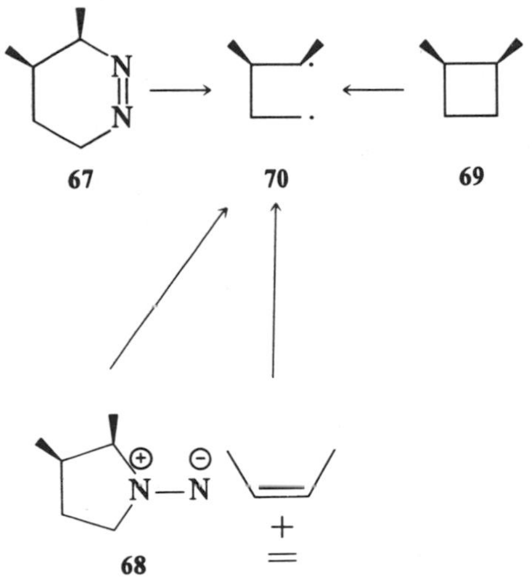

67 70 69

68

common intermediate, the 1,4-diradical **70,** is formed in all these reactions. The reactions of the azo-compounds **67** and **71** are complicated by the fact that a stereospecific and concerted cycloreversion to olefins and nitrogen competes with diradical formation:

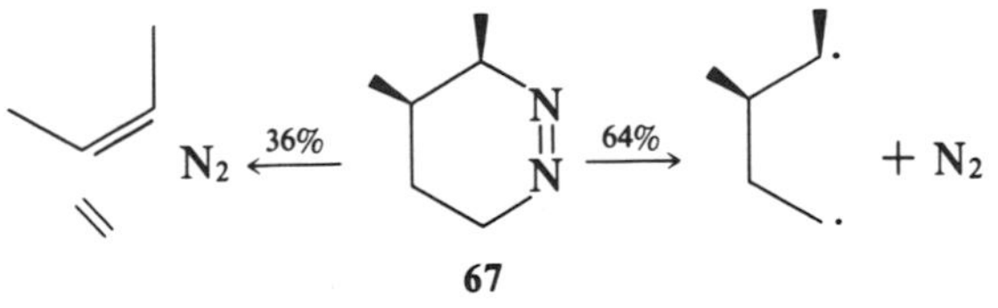

N_2 36% 64% $+ N_2$

67

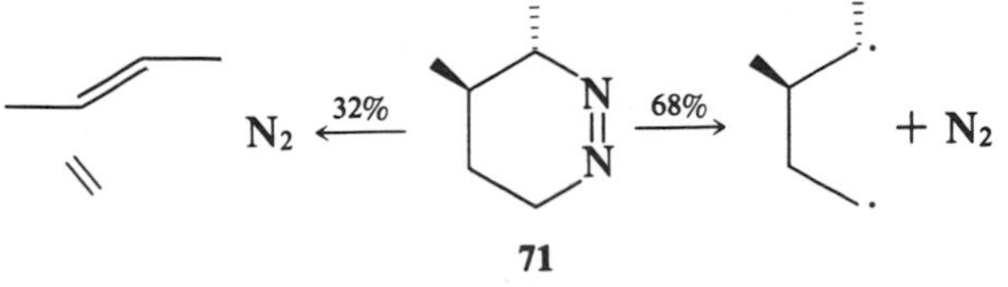

N_2 32% 68% $+ N_2$

71

Ignoring the concerted reactions, the following mechanistic scheme obtains:

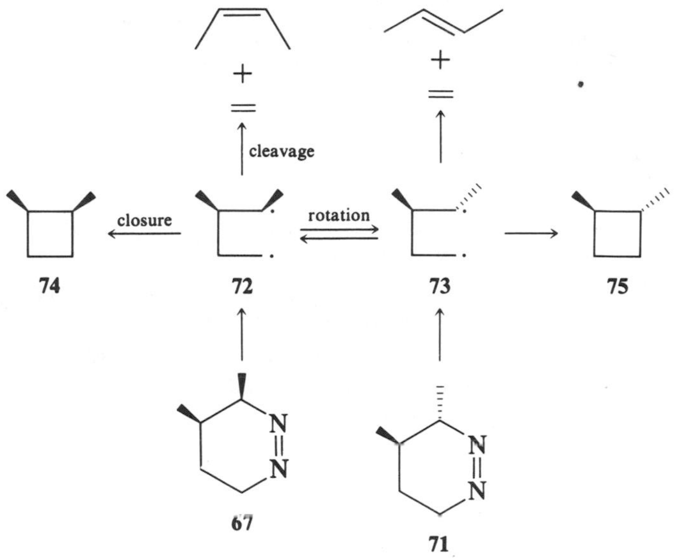

Scheme 5

From a kinetic analysis of the reactions of **67** and **71,** carried out at 439°C in the gas-phase, the relative rate constants for rotation, cleavage, and closure in the 1,4-diradicals (**72** and **73**) can be evaluated. For **72,** k(cleavage)/k(closure) = 1.8 and k(closure)/k(rotation) = 0.7. For **73,** k(cleavage)/k(closure) = 1.6 and k(closure)/k(rotation) = 1.9. Very similar ratios were found in the pyrolyses of the cyclobutanes **74** and **75** at the same temperature, and also in the cycloaddition of ethylene to the butenes at 430°C.

The study of compounds such as **68** (variously described as 1,1-disubstituted diazenes and aminonitrenes) is complicated by the fact that they are reactive intermediates that are usually generated below −120°C; for example:

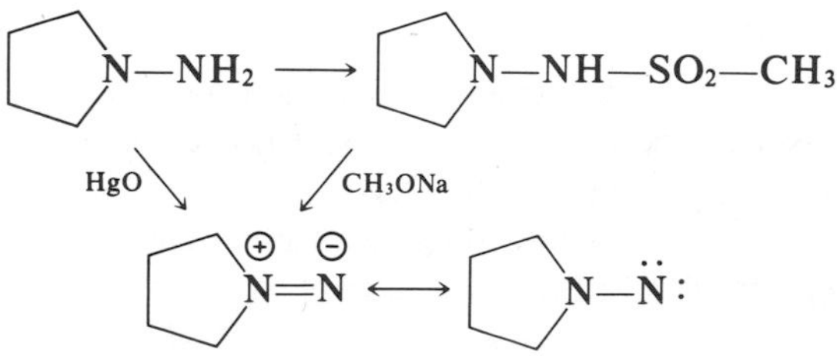

However, the diazene **76** has been generated in solution at −78°C and directly observed by IR and UV spectroscopy:[38]

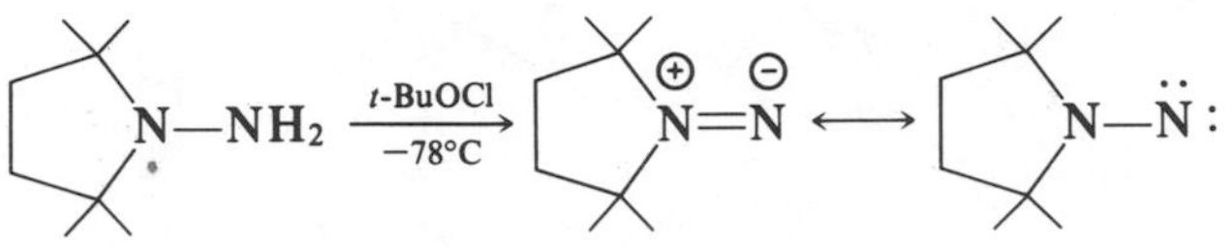

N—NH_2 $\xrightarrow[-78°C]{t\text{-BuOCl}}$ $\overset{\oplus}{N}$=$\overset{\ominus}{N}$ ⟷ N—N̈:

76

A considerable degree of NN double bond character in this "aminonitrene" is indicated by the observed N=N stretching vibration at 1638 cm^{-1} in the infrared spectrum. Less highly substituted diazenes such as **68** have not been directly observed, and a direct comparison with the 439°C pyrolysis of **67** is, therefore, difficult. It was found, however, that the pyrolyses of the *cis*- and *trans*-compounds **77** and **78** in the gas phase at 439°C gave the products expected from Scheme 5. Moreover, the relative rates of rotation, cleavage, and closure in the presumed diradicals agreed well with those cited earlier for compounds **67** and **71**.[39]

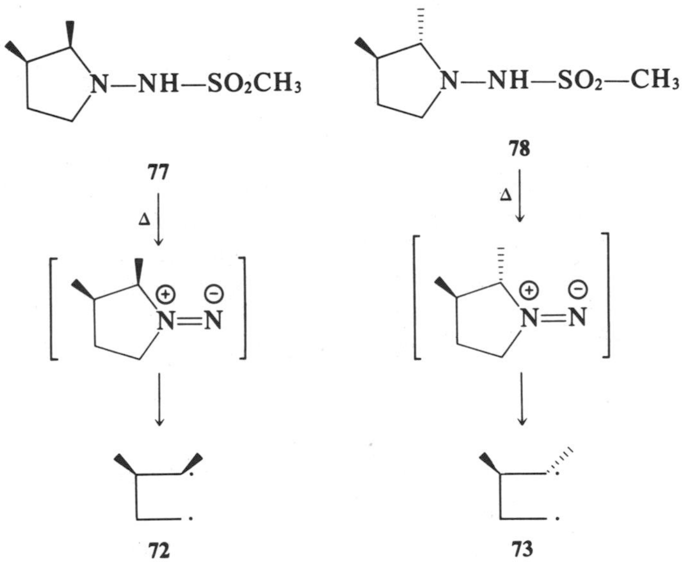

N—NH—SO_2CH_3 (**77**) → Δ → [$\overset{\oplus}{N}$=$\overset{\ominus}{N}$] → **72**

N—NH—SO_2—CH_3 (**78**) → Δ → [$\overset{\oplus}{N}$=$\overset{\ominus}{N}$] → **73**

Thus the involvement of tetramethylene diradicals as common intermediates seems well established. The question remains, however, as to the nature of these diradicals. As mentioned earlier (see Section 3.2), only *trans*-diradicals of type **33** are found to be energy minima according to quantum-chemical calculations.

In the thermal reactions just described, it may be assumed that the diradicals are formed initially in singlet states. Perhaps they also remain in those states and react before intersystem crossing to triplets of presumed lower energy can occur. The opposite sequence of events applies to a class of 1,4-diradicals generated by photochemically induced hydrogen migration in alkyl ketones, known as the Norrish Type II reaction:[40]

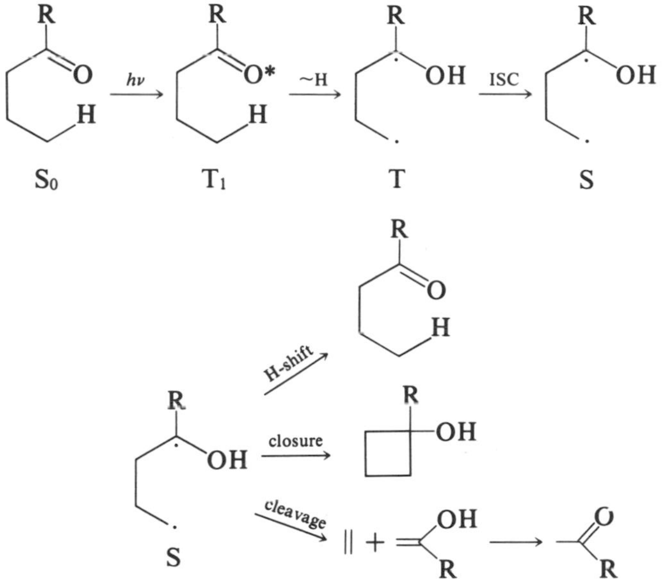

Scheme 6

Direct irradiation of the ketone first produces the excited singlet state, which rapidly undergoes intersystem crossing to the excited triplet. Triplet states of ketones have been directly observed by UV spectroscopy using laser flash photolysis, and they have lifetimes of the order of nanoseconds. The ESR spectrum of the triplet state of benzophenone has also been observed at 77 K.[41] The triplet ketone (Scheme 6) now undergoes a hydrogen shift, giving the triplet diradical. The triplet diradical has also been observed by UV using laser flash photolysis.[40] Its lifetime (30–150 ns at room temperature) is rather insensitive to substituents and to temperature. Intersystem crossing to the singlet diradical appears to be the factor affecting diradical lifetime.[40,42] The singlet diradicals are assumed to be very short-lived (~1 ns) and have never been directly observed. They undergo three rapid types of reaction: hydrogen shift to regenerate the starting material, closure to a cyclobutanol, and cleavage to olefin + ketone. Note that here, too, the triplet diradicals are, presumably, the ground states, but very small singlet–triplet splittings will allow the reactions to occur via the more reactive singlets.

Further evidence for the existence of these 1,4-diradicals is found, for example,

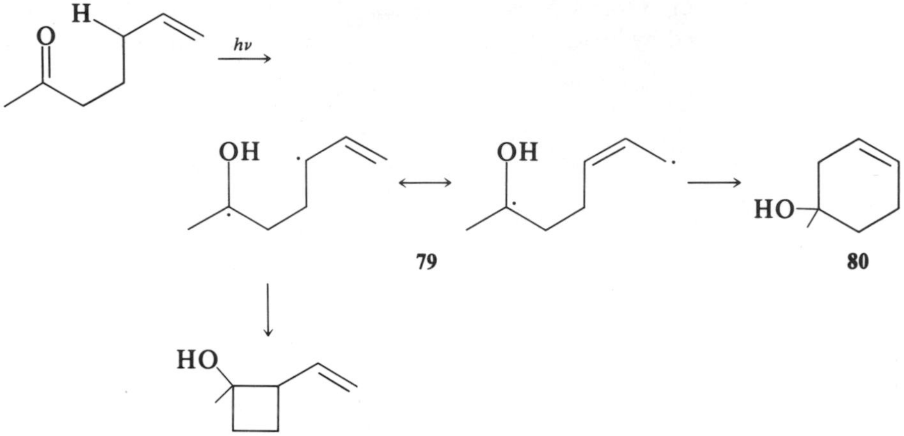

in the isomerization to a cyclohexanol (**80**) via an allyl radical (**79**),[43] and in the trapping reaction with thio- and selenoketones:[40a]

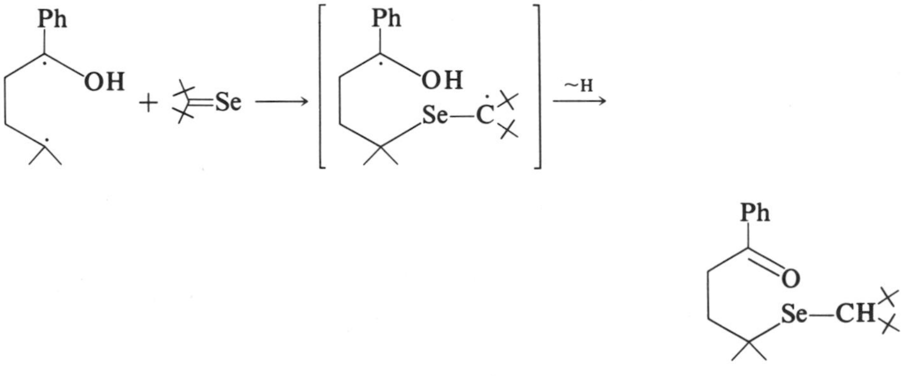

Short-lived singlet diradicals are assumed to be formed during the thermal [2 + 2]-cycloaddition of olefins[44] (reverse of cleavage reaction in Scheme 5). The rates of such addition reactions are enhanced by donor (D) and acceptor (A) substituents in the two olefins, which result in a zwitterionic stabilization of the diradical:

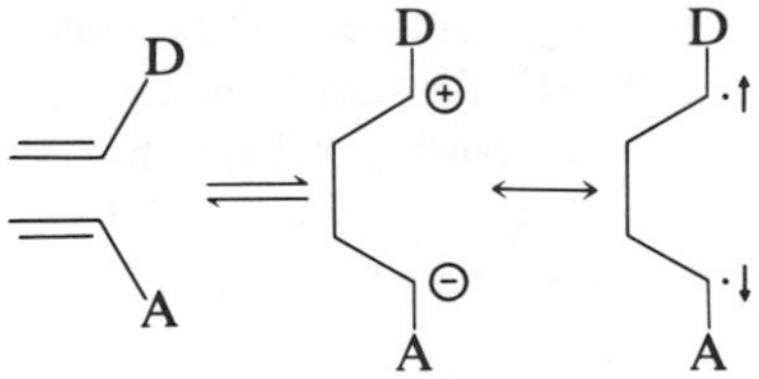

An example is given in the following reaction:

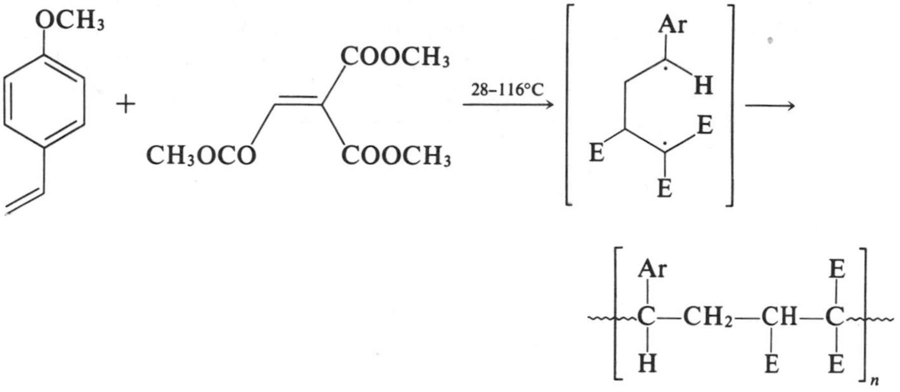

OCH$_3$ + CH$_3$OCO COOCH$_3$ COOCH$_3$ $\xrightarrow{28\text{–}116°C}$ [Ar H E E E] $\longrightarrow$

$$\left[\begin{array}{c} Ar \\ | \\ -C- \\ | \\ H \end{array} -CH_2- \begin{array}{c} \\ \\ CH \\ | \\ E \end{array} - \begin{array}{c} E \\ | \\ C- \\ | \\ E \end{array}\right]_n$$

81

E = COOCH$_3$

in which the 1 : 1 copolymer **81** formed was identical with one produced by deliberate free-radical induced polymerization. The reaction giving **81** was suppressed by free-radical inhibitors.[45]

Triplet 1,4-diradicals (**82**-T) have been directly observed by UV spectroscopy in the laser flash photolysis of benzophenone in the presence of 1,4-dioxene:

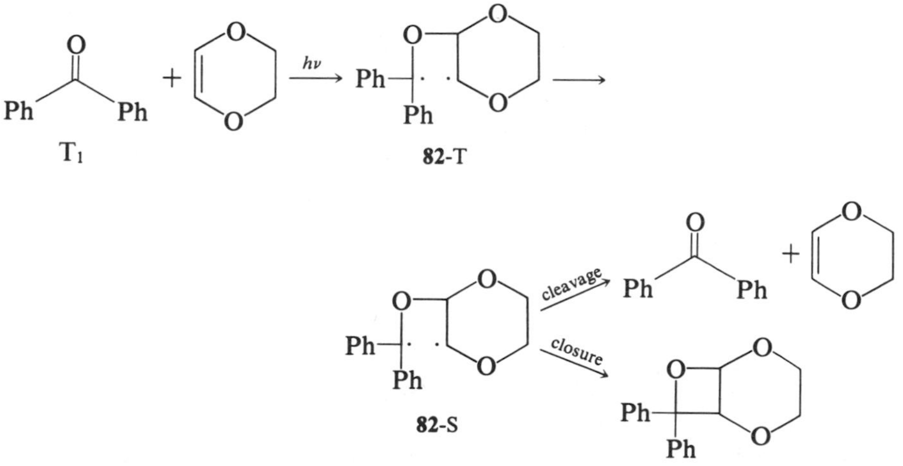

O, Ph, Ph (T$_1$) + O O $\xrightarrow{h\nu}$ Ph, O, Ph, O, O (**82**-T) $\longrightarrow$

Ph, O, Ph, O, O (**82**-S) $\xrightarrow{\text{cleavage}}$ Ph, O, Ph + O O

$\xrightarrow{\text{closure}}$ Ph, O, Ph, O, O

The triplet (**82**-T), absorbing at 535 nm, decays to the singlet **82**-S (not observed). The latter then undergoes cleavage and ring closure reactions analogous to those described in Scheme 6.[46]

A 1,4-diradical stabilized by double allylic resonance (**83**) is formed by photolysis or thermolysis of several precursors (Scheme 7).[47] The ESR spectrum of **83,** obtained at $-180°C$, gives evidence of a delocalized triplet diradical with a low value of D/hc ($0.0204\ cm^{-1}$; compare with the benzylic diradical **18** in Section 3.1). The diradical was generated in argon matrix at 10 K and observed by IR and UV spectroscopy; its dimerization to a mixture of **84** and **85** on warming was monitored spectroscopically. The diradical could be trapped with oxygen, giving the cyclic peroxides **86** and **87,** and when the reactions were carried out thermally in solution or photochemically in the gas phase, ring closure and cleavage to **88** and **89** took place.

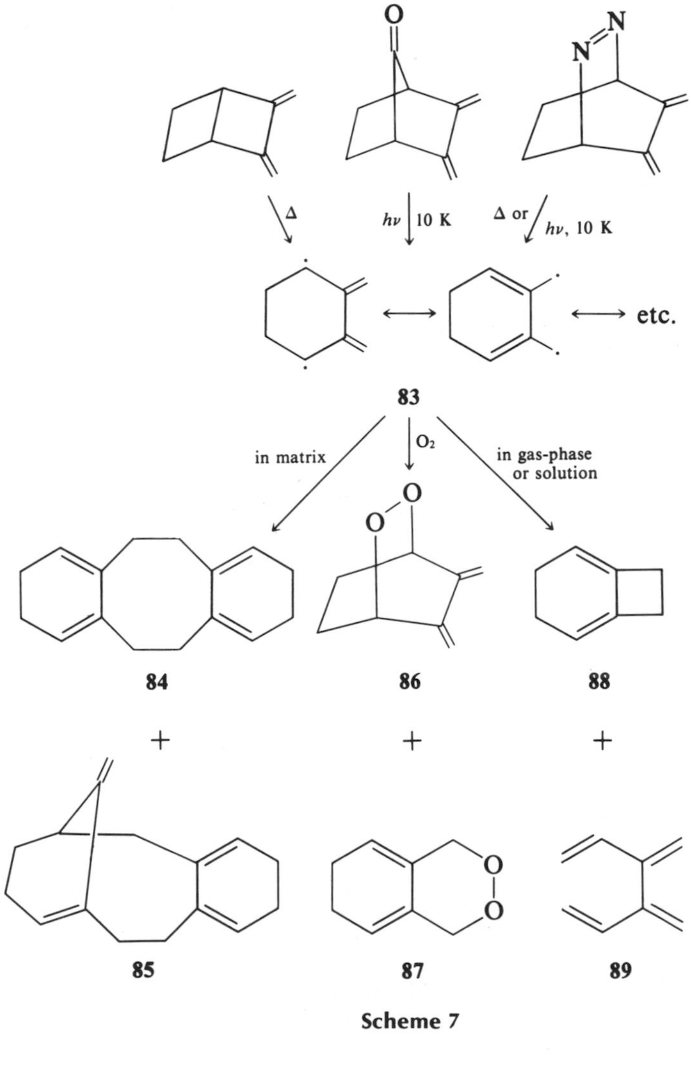

Scheme 7

The same type of diradical is probably formed during the (symmetry-forbidden) dimerization of allenes:[48]

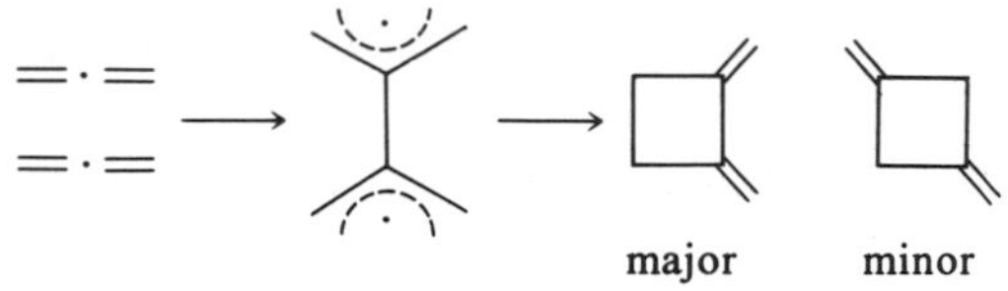

major minor

The introduction of bulky adamantyl substituents hinders ring closure, thereby increasing the lifetimes of such diradicals:

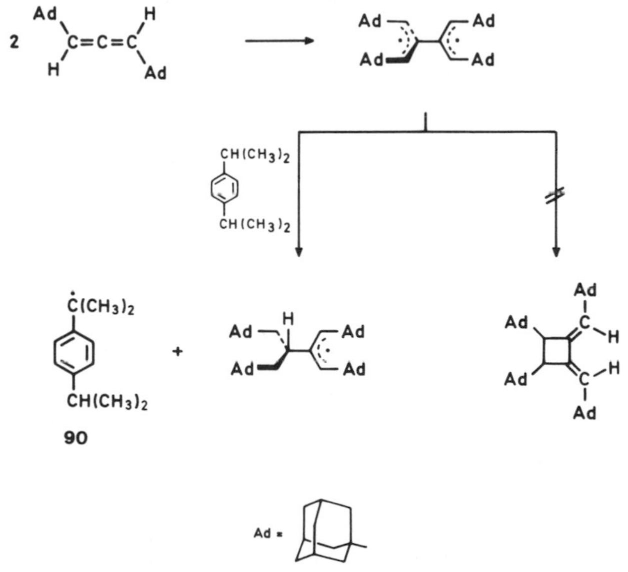

2 Ad(H)C=C=C(H)Ad

$CH(CH_3)_2$ $CH(CH_3)_2$

$\dot{C}(CH_3)_2$ $CH(CH_3)_2$

90

Ad =

The tetraadamantyl diradical was long-lived enough to abstract hydrogen from added *p*-diisopropylbenzene (a common free-radical reaction). The resulting benzylic radicals **(90)** dimerized.[49]

3.7. HIGHER DIRADICALS

Up until now, direct evidence for the existence of singlet diradicals has been scarce. The observation of CIDNP in the reaction products of the 1,10-diradical **91** (Ref. 50) demonstrates that singlet–triplet mixing is occurring. As indicated in Section

2.2.3, the observation of CIDNP requires that the S–T$_0$ splitting be very small, namely, of the order of the hyperfine splitting constant ($\sim 10^{-2}$ cal/mol).

$$\dot{C}H_2(CH_2)_8\dot{C}H_2 \longrightarrow CH_3—(CH_2)_7—CH{=}CH_2 + \text{other products}$$
91

Thermolysis of compound **92** in an NMR probe at 193°C resulted in CIDNP in the starting material:[51]

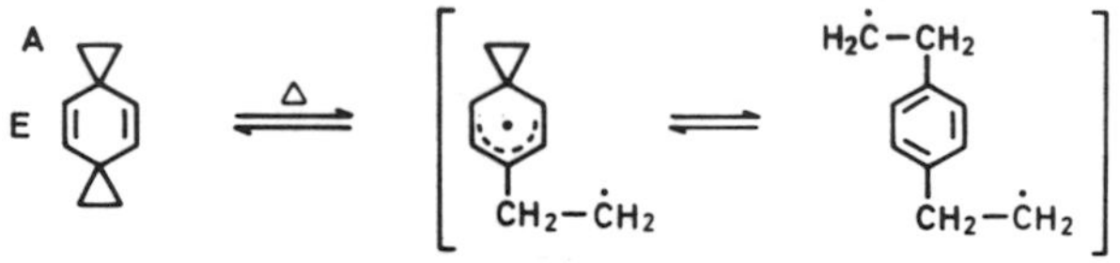

A
E
Δ
$H_2\dot{C}-CH_2$
$CH_2-\dot{C}H_2$
$CH_2-\dot{C}H_2$
92
93
94
H_5C_6-SH
C_2H_5
CH_2-CH_2
$S-C_6H_5$

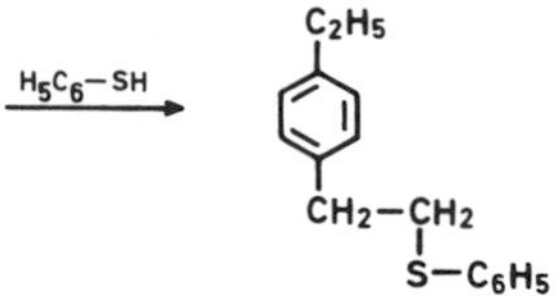

The 1,8-diradical **94** was trapped efficiently in the presence of thiols. Since in **94** $\Delta g = 0$, this diradical cannot give rise to the observed net effect and the polarization was, therefore, ascribed to the diradical **93.**

The photolysis of ketones has long been proposed to involve diradicals (''Norrish Type I'' cleavage):[52]

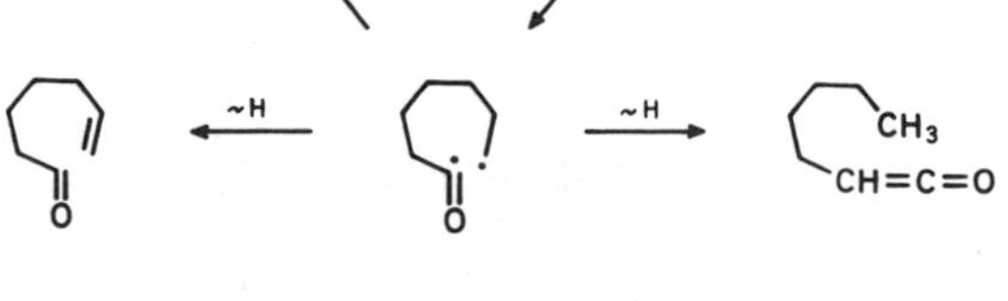

hν
ISC
S_0
$S_1(n,\pi^*)$
$T_1(n,\pi^*)$
''Norrish-I''
~H
~H
CH_3
$CH{=}C{=}O$

By performing the irradiation in an NMR spectrometer, ^{13}C CIDNP was observed for the carbonyl carbon in the starting ketone (S_0) as well as the products. That the largest effect is obtained for the carbonyl carbon is due to the large hyperfine splitting (15.0 mT) with this nucleus. Even 1,6- and 1,5-diradicals have been detected using this technique.[53] In all these diradicals, the singlet–triplet energy separation must be extremely small.

A measure of how far apart two radical centers must be in order *not* to interact can be obtained by ESR spectroscopy of bisnitroxides.[54] Since the nitroxides are extraordinarily stable, measurements can be carried out conveniently at room temperature. In a normal mononitroxide, the free electron couples with the ^{14}N nucleus, giving three lines ($a_N \sim 1.5$ mT). If now the two electrons in a bisnitroxide interact strongly, each electron can couple with both nitrogen nuclei. As a result, a five-line ESR spectrum is obtained. Thus the bisnitroxide **95** ($n = 0$) shows a well-

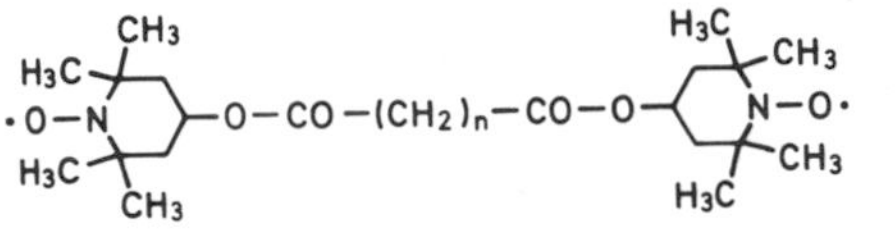

95

defined quintet, but as n increases to 8, the spectrum gradually changes to a triplet. In the latter case, the interaction between the electrons is much smaller than a_N, and the molecule behaves like two monoradicals.

The interaction between the two electrons is related to the singlet–triplet splitting in the diradical. "Strong" interaction implies that the singlet–triplet splitting is large compared to the hyperfine coupling constant a; weak interaction implies a singlet–triplet splitting $\ll a$ (i.e., $\ll 10^{-2}$ cal/mol). A particularly well-studied example[55] is the following bisnitroxide in which the average distance between the two free electrons is 4.8 Å:

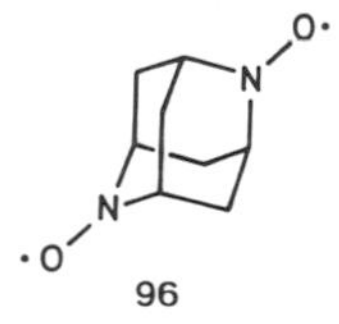

96

singlet–triplet splitting = 0.3 kcal/mol (= 110 cm^{-1})

a_N = 2.0 mT

ground-state triplet, D/hc = 0.0240 cm^{-1}

Note that one obtains quite different kinds of information from CIDNP and ESR. A *reacting* diradical with a singlet–triplet splitting as large as that in **96** could not have been detected by CIDNP.

3.8. PROBLEMS

1. Explain the reaction:[56]

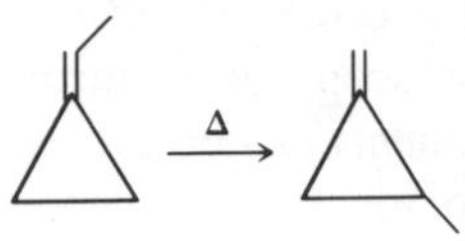

2. The photolysis of **97** at 77 K produces a triplet species with the zero-field splitting parameters $D/hc = 0.0204$ and $E/hc = 0.0052\ \mathrm{cm}^{-1}$. Give a structure for this product.[57]

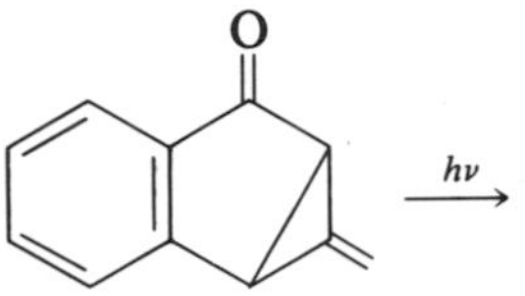

97

3. Write a diradical mechanism for the reaction:[58]

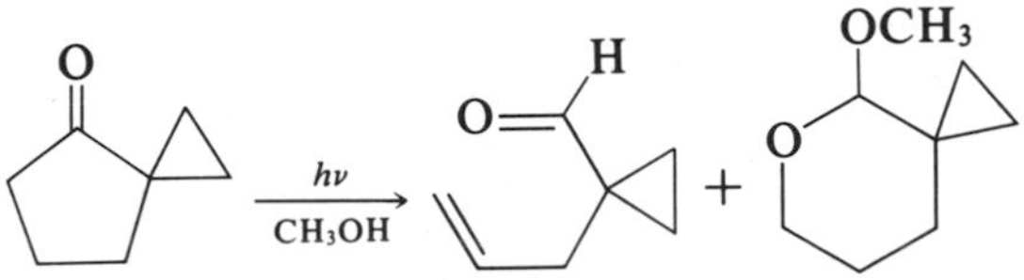

4. Write a mechanism for the reaction:[59]

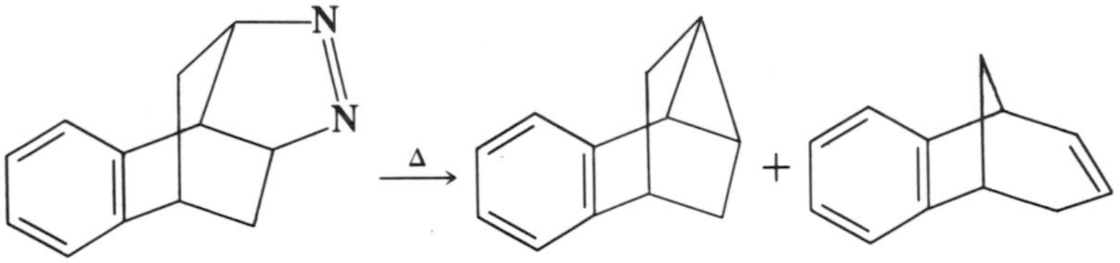

5. The cyclopropane **98** undergoes thermal *cis–trans* isomerization with a remarkably low activation energy of 27.6 kcal/mol.[60] Explain this in terms of substituent effects on the stability of putative radicals.

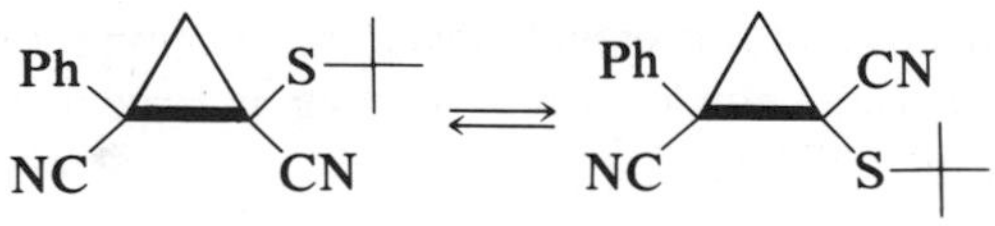

98

6. Use the Norrish Type II reaction to write a mechanism for the following:[61]

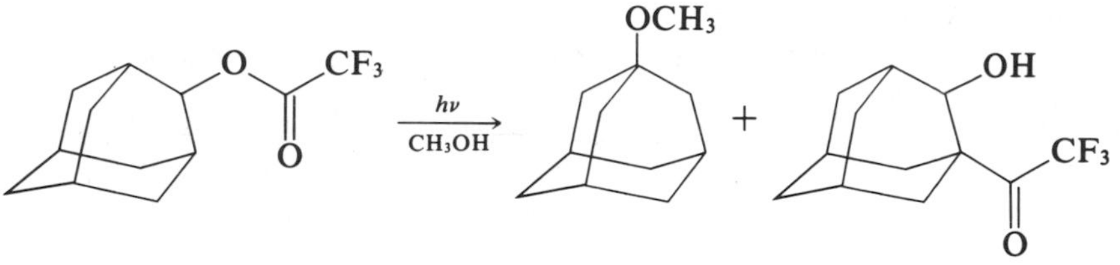

OCH_3 CF_3 $h\nu$ CH_3OH OH CF_3 O O O

7. Cyclic azo-compound **99** gives cyclobutane **100** with >98% stereospecificity on thermolysis or direct photolysis. However, in the sensitized photolysis (using thioxanthone as triplet sensitizer), only 61% retention of configuration is observed.[36a] Explain.

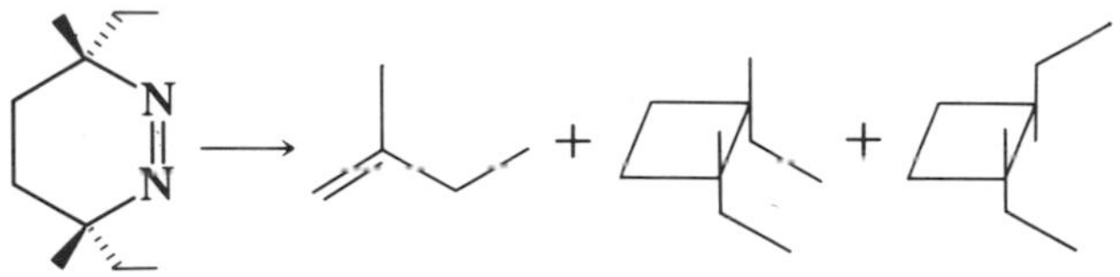

N N

99 **100**

8. Explain the variation in the following relative rate constants for the S_N1 solvolyses of tosylates and brosylates[62]

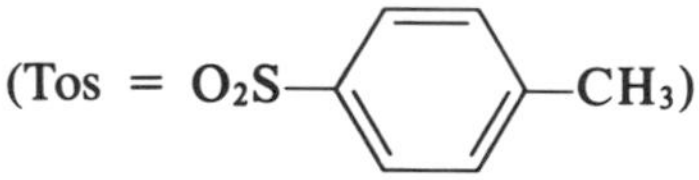

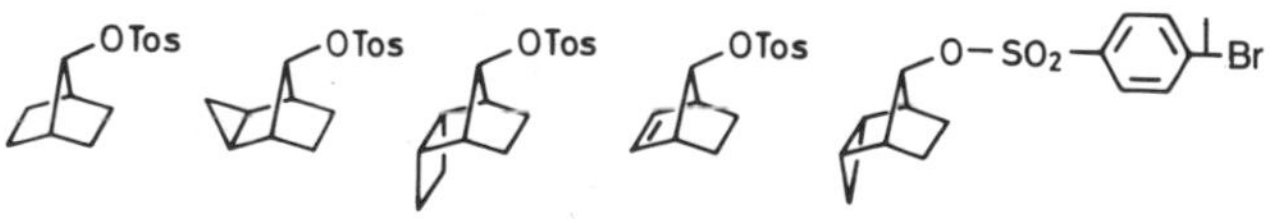

(Tos = O_2S–C_6H_4–CH_3)

	OTos	OTos	OTos	OTos	$O-SO_2-C_6H_4-Br$
k_{rel}	1	0.3	10^4	10^{11}	10^{14}

REFERENCES

1. X. Chapuisat and Y. Jean, *J. Am. Chem. Soc.*, **97**, 6325 (1975); *Top. Curr. Chem.*, **68**, 1 (1976); P. J. Hay, W. J. Hunt, and W. A. Goddard, *J. Am. Chem. Soc.*, **94**, 638 (1972).
2. S. L. Buchwalter and G. L. Closs, *J. Am. Chem. Soc.*, **101**, 4688 (1979).
3. R. Hoffmann, *J. Am. Chem. Soc.*, **90**, 1475 (1968).
4. A. D. Walsh, *Trans. Faraday Soc.*, **45**, 179 (1949).
5. J. J. Gajewski, R. J. Weber, and M. J. Chang, *J. Am. Chem. Soc.*, **101**, 2100 (1979).
6. H. E. O'Neal and S. W. Benson, *J. Phys. Chem.*, **72**, 1866 (1968).
7. R. J. Crawford and A. Mishra, *J. Am. Chem. Soc.*, **88**, 3963 (1966).

8. J. A. Berson, L. D. Pedersen, and B. K. Carpenter, *J. Am. Chem. Soc.*, **98,** 122 (1976).

9. J. A. Berson, in *Rearrangements in Ground and Excited States,* P. de Mayo, Ed., Academic Press, New York, 1980, p. 311.

10. J. E. Baldwin and M. W. Grayston, *J. Am. Chem. Soc.*, **96,** 1629, 1630 (1974); see also P. H. Mazzochi and R. S. Lustig, *ibid.*, **97,** 3714 (1975).

11. J.-F. Muller, D. Muller, H. J. Dewey, and J. Michl, *J. Am. Chem. Soc.*, **100,** 1629 (1978).

12. M. Gisin, E. Rommel, J. Wirz, M. N. Burnett, and R. M. Pagni, *J. Am. Chem. Soc.*, **101,** 2216 (1979); R. M. Pagni, M. N. Burnett, and H. M. Hassaneen, *Tetrahedron,* **38,** 843 (1982).

13. P. Dowd and M. Chow, *Tetrahedron,* **38,** 799 (1982): P. Dowd, *Acc. Chem. Res.*, **5,** 242 (1972).

14. (a) S. B. Auster, R. M. Pitzer, and M. S. Platz, *J. Am. Chem. Soc.*, **104,** 3812 (1982). (b) D. Feller, K. Tanaka, E. R. Davidson, and W. T. Borden, *J. Am. Chem. Soc.*, **104,** 967 (1982).

15. J. A. Berson, *Acc. Chem. Res.*, **11,** 446 (1978); J. A. Berson, in *Diradicals,* W. T. Borden, Ed., Wiley, New York, 1982.

16. M. Rule, J. A. Mondo, and J. A. Berson, *J. Am. Chem. Soc.*, **104,** 2209 (1982).

17. D. F. Kelley, P. M. Rentzepis, M. R. Mazur, and J. A. Berson, *J. Am. Chem. Soc.*, **104,** 3764 (1982).

18. M. R. Mazur and J. A. Berson, *J. Am. Chem. Soc.*, **104,** 2217 (1982).

19. M. Rule, R. F. Salinaro, D. R. Pratt, and J. A. Berson, *J. Am. Chem. Soc.*, **104,** 2223 (1982); R. F. Salinaro and J. A. Berson, *ibid.*, **104,** 2228 (1982).

20. M. J. S. Dewar and S. Kirschner, *J. Am. Chem. Soc.*, **96,** 5246 (1974).

21. G. A. Segal, *J. Am. Chem. Soc.*, **96,** 7892 (1974).

22. C. Doubleday, J. W. McIver, and M. Page, *J. Am. Chem. Soc.*, **104,** 3768 (1982).

23. R. J. Crawford and A. Mishra, *J. Am. Chem. Soc.*, **88,** 3963 (1966).

24. For reviews, see R. G. Bergman, in *Free Radicals,* J. K. Kochi, Ed., Wiley, New York, 1973, Vol. I, p. 191; P. B. Dervan and D. A. Dougherty, in *Diradicals,* W. T. Borden, Ed., Wiley, New York, 1982.

25. M. P. Schneider and H. Bippi, *J. Am. Chem. Soc.*, **102,** 7363 (1980).

26. T. C. Clarke, L. A. Wendling, and R. G. Bergman, *J. Am. Chem. Soc.*, **97,** 5638 (1975).

27. P. C. Hiberty and Y. Jean, *J. Am. Chem. Soc.*, **101,** 2538 (1979).

28. R. J. Crawford and M. H. Chang, *Tetrahedron,* **38,** 837 (1982).

29. D. W. K. Yeung, G. A. MacAlpine, and J. Warkentin, *J. Am. Chem. Soc.*, **100,** 1962 (1978).

30. W. R. Roth and M. Martin, *Justus Liebigs Ann. Chem.*, **702,** 1 (1967).

31. K. Mackenzie, in *The Chemistry of the Hydrazo, Azo, and Azoxy Groups,* S. Patai, Ed., Wiley, New York, 1975, p. 329; C. J. Samuel, *Chem. Commun.*, **1982,** 131.

32. M. H. Chang and D. A. Dougherty, *J. Am. Chem. Soc.*, **104,** 1131 (1982).

33. L. A. Paquette, *J. Am. Chem. Soc.*, **92,** 5765 (1970).

34. N. Rieber, J. Alberts, J. A. Lipsky, and D. M. Lemal, *J. Am. Chem. Soc.*, **91,** 5668 (1969).

35. P. B. Dervan, T. Uyehara, and D. S. Santilli, *J. Am. Chem. Soc.*, **101,** 2069 (1979); P. B. Dervan and D. S. Santilli, *ibid.*, **102,** 3863 (1980).

36. (a) See also P. D. Bartlett and N. A. Porter, *J. Am. Chem. Soc.*, **90,** 5317 (1968). (b) K. R. Kopecky and J. Soler, *Can. J. Chem.*, **52,** 2111 (1974). (c) R. C. Neuman and E. W. Ertley, *J. Am. Chem. Soc.*, **97,** 3130 (1975).

37. G. Scacchi, C. Richard, and M. H. Bach, *Int. J. Chem. Kinet.*, **9,** 513 (1977).

38. W. D. Hinsberg, P. G. Schultz, and P. B. Dervan, *J. Am. Chem. Soc.*, **104,** 766 (1982).

39. P. B. Dervan and T. Uyehara, *J. Am. Chem. Soc.*, **101,** 2076 (1979).

40. (a) J. C. Scaiano, E. A. Lissi, and M. V. Encina, *Rev. Chem. Intermed.*, **2,** 139 (1978). (b) J. C. Scaiano, *Tetrahedron*, **38,** 819 (1982).

41. H. Murai, T. Imamura, and K. Obi, *Chem. Phys. Lett.*, **87,** 295 (1982).

42. See also R. A. Caldwell, T. Majima, and C. Pac, *J. Am. Chem. Soc.* **104,** 629 (1982).

43. N. C. Yang, A. Morduchowitz, and D.-D. Yang, *J. Am. Chem. Soc.*, **98,** 1017 (1963).

44. P. D. Bartlett, *Q. Rev. Chem. Soc.*, **24,** 473 (1970).

45. H. K. Hall and M. Abdelkader, *J. Org. Chem.*, **46,** 2948 (1981).

46. S. C. Freilich and K. S. Peters, *J. Am. Chem. Soc.*, **103,** 6255 (1981).

47. W. R. Roth, M. Biermann, G. Erker, K. Jelich, W. Gerhartz, and H. Görner, *Chem. Ber.*, **113,** 586 (1980).

48. W. R. Roth, M. Heiber, and G. Erker, *Angew. Chem.*, **85,** 511 (1973); *Angew. Chem. Int. Ed. Engl.*, **12,** 504 (1973); J. E. Baldwin and R. H. Fleming, *Top. Curr. Chem.*, **15,** 281 (1970); D. J. Pasto, *J. Am. Chem. Soc.*, **101,** 37 (1979).

49. T. L. Jacobs and R. C. Kammerer, *J. Am. Chem. Soc.*, **94,** 7190 (1972).

50. R. Kaptein, M. Fràter-Schröder, and L. J. Oosterhoff, *Chem. Phys. Lett.*, **12,** 16 (1971).

51. T. Tsuji and S. Nishida, *J. Am. Chem. Soc.*, **96,** 3649 (1974).

52. N. J. Turro et al., *Acc. Chem. Res.*, **5,** 92 (1972).

53. R. Kaptein, R. Freeman, and H. D. W. Hill, *Chem. Phys. Lett.*, **26,** 104 (1974); C. Doubleday, *ibid.*, **77,** 131 (1981); *ibid.*, **79,** 375 (1981).

54. R. Brière, R.-M. Dupeyre, H. Lemaire, C. Morat, A. Rassat, and P. Rey, *Bull. Soc. Chim. Fr.*, **1965,** 3290; E. G. Rozantsev, V. A. Golubev, M. B. Neiman, and Yu. V. Kokhanov, *Bull. Acad. Sci. USSR, Division Chem. Sci.*, **1965,** 559.

55. R.-M. Dupeyre, A. Rassat, and J. Ronzaud, *J. Am. Chem. Soc.* **96,** 6559 (1974).

56. J. P. Chesick, *J. Am. Chem. Soc.*, **85,** 2720 (1963).

57. D. E. Seeger, E. F. Hilinski, and J. A. Berson, *J. Am. Chem. Soc.*, **103,** 720 (1981).

58. J. K. Crandall and R. J. Seidewand, *J. Org. Chem.*, **35,** 697 (1970).

59. W. Adam, N. Carballeira, and O. De Lucchi, *J. Am. Chem. Soc.*, **102,** 2107 (1980).

60. A. De Mesmaeker, L. Vertommen, R. Merényi, and H. G. Viehe, *Tetrahedron Lett.*, **1982,** 69.

61. J. E. Gano and L. Eizenberg, *J. Am. Chem. Soc.*, **95,** 972 (1973); N. A. Marron and J. E. Gano, *ibid.*, **98,** 4653 (1976).

62. See M. J. S. Dewar and J. M. Hargis, *J. Am. Chem. Soc.*, **90,** 4468 (1968) and references therein; G. Ellen and G. W. Klumpp, *Tetrahedron Lett.*, **1974,** 2995.

Chapter 4

CARBENES AND NITRENES

Es wird nun meine nächste Aufgabe sein, das Methylen oder Derivate desselben . . . darzustellen.

—J. U. Nef, *Liebigs Ann. Chem.*, **287,** 359 (1895).

4.1. ELECTRONIC STRUCTURE AND ENERGETICS

Methylene (**1**) and nitrene (**2**) are the simplest representatives of the families of carbenes and nitrenes.†

$$\ddot{C}H_2 \qquad :\dot{N}H$$

1 **2**

Formally, they result from the removal of two hydrogen atoms from methane and ammonia, respectively. They are isoelectronic and exhibit many similarities in chemical behavior. Other isoelectronic first row species are the oxygen atom $:\ddot{O}:$ and $:\ddot{F}:^{\oplus}$. Since the central atoms are surrounded by electron sextets, they are said to be *electron deficient*. Not surprisingly, therefore, they normally react as strong electrophiles, but, as we shall see, the electrophilicity of carbenes and nitrenes is strongly influenced by substituents.

†*Chemical Abstracts* uses the generic name "imidogen" for nitrenes.

If methylene were linear, it would have two degenerate p orbitals, and Hund's first rule would predict a triplet ground state:

Degenerate orbitals in linear CH_2 and NH (nitrogen lone pair ignored)

For NH, there is only one possibility: The molecule is linear and therefore a triplet. If methylene is *not* linear (and in fact, it isn't), the two orbitals will become different: the one in the molecular plane (the σ orbital) will acquire s character and thereby become stabilized. The other orbital (p) remains largely unchanged:

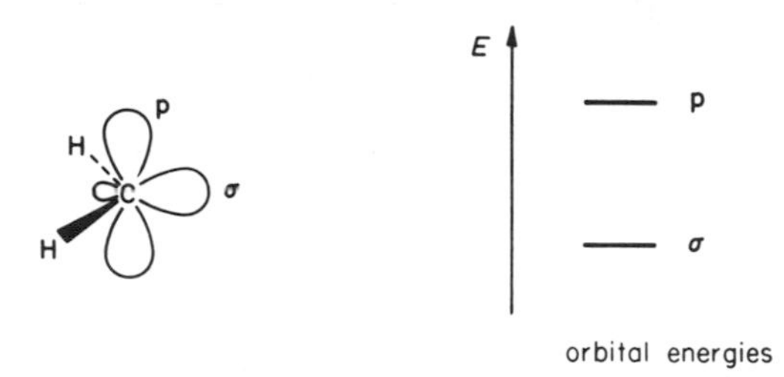

Whether or not the species will now be a singlet depends on the energy separation between the two states. It has to be large enough to overcome the *electron correlation energy*, that is, the energy required to bring two electrons together in one orbital. Therefore, if the energy spacing is small, the species will still have a triplet ground state. The following electronic states can exist:

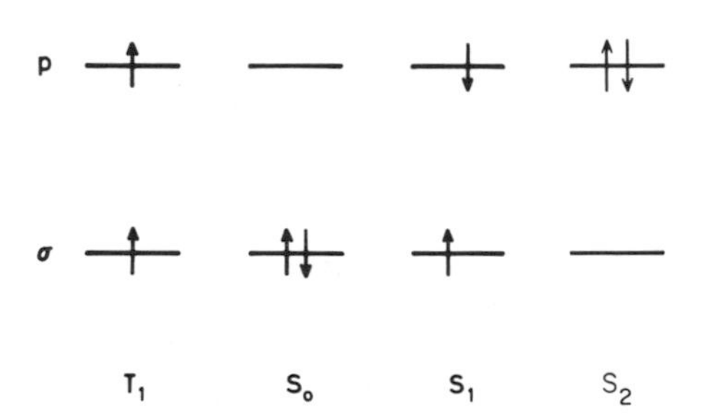

It is now known that T_1 is the ground state. (For the historical development, see Table 4.1.) S_0 is an excited singlet state, and S_1 is a doubly excited singlet state. The ordering of the state energies is given in Figure 4.1. Each species represents a discrete entity.

It has been established in recent years that the energy difference between the lowest triplet and singlet states (the singlet–triplet splitting) is quite small, probably

TABLE 4.1. History of Methylene[a]

Year	Authors	Method	∢HCH in Ground State	Ground State	S–T Splitting (kcal/mol)
1932	Mulliken	Intuitive Qualitative	90–109°	singlet	
1934	Lennard-Jones	Quantum mechanical (QM)	bent	singlet	
1947	Walsh	Thermochemical	(180°)	triplet	*small*
1949	Laidler, Casey	Thermochemical	90°	singlet	−19
1951	Lennard-Jones, Pople	Qualitative QM	120–180°	?	
1951	Linnett, Poë	Qualitative QM	110–134°	?	
1952	Niira, Oohata	Semiempirical QM	140°	triplet	35
1956	Walsh	Qualitative QM		triplet	*small*
1957	Gallup	Extended Hückel	160°	triplet	30
1960	Foster, Boys	Ab initio MO	129°	triplet	24
1960	Padgett, Krauss	SCF–MO	~120°	triplet	*small*
1961[b]	Herzberg	UV spectrum	≈180°	triplet	
1962	Jordan, Longuet–Higgins	Semiempirical MO	180°	triplet	10
	Pedley, Ellison	Semimpirical MO	180°	triplet	~14
1964	Dixon	Valence bond	180°	triplet	~14
1968	Hoffmann, Zeiss, Van Dine	Extended Hückel	155°		
1969[c]	Harrison, Allen	Ab initio	138°	triplet	⩾33.4
1970	Bernheim, Wasserman	ESR spectrum	≃136°	triplet	
1971[d]	Herzberg, Johns	UV spectrum (reconsidered)	136°	triplet	
1971[e]	Hase, Phillips, Simons	Kinetics	—	triplet	8–9
1971	O'Neil, Schaefer, Bender	Ab initio	133.3°	triplet	22.2
	Del Bene	SCF	130°	triplet	33.4
	Pople	SCF	132°	triplet	37
	Harrison	Ab initio	135°	triplet	19

1972[f]	Bender, Schaefer, Allen et al.	Ab initio	134°	triplet	11 ± 2
1972	Hay, Hunt, Goddard	Valence bond	135°	triplet	11.5
1972[g]	Dewar et al.	MINDO/2 (semiempirical)	—	triplet	⩾15
1972[h]	Frey	Kinetics	—	triplet	≃8
1974[i]	Dewar et al.	MINDO/3	134°	triplet	8.7
1974	Staemmler	Ab initio	134.2°	triplet	9.2
1975[j]	Frey, Kennedy	Kinetics	—	triplet	9
1976[k] and 1981	Lineberger et al.	Photoelectron detachment	138 ± 4°	triplet	19.5 ± 0.7
1976[l]	Meadows, Schaefer	Ab initio	—	triplet	19.7
1977[m]	Bauschlicher, Schaefer	Ab initio	—	triplet	11 ± 2
1977–81[n]	Several calculations	Ab initio	—	triplet	10.5 ± 2
1978[o]	Lengel, Zare	Photolysis of ketene	—	triplet	8.1 ± 0.8
1982[p]	Hayden et al.	Molecular beam experiments	—	triplet	8.5 ± 0.8

[a]Older historical details are summarized by P. P. Gaspar and G. S. Hammond in *Carbenes*, R. A. Moss and M. Jones, Jr., Eds., Wiley, New York, 1975, Vol. II, p. 207.
[b]Reference 2.
[c]Reference 3.
[d]Reference 14.
[e]Reference 6a.
[f]Reference 4.
[g]Reference 5.
[h]Reference 6b.
[i]Reference 7.
[j]References 6c and 6d.
[k]Reference 8.
[l]Reference 9.
[m]Reference 10.
[h]Reference 11.
[o]Reference 12a.
[p]Reference 12b.

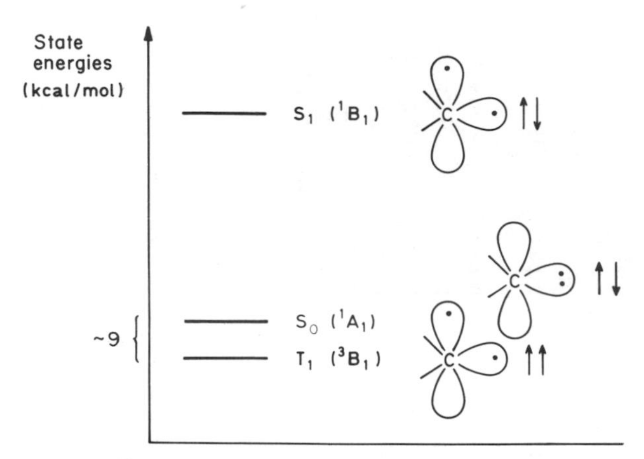

Figure 4.1. Electronic states of methylene.

around 9 kcal/mol, and that the carbene is bent. The H—C—H angle is approximately 136° in the triplet and somewhat smaller in the singlet S_0. Previously, the ordering of the states, their relative energies, and the carbene conformations were matters of much dispute (Table 4.1).

Although there has been argreement since 1952 that the ground state is a triplet, it was only in 1970 that the question of geometry was settled by direct observation of $^3CH_2$ by ESR.[1] In 1961, Herzberg had observed the UV spectrum of methylene and interpreted the rotational fine structure as indicating that CH_2 was probably linear.[2] The methylene was produced by photolysis of diazomethane, which first produced the singlet $^1CH_2$ (the principle of spin conservation):

$$H_2C{=}N_2 \xrightarrow{h\nu} {}^1CH_2 + N_2$$

$${}^1CH_2 \longrightarrow {}^3CH_2$$

Since the singlet so formed was observed to decay to the triplet, this experiment established the triplet ground state.

In the following years (1962–1964), the theoreticians reproduced the linear structure. However, as of 1968, they all agree that CH_2 is bent.

If the structure was in debate, so was the singlet–triplet splitting. As late as 1969, it was stated that this separation must be $\geqslant$33.4 kcal/mol, and Harrison and Allen[3] advised the organic chemists to favor interpretations of reaction mechanisms that were in accord with a large splitting. Two years later, the theoretical separation was down to 19 kcal/mol. Another year, and the theoreticians calculated 11 $\pm$ 2 kcal/mol.[4] Nevertheless, in 1972, it was still stated that the energy difference must be $\geqslant$15 kcal/mol, probably around 25.[5] In 1971–1972, two experimental groups derived a singlet–triplet splitting of 8–9 kcal/mol from kinetic measurements of gas-phase reactions of CH_2.[6] The next year, all theoreticians agreed on a low value: 8–9 kcal/mol.[7]

The story seems to go on endlessly. In 1976, measurements of the electron detachment from the anion $CH_2^{\ominus}$ produced an S–T splitting of 19.5 ± 0.7 kcal/mol.[8] This was immediately confirmed by ab initio calculations,[9] but more refined calculations were unable to maintain the high value.[10] As of 1977, the theoreticians[11] have reinstalled the low value of 9–11 kcal/mol, and experimental support for this figure continues to come forward.[12] Although the discrepancy between this value and the results of the electron detachment experiment is not entirely resolved, we shall adopt the "low" value of 8–10 kcal/mol, which is perfectly reasonable chemically. For substituted carbenes, singlet–triplet splittings of only a few kcal/mol are known, and the singlet may even become the ground state (e.g., in the halocarbenes).

Singlet–triplet splittings in nitrenes are less well known, but they are certainly larger than in the isoelectronic carbenes (e.g., approximately 35 kcal/mol for NH).[40] This will be explained in connection with ESR spectroscopy (Section 4.3.2). Although precise heats of formation of $^3CH_2$ and ^3NH are still the matter of some debate, recent values cluster around 94 ± 1 and 90 ± 5 kcal/mol, respectively.[12a,13]

4.2. GENERATION

The general methods, and a few special ones, of preparing carbenes and nitrenes are collected in Tables 4.2 and 4.3. The most universal methods are the decomposition of diazo-compounds and azides, respectively. Common for the homolytic methods is the breaking of rather weak bonds with expulsion of stable, neutral fragments like N_2 and CO. The breaking of a C=C bond, in contrast, is a strongly endothermic process,

$$CH_2{=}CH_2 \longrightarrow 2\ {}^1CH_2{:}$$

$$\Delta H_f^0 = 12.5 \qquad \sim 2 \times 102 \qquad \Delta H_r^0 \simeq 192 \text{ kcal/mol}$$

which can be achieved only in very special cases where the π-bond is twisted due to steric repulsion between substituents (tetra-1-naphthylethylene).[15] The expulsion of a carbene from a three-membered ring is much less endothermic, and the relatively stable difluorocarbene can be generated in this way:[16]

$$\text{hexafluorocyclopropane} \underset{E_a \sim 8}{\overset{E_a = 38.6}{\rightleftharpoons}} CF_2{=}CF_2 + {:}CF_2$$

$$\Delta H_f^0 \sim -227 \qquad -155 \qquad -44$$

$$\Delta H_r^0 \sim 28 \text{ kcal/mol}$$

TABLE 4.2. Formation of Carbenes[a]

Diazo-compounds

$$R_2C{=}N_2 \xrightarrow{h\nu \text{ or } \Delta} R_2C\!: + N_2$$

Diazirines

$$R_2C\!<\!(N{=}N) \xrightarrow{h\nu \text{ or } \Delta} R_2C{=}N_2 \;\downarrow\; R_2C\!: + N_2$$

Tetrazoles

$$\text{5-R-tetrazole (NH)} \xrightarrow[-N_2]{\Delta} [R{-}C{\equiv}\overset{\oplus}{N}{-}\overset{\ominus}{N}H]$$

$$\xrightarrow{\text{solution}} \text{Dimerization}$$

$$\xrightarrow{\text{gas-phase}} R{-}CH{=}N_2 \xrightarrow[-N_2]{\Delta} R{-}\ddot{C}H + N_2$$

Ketenes

$$R_2C{=}C{=}O \xrightarrow{h\nu \text{ or } \Delta} R_2C\!: + CO$$

Bamford-Stevens Reaction

$$R_2C{=}O + H_3N_2SO_2{-}C_6H_4{-}CH_3 \longrightarrow R_2C{=}N{-}NH{-}Tos$$

$$\xrightarrow{\text{Base (NaH)}} R_2C{=}N{-}\overset{\ominus}{N}{-}Tos\ Na^{\oplus} \xrightarrow{\Delta} R_2C{=}N_2 + Tos^{\ominus}Na^{\oplus}$$

$$R_2C{=}N_2 \xrightarrow{\Delta} R_2C\!: + N_2$$

$$Tos = SO_2{-}C_6H_4{-}CH_3$$

Small Rings

$$(R^1)_2C\!<\!(CR^2_2)_2 \xrightarrow{h\nu \text{ or } \Delta} (R^1)_2C\!: + R^2_2C{=}CR^2_2 \quad (\text{best when } R^1 = R^2 = F)$$

$$R_2C\!<\!(O)C(CH_3)_2 \xrightarrow{h\nu} R_2C\!: + (CH_3)_2C{=}O$$

TABLE 4.2. Formation of Carbenes[a] *(Continued)*

Strained π-Bonds

Tetra(1-naphthyl)ethylene $\overset{\Delta}{\rightleftharpoons}$ di(1-naphthyl)carbene ($\ddot{C}$)

Alkyl Halides

$$CHCl_3 \xrightarrow{NaOH} \overset{\ominus}{C}Cl_3 \xrightarrow[-Cl^{\ominus}]{} :CCl_2$$

$$R_2CH{-}X \xrightarrow{B^{\ominus}} R_2C: + BH + X^{\ominus} \quad \textit{(Carbenoid)}$$

$$R_2CX_2 \xrightarrow{LiR'} R_2C: + LiX + R'X \quad \textit{(Carbenoid)}$$

$$R{-}Hg{-}CCl_3 \xrightarrow{\Delta} RHg{-}Cl + :CCl_2$$

$$R{-}Hg{-}CCl_3 \xrightarrow{\Delta} RHg\cdot + \cdot CCl_3$$

Ylides

$$R_2\overset{\ominus}{C}{-}\overset{\oplus}{S}(CH_3)_2 \xrightleftharpoons{\Delta, h\nu} R_2C: + S(CH_3)_2$$

[a]Further details are given in Reference 18.

Diazo-compounds and azides usually decompose with activation energies around 30 and 35 kcal/mol, respectively. The following reaction[17] is almost thermoneutral.

$$\begin{array}{llll} Ph{-}N_3 \xrightarrow{\Delta} & Ph{-}\ddot{N}: & + N_2 & \Delta H_r^0 \sim 0 \\ \Delta H_f^0 \sim 99 & \sim 100 & 0 & E_a \sim 35 \text{ kcal/mol} \end{array}$$

Although phenyl isocyanate is isoelectronic with phenyl azide, it does *not* give nitrenes thermally[20] due to the fact that the reaction would be endothermic by about 45 kcal/mol. A correspondingly higher activation energy may, therefore, be expected. This energy can be supplied by *photolysis*, and the low-temperature irradiation of isocyanates is, in fact, a useful means of producing nitrenes.[21]

$$\begin{array}{lllll} Ph{-}N{=}C{=}O & \underset{h\nu}{\overset{\Delta}{\not\rightarrow}} & Ph{-}\ddot{N}: & + CO & \Delta H_r^0 \sim 45 \\ \Delta H_f^0 \sim 29 & & \sim 100 & -26 & \text{kcal/mol} \end{array}$$

TABLE 4.3. Formation of Nitrenes[a]

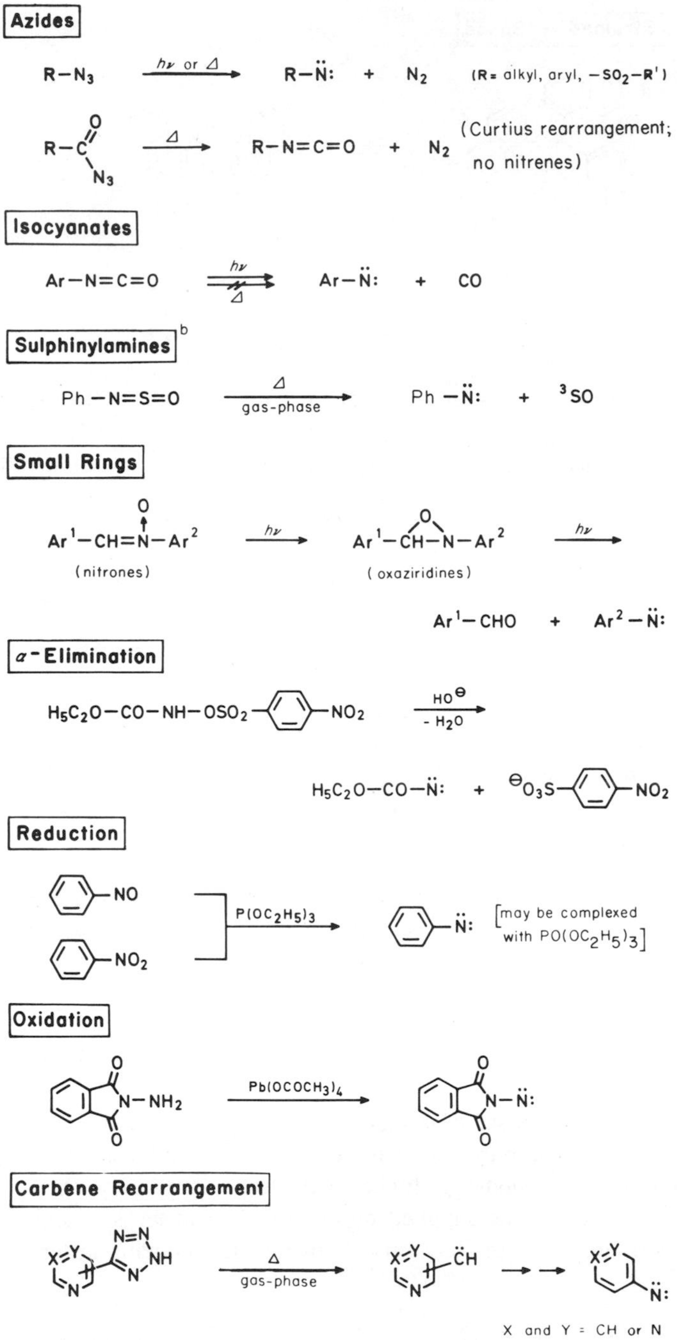

Azides

$$R{-}N_3 \xrightarrow{h\nu \text{ or } \Delta} R{-}\ddot{N}: + N_2 \quad (R = \text{alkyl, aryl, } {-}SO_2{-}R')$$

$$R{-}C(=O){-}N_3 \xrightarrow{\Delta} R{-}N{=}C{=}O + N_2 \quad \text{(Curtius rearrangement; no nitrenes)}$$

Isocyanates

$$Ar{-}N{=}C{=}O \underset{\Delta \text{ (not)}}{\overset{h\nu}{\rightleftharpoons}} Ar{-}\ddot{N}: + CO$$

Sulphinylamines[b]

$$Ph{-}N{=}S{=}O \xrightarrow[\text{gas-phase}]{\Delta} Ph{-}\ddot{N}: + {}^3SO$$

Small Rings

$$\underset{\text{(nitrones)}}{Ar^1{-}CH{=}N(\rightarrow O){-}Ar^2} \xrightarrow{h\nu} \underset{\text{(oxaziridines)}}{Ar^1{-}\overset{O}{CH{-}N}{-}Ar^2} \xrightarrow{h\nu} Ar^1{-}CHO + Ar^2{-}\ddot{N}:$$

α-Elimination

$$H_5C_2O{-}CO{-}NH{-}OSO_2{-}C_6H_4{-}NO_2 \xrightarrow[-H_2O]{HO^{\ominus}} H_5C_2O{-}CO{-}\ddot{N}: + {}^{\ominus}O_3S{-}C_6H_4{-}NO_2$$

Reduction

$$C_6H_5{-}NO,\; C_6H_5{-}NO_2 \xrightarrow{P(OC_2H_5)_3} C_6H_5{-}\ddot{N}: \quad \text{[may be complexed with } PO(OC_2H_5)_3\text{]}$$

Oxidation

$$\text{N-aminophthalimide } (N{-}NH_2) \xrightarrow{Pb(OCOCH_3)_4} \text{phthalimido-}N{-}\ddot{N}:$$

Carbene Rearrangement

$$\text{(X=Y azinyl)-tetrazole (N=N, NH)} \xrightarrow[\text{gas-phase}]{\Delta} \text{(X=Y azinyl)-}\ddot{C}H \rightarrow \rightarrow \text{(X=Y ring)-}\ddot{N}:$$

X and Y = CH or N

[a]Further details are given in Reference 19.
[b]Reference 22.

Carbenoids

Special attention should be drawn to the so-called carbenoids, that is, compounds that behave like carbenes, although often they are complexed rather than free. For example,

$$CHCl_3 \xrightarrow[-BuH]{BuLi} \underset{\text{Carbenoid}}{Li{-}CCl_3} \longrightarrow \text{(dichlorocyclopropane)} + LiCl$$

$$Li{-}CCl_3 \longrightarrow LiCl + :CCl_2\ (\text{free})$$

This carbenoid can be considered a complex between $:CCl_2$ and LiCl. On heating, it dissociates into the free carbene and LiCl:[23]

$$Cl_3C{-}Li \xrightarrow{\Delta} (CCl_3^{\ominus}Li^{\oplus})_{solv.} \quad (\text{not formed})$$

$$Cl_3C{-}Li \xrightarrow{\Delta} :CCl_2 + LiCl$$

Compounds such as **3** are obtained in metal–halogen exchange reactions and can be directly observed by low-temperature ^{13}C NMR spectroscopy.[24] Compound **3** may be regarded as the carbenoid derived from the corresponding cyclopropylidene **4** and lithium bromide.

Br Br $\xrightarrow[-100°C]{BuLi}$ Li Br

3 **4**

The carbenoids probably exist as higher aggregates in solutions containing strong bases. In the presence of *N*,*N*-tetramethyl-1,2-diaminoethane, CCl_3Li exists as the complex[25]

H_3C CH_3 N CCl_3Li N H_3C CH_3

which is stable in hexane solution at −90°C.

The distinction between free and complexed carbenes (carbenoids) can be made by means of competition experiments when the free carbene can be generated from an independent source, for example, a diazo-compound or a diazirine. If the same carbene intermediates are formed from the diazo-compound and from the base-induced reaction, they must show identical reactivities toward various substrates. An elegant method of converting carbenoids into free carbenes has been developed by Moss.[26] If the carbenoid is in fact a carbene complexed with a salt, as described above, it should be possible to liberate the carbene by removing the salt. This can be achieved by taking advantage of the fact that potassium ions are strongly solvated by cyclic polyethers (crown-ethers). Effectively, potassium ions are shielded from other molecules and ions through complexation with the crown-ether:

"18-crown-6"-potassium ion complex

Thus, if phenylbromocarbene, for example, is generated with the aid of a potassium-containing base in the presence of "18-crown-6," the potassium ions are not available for complexation with the carbene, and the *free* carbene results:

$$Ph{-}CHBr_2 \xrightarrow[-t\text{-BuOH}]{KO\text{-}t\text{-Bu}} \underset{\text{carbenoid}}{Ph{-}CBr_2K} \xrightarrow[-KBr]{\text{18-crown-6}} Ph{-}\ddot{C}{-}Br$$

Ph(Br)C(N=N) (phenylbromodiazirine) $\xrightarrow[-N_2]{h\nu}$ Ph—C̈—Br

The *same* carbene is obtained directly by photolysis of phenylbromodiazirine.

If the crown-ether is omitted in the base-induced reaction, only the carbenoid is obtained, and its reactivity is found to differ widely from that of the free carbene. Carbenoids are more stable, less reactive, and more selective than free carbenes.

4.3. DIRECT OBSERVATION

Since carbenes and nitrenes are short-lived and highly reactive species, a direct physical detection requires special techniques. The lifetimes can be prolonged by removal of reaction partners. Spectroscopic observations are therefore made in the gas-phase either at low pressure or with an inert gas as diluent, or at low temperatures, typically 4–77 K, in a matrix or glassy solution. The reactive species are often generated by flash photolysis. For example,

5 $H_2C{=}N_2 \xrightarrow{h\nu} :CH_2 + N_2$

6 $H_2C{=}C{=}O \xrightarrow{h\nu} :CH_2 + CO$

7 $CH_2\langle N{=}N \rangle$ (diazirine) $\xrightarrow{h\nu} :CH_2 + N_2$

Naturally, one has to ensure that it is not an excited state of the starting material, a radical derived therefrom, or a reaction product that is being observed. For such purposes, it is useful to generate the reactive intermediate from several different precursors.

4.3.1. Optical Spectroscopy

Herzberg[2,14] was able to prove that the triplet species formed by photolysis of diazomethane **(5)** contained two and only two hydrogen atoms by observing the following isotopic shifts in the UV spectrum of the intermediate:

Precursor	λ (nm)
$CH_2{=}N_2$	141.45
$CHD{=}N_2$	141.55
$CD_2{=}N_2$	141.58

This does not prove that the species is CH_2; it could be an isomer of diazomethane, for example, diazirine **(7)**. Proof that the species had the symmetrical structure H—X—H came from an analysis of the changes in the rotational fine structure due to substitution of D for H. This originally suggested a H—C—H angle of 180° (Table 4.1). Even the revised angle of ~136° is much too large for the molecule to be diazirine. The H—C—H angle in cyclopropane is ~118°.

Proof that the species contained one and only one carbon atom was derived from the observation of an isotopic shift in the absorption band when $^{13}CH_2{=}N_2$ was employed. The species must, therefore, be $:CH_2$.

Additional absorption lines at longer wavelengths (550–950 nm) were ascribed to a short-lived asymmetrical top molecule, the first singlet state of $:CH_2$ ($\sphericalangle$HCH ≅ 104°).

In order to observe carbenes in reacting systems in solution, a much more stringent time resolution is required. Using laser flash photolysis with a time resolution in the nanosecond or picosecond range, a number of arylcarbenes have been directly observed by UV spectroscopy. Such studies are of the utmost importance since they allow a direct determination of the rate constants for carbene reactions. Some triplet (T) and singlet (S) arylcarbenes detected in this manner are given below together with some absorption (A) and emission (E) maxima.

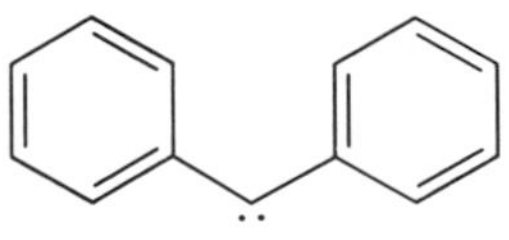

Cl

T — A 301, 465 nm (Refs. 27 and 28); E 480 nm (Ref. 27)

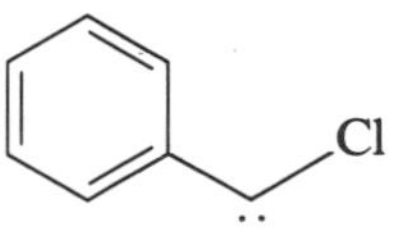

S — A 300 nm (Ref. 29)

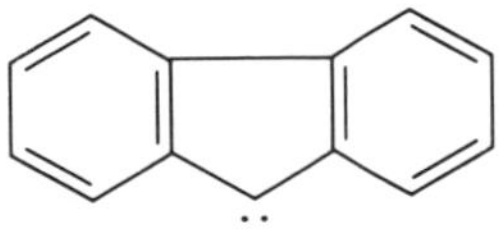

T — A 470 nm (Ref. 30)

Of course, the mere observation of an absorption (or emission) line does not prove that a carbene has been detected. Elaborate studies were necessary before the absorption maximum of triplet fluorenylidene could be unambiguously assigned.[30–32]

Nitrenes, too, have been observed by UV spectroscopy following photolysis, usually of azides, in the gas-phase or in glassy solution.[33] The UV spectra of the triplet arylnitrenes at 77 K resemble those of the electronically related benzyl radicals:[34]

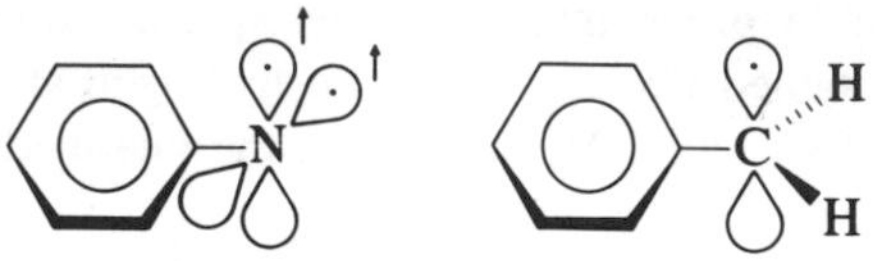

λ_{max} 314, 402 318, 463 nm

The first *singlet* arylnitrene to be observed directly was generated by nanosecond laser flash photolysis of 1-azidopyrene (**8**):[35]

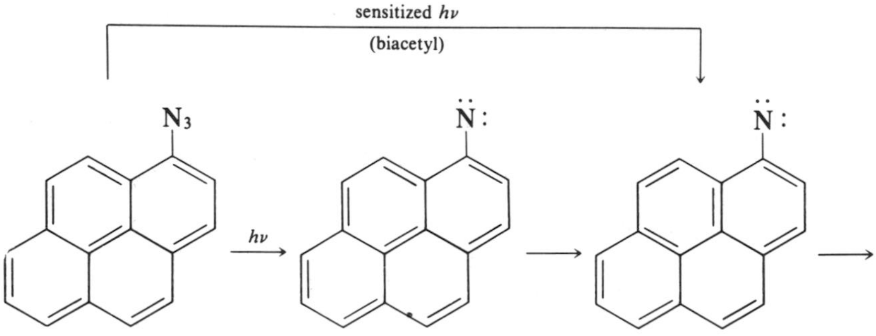

sensitized $h\nu$ (biacetyl)

8 $\xrightarrow{h\nu}$ S_0 $\longrightarrow$ T_1 $\longrightarrow$

S_0: λ_{max} 450 nm, lifetime 22 ns (25°C), 34 ns (77 K)

T_1: λ_{max} 415 nm

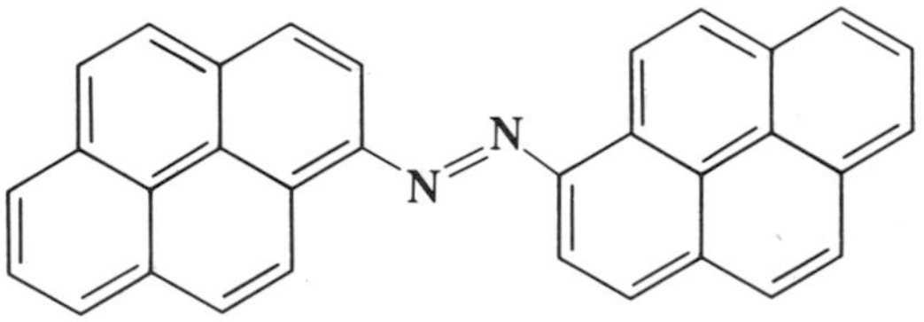

9

The singlet (S_0), absorbing at 450 nm, decayed to the ground state triplet (T_1) (rate constant ~ 10^7 s^{-1} at 25°C), which was also formed without intervention of S_0 in the biacetyl sensitized (triplet state) photolysis of **8**. The electronic excitation of T_1 at 415 nm produced the excited triplet (T_2) which had a lifetime of approximately 7 ns before reverting to T_1. T_1 is eventually consumed by dimerization to azopyrene (**9**), one of the most widespread reactions of arylnitrenes.

In order to obtain infrared spectra of carbenes or nitrenes it is necessary both to have a higher concentration of the species and to preserve it for a longer time. Therefore, low-temperature matrix isolation is used. The most convenient means of producing carbenes is again photolysis of a diazo-compound, but thermal reactions may also been employed. An example is the pyrolytic decomposition of trichloromethylmercury compounds. For example (Ref. 36):

$$\text{Ph—Hg—CCl}_3 \xrightarrow{350°\text{C}} \text{Ph—Hg—Cl} + :\text{CCl}_2$$

Since high temperature is required to decompose the mercury compound, it is subjected to low-pressure pyrolysis (flash vacuum pyrolysis), the products being immediately condensed, together with argon gas, on a KBr disk at temperatures near 10 K:

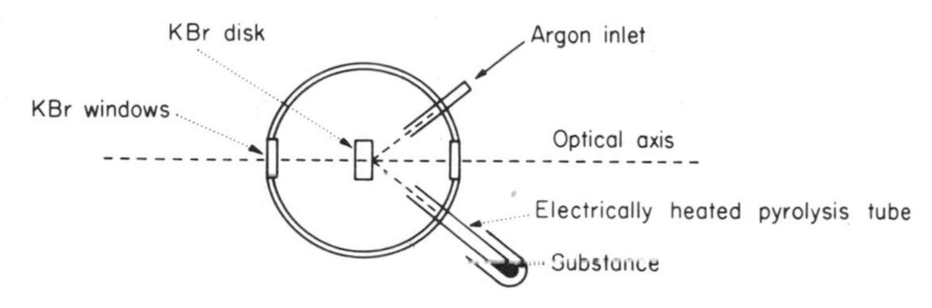

The IR spectrum of the matrix thus obtained shows, apart from signals due to $PhHgCCl_3$ and PhHgCl, peaks at 720 and 746 cm^{-1} ascribed to $:CCl_2$. The same peaks were observed when $:CCl_2$ was generated from other sources.[37] After warming to 40 K these peaks disappeared, and new peaks due to the dimer C_2Cl_4 developed in their place. Note that while $:CH_2$ has a triplet ground state, the *halocarbenes are singlets.*[10,38] This will be discussed in Section 4.6.2. Because they are singlets, direct observation by means of ESR spectroscopy is not possible.

Cyclopentadienylidene (**10**) has been obtained by matrix photolysis at 12–20 K and identified by its UV (λ_{max} 296 nm) and IR spectra (strongest band at 703 cm^{-1}). The carbene dimerized to fulvalene (**11**) above 20 K, or, when generated in a CO matrix, reacted with the host to give the ketene **12**.[39]

N_2 —($h\nu$, N_2, 20 K)→ **10** —(20–35 K)→ **11**

10 —(20 K, CO)→ **12**

The observation of matrix IR spectra of arylcarbenes will be described in connection with carbene and nitrene rearrangements (Section 4.4.5).

4.3.2. ESR Spectroscopy

The basic principles of ESR spectroscopy were presented in Section 2.2.2. In order for a carbene or nitrene to be observable by ESR, it must be in the triplet state. As mentioned earlier, the halocarbenes have singlet ground states. The same appears to be true for phenylhalocarbenes,

Cl and Br

which cannot be observed by ESR spectroscopy.

Some D and E values for triplet carbenes and nitrenes are given in Tables 4.4 and 4.5. In a truly degenerate system such as H—N, $E = 0$. In most nitrenes, E is close to zero. In most carbenes the σ and p orbitals are different, thus rendering E significantly different from zero. In conjugated systems (e.g., phenylnitrene and phenylcarbene), the out-of-plane p electron will be delocalized over the molecule:

N ⟷ N

This is a normal benzyl radical resonance, which would be expected to lower the energy of the molecule by approximately 12.5 kcal/mol, as in benzyl itself. Since the two radical electrons are now further apart, the D-value will decrease. Consequently, D-values for aromatic systems are much smaller than for aliphatic ones.

Nitrogen is more electronegative than carbon, so that we may expect a nitrene to hold its electrons closer to the nucleus than a carbene does. The average distance between the two nitrene electrons is therefore smaller, the interaction between them stronger, and the D-values larger than in carbenes. For the same reason, singlet–triplet splittings are larger in nitrenes than in carbenes: The electron correlation energy is larger. The singlet–triplet splitting in NH is about 35 kcal/mol.[40]

If the nitrene is isolated from the benzene ring by an SO_2 group, the D-factor rises sharply; however, not as much as when the phenyl group is replaced by alkyl (compare Ph—SO_2—N and CH_3—SO_2—N). This indicates that a nitrene p electron is able to delocalize partly through the SO_2 group.

If the carbene or nitrene can delocalize over several rings, the D-factor decreases. α-Naphthylnitrene is more highly delocalized than β-naphthynitrene; the same holds for the carbenes. Although the resonance energies of the naphthylmethyl radicals are only approximately known, simple MO calculations, resonance theory, and experiment (bond dissociation energies) indicate a higher resonance energy for α-naphthylmethyl.

In diphenylcarbene (**13**), one could expect a low D-value due to conjugation of one p electron with each benzene ring. However, comparison with fluorenylidene (**15**) (Table 4.4), which certainly is bent and in which one electron is forced to occupy the in-plane σ orbital, indicates almost the same degree of delocalization in the two carbenes. ESR measurements on diphenylcarbene in a host matrix of 1,1-diphenylethylene have shown that the carbene is not only bent, but the phenyl groups are twisted at a dihedral angle of 36°:[41]

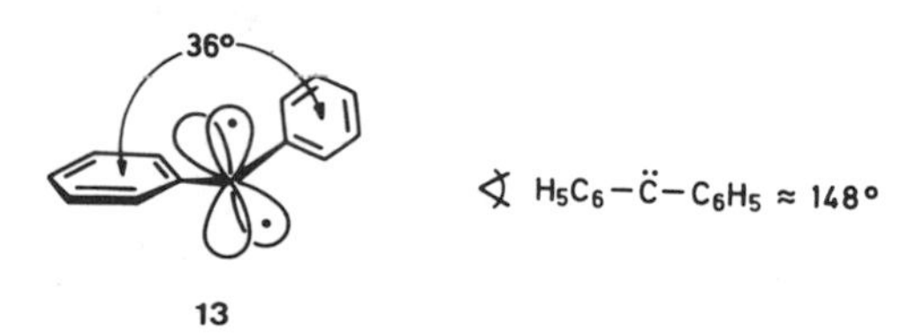

A linear carbene is obtained in the 9,9-dianthrylcarbene[42] (**14**) (Table 4.4); $E = 0.0$. In this case, a bent structure would induce severe sterical hindrance between the two large substituents.

4.4. REACTION TYPES

4.4.1. Investigational Techniques

The only rigorous way to ascertain that a *particular* reactive intermediate is involved in a given reaction is to observe it directly and monitor its reaction with a spectroscopic method. Kinetics alone may, of course, give evidence for the involvement of an intermediate, but only the combination with chemical intuition will define its nature. Since the rate constants for carbene reactions in solution typically fall in the range 10^6–10^9 liter mol^{-1} s^{-1}, the upper limit being the diffusion-controlled rate, very fast spectroscopic methods are required. Nanosecond and picosecond flash photolysis have been mentioned. The decay of an intermediate may be followed by absorption[28–31] or emission[43] UV or by ESR spectroscopy.[44] An intermediate may be identified by static spectroscopy of low-temperature matrices or glassy solutions, but this does not necessarily prove that the same intermediate is involved in a chemical reaction at, say, room temperature. A carbene observed by ESR at low temperature will be the (ground state) triplet, but the corresponding singlet and other intermediates possibly involved in a solution reaction will escape detection.

A general reaction scheme, exemplified for the decomposition of a diazo-compound, can be formulated as follows:

$$\begin{array}{ccccccc}
CR_2{=}N_2 & \xrightarrow{h\nu} & {}^1CR_2{=}N_2^* & \xrightarrow[-N_2]{\Delta \text{ or } h\nu} & {}^1{:}CR_2 & \xrightarrow{\text{fast}} & \text{singlet reaction} \\
 & \searrow^{h\nu}_{\text{sensitized}} & \big\updownarrow \text{ISC} & & \big\updownarrow \text{ISC} & & \\
 & & {}^3CR_2{=}N_2^* & \xrightarrow{-N_2} & {}^3{:}CR_2 & \xrightarrow{\text{(fast)}} & \text{triplet reaction}
\end{array}$$

Scheme 1

TABLE 4.4. ESR Parameters for Triplet Carbenes

Carbene	$D/hc(\text{cm}^{-1})$	$E/hc(\text{cm}^{-1})$	Reference
$\ddot{C}H_2$	~0.6881	~0.00346	a
$(t\text{-}C_4H_9)_2C$:	0.689	0.039	b
$F_3C-\ddot{C}H$	0.712	0.021	c
$(F_3C)_2\ddot{C}$	0.744	0.0437	c
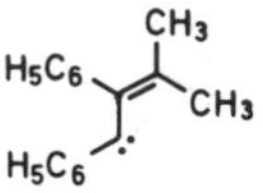 CH_3, H_5C_6, CH_3, H_5C_6	0.3580	0.0206	d
$HC\equiv C-\ddot{C}H$	0.628	0.000	e
$Ph-C\equiv C-\ddot{C}H$	0.541	0.0035	e
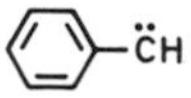 $-\ddot{C}H$	0.5098	0.0249	f
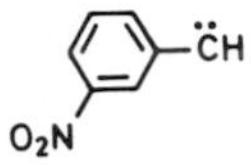 $\ddot{C}H$, O_2N	0.5230	0.0243	f
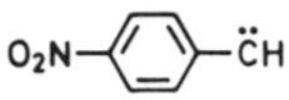 O_2N- $-\ddot{C}H$	0.4859	0.0224	f
$\ddot{C}H$	*anti:* 0.4555 *syn:* 0.4347	0.0202 0.0208	f
$\ddot{C}H$	*anti:* 0.4711 *syn:* 0.4926	0.0243 0.0209	f
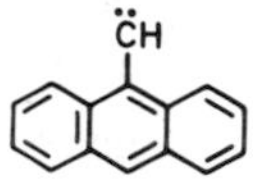 $\ddot{C}H$	0.3008	0.0132	f
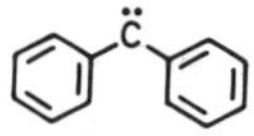 $\ddot{C}$ 13	0.4050	0.0194	f
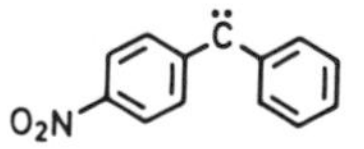 $\ddot{C}$, O_2N	0.3765	0.0175	f
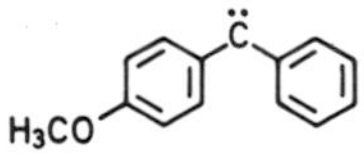 $\ddot{C}$, H_3CO	0.4042	0.0194	f
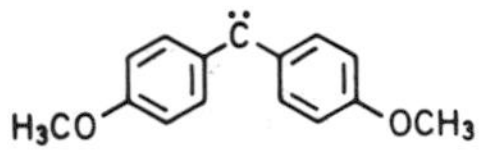 $\ddot{C}$, H_3CO, OCH_3	0.4065	0.0193	f,g
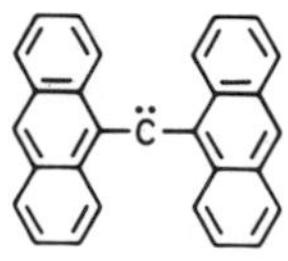 $-\ddot{C}-$ 14	0.3008	0.0	f
 10	0.4089	0.0120	f
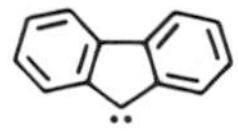 15	0.4078	0.0283	f

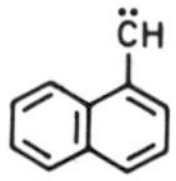

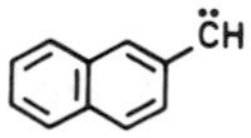

TABLE 4.4. ESR Parameters for Triplet Carbenes *(Continued)*

Carbene	$D/hc(\mathrm{cm}^{-1})$	$E/hc(\mathrm{cm}^{-1})$	Reference
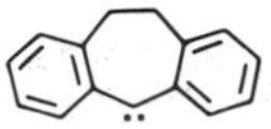	0.3932	0.0170	h
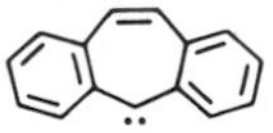	0.3787	0.0162	h
$(H_3C)_3Si-\ddot{C}-H$	0.613	0.00145	i

[a] E. Wasserman, W. A. Yager, and V. J. Kuck, *Chem. Phys. Lett.*, **7**, 409 (1970).
[b] J. E. Gano, R. H. Wettach, M. S. Platz, and V. P. Senthilnathan, *J. Am. Chem. Soc.*, **104**, 2326 (1982).
[c] E. Wasserman, L. Barash, and W. A. Yager, *J. Am. Chem. Soc.*, **87**, 4974 (1965).
[d] G. E. Palmer, J. R. Bolton, and D. R. Arnold, *J. Am. Chem. Soc.*, **96**, 3708 (1974).
[e] R. A. Bernheim, R. J. Kempf, J. V. Gramas, and P. S. Skell, *J. Chem. Phys.*, **43**, 196 (1965).
[f] Reference 42.
[g] For further mono- and disubstituted diphenylcarbenes, see R. W. R. Humphreys and D. R. Arnold, *Can. J. Chem.*, **55**, 2286 (1977).
[h] I. Moritani, S. Murahashi, M. Nishino, Y. Yamamoto, K. Itoh, and N. Mataga, *J. Am. Chem. Soc.*, **89**, 1259 (1967).
[i] M. R. Chedekel, M. Skoglund, R. L. Kreeger, and H. Shechter, *J. Am. Chem. Soc.*, **98**, 7846 (1976).

The rate determining step is the formation of the carbene. It may reasonably be expected, therefore, that the rate of decomposition of the diazo-compound is essentially independent of solvent and additives. In many instances, the diazo-compound can react directly with the substrate, leading to the same product as that expected from a carbene reaction.[45] This may be detected kinetically, or, in favorable cases, by isolation of the primary reaction product. As an example, the photolysis of diazofluorene **(16)** in norbornene results in the precipitation of the pyrazoline **17** which is thermally converted to the cyclopropane **18**, the expected product of

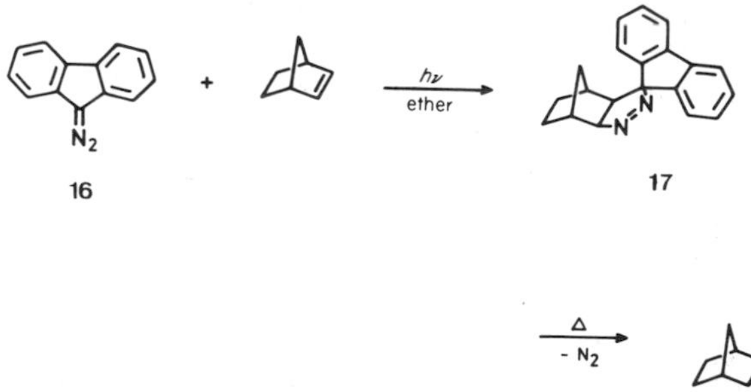

N_2 + → hν, ether →

16 17

→ Δ, $-N_2$ →

18

carbene addition to a double bond.[45a] The N_2 expulsion from pyrazolines like **17** was discussed in Chapter 3 and proceeds with activation energies between 30 and 40 kcal/mol, that is, about the same as required to decompose diazo-compounds and azides.

TABLE 4.5. ESR Parameters for Triplet Nitrenes[a]

Nitrene	$D/hc(\mathrm{cm}^{-1})$	$E/hc(\mathrm{cm}^{-1})$	ρ_N[b]
$\cdot\ddot{N}H$ (calc.)	1.86	0	
$H_3C-\ddot{N}\cdot$	1.595	<0.003	
$(H_3C)_3C-\ddot{N}\cdot$	1.625	<0.002	
$H_5C_2-O-C(=O)-\ddot{N}\cdot$	1.603	0.0215	
$C_6H_5-SO_2-\ddot{N}\cdot$	1.428	<0.002	
$C_6H_5-\ddot{N}\cdot$[c]	0.9978	<0.002	0.63
$m\text{-}O_2N-C_6H_4-\ddot{N}\cdot$	0.989	<0.002	0.62
$p\text{-}O_2N-C_6H_4-\ddot{N}\cdot$	1.065	<0.002	0.67
$p\text{-}H_3CO-C_6H_4-\ddot{N}\cdot$	0.9978	0.0039	0.63
1-naphthyl$-\ddot{N}\cdot$	0.7890	<0.002	0.50
2-naphthyl$-\ddot{N}\cdot$	1.0083	0.003	0.63
3-pyridyl$-\ddot{N}\cdot$	1.0048	<0.003	0.64

[a]Data from E. Wasserman, *Prog. Phys. Org. Chem.*, **8,** 319 (1971).
[b]*p*–π spin density on nitrogen.
[c]For substituted phenylnitrenes, see also J. H. Hall, J. M. Fargher, and M. R. Gisler, *J. Am. Chem. Soc.*, **100,** 2029 (1978).

As another example, phenyl azide decomposes in hydrocarbon solvents with an activation energy of approximately 35 kcal/mol.

$$\mathrm{Ph{-}N_3} \longrightarrow \mathrm{Ph{-}N{:}} + \mathrm{N_2}$$

In solvents containing olefinic double bonds, the azide is consumed much more rapidly; that is, E_a is smaller. The isolable product is the aziridine **20**, which is formed in a 1,3-dipolar cycloaddition of the azide to the olefin. The triazoline **19** so produced usually decomposes under the reaction conditions, giving **20**. In certain cases, triazolines like **19** may be detected or isolated.[46]

The production of the *same* carbene or nitrene from different precursors and the observation that the *same* products are formed in each case constitutes good evidence

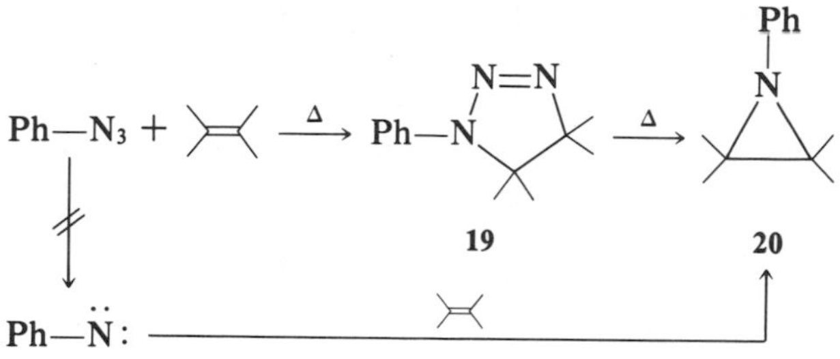

$$Ph{-}N_3 + \text{alkene} \xrightarrow{\Delta} \mathbf{19} \xrightarrow{\Delta} \mathbf{20}$$

$$Ph{-}\ddot{N}: \xrightarrow{\text{alkene}} \mathbf{20}$$

Scheme 2

that the presumed intermediate is indeed involved. The best evidence is obtained from competition experiments (i.e., experiments in which several substrates—or different positions within a substrate—can compete for the reactive intermediate). The identity of phenylcarbene derived from different sources was established in this manner (Scheme 3).[47]

Having established that the desired reactive intermediate is formed, it is of interest to establish its spin, singlet, or triplet. This may be achieved through an examination of the products and their stereochemistry since singlets and triplets often react differently, as will be discussed later. However, since the interpretation of this difference is based on some preconceived ideas of how singlets and triplets should react, it is necessary to have some independent evidence as to the spin state.

The principle of spin conservation (the Wigner–Witmer rules[48]) demands that thermolysis or direct photolysis of a diazo-compound give a singlet carbene plus (singlet) nitrogen (cf. Scheme 1). By triplet sensitized photolysis (e.g., with benzophenone), it is possible to populate the triplet excited diazo-compound, and this may then give a triplet carbene directly. One cannot exclude that some direct photolysis also takes place, giving the singlet carbene. The singlet-excited diazo-compound may in turn be specifically populated by singlet energy transfer, which can be achieved using aromatic hydrocarbons as sensitizers. These techniques should then provide opportunities for more or less specific generation of singlet and triplet intermediates.

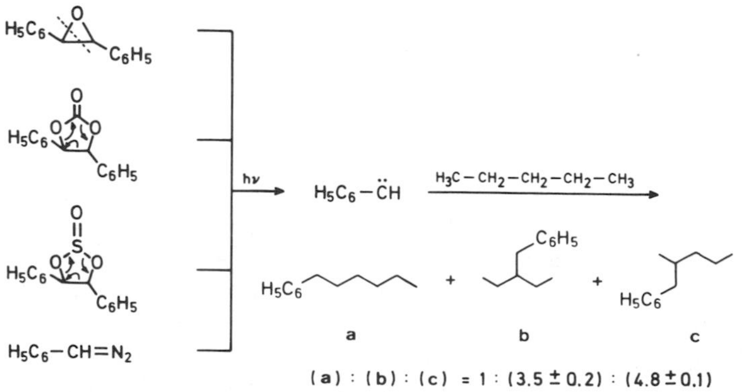

$$H_5C_6{-}CH{=}N_2 \text{ (and other precursors)} \xrightarrow{h\nu} H_5C_6{-}\ddot{C}H \xrightarrow{H_3C-CH_2-CH_2-CH_2-CH_3} \mathbf{a} + \mathbf{b} + \mathbf{c}$$

(a) : (b) : (c) = 1 : (3.5 ± 0.2) : (4.8 ± 0.1)

Scheme 3

Fortunately, this hope is often borne out by experiment, but it is not foolproof. Singlet excited diazo-compounds may undergo intersystem crossing to the triplet excited states before decomposition to carbenes. Furthermore, the singlet and triplet carbenes can, themselves, interconvert. If the singlet–triplet splitting is small, this interconversion will be fast.

If the singlet is the ground state (e.g., in the halocarbenes), there is less complication: Only singlet reaction is expected.[57] If the triplet is the ground state, an initially formed singlet carbene will in time convert to the triplet. If one dilutes the substrate with an inert solvent, the singlet carbene will undergo several unproductive collisions before encountering a molecule with which it can react. Therefore, the probability of intersystem crossing to the triplet ground state is increased. If a difference in product distribution is observed simply as a consequence of dilution, this may be related to singlet–triplet conversion.

Using this technique, it was concluded that the following carbenes have triplet ground states:[49–51]

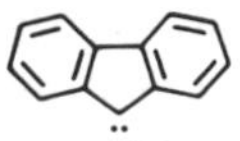

15

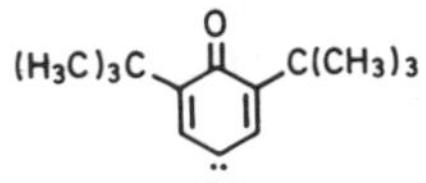

$(H_3C)_3C$ ... $C(CH_3)_3$

21

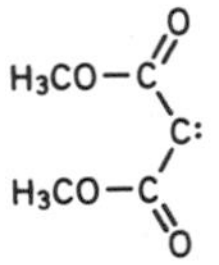

$H_3CO-C(=O)-\ddot{C}-C(=O)-OCH_3$

22

$H-\ddot{C}-CO-OC_2H_5$

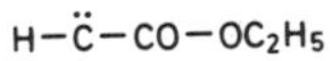

23

This is not the only evidence: The ESR spectra of **15** (Table 4.4) and **23** (Ref. 52) have been recorded, and calculations[53] support a triplet ground state for **23** with a singlet–triplet splitting of about 7 kcal/mol. Yet, when **23** is generated in solution, without employing the dilution technique, the reactions observed are those expected of a singlet.[54,55]

The manifestation of the dilution effect requires that the rates of intersystem crossing and of the product-forming reactions be of the same order of magnitude. The effect is due not only to a delaying of the singlet reaction, but also to catalysis of intersystem crossing: This spin-forbidden reaction is "catalyzed" by spin–orbit coupling (i.e., coupling between the electron spin and orbital momenta).[56a] This coupling mixes the two states. In order that the spin inversion be successful, however, there must be a continuum of vibrational levels through which one electronic state can go over into the other. Such a continuum is not available for small molecules like CH_2, but it may well be for larger ones. The necessary continuum can easily be provided, however, by collision with another molecule, in gas or liquid phases. Accordingly, the conversion of triplet excited ethylene to the ground state obeys second-order kinetics:[56b]

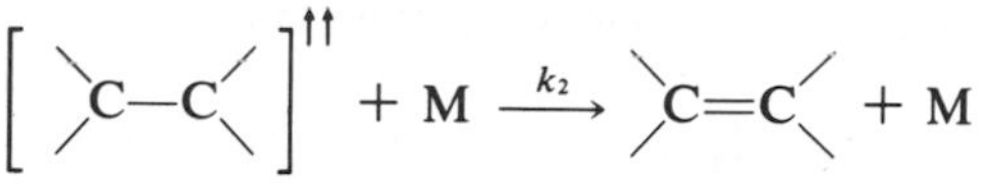

$$\left[\rangle C{-}C\langle \right]^{\uparrow\uparrow} + M \xrightarrow{k_2} \rangle C{=}C\langle + M$$

T_1 S_0

where M is an inert molecule or the wall of the reactor.

Similarly, Herzberg's triplet methylene was produced by collisions between singlet CH_2 and nitrogen. By decreasing the nitrogen pressure, this process was retarded, and it became possible to observe the singlet CH_2 directly.[2]

The collision-induced intersystem crossing is more effective with large molecules M, for which reason it is also referred to as "the heavy-atom effect." Methylene bromide and hexafluorobenzene are particularly popular in solution phase studies.

Singlet carbenes or nitrenes may be selected for chemical study by rapid removal of any triplets present: Since triplet carbenes may be regarded as 1,1-diradicals, they are expected to be rapidly trapped by free radical scavengers such as butadiene or 1,1-diphenylethylene. Thus the stereospecificity of cyanocarbene **(24)** in the addition to *cis*-2-butene increased on addition of 1,1-diphenylethylene (cf. Section 4.4.3).[58]

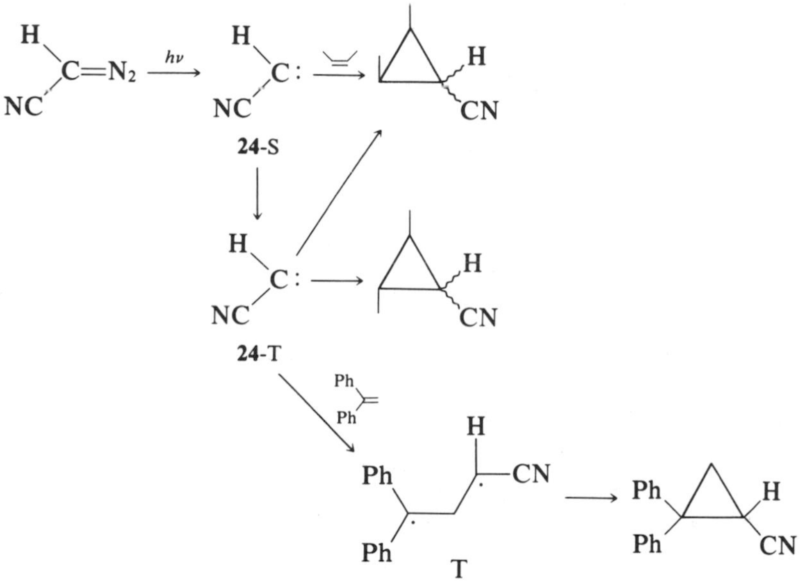

H(NC)C=N2 —hν→ H(NC)C: (**24**-S) → cyclopropane (H, CN)

24-S ↓ H(NC)C: (**24**-T) → cyclopropane (H, CN)

24-T —Ph2C=CH2→ Ph2C•–CH2–C•H(CN) (T) → 1,1-diphenyl-2-cyanocyclopropane (Ph, Ph, H, CN)

Triplet carbenes may also be selectively trapped by nitrous oxide, giving iminoxyls **(25)** observable by ESR spectroscopy:[59]

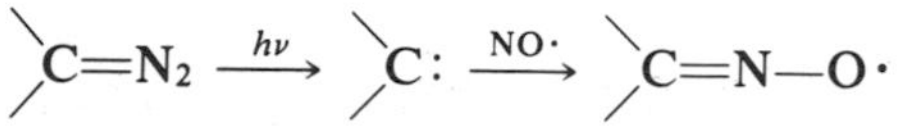

$$\rangle C{=}N_2 \xrightarrow{h\nu} \rangle C: \xrightarrow{NO\cdot} \rangle C{=}N{-}O\cdot$$

25

Some carbenes detected in this (indirect) manner are

$$CH_3—\ddot{C}—CH_3,\ Ph—\ddot{C}H,\ Ph—\ddot{C}—C_2H_5,$$

$$Ph—CO—\ddot{C}H,\ EtOCO—\ddot{C}—COOEt,$$

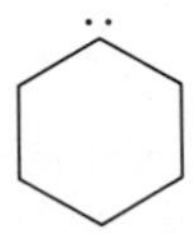

The ESR intensities of the iminoxyls increased sharply when sensitized photolysis was used, whereby triplet carbenes can be formed directly. It was not possible to trap triplet nitrenes with NO under these conditions, however.[59] This is in line with other indications that nitrenes are less reactive and, at least in some cases, more stable than the corresponding carbenes[60] (see also Section 4.6.1).

4.4.2. Dimerization

Triplet carbenes, like radicals, dimerize in a very fast and exothermic reaction:[61]

$$^3CH_2 + {}^3CH_2 \longrightarrow CH_2{=}CH_2$$

$$\Delta H_r^0 \simeq -176 \text{ kcal/mol}$$

$$k = (3.2 \pm 0.9) \times 10^{10} \text{ liter mol}^{-1} \text{ s}^{-1}$$

The recombination of two triplets to give a singlet product is spin-allowed.

The efficiency of this process obviously depends on the concentration of the carbene. In solution, "dimers" are often formed by attack of the carbene on the carbene precursor, whose concentration is much higher:

$$R_2C\!: + R_2C{=}N_2 \longrightarrow R_2C{=}CR_2 + N_2$$

$$\searrow R_2C{=}N—N{=}CR_2$$

This reaction also leads to azines. The azines themselves decompose at much higher temperatures with predominant N—N bond cleavage.

Triplet ground-state phenylnitrene dimerizes in solution in two ways:[63]

$$Ph\text{-}N_3 \xrightarrow[or\ h\nu]{\Delta} {}^1Ph—\ddot{N}\!: \longrightarrow {}^3Ph—\ddot{N}\!: \qquad (1)$$

$$2\ {}^3Ph—\ddot{N}\!: \longrightarrow Ph—N{=}N—Ph \qquad (2)$$

$$Ph—\ddot{N}\!: + Ph—N{=}\overset{\oplus}{N}{=}\overset{\ominus}{N} \longrightarrow Ph—N{=}N—Ph + N_2 \qquad (3)$$

Azobenzene is the main product of decomposition of phenyl azide, both in solution and in the gas-phase. Even at low pressure (0.05 mm) yields up to ~90% of

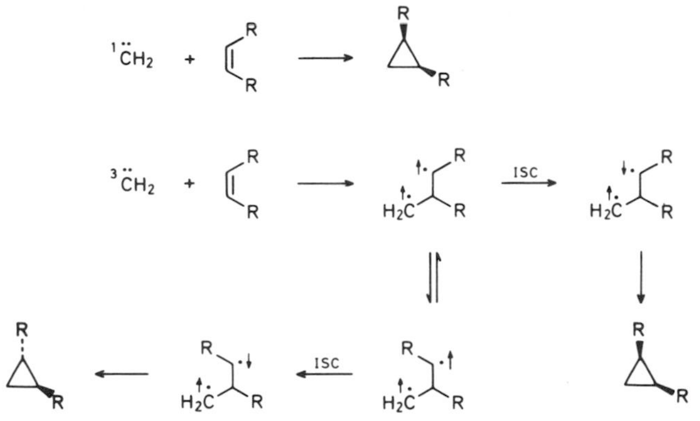

Scheme 4

azobenzene can be obtained,[20,60] indicating that phenylnitrene is relatively stable and unreactive. The nitrenes can be generated by multiple rearrangement from nonazide precursors, so that only Eq. (2) can contribute to the dimerization. The yield of azobenzene remains high.[64]

4.4.3. Addition to Double Bonds

Skell and Woodworth[65] argued that the addition of a singlet carbene to an olefin could be a one-step reaction, two new bonds being formed simultaneously. A triplet, in contrast, cannot give a cyclopropane in a one-step reaction, since this process is spin-forbidden. Therefore, it was postulated that the triplet would add in two steps, leading first to a triplet diradical intermediate (Scheme 4). Since rotations can take place in this intermediate prior to intersystem crossing, loss of stereochemistry would result. Thus the Skell rule can be formulated: *Singlet carbenes add to olefins stereospecifically; triplet carbenes, nonstereospecifically.* If, however the rate of spin inversion in the 1,3-diradical is much higher than the rate of rotation, the triplet may still add stereospecifically. Likewise, there is no rule that forbids the *singlet* carbene to add in a two-step manner. We have seen, however, that the singlet 1,3-diradical is *not* a stable intermediate, but rather a transition state (Chapter 3). Therefore, the addition of singlet carbenes to double bonds should indeed be energetically concerted one-step reactions.

It is clear that, in spite of its concertedness, the singlet carbene addition cannot by synchronous. The direct approach of $:CH_2$ to an olefin in a transition state that has the geometry of the product cyclopropane is forbidden by orbital symmetry: The transition state contains four electrons and is antiaromatic:

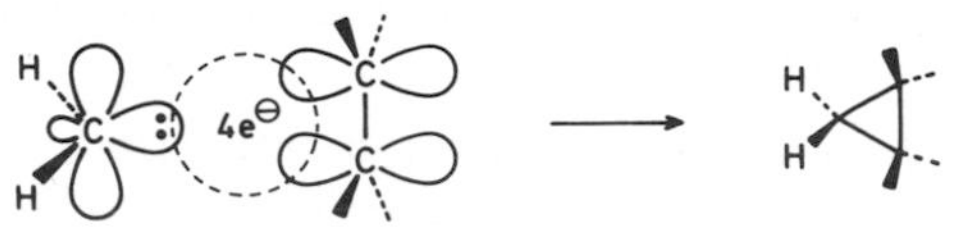

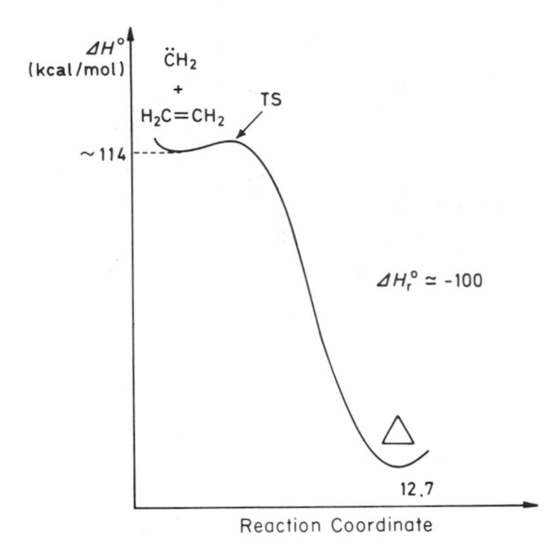

Figure 4.2. Addition of singlet methylene to ethylene.

It is not necessary, however, that the transition state be symmetrical. The reaction is highly exothermic, has a very low activation energy, if any, and hence the transition state is very reactantlike (Hammond postulate, see Figure 4.2). This means that the transition state geometry is very loose; the cyclopropane geometry is only reached later. Calculations indicate that the reactants approach each other initially with the geometry **26**, which has only two electrons[66] in the transition state and is, therefore, "allowed." Since the carbene in **26** "attacks" with its empty *p* orbital, this approach reflects the electrophilic nature of the carbene. Note, however, that with the *same* geometry, the filled carbene σ orbital can overlap with the vacant π^* orbital or LUMO of the alkene (see **27**).[67] This is synonymous with a *nucleophilic*

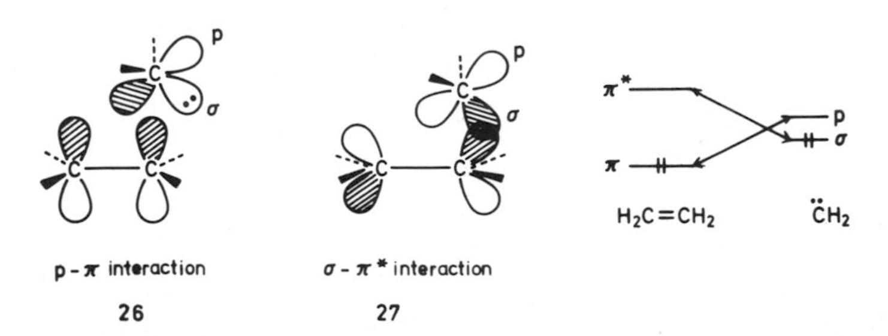

attack by the carbene. Whether the latter will be important or not depends on the π^*-energy of the double bond (see level correlation diagram). If this is low enough, the carbene can show nucleophilic properties (for further discussion, see Section 4.4.5).

Experimentally, the Skell hypothesis has been confirmed. The halocarbenes, which are singlets, add stereospecifically. Fluorenylidene (**15**), generated by photolysis of 9-diazofluorene, adds mainly stereospecifically, the first formed species being a singlet. Dilution with hexafluorobenzene causes demotion to the triplet ground state and nonstereospecific addition.[49]

Further experiments[30] showed that triplet fluorenylidene (**15**-T) could be directly observed by UV spectroscopy using laser flash photolysis [λ_{max}(**15**-T) = 470 nm].

1 :⇅ 15 - S dilution 3 :⇈ 15 - T

Its lifetime at room temperature was 95 ns in hexafluorobenzene solution. The singlet (**15**-S) could not be observed under these conditions, and it was concluded that its lifetime must be less than 5 ns.

Another transient intermediate absorbing at 400 nm was observed upon photolysis of 9-diazofluorene in acetonitrile solvent. This species, first believed to be triplet fluorenylidene (**15**-T),[32] was later shown[30,68] to be instead the nitrile ylide formed in a reaction with acetonitrile:

15 (S or T) $\xrightarrow{CH_3CN}$ N C CH_3 400 nm

Nitrenes behave similarly (Scheme 5). Ethoxycarbonylnitrene generated thermally (singlet) adds mainly stereospecifically. If the olefin concentration is lowered, more intersystem crossing to the nonstereospecific triplet in encountered. Photolysis produces approximately 30% of the nitrenes directly in the triplet state with correspondingly more nonstereospecific addition. The triplet can be removed by rapid reaction with radical scavengers, for example, butadiene. Note that part of the reaction is a non-nitrene route via addition of the azide to the alkene, giving a triazoline as intermediate product. The triazoline undergoes stereospecific nitrogen extrusion (Scheme 5).[69a]

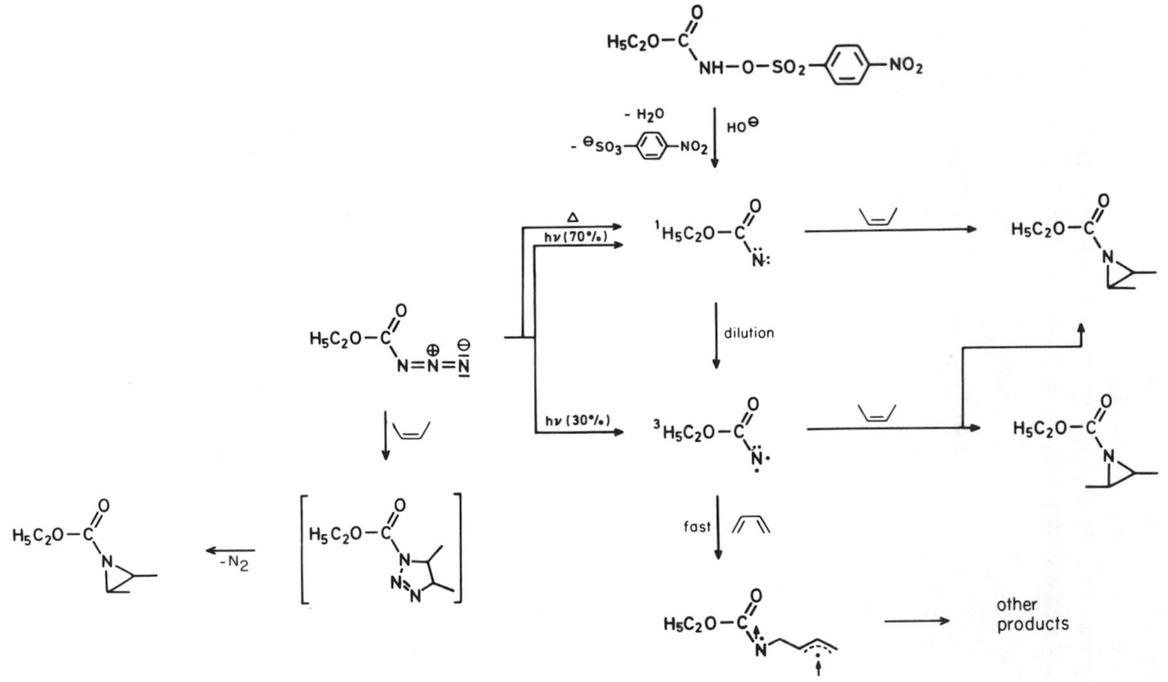

Scheme 5

In contrast to alkoxycarbonylnitrenes, arylnitrenes do not as a rule undergo addition reactions unless they are made sufficiently electrophilic by the introduction of electron-withdrawing substituents.[69b] Instead, they dimerize to azo-compounds, abstract hydrogen from solvent to give anilines, or insert into C—H bonds.[19,69b] Triazinylnitrene is an example of an aromatic nitrene adding to a C=C bond in benzene to give an azepine derivative:[69c]

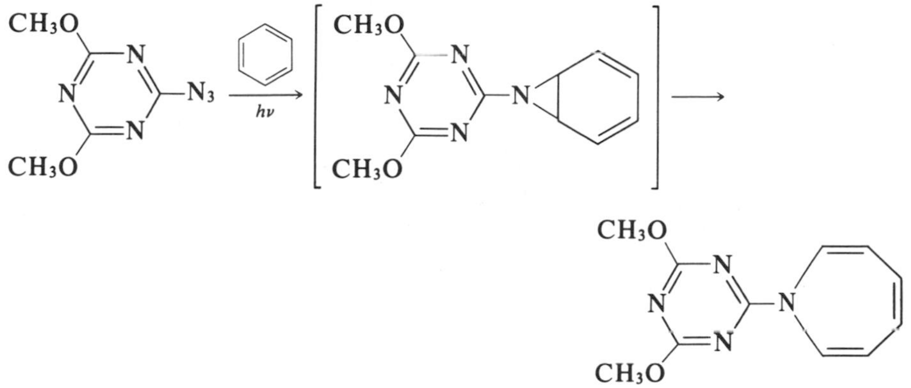

The corresponding reactions of ethoxycarbonylnitrene and of many carbenes are also known.[18,19]

The observation of stereospecific addition of a carbene (or nitrene) is not in itself sufficient evidence for singlet reaction. The carbene **28**, which has a triplet ground state, undergoes stereospecific addition. Neither dilution nor triplet-sensitized photolysis changes this.[70] The UV spectra of **28**, produced by flash photolysis at 77 K and at room temperature were identical. It was concluded that it is *triplet* **28** that adds stereospecifically. This was explained in terms of formation of a charge-transfer complex between the carbene and the olefin. Evidence for this hypothesis was found

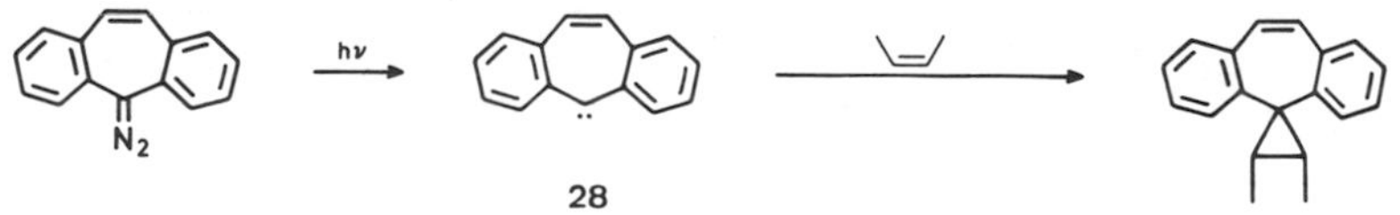

when the diazo-compound was irradiated in the presence of tetracyanoethylene (an electron acceptor). The absorption spectrum of the tetracyanoethylene radical anion was observed. In this reaction, the carbene is acting as a nucleophile toward the alkene.

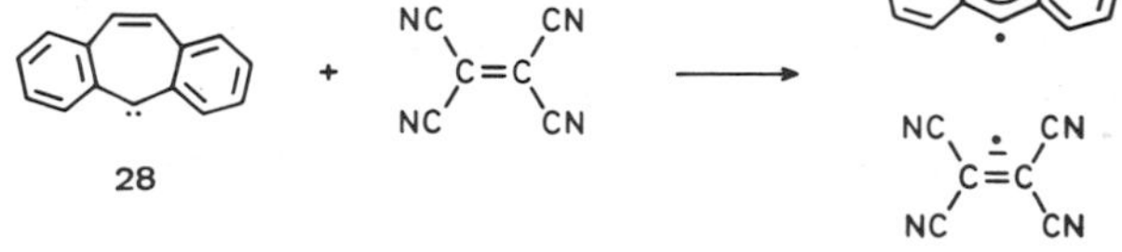

This experiment does not reveal the spin multiplicity of the reacting carbene. It is possible that the singlet–triplet splitting in **28** is very small, resulting in rapid intersystem crossing in both directions. The concentration of the *singlet* will therefore always be high enough that it dominates the reaction. This would explain the stereospecific addition. Such a rapid singlet–triplet interconversion has been postulated for phenyl- and diphenylcarbenes.[28]

$$^{1}Ph_2C: \rightleftharpoons {}^{3}Ph_2C: \qquad (4)$$

13-S **13**-T

The latter adds nonstereospecifically to *cis*-2-butene, and the stereospecificity cannot be altered by dilution or even by the addition of oxygen as a radical trap. If one could accelerate the rate of addition of one of the carbenes, the triplet say, it might be possible to bring this rate into the range of the singlet–triplet interconversion, Eq. (4). Indeed, this can be done by letting the two carbenes compete for styrene, which, being a conjugated alkene, acts as a "radical trap," preferentially scavenging the triplet carbene:

$$C_6H_5CH{=}CH_2 + {}^{3}R_2\dot{C}\cdot \longrightarrow C_6H_5\dot{C}H{-}CH_2{-}\dot{C}R_2$$

29

The resulting diradical **29** should collapse to a phenylcyclopropane in a nonstereospecific manner. The singlet Ph_2C: continues to add stereospecifically. The following results were obtained on dilution with hexafluorobenzene:[71]

$$C_6H_5CH{=}CHD + (H_5C_6)_2\ddot{C} \longrightarrow \text{cis-1,1,2-triphenyl-3-deuteriocyclopropane} + \text{trans isomer}$$

mol% C_6F_6	% *cis*	% *trans*
0	65	35
36	60	40
50	59	41
80	56	44
90	55	45
95	56	44

This *trans/cis* ratio, that is, the degree of nonstereospecificity, increases with the extent of dilution. *This indicates that the triplet is the ground state* and that it is *a little* more stable than the singlet. In the presence of ~90 mol % C_6F_6, there can be no more than 8–12% singlet addition still taking place, that is, $100 - [2 \times (45 \pm 1)]$.

In the case of $:CH_2$ itself, the singlet lies $\geqslant 8$ kcal/mol above the triplet (see Section 4.1). From the relation

$$-\Delta G^0 = RT\ln K$$

one calculates an equilibrium constant $K \geqslant 7.4 \times 10^5$ at room temperature. Therefore, there is little chance that an initially formed triplet $:CH_2$ will give any observable singlet reaction. For diphenylcarbene, it is quite different. The singlet–triplet splitting derived from direct spectrophotometric measurements of the rates of interconversion of the two spin states is $\Delta G^0 \sim 3$ kcal/mol.[28] This gives $K \sim 160$ at room temperature. Thus the singlet–triplet equilibrium can give rise to significant amounts of singlet product even though the triplet is the ground state.

Apparently, an even smaller singlet–triplet separation obtains in *p,p'*-dianisylcarbene (**30**). Here the singlet state would be stabilized by resonance between the

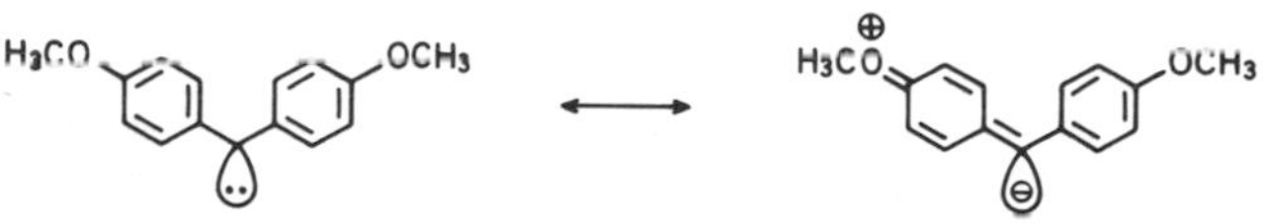

30

methoxy groups and the empty carbene *p* orbital. At low concentrations, the ESR spectrum of **30** is not observable. At higher concentrations, the triplet ESR spectrum *is* observed.[42] This indicates that the singlet and triplet states are almost degenerate, and the singlet may, in fact, be the ground state. If this is in equilibrium with a small amount of the triplet, the latter will not be observable at low concentrations.

Before leaving the addition reactions, another type of addition must be mentioned: the 1,4- or homo-1,4-addition. One may wonder why there are not many examples of this formally allowed concerted cycloaddition reaction:

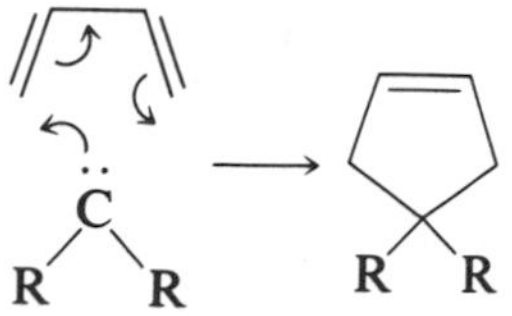

Apparently, the reaction is indeed perfectly allowed, but the thermodynamics are even more favorable for the normal 1,2-addition, so that special steric requirements must be met in order to disfavor 1,2-addition and hence make 1,4-addition experimentally observable. The reaction takes place, in competition with 1,2-addition, between halocarbenes and norbornadienes:[62]

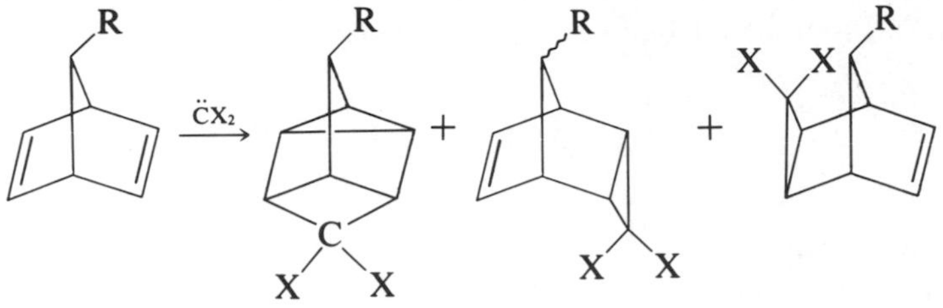

Another highly interesting case is the intramolecular 1,4-addition of cyclopentadienylcarbene, which takes place to the exclusion of 1,2-addition:[62c]

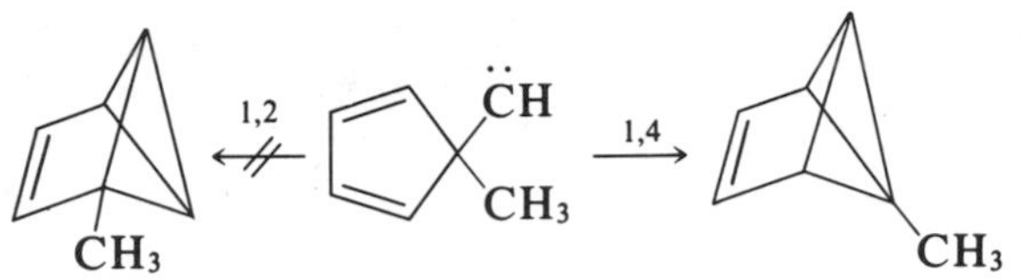

Here, apparently, the transition state for 1,2-addition would be excessively strained, but the carbene has the proper orientation for 1,4-addition to the diene system.

4.4.4. Insertion into C—X bonds

Carbenes and nitrenes insert into single bonds, the most important being the C—H insertion. As with the addition to double bonds, one can imagine a direct insertion of a singlet carbene (Scheme 6, Path *a*); this has come to be known as the Doering–Prinzbach mechanism. It has also been suggested that the singlet carbene abstracts a hydrogen atom to give a singlet radical pair, which then recombines (Benson–DeMore mechanism, Path *b*).[18] For triplet reactions, there is general agreement that an abstraction–recombination path (Path *c*) is followed:

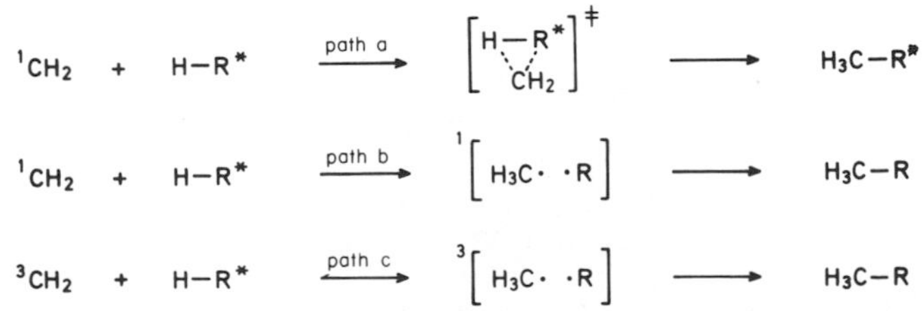

$$^1CH_2 + H{-}R^* \xrightarrow{\text{path a}} \left[\begin{array}{c} H{-}R^* \\ CH_2 \end{array}\right]^{\ddagger} \longrightarrow H_3C{-}R^*$$

$$^1CH_2 + H{-}R^* \xrightarrow{\text{path b}} {}^1[H_3C\cdot \ \cdot R] \longrightarrow H_3C{-}R$$

$$^3CH_2 + H{-}R^* \xrightarrow{\text{path c}} {}^3[H_3C\cdot \ \cdot R] \longrightarrow H_3C{-}R$$

Scheme 6

*denotes optical activity.

Many attempts to distinguish Paths *a* and *b* for the singlet reaction have been reported, and it now seems clear that Path *a* is the normal course. If Path *a* is followed, the reaction should proceed with retention of configuration; that is, if RH is optically active, the product should maintain the activity. Indeed, this is usually the case:[72]

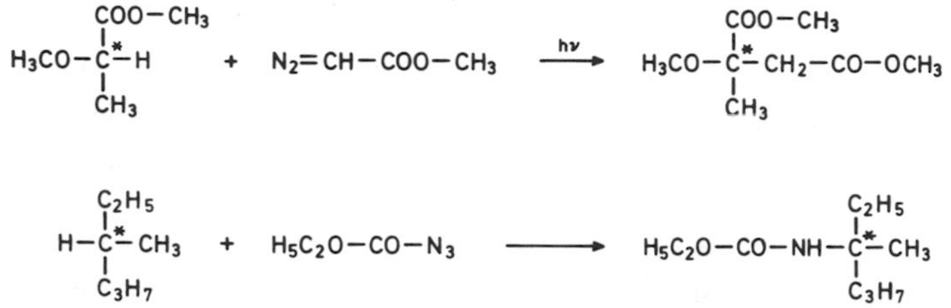

$$H_3CO-\overset{COO-CH_3}{\underset{CH_3}{C^*}}-H \;+\; N_2{=}CH-COO-CH_3 \xrightarrow{h\nu} H_3CO-\overset{COO-CH_3}{\underset{CH_3}{C^*}}-CH_2-CO-OCH_3$$

$$H-\overset{C_2H_5}{\underset{C_3H_7}{C^*}}-CH_3 \;+\; H_5C_2O-CO-N_3 \longrightarrow H_5C_2O-CO-NH-\overset{C_2H_5}{\underset{C_3H_7}{C^*}}-CH_3$$

A substituted phenylnitrene was found to insert into an optically active side-chain with complete retention of configuration in the gas-phase (singlet reaction), but only 65% retention in solution,[73] the latter result being ascribed to collisional deactivation to the ground-state triplet nitrene.

One further piece of evidence for the Path *a* mechanism is the C—H insertion of the arylcarbene **31**:

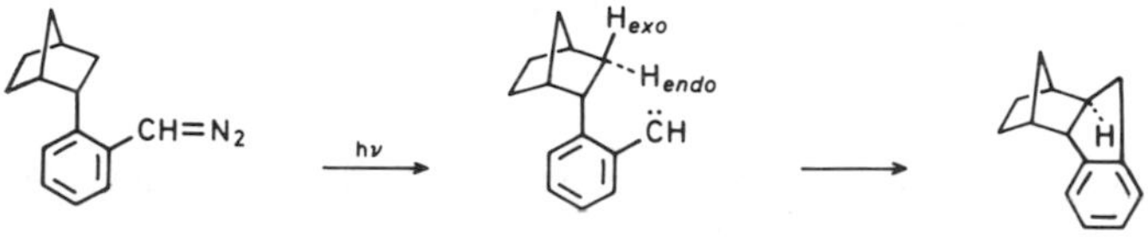

31 32

Insertion occurs predominantly into the *exo*-C—H bond, giving the strained product **32**.[74] A molecular model will show that the carbene can only approach the *exo*-C—H bond *from the side*, not from the end.

Recent ab initio quantum-chemical calculations[75] on the simplest singlet carbene

$$^1H_2C: \;+\; H-H \longrightarrow CH_4$$

reaction fully support the mechanism given in Path *a*, that is, three-center (but unsymmetrical) reaction which proceeds without activation energy. The bimolecular rate constant for this reaction is known[61] to be about 10^9 liter mol^{-1} s^{-1}, which is one order of magnitude lower than the recombination rate constant for methyl radicals (cf. Table 2.1).

Analogous calculations[75] for the *triplet* insertion favored the *linear* transition state shown. This will lead to abstraction–recombination in accord with Path *c*

$$^3\dot{\underset{\cdot}{C}}H_2 + H{-}H \longrightarrow \left[H \overset{0.90\,Å}{—} H \overset{1.40\,Å}{—} C\overset{H}{\underset{H}{)}}\, 126.5° \right]^{\ddagger} \longrightarrow \cdot CH_3 + \cdot H$$

above. The calculated activation energy was 15 kcal/mol. Although the rate constant for this reaction has not been measured experimentally, it is known[61] that it is less than 10^7 liter mol^{-1} s^{-1}, that is, at least two orders of magnitude below that for the singlet reaction.

CIDNP is a very powerful tool for establishing the existence of radical pair intermediates and their spin (Section 2.2.3). Triplet diphenylcarbene abstracts a hydrogen atom from toluene, giving the triplet radical pair **34**.[76] The radical pair can recombine to give **35**, or each fragment can, after diffusion, dimerize to yield **36** and **37**.

$$\underset{\mathbf{33}}{Ph_2C{=}N_2} \xrightarrow[\text{or } \Delta]{h\nu} \underset{\mathbf{13\text{-}T}}{Ph_2C{:}\uparrow\uparrow} \xrightarrow{Ph{-}CH_3} \underset{\mathbf{34}}{{}^3[Ph_2\dot{C}{-}H \quad H_2\dot{C}{-}Ph]}$$

$$\underset{\mathbf{35}}{Ph_2CH{-}CH_2Ph} \qquad \underset{\mathbf{36}}{Ph_2CH{-}CHPh_2} \qquad \underset{\mathbf{37}}{PhCH_2{-}CH_2Ph}$$

$$\underset{\mathbf{38}}{Ph_2CH{-}N{=}N{-}CH_2Ph} \xrightarrow[\text{or } h\nu]{\Delta} \underset{\mathbf{39}}{{}^1[Ph_2\dot{C}H \quad \dot{C}H_2{-}Ph]}$$

The proton–NMR spectrum of the coupling product **35** shows *AE* multiplets (Figure 4.3). Application of Kaptein's rules gives:

$$\Gamma_m = \mu\epsilon a_i \cdot a_j \cdot J_{ij} \cdot \sigma_{ij} = (+)(+)(-)(-)(+)(-) = (-) \Rightarrow AE$$

recalling that the hyperfine splitting constants a are negative for α-protons and that the vicinal coupling constant J_{ij} is positive.

We have seen that azo-compounds of type **38** can decompose thermally to *singlet* radical pairs **39**, and in the present example, the same products **35–37** were formed, although their ratios were different from those obtained from **33**. The only difference between the radical pairs **34** and **39** is their spin, and this is nicely borne out in the CIDNP spectra (Figure 4.3). Since the triplet pair (**34**) gives an *AE* multiplet, the singlet pair (**39**) must give an *EA* multiplet.

In the preceding section, it was mentioned that diphenylcarbene exists as an equilibrium mixture of singlet and triplet states. We must then conclude that *only triplet Ph_2C abstracts hydrogen* to give a radical pair intermediate. If the singlet

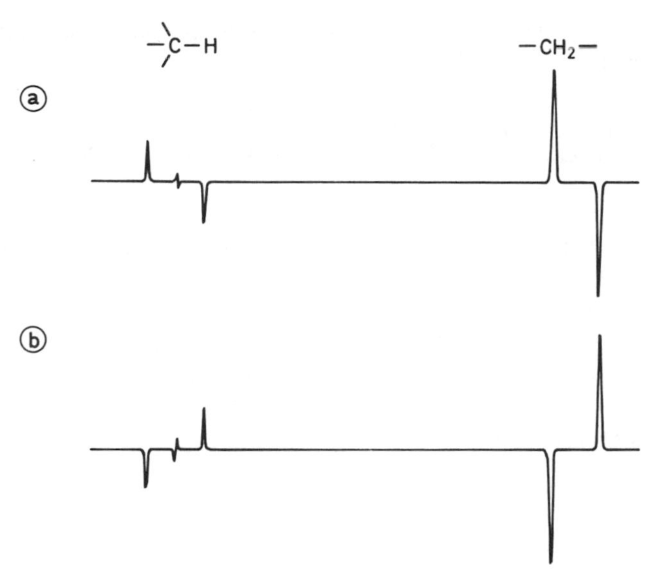

Figure 4.3. CIDNP spectra of triphenylethylene (**35**), produced by (a) photolysis of diphenyldiazomethane (**33**) in toluene; (b) thermolysis of azo-compound **38** in toluene. (Reprinted with permission from Ref. 76b. Copyright 1969 American Chemical Society.)

Ph_2C inserts into a C—H bond, it will not be detected by the CIDNP experiment since it is a one-step reaction.

Similar CIDNP experiments with methylene itself[54] showed that triplet $\ddot{C}H_2$ (produced by sensitized photolysis) gives *AE* polarized ethylbenzene with toluene:

$$^3\ddot{C}H_2 + Ph{-}Ch_3 \longrightarrow {}^3[\dot{C}H_3 \quad \dot{C}H_2{-}Ph] \longrightarrow \underset{AE}{Ph{-}CH_2{-}CH_3}$$

The singlet methylene (formed by direct photolysis) gave no polarized spectrum, supporting the direct insertion mechanism:

$$^1\ddot{C}H_2 + Ph{-}CH_3 \longrightarrow Ph{-}CH_2{-}CH_3$$

Although it seems established that singlet carbenes *insert* directly, triplet carbenes indirectly into C—H bonds, it need not be so for other single bonds. Singlet $:CH_2$ abstracts chlorine atoms from $CDCl_3$; the triplet, in contrast, abstracts D atoms.[54,77]

$$^1{:}CH_2 + CDCl_3 \longrightarrow {}^1[CH_2Cl\cdot \ \cdot CDCl_2] \longrightarrow \underset{E}{CH_2Cl{-}CDCl_2}$$

$$^3{:}CH_2 + CDCl_3 \longrightarrow {}^3[CH_2D\cdot \ \cdot CCl_3] \longrightarrow \underset{A}{CH_2D{-}CCl_3}$$

The reason why the singlet *abstracts* chlorine is, most probably, that an intermediate ylide is formed:[78]

$$^{1}{:}CH_2 + CDCl_3 \rightleftharpoons H_2\overset{\ominus}{C}—\overset{\oplus}{Cl}—CDCl_2 \longrightarrow {}^{1}[CH_2Cl\cdot \;\cdot CDCl_2]$$

Ylide formation between carbenes and lone-pair molecules is widespread and often reversible. The same type of reaction takes place with ethers and sulfides (cf. Table 4.2) and also with nitrogen compounds such as pyridine[79] and acetonitrile.[68]

The probable formation of a similar ylide (**40**) during the reaction between bis(methoxycarbonyl)carbene and dibenzyl ether is supported by the observation of CIDNP ascribed to the singlet radical pair **41**.[80] The outcome of the reaction corresponds to a C—O bond insertion (**42**).

$$R—CH_2—O—CH_2—R + {:}CZ_2 \longrightarrow {}^{1}\left[R—CH_2—\overset{\oplus}{O}(—CH_2—R)—\overset{\ominus}{C}Z_2\right] \longrightarrow$$

40

$$\overline{{}^{1}\;R—CH_2—O—\dot{C}Z_2 \quad \dot{C}H_2—R} \xrightarrow{\text{cage}} R—CH_2—O—CZ_2—CH_2—R$$

41 42

$(R = C_6H_5;\ Z = COOCH_3)$

It should be stressed at this point that the outcome of a carbene reaction may be entirely dependent on the *phase* in which the experiment is carried out. For example,[81] phenyl carbene (**43**) generated by photolysis of phenyldiazomethane in diethyl ether solution at 0°C afforded the ether **45** (30%), presumably arising from ethylene elimination in the ylide **44**, the latter being formed in a reaction of the *singlet* carbene as just described (Scheme 7). The two products of C—H insertion, **46** and **47**, were also formed, presumably from the *triplet* carbene. As expected since the triplet behaves like a diradical, the ratio of secondary to primary C—H insertion (**46**/**47**) was large (148 on a per-bond basis, cf. also Scheme 3).

However, when the same experiment was performed in frozen solution at −196°C, the product **45** was absent (i.e., only triplet carbene reacting), and the secondary/primary C—H insertion ratio dropped to 17. Even more remarkably, the tertiary/primary C—H insertion ratio in the reaction with 2,3-dimethylbutane was 69 at 0°C and decreased to 6.5 at −196°C. From these and other experiments,[81] it may be concluded that the propensity of a triplet carbene toward *primary* C—H abstrac-

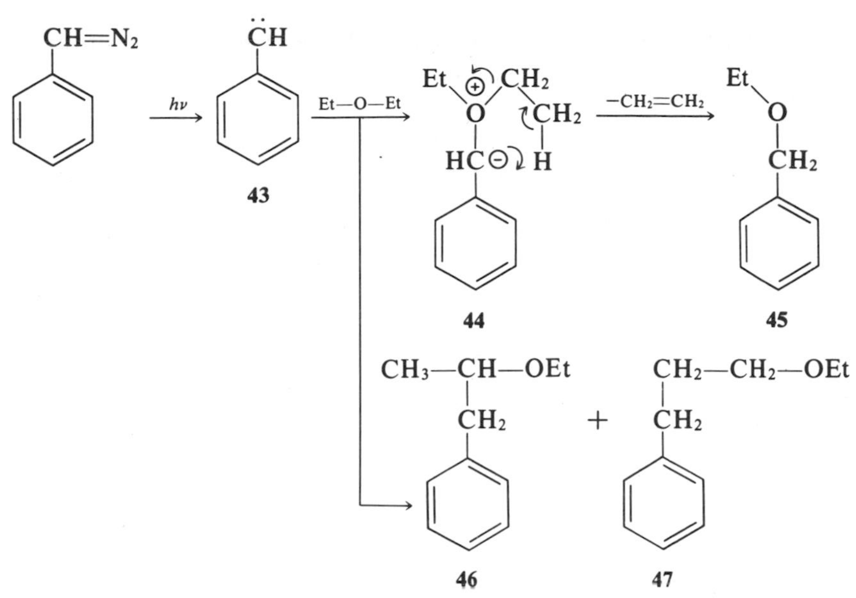

$CH{=}N_2$ $\xrightarrow{h\nu}$ $\ddot{C}H$ (43) $\xrightarrow{Et-O-Et}$ 44 $\xrightarrow{-CH_2{=}CH_2}$ 45

$CH_3-CH(CH_2Ph)-OEt$ (46) + $PhCH_2-CH_2-CH_2-OEt$ (47)

Scheme 7

tion increases dramatically in frozen matrices. The cause is not a sudden decrease in the activation energy for primary C—H abstraction, but, more likely, a proximity effect. In the frozen matrix, mobility is low, and the carbene finds itself in closer proximity to primary than to secondary or tertiary C—H bonds.

Furthermore, it has been found that hydrogen abstraction also dominates over addition to C═C bonds in low-temperature matrices.[82] The rates of decay of the ESR signals of triplet diphenylcarbene in glassy or crystalline hosts have been measured.[44,83] These decay reactions have extremely low Arrhenius activation parameters. In particular, the low *A*-factors demonstrate that, at temperatures around −196°C, the predominant reaction is hydrogen abstraction proceeding by tunneling.[83]

In analogy with the reaction with ethers illustrated in Scheme 7, *singlet* carbenes insert into O—H bonds of alcohols, whereas *triplet* carbenes undergo C—H bond insertion.[18] The singlet carbene O—H insertion may involve either an ylide analogous to **44** or proton transfer from the alcohol to the carbene, giving a carbenium ion.[84]

It should also be noted that the formation of a "C—H insertion product" is not sufficient evidence that it is derived from a carbene. In connection with the Wolff rearrangement, it will be shown (Section 4.4.5) that, apparently, an excited state of a diazo-compound can undergo direct C—H insertion, bypassing the carbene. Furthermore, the triplet 1,5-diradical **49** is produced by matrix photolysis of the diazo-compound **48**.[85] One might reasonably expect that this is due to an intramolecular hydrogen abstraction by the naphthylcarbene **50**; however, **50** is not detectable even at 4 K.

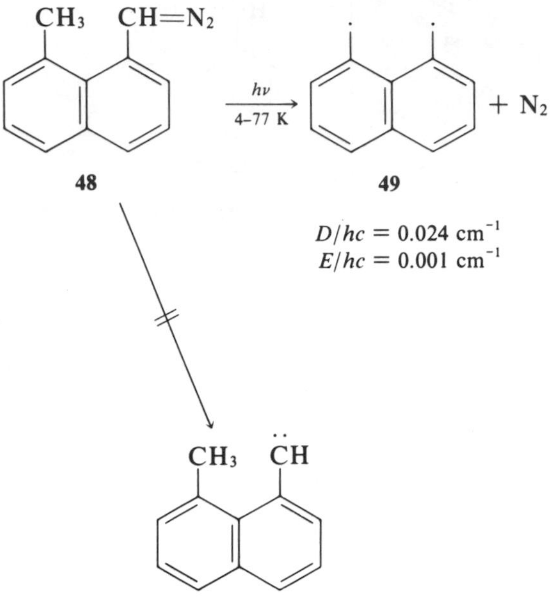

48 $\xrightarrow[4\text{–}77\text{ K}]{h\nu}$ **49** + N_2

$D/hc = 0.024\ \text{cm}^{-1}$
$E/hc = 0.001\ \text{cm}^{-1}$

48 ⇸ **50**

In contrast, the azide **51** gave both the triplet diradical **52** and the triplet nitrene **53**, both observable by ESR spectroscopy.[86] However, the nitrene could not be induced to convert to the diradical, either thermally or photochemically, thus suggesting that the two species are formed simultaneously, not sequentially. Likely precursors of **52** are an excited state of the azide or, possibly, the ESR-inactive singlet nitrene.

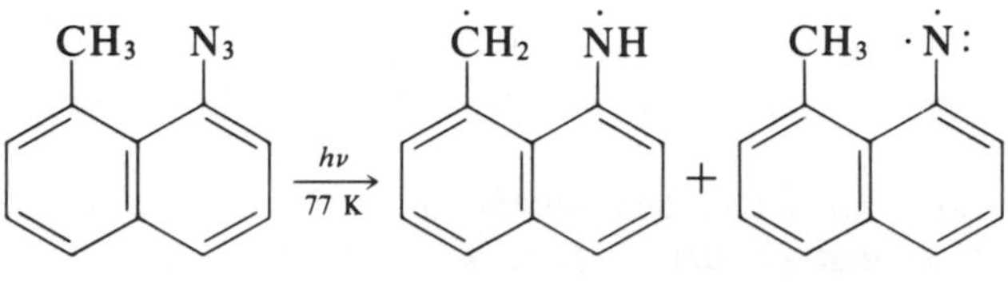

51 $\xrightarrow[77\text{ K}]{h\nu}$ **52** + **53**

$D/hc = 0.0255 \qquad 0.79\ \text{cm}^{-1}$
$E/hc = 0.0008 \qquad <0.003\ \text{cm}^{-1}$

4.4.5. Rearrangements

Hydrogen Shifts. The 1,2-hydrogen shift in a singlet carbene is one of the most general and versatile carbene reactions:

$$\longrightarrow \quad R{-}CH{=}CH_2 \qquad (5)$$

It constitutes the crucial step in the Bamford–Stevens reaction, whereby a ketone (or aldehyde) possessing at least one α-hydrogen atom is converted to an alkene:

$$R^1{-}CH_2{-}C(=O){-}R^2 \xrightarrow{Tos{-}NH{-}NH_2} R^1{-}CH_2{-}C(=N{-}NH{-}Tos){-}R^2 \xrightarrow{Base}$$

$$R^1{-}CH_2{-}C(=N{-}\overset{\ominus}{N}{-}Tos){-}R^2 \xrightarrow[\substack{-Tos^{\ominus} \\ -N_2 \\ \sim H}]{\Delta} R^1{-}CH{=}CH{-}R^2$$

The hydrogen shift can be regarded as an insertion into an α-C—H bond and is analogous to the Wagner–Meerwein rearrangement in carbocations. Alkyl and aryl groups migrate too. The migratory aptitude increases in the series

$$H > aryl > alkyl$$

From the known enthalpy of formation of methylene ($:CH_2$), one can estimate the enthalpy of formation of methylcarbene (ethylidene, $CH_3—\ddot{C}H$) as about 90 kcal/mol. The 1,2-hydrogen shift is, therefore, highly exothermic (Figure 4.4), and in keeping with the Hammond postulate, the transition state should be very early; that is, it will resemble the carbene more than the olefin, and a low energy of activation may be expected. This is borne out by theoretical calculations, which predict an activation energy ≤ 5 kcal/mol.[88]

Since this process occurs so readily, it is difficult to observe *intermolecular* reactions of alkylcarbenes. However, if the *triplet* carbene is generated directly by sensitized photolysis, the hydrogen shift will be much less favored: It now resembles the 1,2-shift in a free radical, a very rare process (see Section 2.5.3). Under such conditions, C—H abstraction and additions to olefins can be observed.[89]

The singlet carbene in Eq. (5) resembles a carbenium ion inasmuch as both possess a vacant *p* orbital. Since carbenium ions are highly stabilized by hyperconjugation, one could expect this for singlet carbenes also. Taken with the small singlet–triplet splitting in methylene itself, hyperconjugation could then lead to singlet ground states in alkylcarbenes. It is highly interesting, therefore, that the first ESR spectrum of an alkylcarbene was reported recently: di(tertiary-

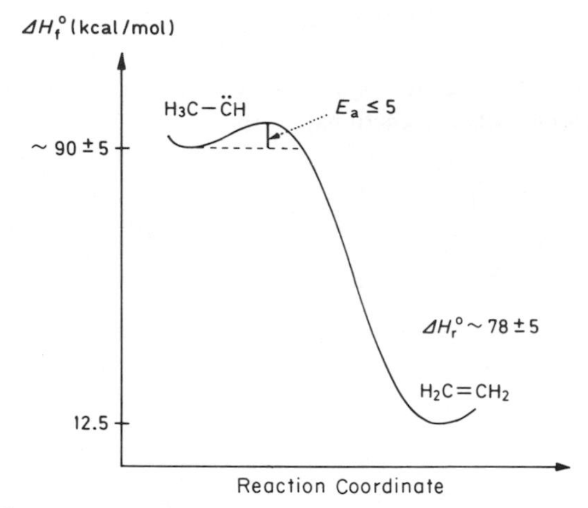

Figure 4.4. Energy profile for methylcarbene rearrangement.

butyl)carbene (**54**) was obtained by photolysis of the diazo-compound at 14 K.[90] The ESR spectrum disappeared irreversibly on warming above 70 K. The results suggest that either the triplet is the ground state, or it lies only a few cal/mol above a ground state singlet.

$$C{=}N_2 \xrightarrow[14\ K]{h\nu} C: \quad \mathbf{54}$$

$$D/hc = 0.689\ \text{cm}^{-1}$$
$$E/hc = 0.039\ \text{cm}^{-1}$$

As a further consequence of the comparison with carbenium ions, one would expect preferred migration of the hydrogen atom that is eclipsed with the vacant *p* orbital [Eq. (5)]. This has been confirmed not only for the 1,2-hydrogen shift, but also for the 1,3-shift as shown below using conformationally rigid systems:[91]

H, D $\xrightarrow{\Delta}$ D

H, D $\xrightarrow{\Delta}$ H, D

$k_{exo}/k_{endo} \simeq 138$

In conformationally flexible open-chain systems, smaller preferences can be expected because rotational barriers are low. As an example, the thermal or photochemical decomposition of diazosuccinic ester **(55)** gives predominantly the maleic ester **(56),** even though the fumaric ester **(57)** is thermodynamically more stable; the preference is low, however:[92]

$$\underset{\mathbf{55}}{\text{EtOCO—}\overset{\overset{\Large N_2}{\|}}{C}\text{—CH}_2\text{—COOEt}} \xrightarrow{\Delta} \text{EtOCO—}\ddot{C}\text{—CH}_2\text{—COOEt} \longrightarrow$$

$$\mathbf{56} + \mathbf{57} \qquad \mathbf{56/57} \geqslant 2.5$$

This indicates that the preferred conformation of the carbene is the *gauche*-form **58:**

$$\mathbf{58} \longrightarrow \mathbf{56}$$

In the presence of Cu^{2+} ions the reverse stereochemistry is obtained: the maleate/fumarate ratio decreases tenfold.[93] Most probably, a bidentate carbene–copper complex now forms, in which the carbene is locked in a conformation favorable for fumarate production:

$$\mathbf{55} \xrightarrow{Cu^{II}} [\text{carbene–}Cu^{II}\text{ complex}] \longrightarrow \mathbf{57} + \mathbf{56}$$

$$\mathbf{56/57} = 0.25$$

Aryl groups undergo ready 1,2-migrations in both radicals and carbenium ions, so one might expect this to be the case in both singlet and triplet carbenes as well.

The photolysis of the diazo-compound **59** at room temperature gives the products of both hydrogen and phenyl migration:[94]

$$\underset{\mathbf{59}}{H-C(Ph)(CH_3)-C(Ph)=N_2} \xrightarrow{h\nu} \underset{\mathbf{60a}}{(Ph)(H_3C)C=C(Ph)(H)} + \underset{\mathbf{60b}}{(Ph)(H_3C)C=C(H)(Ph)} + \underset{\mathbf{60c}}{(H)(H_3C)C=C(Ph)(Ph)}$$

ratio ~ 1 : 1 : 1

When this reaction is carried out at progressively lower temperatures, the extent of phenyl migration (giving **60c**) increases until the solution solidifies (−110°C). Below this temperature, phenyl migration decreases and is virtually absent at −196°C; the only products are now **60a** and **60b**, formed by hydrogen migration. The likely explanation is that low temperatures favor the *triplet* carbene; hence phenyl migration will become increasingly important. However, once the matrix solidifies, steric constraints will make phenyl migration more difficult but hardly influence the shift of a hydrogen atom. Although the hydrogen atom might be reluctant to migrate in the triplet (as in a radical), a small singlet–triplet splitting would allow for a singlet–triplet equilibrium and thus permit the hydrogen shift to occur in the singlet manifold.

$$\mathbf{61} \xrightarrow[-CH_3CN,\ -CO_2]{\Delta} \underset{\mathbf{62}}{Ar-CH=C:} \xleftarrow{\Delta} \underset{\mathbf{63}}{Ar-CH=C=C=O}$$

$$\mathbf{62} \longrightarrow Ar-C\equiv C-H \qquad \mathbf{64} \xrightarrow[-CO_2,\ -(CH_3)_2C=O]{\Delta} \mathbf{63}$$

A special class of carbenes, the vinylidenes (**62**), can be generated by flash vacuum pyrolysis of either 4-methyleneisoxazol-5(4*H*)-ones (**61**)[95] or 5-methylene-1,3-dioxan-4,6-diones (**64**),[96] in the latter case via unstable but detectable methyleneketenes (**63**). (For other methods of producing vinylidenes, see Scheme 3 in Section 3.1. and the reviews.[97,98]) The vinylidenes undergo extremely rapid shifts of hydrogen[99] (and aryl) and the reactions therefore constitute very versatile syntheses of acetylenes, particularly arylacetylenes. Interestingly, a ^{13}C-labeling study has shown[96] that (a) both hydrogen and phenyl migrate and (b) the vinylidene–acetylene rearrangement is reversible at high temperature:

$$\mathrm{Ph{-}\overset{*}{C}{\equiv}C{-}H} \overset{\Delta}{\rightleftharpoons} \mathrm{(Ph)_2\overset{*}{C}{=}C{:}} \overset{\Delta}{\rightleftharpoons} \mathrm{H{-}\overset{*}{C}{\equiv}C{-}Ph}$$

Other examples, mostly photochemical ones, of retro-1,2-rearrangements are known.[100] Inspection of Figure 4.4 indicates that, in order to go from the olefin to the carbene, a large amount of energy must be supplied. This can be achieved by photolysis, the triplet excited state of ethylene lying approximately 100 kcal/mol above the ground state:

$$S_0 \xrightarrow{h\nu} [S_1]^{\uparrow\downarrow} \xrightarrow{\text{ISC}} [T_1]^{\uparrow\uparrow}$$

The triplet, or the singlet if it is sufficiently long-lived, can rearrange to an alkylcarbene in an exothermic reaction:

Evidence for such processes is presented in the reaction schemes below.[100,101]

Ph D Me Ph CH2—C—Ph Me $\underset{\Delta}{\overset{h\nu}{\rightleftharpoons}}$ D Ph Me Ph CH2—C—Ph Me → D H Ph Me Ph C—Ph H Me

Ph H Ph DO $\overset{h\nu}{\rightleftharpoons}$ Ph H Ph DO → Ph H Ph D O

Although formally analogous to the carbene rearrangements, the 1,2-shifts of hydrogen and alkyl groups in azides are not so clear-cut:

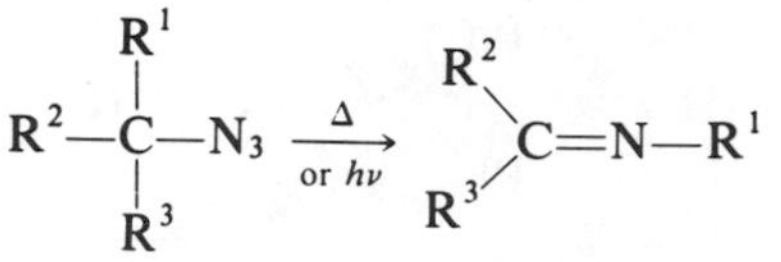

$$R^2\text{—}C(R^1)(R^3)\text{—}N_3 \xrightarrow[\text{or } h\nu]{\Delta} (R^2)(R^3)C\text{=N—}R^1$$

There is much evidence that the *thermal* reaction takes place via *nitrenes*, and these can be trapped in the form of intramolecular C—H insertions into neighboring phenyl groups, for example, **65** → **66**.[102] The product **66** is not formed on *photolysis*, however, thereby suggesting that the photolysis does not proceed via free nitrenes but rather through an excited state of the azide (**65**), probably the singlet

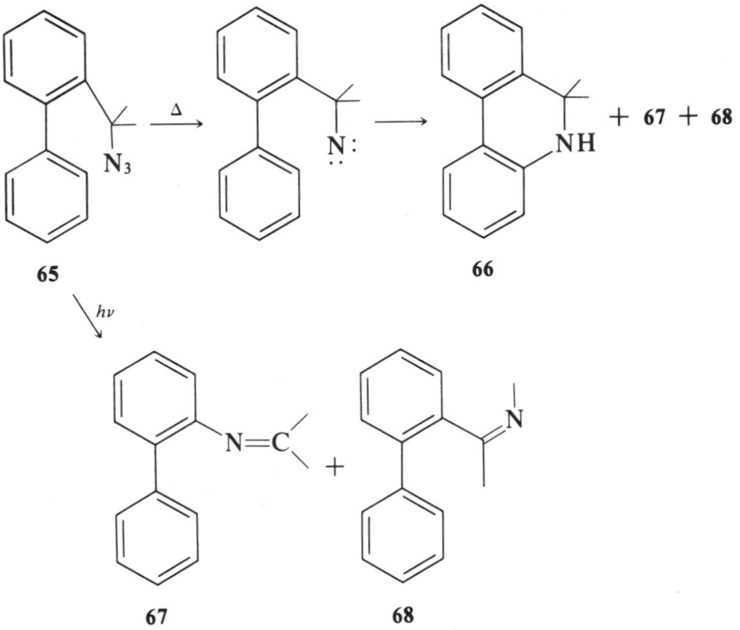

N₃ —Δ→ N: ⟶ NH + **67** + **68**

65 **66**

hν

N=C + N

67 **68**

excited state.[102,103] The triplet excited states of azides do give rise to triplet nitrenes, which can be observed by ESR spectroscopy (see methylnitrene, Table 4.5). Like radicals, the triplet nitrenes are not expected to rearrange,[104] and in fact, triplet quenchers have no effect on the photochemical reactions **65** → **67** + **68**, thereby confirming the singlet excited azide as the likely reactive species.

The 1,2-hydrogen shift in an azide (or nitrene) is complicated by the fact that the resulting imines are unstable. The direct irradiation of methyl azide in an argon matrix at 4 K gave methyleneimine (**69**), identified by its IR spectrum, but no spectroscopic evidence for the formation of methylnitrene (singlet) was obtained.[105]

Either the hydrogen migration is fast even at 4 K, or the reaction bypasses the nitrene, as in the photolysis of **65**.

$$CH_3{-}N_3 \xrightarrow{\Delta \text{ or direct } h\nu} {}^1[CH_3{-}\ddot{N}:] \longrightarrow \underset{\mathbf{69}}{CH_2{=}NH}$$

$$CH_3{-}N_3 \xrightarrow{\text{sensitized } h\nu} {}^3CH_3{-}\dot{N}:$$

As already mentioned, the triplet sensitized photolysis does give the triplet methylnitrene.

The analogy with tertiary alkylnitrenes such as **65** suggests that singlet methylnitrene should be formed on pyrolysis of methyl azide. The singlet nitrene, in turn, should rearrange to methyleneimine (**69**). Although the gas-phase pyrolysis of methyl azide has been interpreted in these terms since 1933,[106] evidence for the actual formation of **69** was obtained only recently through direct observation with the aid of millimeter-wave[107] and photoelectron spectroscopy.[108] It will be very difficult, if not impossible, to prove that singlet methylnitrene is an *intermediate* in this reaction, for calculations[104,108] suggest that there is hardly any activation energy toward its isomerization to methyleneimine.

Wolff-Type Rearrangements. The thermal or photochemical reorganization of an α-diazo-carbonyl compound to give a ketene is known as the Wolff rearrangement. For the sake of convenience, we include under this heading the analogous rearrangements of thioxo-, imidoyl-, and vinylcarbenes.

$$R^1{-}C({=}O){-}C({=}N_2){-}R^2 \xrightarrow[\text{or } h\nu]{\Delta} R^1{-}C({=}O){-}\ddot{C}{-}R^2 \longrightarrow O{=}C{=}C(R^2)(R^1) \quad (6)$$

The reaction [Eq. (6)] is usually formulated as a carbene reaction, but, as shown further on, this may not always be correct. The ketene formed as a product is rarely isolated as such. Rather, it is converted to a carboxylic acid or an ester by performing the reaction in aqueous or alcoholic solution, as in the well-known Arndt-Eistert homologization:[109]

$$R{-}C({=}O){-}Cl + H_2C{=}N_2 \xrightarrow{-\,HCl} R{-}C({=}O){-}CH{=}N_2 \xrightarrow[\text{or } h\nu]{\Delta}$$

$$[R{-}CH{=}C{=}O] \xrightarrow{H_2O} R{-}CH_2{-}COOH$$

The most exciting aspect of the Wolff rearrangement is the demonstration of carbon scrambling, which implies oxirenes **(71)** as intermediates in a carbene–carbene interconversion:[60,110–112]

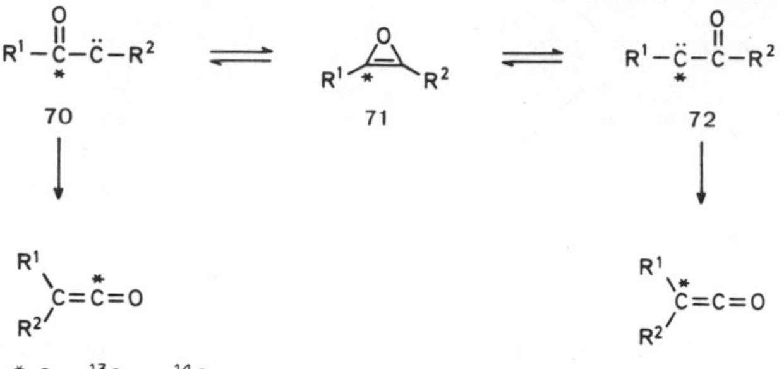

$$R^1-\overset{O}{\overset{\|}{C^*}}-\ddot{C}-R^2 \rightleftharpoons \text{71} \rightleftharpoons R^1-\ddot{C}^*-\overset{O}{\overset{\|}{C}}-R^2$$

70 71 72

$R^1R^2C=\overset{*}{C}=O$ (from 70) $R^1R^2\overset{*}{C}=C=O$ (from 72)

* C = ^{13}C or ^{14}C

Although no oxirene has yet been detected with certainty, even in argon matrices at 8 K,[119] the evidence from ^{13}C and ^{14}C labeling studies overwhelmingly suggests that oxirenes can be involved in both thermal and photochemical reactions, the yields of carbon scrambling approaching 100% in some photochemical cases. Conspicuously, however, rigid cyclic diazoketones such as **73** and **74** rearrange without oxirene participation:[113]

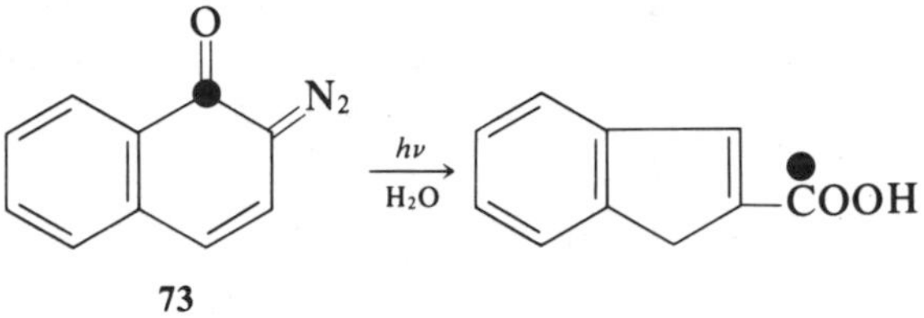

73 $\xrightarrow[H_2O]{h\nu}$ COOH

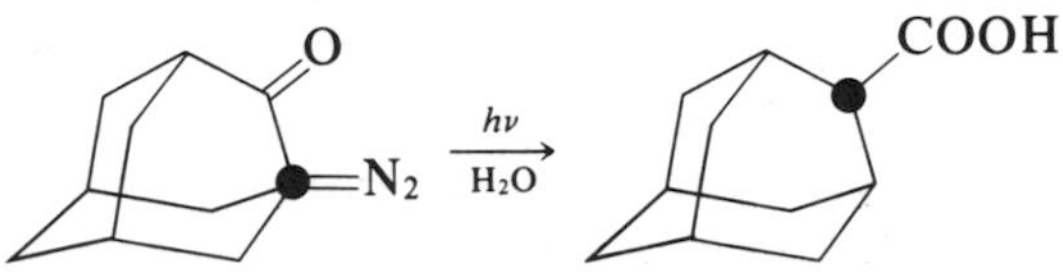

74 $\xrightarrow[H_2O]{h\nu}$ COOH

In contrast, rearrangement via the oxirene **76** was established for the large and flexible ring **75**:[114]

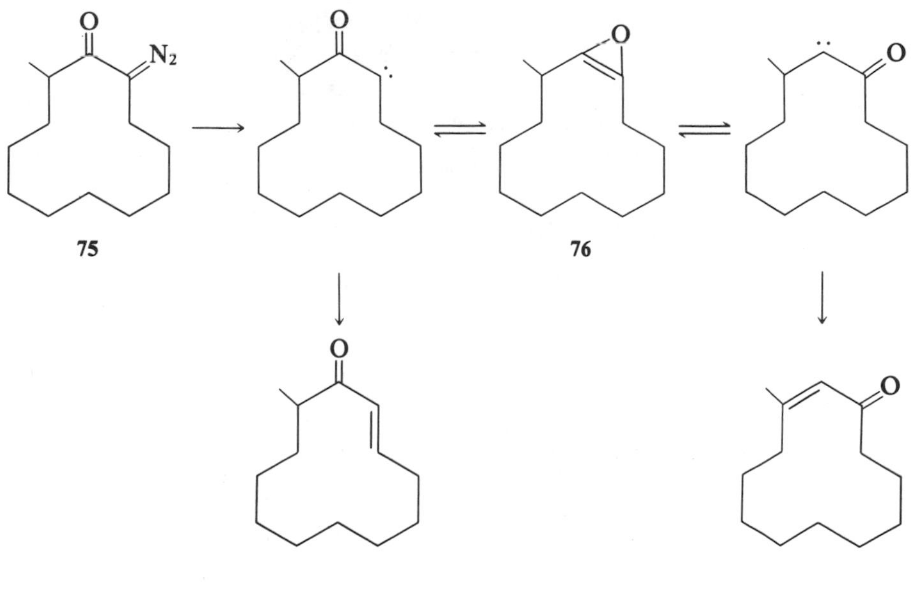

75

76

These and many other observations can be understood in terms of the following reaction chart:

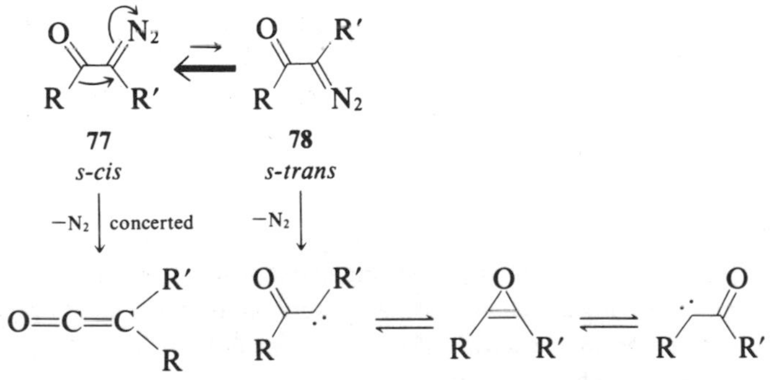

77
s-cis

78
s-trans

$-N_2$ concerted

$-N_2$

It is known from spectroscopic studies and dipole moment measurements that acyclic α-diazoketones and α-diazoesters exist predominantly in the *s-cis* form, although the barrier toward interconversion (**77** ⇌ **78**) is low.[115,116] A rationale[115] would be, then, that the *s-cis* form **77** undergoes migration of the group R concerted with loss of N_2. Negative evidence for this proposal has been adduced through the failure to observe CIDNP during the decomposition of diazoacetaldehyde in solution; that is, evidence for carbene intermediates was not obtained.[54] Moreover, it was shown[117] that the α-diazoamide **79** undergoes an intramolecular photochemical C—H insertion to give the β-lactam **81** together with the ketene **82** *in a noncarbene*

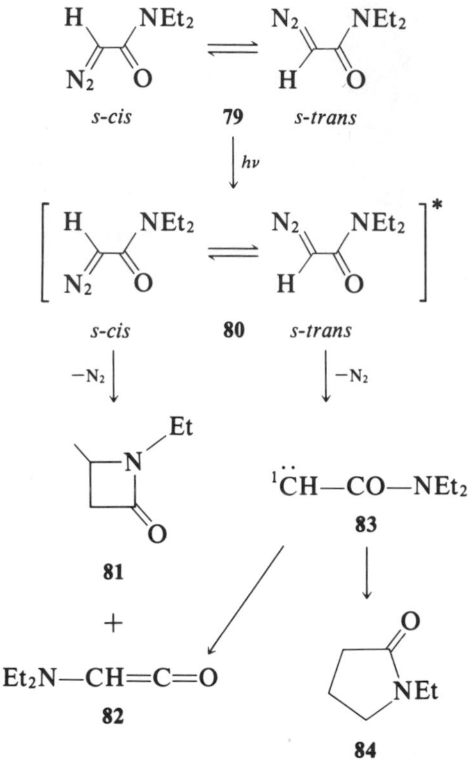

Scheme 8

reaction. From the effects of sensitizers and quenches, it was concluded that **81** is formed directly from the *s-cis* form of the singlet excited state of the diazo-compound **(80)**. The accompanying γ-lactam **84** was ascribed to the singlet carbene **(83)**, presumably derived from the *s-trans* form. In a related study,[118] it was concluded that the ketene (**82** in Scheme 8) may be formed both from the singlet carbene and from the concerted rearrangement.

The labeling experiments cited earlier constitute almost irrefutable evidence for the existence of free ketocarbenes in all cases where oxirene participation has been demonstrated. This has been the case particularly in photochemical reactions, and in thermal gas-phase reactions, that is, under conditions of relatively high energy. Consequently, the concentrations of the *s-trans* forms will be relatively high under these conditions. Some concerted Wolff rearrangement can take place from the *s-cis* forms, but the *s-trans* forms will allow the formation of free carbenes and hence carbene–carbene interconversion via oxirenes. The cyclic diazoketones such as **73** and **74** are unable to adopt an *s-trans* form; hence no oxirene participation is found. A molecular model will show, however, that the preferred conformation of the larger ring **75** is one in which the diazoketone is oriented *s-trans*. Accordingly, this compound produces free carbenes, and hence oxirene, both thermally and photochemically.

Although the existence of oxirene cannot yet be said to be proven,[119] both thiirene (**85**) and the corresponding selenirene (**86**) and substituted derivatives thereof have been matrix-isolated and identified by their IR spectra:[120,121]

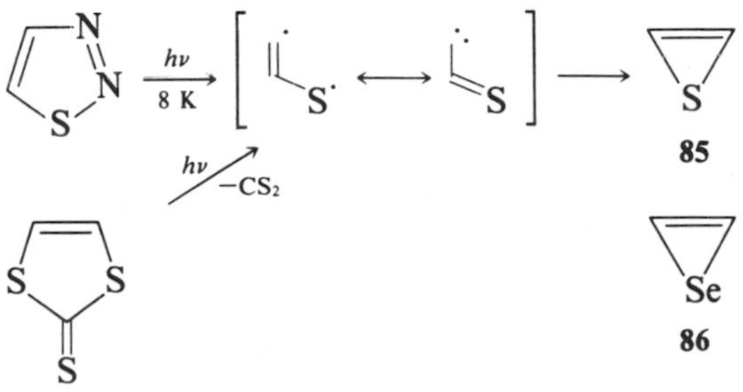

N N S —hν, 8 K→ [S· ⟷ S] → S

85

hν, $-CS_2$

S S S

Se

86

Even benzothiirene (**87**) may have been observed by IR spectroscopy during the following reaction:[121b]

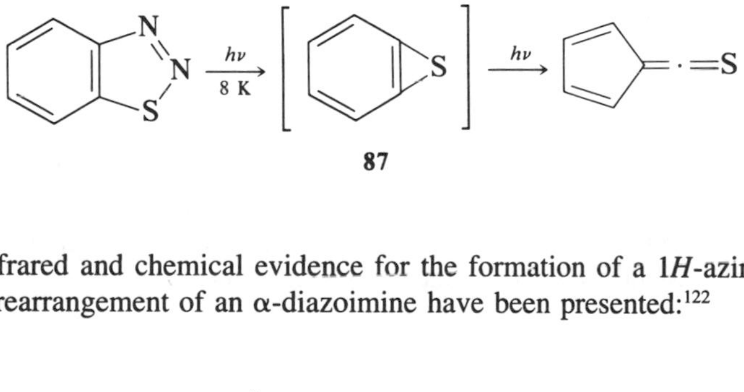

N N S —hν, 8 K→ [S] —hν→ =·=S

87

Both infrared and chemical evidence for the formation of a 1*H*-azirine (**88**) in the Wolff rearrangement of an α-diazoimine have been presented:[122]

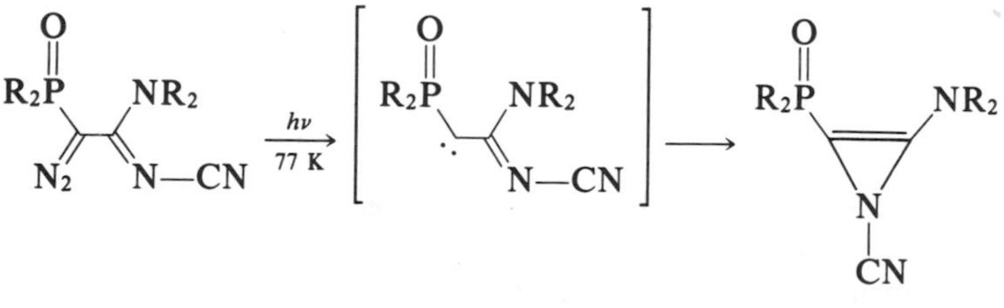

O, R_2P, NR_2, N_2, N—CN —hν, 77 K→ [O, R_2P, NR_2, N—CN] → O, R_2P, NR_2, N, CN

88

2200 cm^{-1} (C≡N)
1867 cm^{-1} (C=C)

Although 1*H*-benzazirine **(89)** has not yet been isolated, a ^{13}C-labeling study of the gas-phase pyrolysis of isatin demonstrates the *quantitative* involvement of **89** prior to ring contraction to cyanocyclopentadiene **(92)**:[123]

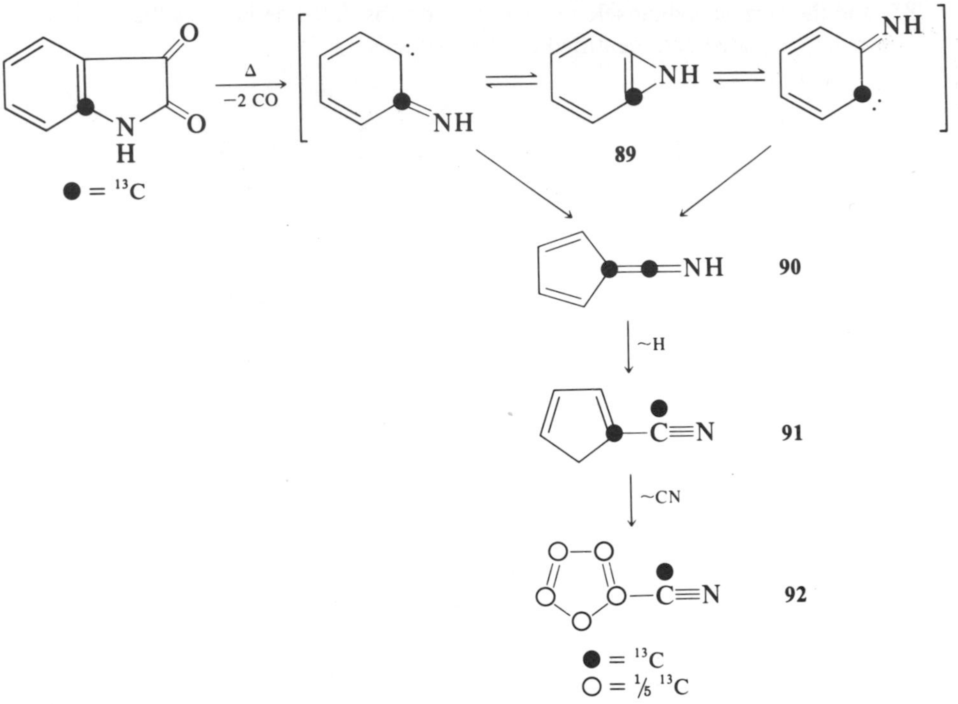

O N H O $\xrightarrow[-2\,CO]{\Delta}$ [NH ⇌ NH **89** ⇌ NH]

● = ^{13}C

=NH **90**

~H

C≡N **91**

~CN

C≡N **92**

● = ^{13}C
○ = ⅕ ^{13}C

The immediate Wolff rearrangement product is the unstable ketenimine **90**, which tautomerizes to the nitrile **91** already during the gas-phase reaction. It was shown that the cyano group in cyanocyclopentadiene undergoes rapid sigmatropic shifts around the five-membered ring under the reaction conditions:

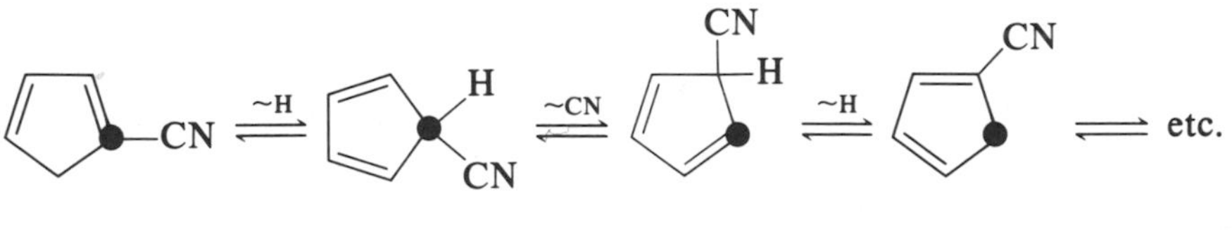

CN $\overset{\sim H}{\rightleftharpoons}$ H CN $\overset{\sim CN}{\rightleftharpoons}$ CN H $\overset{\sim H}{\rightleftharpoons}$ CN ⇌ etc.

Therefore, the final product is **92**, in which the total ^{13}C content of the ring equals that of the cyano group.

The ring opening of cyclopropenes to vinylcarbenes belongs logically to this class of reactions:[60]

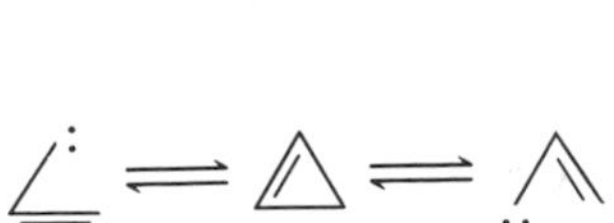

A Wolff-type rearrangement of a vinylcarbene is observed in the very efficient thermal rearrangement of benzocyclopropene **(93)** to vinylidenecyclopentadiene (fulvenallene) **(94)**:[124]

$$\mathbf{93} \xrightarrow[\text{gas-phase}]{\Delta} [\text{carbene}] \longrightarrow \mathbf{94}\ (=C=CH_2)$$

The same type of reaction can be used to prepare the highly unstable isoindene derivative **95**:[125]

$$\xrightarrow[\text{gas-phase}]{\Delta} \quad \longrightarrow \quad \mathbf{95}$$

The most familiar "nitrene" analogs of the preceding carbene rearrangements are the Curtius, Hofmann, and Lossen rearrangements. However, there is much evidence that these are concerted reactions, not involving nitrenes:[19,112]

$$R{-}C(=O)N_3 \xrightarrow{\Delta} R{-}N{=}C{=}O$$
$$R{-}C(=O)N_3 \xrightarrow{h\nu} R{-}C(=O)\ddot{N}: \longrightarrow \text{trapping reactions}$$

Isocyanates are formed thermally, but acylnitrenes cannot be trapped under these conditions. In contrast, photolysis apparently does give nitrenes that undergo certain trapping reactions, but little or no isocyanate is formed in this case.

There are problems with this simple reaction scheme, however.[112] For example, nitrile oxides rearrange thermally and, in some cases, photochemically to isocyanates. The reaction is most simply formulated as an analog of the Wolff rearrangement,

$$R{-}C{\equiv}N{\rightarrow}O \longleftrightarrow R{-}\ddot{C}{-}N{=}O \xrightarrow{\Delta} R{-}C{=}N\ (\text{O-bridged}) \longrightarrow R{-}C(=O){-}\ddot{N}: \longrightarrow R{-}N{=}C{=}O$$

thus implying that (singlet) acylnitrenes *do* rearrange to isocyanates. One way out of the dilemma would be to postulate that thermolysis of acyl azides and nitrile oxides gives very short-lived singlet nitrenes that rapidly rearrange. Photolysis, in contrast, gives triplet nitrenes that do not rearrange but undergo inter- or intramolecular C—H insertion reactions. The latter process could also take place from excited states of the carbonyl azides. These possibilities have not yet been sufficiently explored. Also, ESR spectra of simple carbonylnitrenes are conspicuously absent from the literature.[87]

Interconversion of Carbenes and Nitrenes (Refs. 60 and 112). The discovery in 1968 that arylcarbenes and arylnitrenes can interconvert[126]

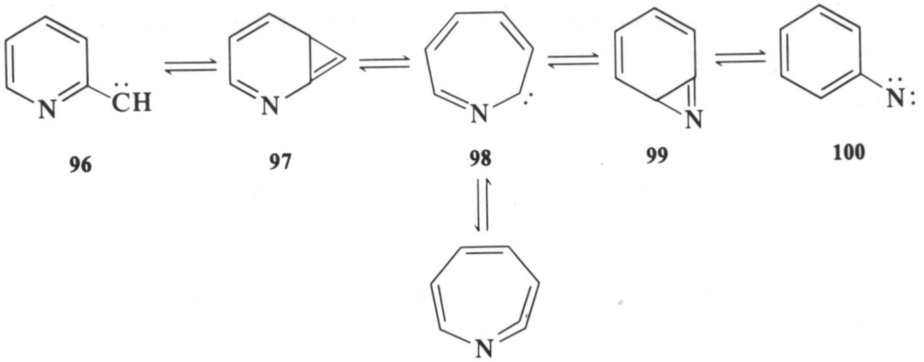

96 97 98 99 100

101

Scheme 9

was based on the observation that several *v*-triazolo[1,5-*a*]pyridines (e.g., **102**)—which are precursors of the diazo-compounds **103**—on flash vacuum pyrolysis gave high yields of *nitrene* products, particularly azobenzenes and anilines, together with smaller amounts of cyanocyclopentadienes. The *same* products were obtained from the appropriately substituted azides, for example, **104**:

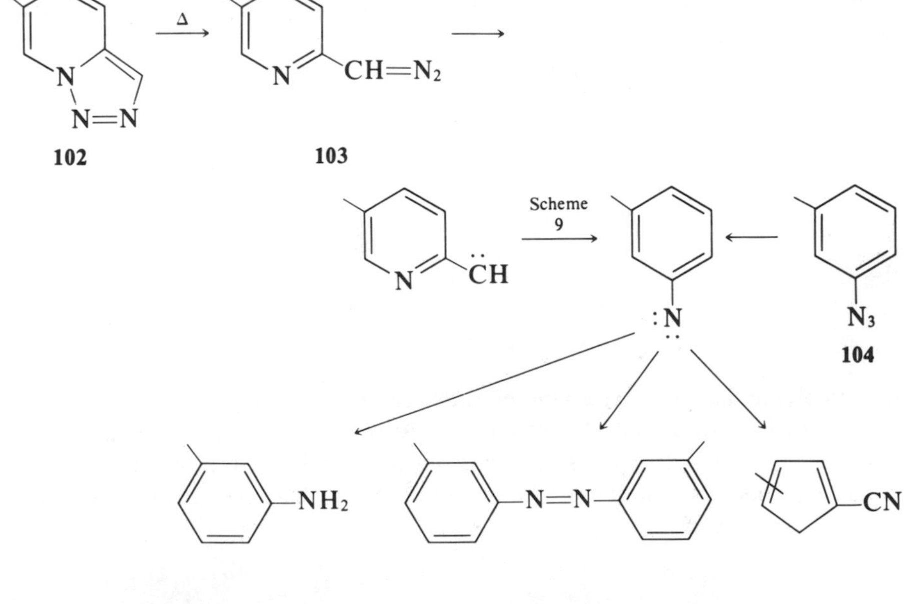

102 103

104

When the 2-pyridylcarbene so generated carries a methyl group (**105**), the 1,2-migration of hydrogen is the only reaction observed:

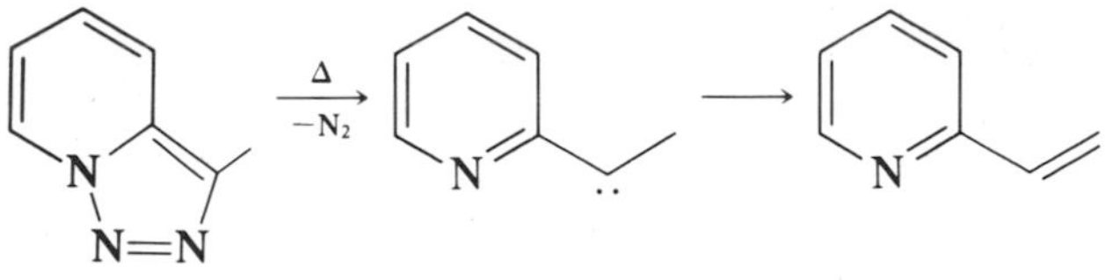

Δ
$-N_2$

N
N=N

N

N

105

In contrast, a phenyl-substituted carbene (**106**) undergoes the carbene–nitrene rearrangement, giving a carbazole (**108**) in quantitative yield and with the methyl group specifically in the 2-position in conformity with the mechanism given in Scheme 9. The same product is obtained from the azide **109** via the nitrene **107**.

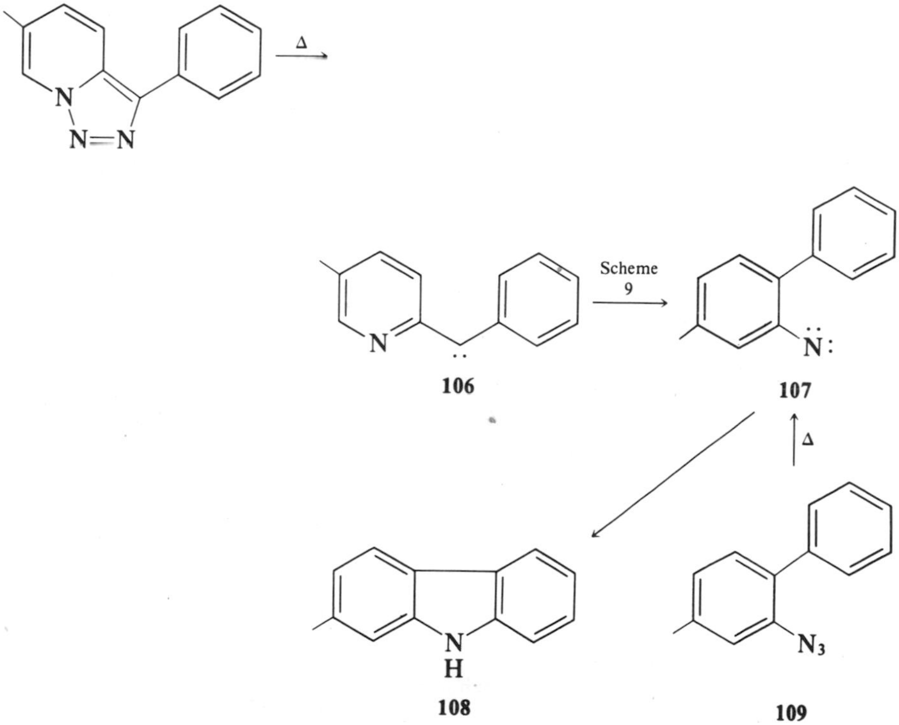

Δ

N
N=N

Scheme
9

N

N:

106

107

Δ

N
H

N_3

108

109

Scheme 10

The rearrangement shown in Scheme 10 specifically involves the pyridine ring—the phenyl ring acts merely as a substituent. Nevertheless, a phenyl ring can undergo the same type of rearrangement, as shown by the ^{13}C-labeling of the phenyl(3-pyridyl)carbene (**110**):[67]

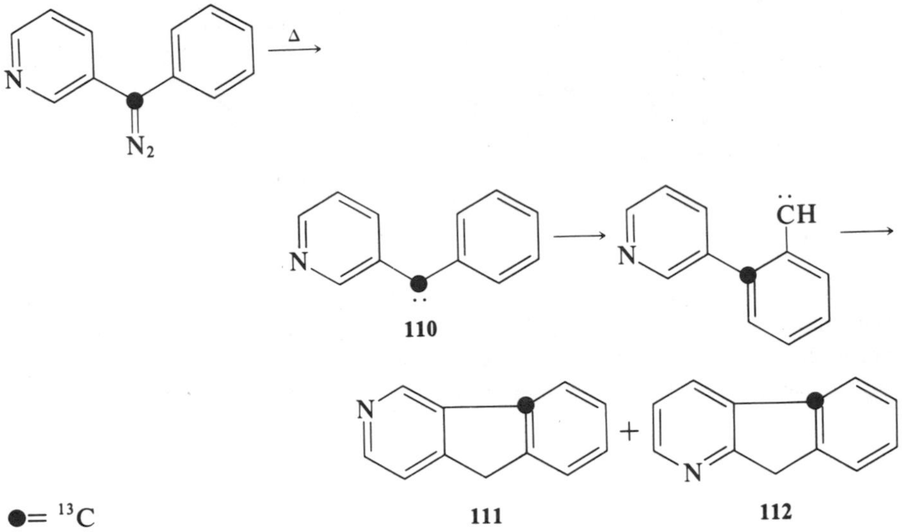

N
N_2
Δ
110
ĊH
111
112
● = ^{13}C

Scheme 11

The exclusive labeling of the benzene rings in the products **111** and **112** requires a process of the type:

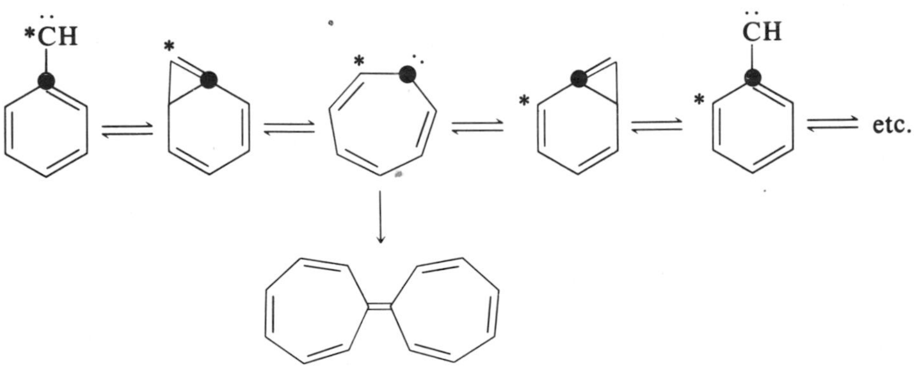

*ĊH
*
*
*
*
ĊH
etc.
113

Scheme 12

Evidence for the occurrence of this reaction is found in the formation of heptafulvalene (**113**) as the major product of the gas-phase pyrolysis of phenyldiazomethane and in the formation of 2,7-dimethylfluorene (**115**) in the pyrolysis of *p,p'*-dimethyldiphenyldiazomethane (**114**).[127,128]

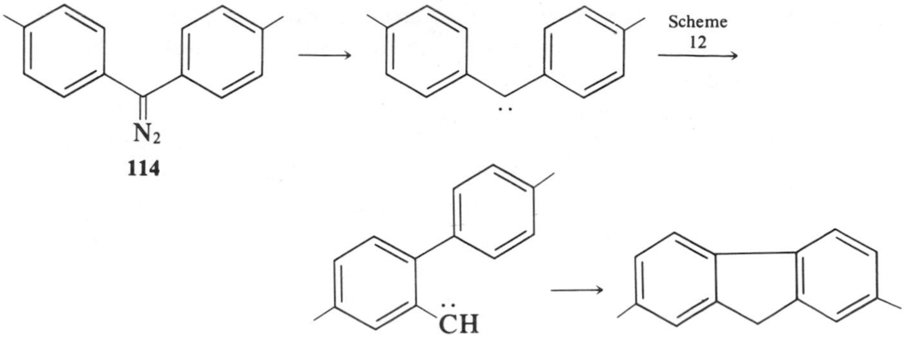

114

115

The phenyl(4-pyridyl)carbene **116a** undergoes *both* carbene–nitrene rearrangement (~91%) and carbene–carbene rearrangement (~9%), as shown by ^{14}C-labeling:[67]

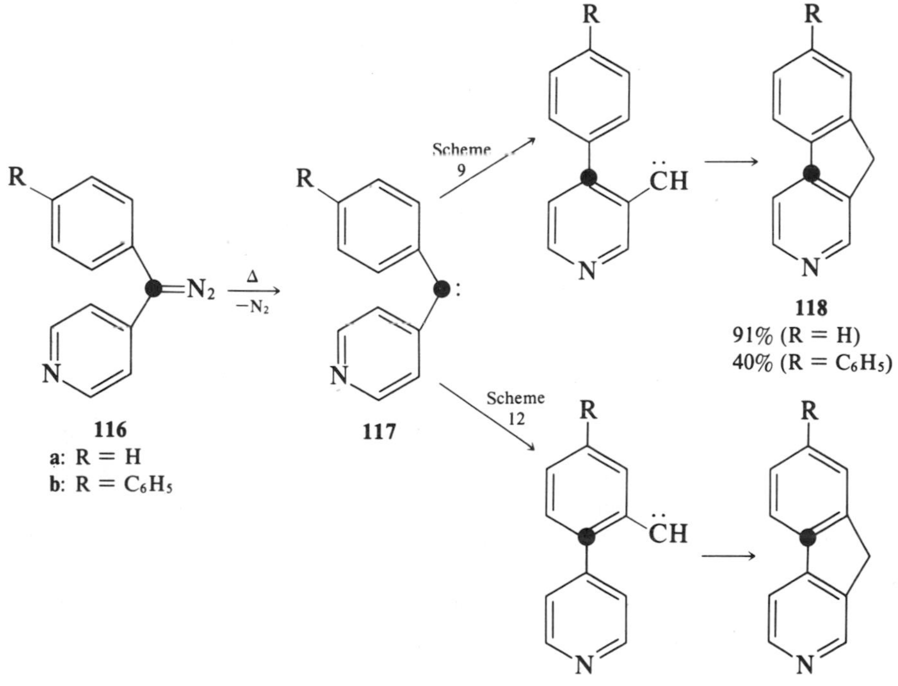

116
a: R = H
b: R = C_6H_5

117

118
91% (R = H)
40% (R = C_6H_5)

119
9% (R = H)
60% (R = C_6H_5)

Scheme 13

The extent of the rearrangement involving the benzene ring (giving **119**) can be *increased* by placing a *p*-phenyl substituent in this ring **(116b)**.[129]

All these results, and several other observations, can be understood in terms of a simple frontier MO analysis.[67] The initial phase of the rearrangements requires bonding between the carbene center and the aromatic ring (e.g., **96** → **97**). Even if **97** is not an intermediate, a concerted reaction **96** → **98** also requires such bonding to develop along the reaction coordinate.

We now dissect the carbene into a free, noninteracting singlet methylene and the affixed rings. The HOMO and LUMO of the singlet carbene are the occupied σ and the vacant *p* orbitals, respectively (Figure 4.5). The σ orbital can interact with the LUMO of the aromatic ring in the *ipso*-position (the position to which the carbene is attached). The *p* orbital can interact with the aromatic HOMO in the *ortho*-position, leading eventually to the formation of a new C—C bond as in **97**. These two interactions can also be illustrated as follows:

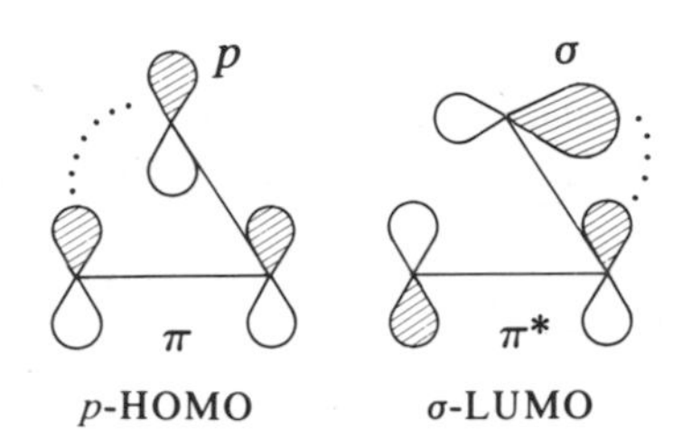

The two interactions take place simultaneously, and they reinforce each other—they are synergic. The *p*-HOMO interaction is "electrophilic" from the point of view of the carbene. It will be favored by high-energy HOMOs with large coefficients in the *ortho*-positions (see Figure 4.5, right-hand side), that is, the same factors that favor normal electrophilic aromatic substitution. The σ-LUMO interaction is "nucleophilic" from the point of view of the carbene; it is favored by low-lying LUMOs possessing large coefficients in the *ipso*-position. A pyridine ring is electron deficient and undergoes electrophilic aromatic substitution with difficulty, and when at all, then in the 3-position. However, pyridine readily undergoes nucleophilic substitution in the 2- and 4-positions. Hence the *p*-HOMO interaction in our carbenes will be strongest with the benzene rings; the σ-LUMO interactions with the pyridine rings (Figure 4.5). Therefore, the outcome of the reaction will be determined by the orbital coefficients. The pyridine LUMO has large coefficients in positions 2 and 4, and indeed, the carbenes **106** and **117a** (Schemes 10 and 13) undergo preferred or exclusive rearrangement involving these rings. The LUMO-coefficient in the 3-position in pyridine is very small, however, and no nucleophilic substitution in this position is known. Despite the low LUMO energy, the σ-LUMO interaction with the pyridine ring in the 3-pyridylcarbene **110** (Scheme 11) is therefore disfavored. The *p*-HOMO interaction with the benzene ring is unaffected, and leads to the observed reaction.

The *p*-phenyl substituent in **117b** *lowers the LUMO* energy of the benzene ring, which can now compete successfully with pyridine in the σ-LUMO interaction; both processes are observed (Scheme 13). Likewise, the naphthyl substituent in the carbene **120** lowers the benzenoid LUMO and, at the same time, increases the *p*-HOMO interaction due to the large coefficient at the 1-position in naphthalene. As a result, the exclusive pyridine ring rearrangement shown in Scheme 10 is no longer observed; instead, both rings undergo rearrangement (Scheme 14).[67]

E
N
LUMO
p
σ
HOMO
LUMO
HOMO
p
C
σ

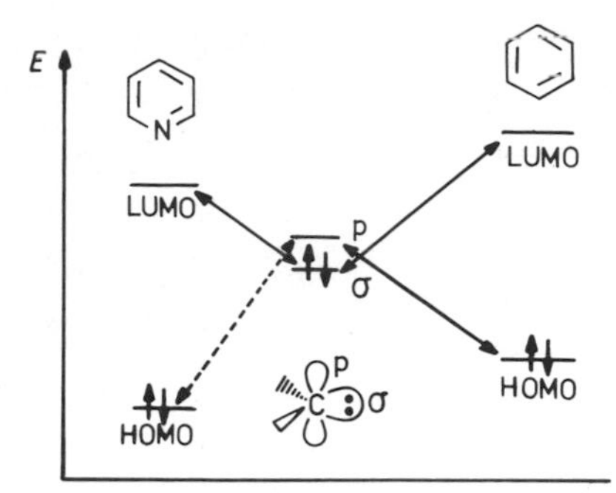

Figure 4.5. Frontier MO interactions in phenyl(pyridyl)carbenes, approximated as the interactions between methylene and the isolated aromatic rings.

N
N=N
Δ
$-N_2$
N
120
N:
HC
N
N
H
30%
N
16%
+
N
48%

Scheme 14

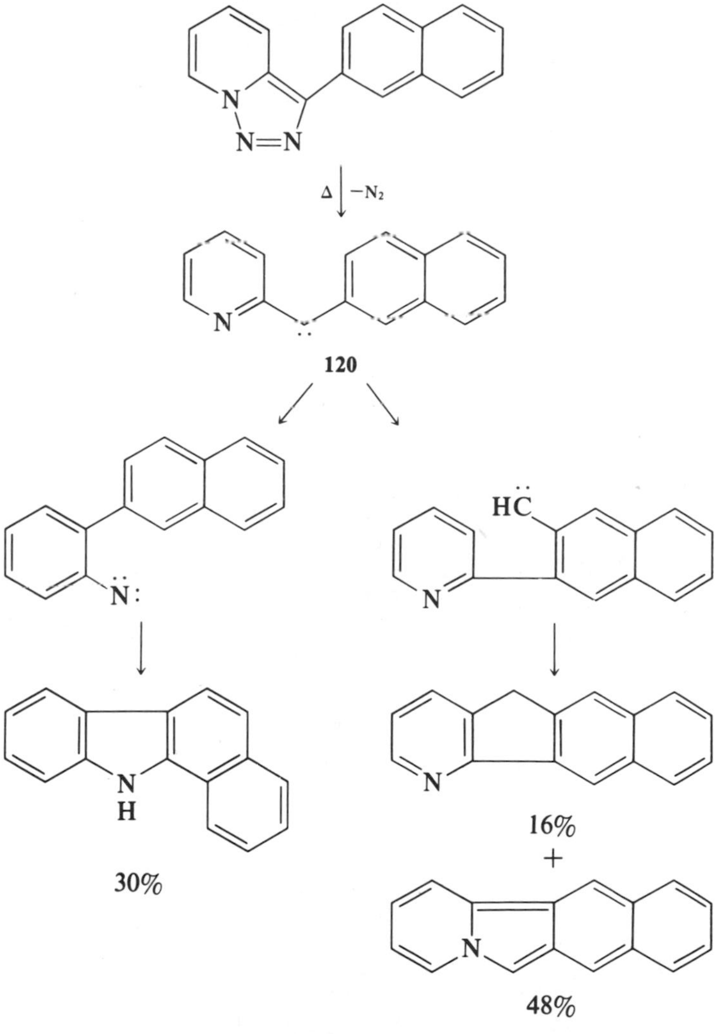

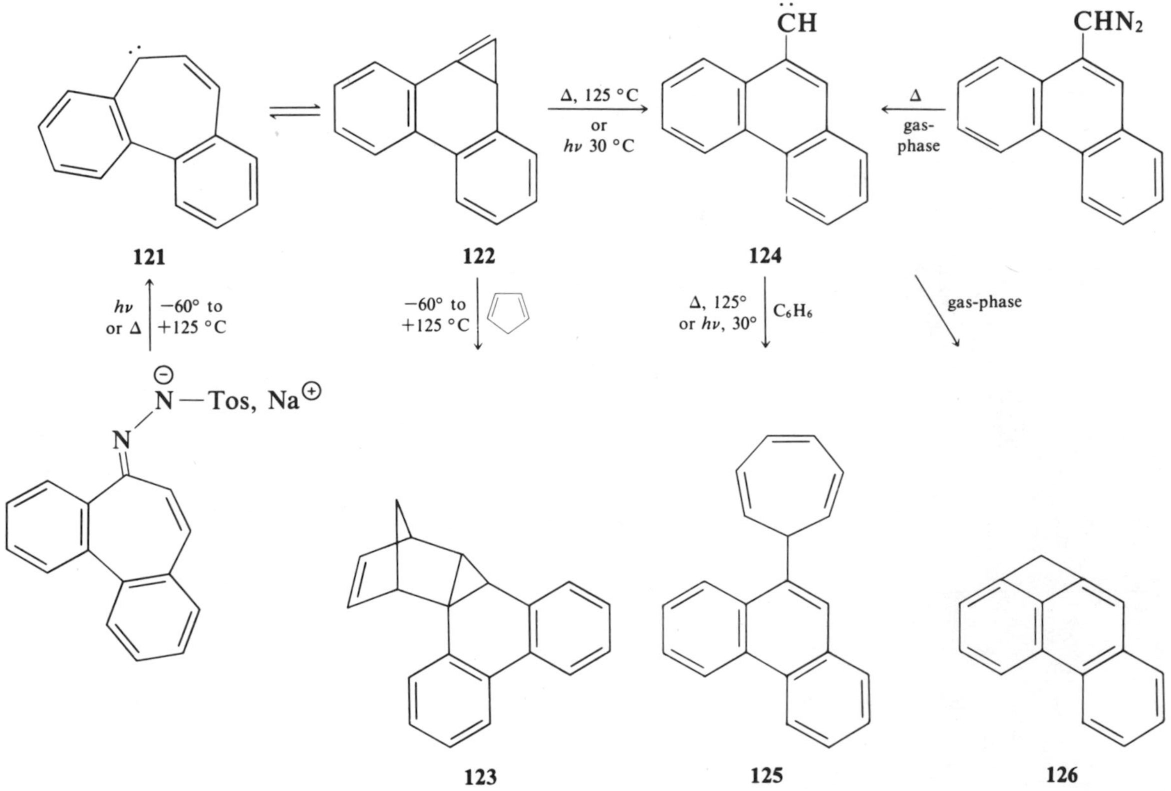

CHN_2

Δ gas-phase

$\ddot{C}H$

Δ, 125 °C or $h\nu$ 30 °C

121

$h\nu$ or Δ, −60° to +125 °C

$\overset{\ominus}{N}$—Tos, $Na^{\oplus}$

122

−60° to +125 °C

123

124

Δ, 125° or $h\nu$, 30°, C_6H_6

125

gas-phase

126

Scheme 15

There is some uncertainty as to the existence, energetics, and lifetimes of the bicyclic intermediates **97** and **99** in Scheme 9 and the corresponding cyclopropenes in Scheme 12. There is no doubt, however, that such fused cyclopropenes are true intermediates in larger systems since they can be trapped in [2 + 4]-cycloaddition reactions in solution (Scheme 15).[130] Thus the carbene **121** isomerizes to the fused cyclopropene **122**, which can be trapped with several dienes, including cyclopentadiene, giving **123**. In the absence of a trapping agent and using benzene as a solvent, further rearrangement to phenanthrylcarbene **(124)** takes place; the latter adds to the solvent to give the cycloheptatriene **125**.[130] When phenanthrylcarbene is generated in the gas phase, it undergoes another rearrangement, namely, insertion into the *peri*-C—H bond, giving the cyclobutaphenanthrene **126**.[131]

There is also some uncertainty as to the nature of the seven-membered ring carbenes, such as **98** in Scheme 9. The results of all the carbene–carbene and carbene–nitrene rearrangements described in Schemes 9–15 require that intermediates of this overall structure be formed, but they do not tell whether the intermediate is a carbene (e.g., **98**) or a ketenimine (e.g., **101**), or whether both types

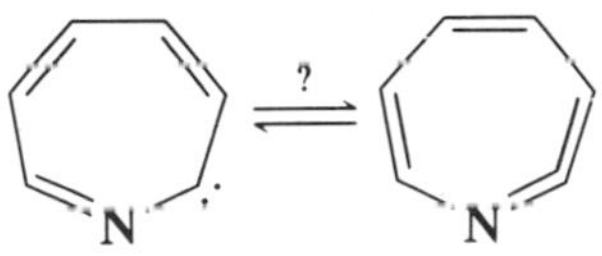

?

98 **101**

of intermediate can coexist in an equilibrium. However, matrix-isolation experiments have successfully demonstrated that the most stable form of this intermediate is the ketenimine **101**, as evidenced by IR spectroscopy.[132] The same intermediate

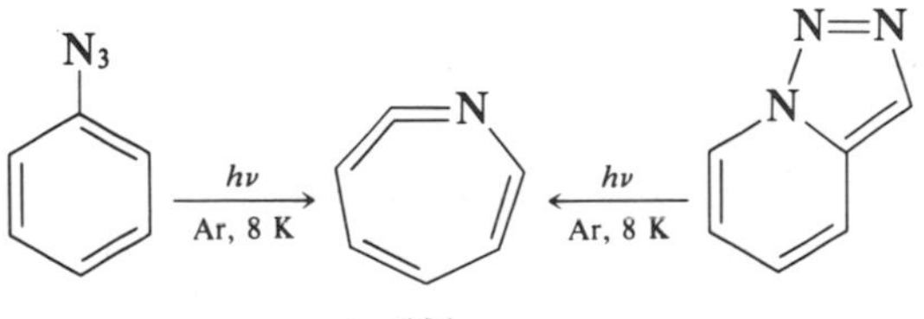

N_3 $h\nu$ Ar, 8 K N $h\nu$ Ar, 8 K N=N N

101
1895 cm^{-1}

is probably responsible for the formation of azepines (e.g., **127**) in the solution phase photolysis (and, more rarely, thermolysis) of aryl azides:[112]

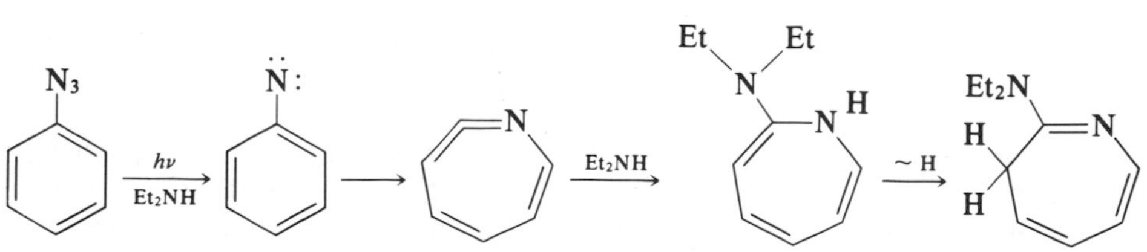

N_3 $h\nu$ Et_2NH N: N Et_2NH Et Et N N H ~ H Et_2N H N H

101
and/or **98**

127

A corresponding allene **(129)** has also been identified by its matrix IR and UV spectra following photolysis or gas-phase pyrolysis of phenyldiazomethane with trapping in argon at 23 K:[133]

CH=N2 — hν, Ar, 10 K → CH (128) — CO, Ar, 40 K → CH=C=O (130)

128 — hν, 10 K → 129

1) 500 °C 2) Ar, 23 K

1816, 1824 cm^{-1}

129

Phenylcarbene itself **(128)** was also directly observed by IR and UV spectroscopy in this study and its identity confirmed by reaction with CO in the matrix, whereby phenylketene **(130)** was formed.

Not only are carbene–carbene and carbene–nitrene rearrangements known: Numerous heteroaromatic nitrenes interconvert with each other, for example, the 2-pyridylnitrenes **133** and **135**.[134,135] First of all, the sublimation of the tetrazole **131** gives the azide **132**. The azide loses nitrogen above 380°C under flash vacuum pyrolysis conditions, and the cyclic carbodiimide **134** can be isolated at −196°C and identified by its IR spectrum. At higher pyrolysis temperatures, **134** disappears, and the stable products **136** and **137** are formed instead. Using labeling, both with ^{15}N and with substituents, it was shown that the products **136** and **137** must be formed *after* complete equilibration of the nitrenes, **133** ⇌ **134** ⇌ **135**. For example, both the 2-amino-pyridine **(136)** and the 2-cyanopyrrole **(137)** had undergone complete nitrogen scrambling (Scheme 16). The carbodiimide **134** was stable at −196°C and dimerized on warming to about −70°C. An almost quantitative yield of the carbodiimide dimer can be obtained from the analogous tetrazolophenanthridine (**138**, Scheme 17). The carbodiimide **139** was formed cleanly at 490°C and dimerized on warming to −40°C. The structure of the dimer **(140)** was established by X-ray crystallography and proves that the intermediates in these reactions are indeed seven-membered ring carbodiimides.

Skattebøl Rearrangements. In the Wolff rearrangements and the carbene–nitrene rearrangements just mentioned, the electron deficiency moves from atom to atom. One further carbene–carbene rearrangement in which the electron-deficient carbon atom remains the same is named after its discoverer[136] and involves a rather unique rearrangement of a vinylcyclopropylidene **(141)** to a cyclopentenylidene **(142)**. The

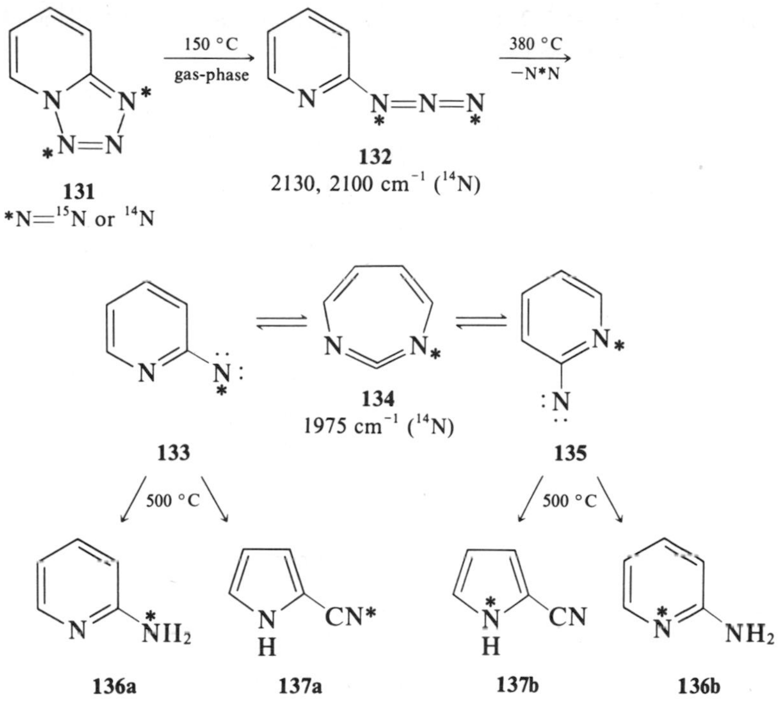

150 °C gas-phase

380 °C −N*N

131
*N = ^{15}N or ^{14}N

132
2130, 2100 cm^{-1} (^{14}N)

133

134
1975 cm^{-1} (^{14}N)

135

500 °C

500 °C

136a **137a** **137b** **136b**

Scheme 16

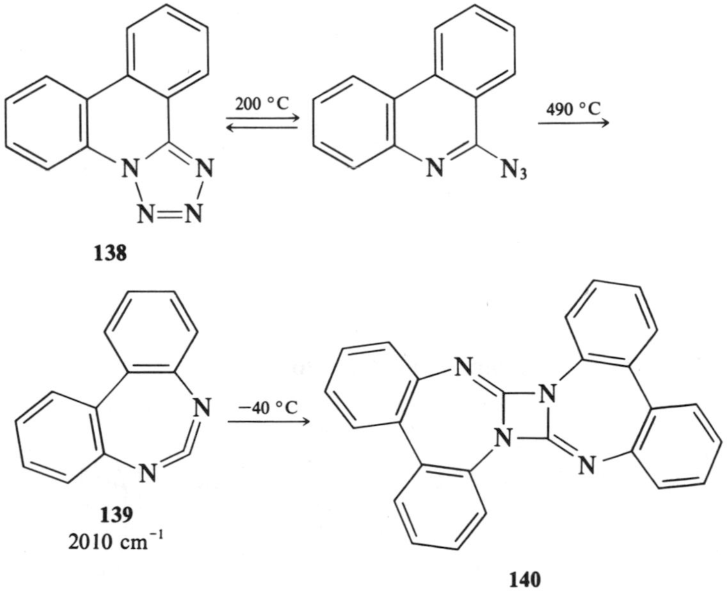

200 °C

490 °C

138

−40 °C

139
2010 cm^{-1}

140

Scheme 17

latter undergoes a rapid 1,2-shift of hydrogen to give a cyclopentadiene **(143)** as the final product (Scheme 18).

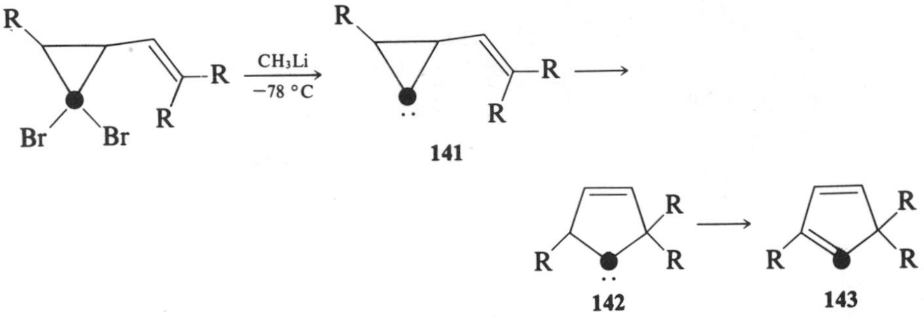

CH$_3$Li
−78 °C

141

142 143

Scheme 18

The mechanism implied in Scheme 18 is supported by carbon labeling and the substituent pattern obtained. The reaction can be understood as proceeding through a π-complex **(144)** in which the vacant carbene *p* orbital interacts particularly with the terminal *p* orbital of the olefinic double bond (see **145**). Steric constraints make the intramolecular addition of the carbene to the double bond unfavorable. The reaction proceeds in a concerted manner (see **146**) to product.[137,138]

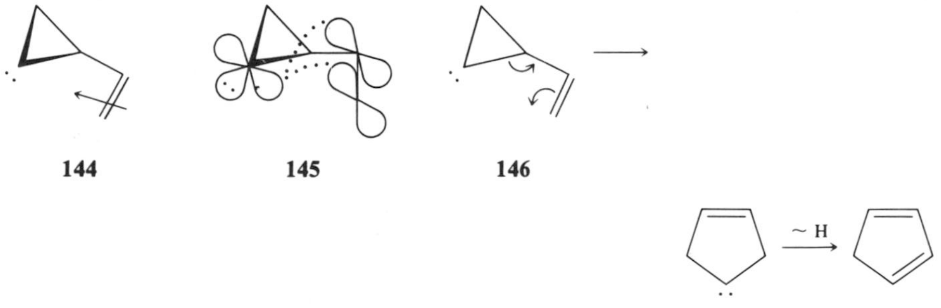

144 145 146

~H

Further Examples of Intramolecular Reactions

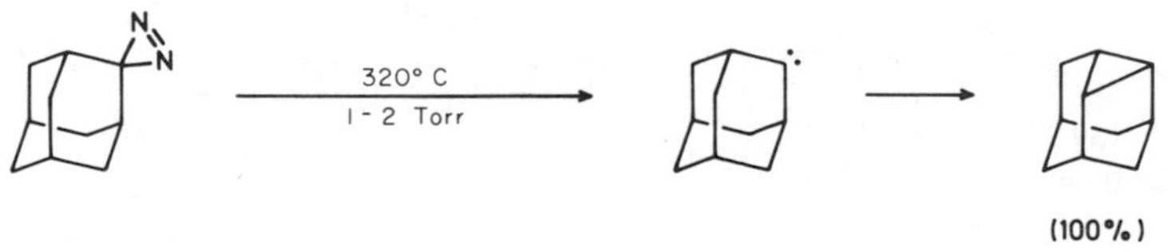

320° C
1 - 2 Torr

(100%)

(Ref. 139)

N−NH−Tos

Δ
NaOMe

=CH$_2$ +

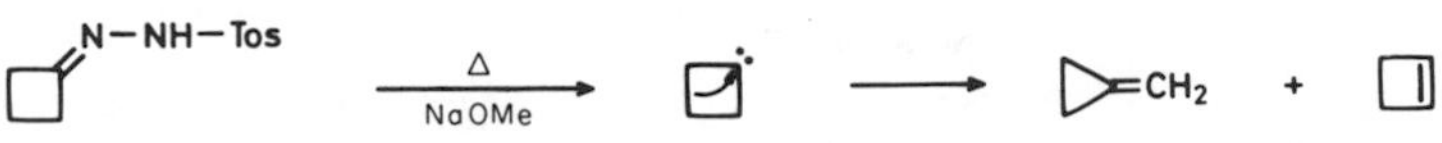

(Ref. 140)

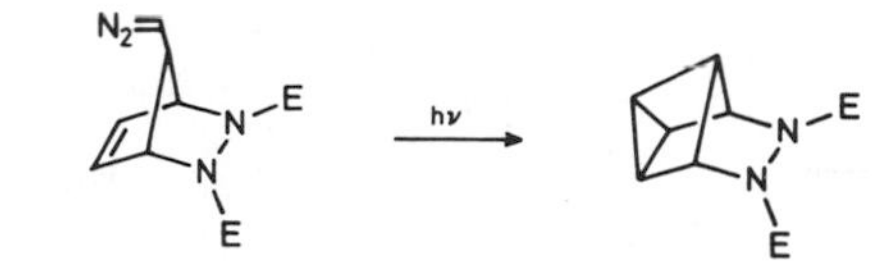

$E = COOC_2H_5$

(Ref. 141)

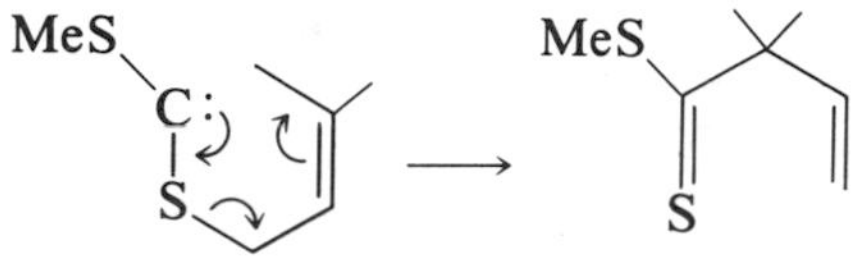

(Ref. 142)

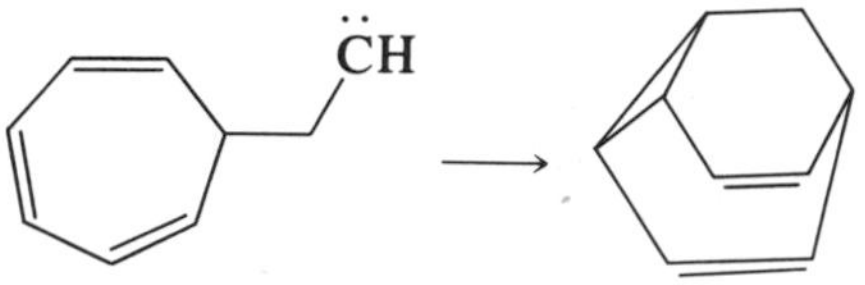

(Ref. 143)

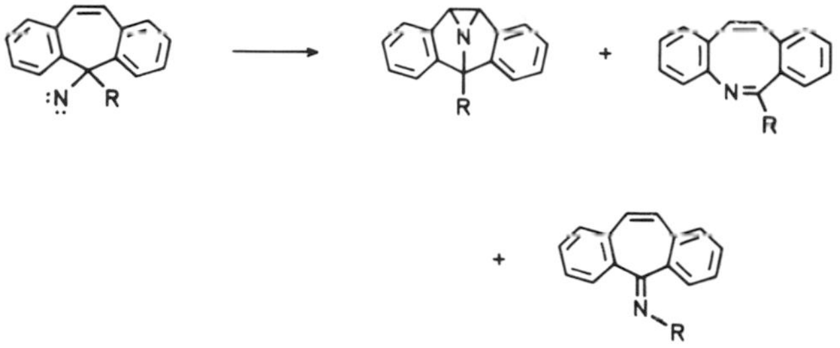

(Ref. 144)

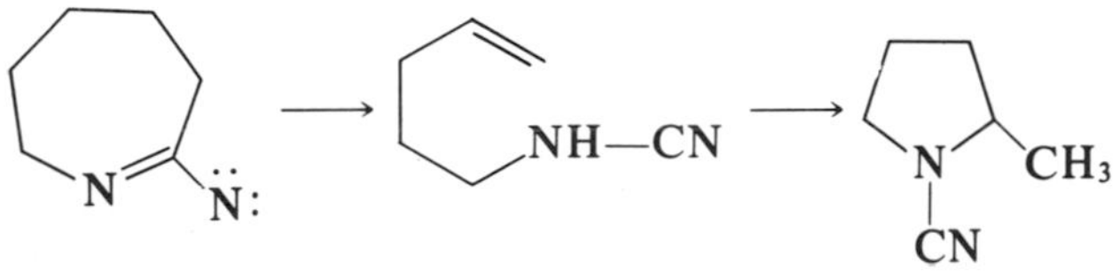

(Ref. 145)

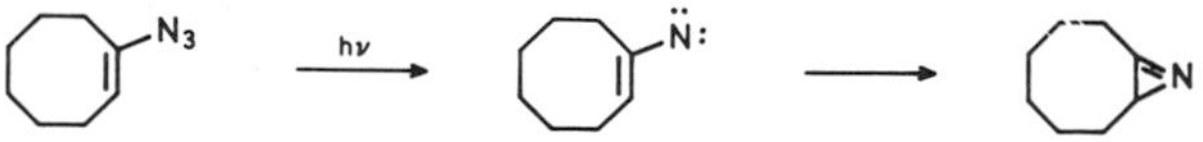

(Ref. 146)

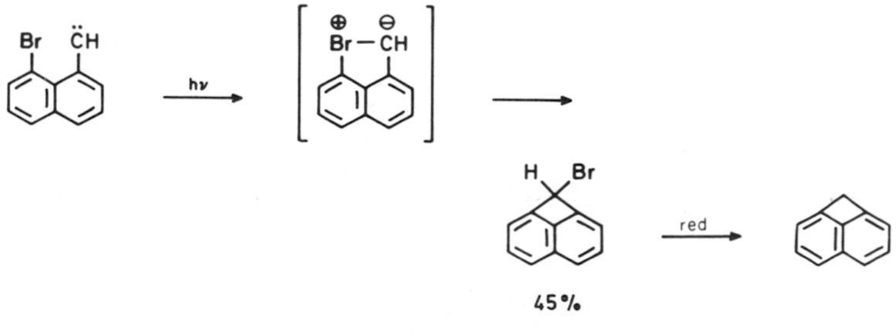

(Ref. 147)

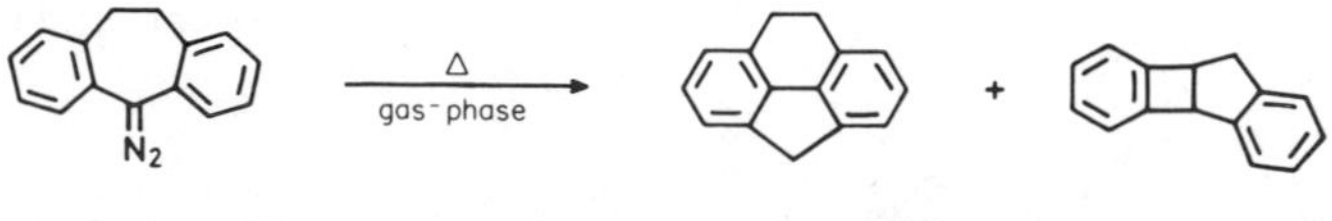

45%

(Ref. 60)

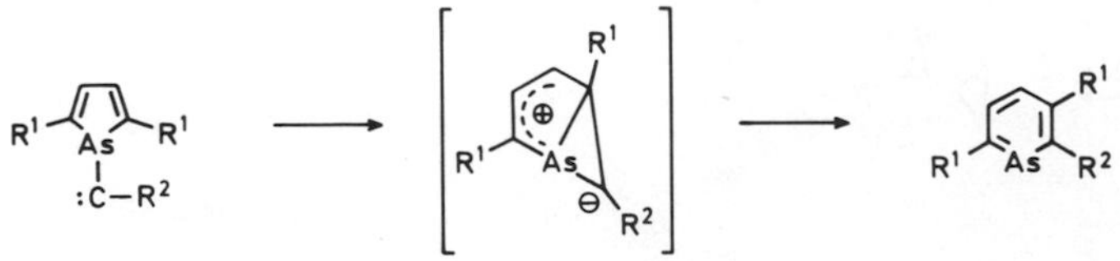

(Ref. 148)

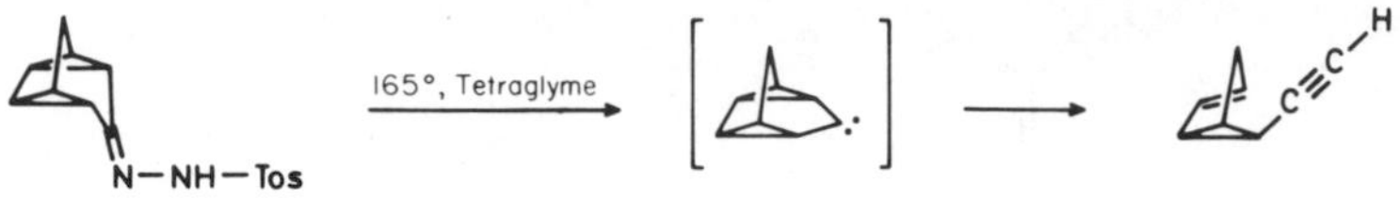

165°, Tetraglyme

N—NH—Tos

(Ref. 149)

Δ

gas-phase

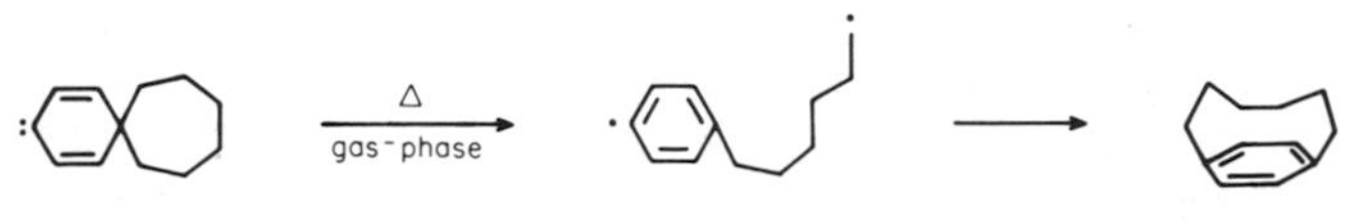

[6]-Paracyclophane

(Ref. 150)

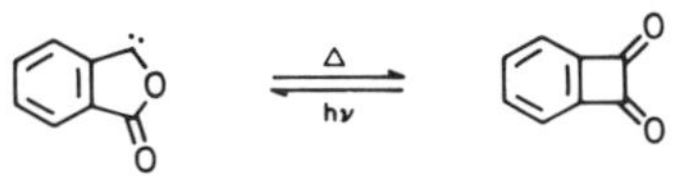

Δ

hν

(Ref. 151)

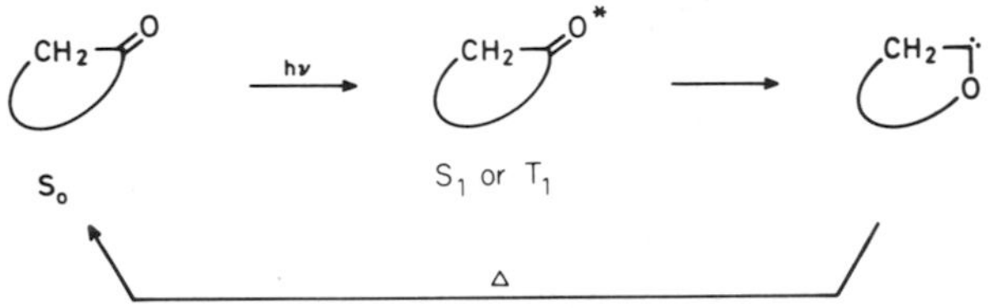

hν

S_0

S_1 or T_1

Δ

(Ref. 152)

Dimer

(Ref. 153)

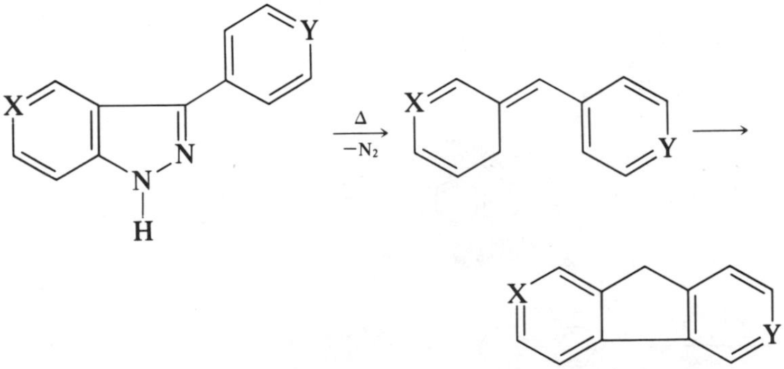

Δ

$-N_2$

(Ref. 154)

4.5. CHEMICAL ACTIVATION

Whenever an exothermic reaction occurs, the enthalpy of reaction will initially be contained as vibrational, rotational, and translational energy in the products. In the condensed phases, this energy is rapidly dissipated through collisions. In the gas-phase, in contrast, the collision frequency depends on the pressure, and if this is sufficiently low, the heat of reaction may be channeled into secondary intramolecular reactions.

The addition of methylene to ethylene was depicted in Figure 4.2. Since the reaction is exothermic by approximately 100 kcal/mol, and since the E_a for the isomerization of cyclopropane to propene is only ~64 kcal/mol (Chapter 3), this isomerization will take place if the cyclopropane is not rapidly deactivated.

$$\ddot{C}H_2 + H_2C{=}CH_2 \longrightarrow \triangle \qquad \Delta H_r^0 \simeq -100 \text{ kcal/mol}$$

$$\triangle \longrightarrow \text{propene} \qquad E_a \cong 64 \text{ kcal/mol}$$

Consequently, the product ratio cyclopropane/propene increases with the pressure.[155]

In general, the energy available to the chemically activated molecule, E^*, is the sum of the enthalpy of the reaction, the activation energy E_a' for the step leading to the chemically activated molecule, the thermal energy at the reaction temperature E_{therm} ($\sim nRT$ where n is the number of internal degrees of freedom), plus any other excess energy E_{excess} that may have been carried over from preceding reaction steps.

$$E^* = \Delta H_r^0 + E_a' + E_{therm} + E_{excess}$$

E_{excess} is particularly important in photochemical processes. The photolysis of $CH_2{=}N_2$ at 3660 Å leads to $:CH_2$ with an excess energy of approximately 16 kcal/mol. At 4358 Å, the excess energy is about 12.5 kcal/mol;[156] that is, the excess energy increases with the energy of the irradiating photons.

Consider the reaction scheme

$$\ddot{C}H_2 + H_2C{=}CH_2 \longrightarrow \triangle^* \begin{cases} \xrightarrow{k_i(E)} \text{propene} \\ \xrightarrow[M]{k_d} \triangle \end{cases}$$

where the star (*) denotes chemical activation, $k_i(E)$ is the rate constant for isomerization, and k_d is the bimolecular rate constant for collisional deactivation. If it is assumed that deactivation takes place on every collision, k_d equals the collision frequency ω, which is obtained from the kinetic theory of gases.†

†This is the "strong collision" hypothesis. It is now known that only 1–10 kcal/mol are removed per collision. Big molecules are more effective in removing excess energy than small ones.

From the observed ratio k_i/k_d as a function of pressure, it is now possible to compute $k_i(E)$. If E^* is strictly monoenergetic, $k_i/k_d = k_i/\omega = (\wedge)/(\triangle)$, and k_i will be pressure independent.

The value for $k_i(E)$ can now be combined with the Rice–Ramsperger–Kassel expression for the rate of unimolecular reactions,†

$$k_i(E) = A\left(\frac{E^* - E_i}{E^*}\right)^{s-1} \tag{7}$$

where A is the high-pressure Arrhenius factor, obtained from conventional kinetics (in the case of pyrolysis of cyclopropane, $A \cong 10^{15}\ s^{-1}$); E^* is the total excess energy, E_i is the activation energy for the isomerization to propene (~64 kcal/mol), and s is the number of effective oscillators, that is, the number of vibrational modes involved in the isomerization. These are $3N$–6 in number in nonlinear N atomic molecules. Normally, it is assumed (and borne out in practice) that only two-thirds of the modes are effective. This gives $s = 2N - 4$, and $s - 1 = 2N - 5$.

Solving Eq. (7) now gives E^*, the total excess energy. The singlet–triplet energy separation in methylene, E_{ST} was elucidated in this way. With a knowledge of $\Delta H_f^0(^3CH_2) = 93 \pm 2$ kcal/mol and of E^* for the 4358–Å photolysis of $CH_2{=}N_2$, one computes for cyclopropane

$$E^* \simeq \Delta H_r^0 + E_{excess} = \Delta H_f^0(\triangle) - \Delta H_f^0(^1CH_2) - \Delta H_f^0(C_2H_4) + E_{excess} \tag{8}$$

where

$$\Delta H_f^0(^1CH_2) = \Delta H_f^0(^3CH_2) + E_{ST}.$$

Here E_{ST} and E_{excess} are still unknown quantities, but their sum can be determined as 112.6 kcal/mol. This means that E_{ST} cannot be more than $112.6 - 93 \pm 2 \simeq 20$ kcal/mol. In order to extract the value of E_{ST}, the singlet methylene is deactivated with increasing amounts of moderator gases, CF_4 or C_8F_{18}. The singlet methylene will now undergo many collisions with the fluorocarbons and lose its excess energy before reacting with ethylene to yield cyclopropane. E^* for cyclopropane is now measured as a function of moderator gas pressure. When no further decrease in E^* can be achieved, the singlet methylene is presumably vibrationally relaxed, and E_{ST} is obtained from Eq. (8) by setting $E_{excess} = 0$. Since about 11 kcal/mol of excess energy could be removed in this way, the value of E_{ST} was determined to be approximately 9 kcal/mol.[6a]

When CH_2 was produced by photolysis of ketene at 3200–3800 Å, a value of $E_{ST} + E_{excess} \simeq 10$ kcal/mol was obtained, and only 1 kcal/mol could be removed by deactivation. Thus ketene gives a less highly excited carbene than does diazo-

†For a more advanced treatment, see Ref. 157.

methane. Let us calculate why this is so. The dissociation energy of ketene is obtained from

$$\begin{array}{cccc} CH_2{=}C{=}O & \longrightarrow & {}^1CH_2 & + \quad CO \\ \Delta H_f^0 = -11.4 & & 93 \pm 2 + E_{ST} & -27.2\ \text{kcal/mol} \end{array}$$

that is,

$$\Delta H_r^0 = +77\ \text{kcal/mol} + E_{ST}$$

Light at 3300 Å corresponds to a photon energy of 86.7 kcal/mol. Thus the CH_2 and CO formed can carry no more than $86.7 - 77 - E_{ST} = (10 - E_{ST})$ kcal/mol of excess energy between them. If all this goes to CH_2, the singlet–triplet gap can be no more than ~ 10 kcal/mol. With $E_{ST} \sim 9$, the singlet methylene will carry about 1 kcal/mol of excess energy.

Similarly, the C—H insertion reaction of $^1CH_2$ with cyclopropane is strongly exothermic:

$$^1CH_2: + \triangle \longrightarrow [\triangleright\!\!-CH_3]^*$$

$$\Delta H_f^0 \sim 102 \qquad 12.7 \qquad 5.7\ \text{kcal/mol}$$
$$\Delta H_r^0 \sim -109\ \text{kcal/mol}$$

Consequently, the excited methylcyclopropane (**147**) rearranges:[158]

147: ▷—CH₃* → (M) ▷—CH₃ ; → CH₂=⟨ + =╲_ + ╲=╲ + /=╲

The isomerization of methylcarbene (**148**) to ethylene was shown (Figure 4.4) to be exothermic by ~ 78 ± 5 kcal/mol. We have already seen (p. 137) that ethylene undergoes *cis–trans* isomerization with $E_a \simeq 64$ kcal/mol. Accordingly, the stereochemistry criterion of Skell cannot be used to deduce anything about the nature of the initial carbene (i.e., singlet or triplet) under gas-phase conditions:

$$\underset{\mathbf{148}}{H_3C-\ddot{C}H} \xrightarrow[\sim -78 \pm 5]{\Delta H_r^0} (H)(H_a)C{=}C(H)(H_a) \underset{}{\overset{E_a = 64}{\rightleftharpoons}} (H)(H_a)C{=}C(H_a)(H)$$

$$\downarrow E_a = 78\ \text{kcal/mol}$$

$$HC{\equiv}CH + H_2$$

Furthermore, ethylene can decompose by elimination of H_2 with an activation energy of 78 kcal/mol,[159] which may be just a little too much to make this an efficient process. If the methylcarbene contains excess energy, however, from the photodecomposition of diazoethane, the decomposition to acetylene takes place.[160,161]

For the same reason, methyleneimine (**69**) generated by flash vacuum pyrolysis of methyl azide experiences extensive decomposition into hydrogen cyanide and H_2:[107]

$$CH_3{-}N_3 \xrightarrow{E_a \sim 40} CH_2{=}NH\ (\mathbf{69}) \longrightarrow HCN + H_2$$

$$\Delta H_f^0:\ 67 \qquad 26 \qquad 32 \qquad 0\ \text{kcal/mol}$$

From the enthalpy of formation of methyl azide (67 kcal/mol), the activation energy for nitrogen loss (~40 kcal/mol) and the enthalpy of formation of methyleneimine[162] (~26 kcal/mol), **69** can be activated by as much as 81 kcal/mol, when produced at low pressure.

When a chemically excited molecule is formed in a bimolecular process, one can imagine that the total excess energy E^* is located primarily in the newly formed part of the molecule. For example,

F F F CF=CF2 + C̈D2 ⟶ F F F F F F D D

149

The molecule **149** loses its excess energy by expulsion of a stabilized carbene $\ddot{C}F_2$. If the excess energy is localized, only the CF_2 group on the right-hand side of the molecule should be expelled, giving **150**. If the energy is distributed over the whole molecule, **151** should also be formed:

149 ⟶ F F F CF=CD2 + C̈F2

150

149 ⟶ F F F H2C=CF D D + C̈F2

151

From the variation of the product ratio **150/151** with pressure it was possible to evaluate the rate constant for energy exchange between the two halves of **149** as ~ 10^{12} s; that is, the molecule distributes the energy within approximately 10 molecular vibrations.[163]

Chemical activation is not only important in carbene chemistry. Many free-radical reactions are exothermic, for example:

RH; $-R\cdot$; M; H; + H· ⟶; *; $k_i(E)$; $E_a \sim 30$; $\cdot CH_3$ +

ΔH_f^0: −1.7 52.1 ~13 34 4.9 kcal/mol

The exothermicity of the first step (37.4 kcal/mol) is sufficient to overcome the activation energy for the fragmentation to a methyl radical and propene.[164]

A similar effect is seen in excited triplet 1,3-diradicals formed from $^3CH_2$ and olefins:[165]

M; $^3CH_2$ +; ⟶; $\cdot CH_3$ +

ΔH_f^0: ~93 ± 2 −1.7 $\Delta H_f^0 \sim 50$, $E^* \sim 40$ 34 ~30 kcal/mol

Chemical activation can be used to practical advantage to make difficult reactions proceed more readily. For example, cyanocyclopentadiene **(153)** is formed in low yield by rearrangement of phenylnitrene **(100)** in the gas-phase pyrolyses of phenyl azide **(152)** and triazolopyridine **(155)** (Scheme 19; see also Scheme 9). A much higher yield of **153** is obtained from the tetrazolylpyridine **156** *at the same temperature,* even though this reaction passes through the same intermediates **154** and **155.** The reason is a higher heat of formation of **156,** together with a higher energy of activation for its decomposition, which leads to chemical activation of the intermediates. If the pressure is increased, excess energy is removed, and the yield of **153** diminishes.[17]

The possibility of chemical activation should always be kept in mind in mechanistic studies of gas-phase reactions, particularly at low pressure, for example, flash vacuum pyrolysis experiments. So-called control experiments purporting to show that a particular reaction does or does not take place under the ''reaction conditions'' require definition of these conditions, including possible chemical activation.[17,166]

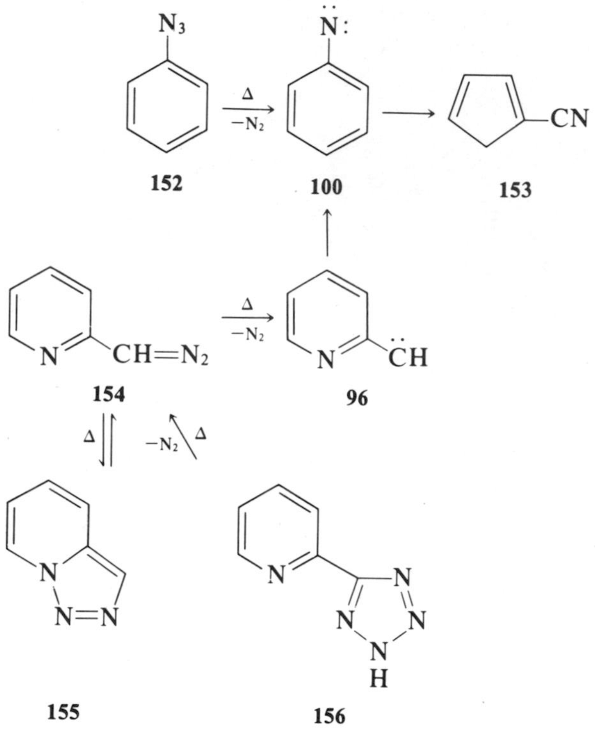

152 100 153

154 96

155 156

Scheme 19

4.6. STRUCTURE AND REACTIVITY

4.6.1. Carbenes Versus Nitrenes

There is much evidence that nitrenes (both singlets and triplets) are more stable, less reactive, and more selective than carbenes.[60] The almost universal rearrangement of six-membered heteroarylcarbenes to more stable and/or less reactive

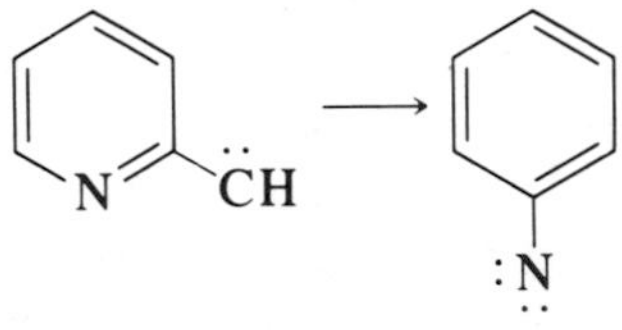

arylnitrenes[112,166] is an example (see Section 4.4.5). This reaction most probably occurs in the singlet manifold. The failure to trap triplet nitrenes with nitrous oxide (see Section 4.4.1) also indicates a lower reactivity of the nitrenes compared with carbenes.

Some selectivity data for intermolecular C—H insertion reactions of carbenes and nitrenes are collected in Table 4.6. Although the data should not be compared too closely because of differing reaction conditions, higher selectivities of nitrenes are apparent.

TABLE 4.6. Selectivity of C—H Insertion Reactions with Hydrocarbons

	tert-CH	*sec*-CH	*prim*-CH	Reference
MeO—CO—$\ddot{C}$H (singlet)	3.1	2.3	1	*c*
EtO—CO—$\ddot{N}$ [a]	34	10	1	*d*
C_6H_5—$\ddot{C}$H[b]		6.3	1	*e*
C_6H_5—$\ddot{N}$ [b]	≥140–280	>7	1	*f*
CH_2	1.40	1.22	1	*g*

[a]Probably singlet.
[b]Probably triplet.
[c]W. v. E. Doering and L. H. Knox, *J. Am. Chem. Soc.*, **83,** 1989 (1961).
[d]W. Lwowski and T. J. Maricich, *J. Am. Chem. Soc.*, **87,** 3630 (1965).
[e]C. D. Gutsche, G. L. Bachman, and R. S. Coffey, *Tetrahedron,* **18,** 617 (1962).
[f]J. H. Hall, J. W. Hill, and H.-C. Tsai, *Tetrahedron Lett.*, **1965,** 2211.
[g]B. M. Herzog and R. W. Carr, *J. Phys. Chem.*, **71,** 2688 (1967).

In a direct comparison of the reactivities of thermally generated (singlet) ethoxycarbonylcarbene **(157)** and ethoxycarbonylnitrene **(158)** with substituted benzenes, it was found[167] that the carbene was approximately 3.33 times more reactive than the nitrene. Both the carbene and the nitrene behave as *electrophiles* in this reaction, giving Hammett ρ values of −0.38 and −1.32, respectively. Thus, in spite of its lower overall reactivity, the nitrene is more electrophilic than the carbene, as expected from the higher electronegativity of nitrogen compared with carbon.† The

$$\text{EtO—C(=O)—}\ddot{\text{C}}\text{H}\ (\mathbf{157}) + C_6H_5\text{—R} \longrightarrow [\text{EtOC(=O)—CH}\langle\text{norcaradiene—R}] \longrightarrow \text{EtO—C(=O)—cycloheptatriene—R} \quad (9)$$

$$\text{EtO—C(=O)—}\ddot{\text{N}}\text{:}\ (\mathbf{158}) + C_6H_5\text{—R} \longrightarrow [\text{EtO—C(=O)—N}\langle\text{azanorcaradiene—R}] \longrightarrow \text{EtO—C(=O)—N(azepine)—R} \quad (10)$$

†It is dangerous to talk about degrees of electrophilicity. If the nitrene is inherently more stable than the carbene, the transition state will be later in the nitrene reaction (Hammond postulate) and the "electrophilicity" will be more strongly felt.

nitrene is more selective than the carbene, showing a larger spread in the relative rates of addition to anisole, toluene, benzene, fluorobenzene, chlorobenzene, and trifluoromethylbenzene.

Whereas the addition of carbenes to double bonds is a rather general reaction of practical use, for example, in the synthesis of cycloheptatrienes [Eq. (9); see also Scheme 15], the corresponding reactions of nitrenes to give aziridines or azepines [Eq. (10)] take place only with electronegatively substituted nitrenes such as **158,** the sulfonylnitrenes **(159),**[168] cyanonitrene **(160),**[169] or *s*-triazinylnitrene[69c]

$$R{-}SO_2{-}\ddot{N}: \qquad N{\equiv}C{-}\ddot{N}:$$

159 **160**

(see Section 4.4.3). Phenylnitrene is not sufficiently electrophilic to undergo this reaction,[170] but the introduction of electron-withdrawing groups increases the electrophilicity so much that these nitrenes **(161)** react with electron-rich aromatics.[69b] Although the products are not azepines [cf. Eq. (10)], they are probably formed via initial addition to an aromatic double bond:

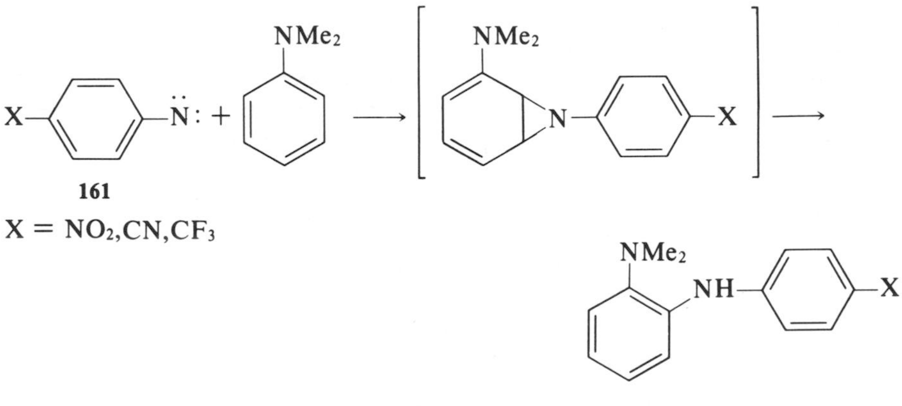

NMe2 NMe2

X—C6H4—N̈: + C6H5NMe2 ⟶ [aziridine intermediate N—C6H4—X] ⟶

161

$X = NO_2, CN, CF_3$

NMe2

NH—C6H4—X

A likely rationale for these observations is that singlet phenylnitrene relaxes to the triplet ground state faster than it undergoes intermolecular addition to C=C double bonds. The triplet is relatively unreactive and gives azobenzene, aniline, and tars as major products. Electronegative substituents increase the electrophilicity and thus the reactivity of the singlet nitrene so that addition reactions now compete with intersystem crossing. The singlet reactivity can also be increased by augmenting the concentration of the reaction partner, as in the *intramolecular* azepine formation shown in Eq. (11).[112,171] Carbazole formation from 2-biphenylylnitrenes (**107;** Scheme 10) is another very efficient and versatile intramolecular cyclization of an arylnitrene, in all probability proceeding in the singlet manifold.[112]

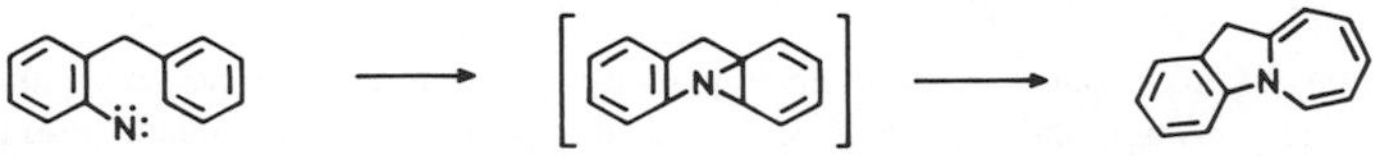

N̈: ⟶ [N] ⟶ N (11)

Arylcarbenes also have triplet ground states, but they show a richer intermolecular chemistry for two reasons: (a) the higher intrinsic reactivity of carbenes as compared with nitrenes and (b) the smaller singlet–triplet splitting, which grants an equilibrium concentration of the reactive and electrophilic singlet (see also Section 4.4.1).

4.6.2. Lone-Pair Stabilization and Philicity

It was mentioned earlier that the halocarbenes and phenylhalocarbenes appear to have singlet ground states. This contention derives from the fact that ESR spectra cannot be obtained nor can the carbenes be "deactivated" to triplets by dilution. Quantum-chemical calculations also indicate singlet ground states.[10,38,57]

This can be accounted for in terms of the resonance structures

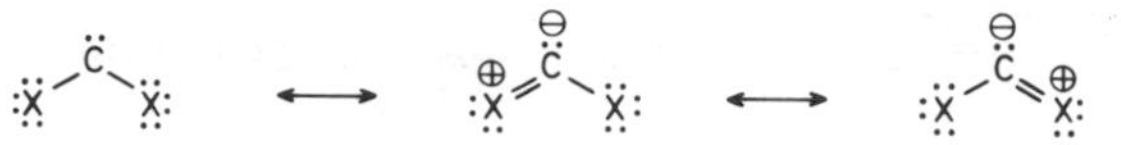

in which the halogen atoms donate electrons into the empty carbene p orbital. Because the halogens are more electronegative than carbon, there will be a strong concomitant σ polarization in the opposite direction:

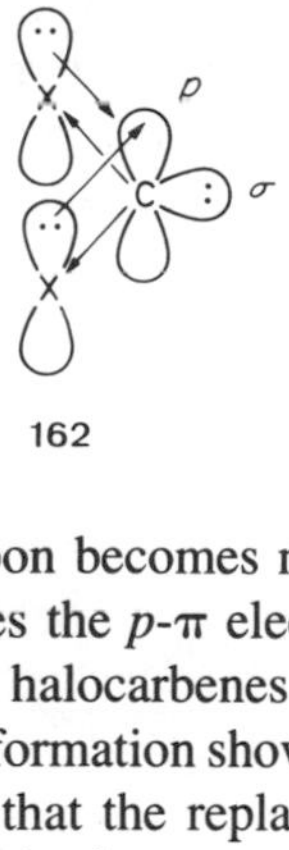

162

Thus the σ core of the carbene carbon becomes more positive, but back-donation from the halogen lone pairs increases the p-π electron density on carbon.

We may estimate how much the halocarbenes are stabilized relative to a non-interacting model. The enthalpies of formation shown in Table 4.7 have been derived from mass spectrometry. It is seen that the replacement of the two hydrogens in $^1\ddot{C}H_2$ by fluorine causes a decrease in ΔH_f^0 by approximately 146 kcal/mol; a similar substitution in methane,

$$CH_4 \longrightarrow CH_2F_2$$

causes a decrease by only 89.3 kcal/mol in ΔH_f^0. On this basis, $\ddot{C}F_2$ appears to be stabilized by ~57 kcal/mol, a huge amount of energy. Similar data can be derived for the other halocarbenes (Table 4.7).

Some relative reactivities of carbenes toward different alkenes are reported in Table 4.8. For comparison, data are also given for the Priležajev epoxidation, which involves the electrophilic transfer of an oxygen atom from a peracid. The more

TABLE 4.7. Enthalpies of Formation of Halocarbenes and Related Compounds

ΔH_f^0 (kcal/mol)[a]		Stabilization[b]
CH_4 (−17.9)	$^1\ddot{C}H_2$ (102)[c]	0
CH_2F_2 (−107.2)	$\ddot{C}F_2$ (−44 ± 2)	~57
CH_2Cl_2 (−22.8)	$\ddot{C}Cl_2$ (53 ± 3)	~44
CH_2Br_2 (+5)	$\ddot{C}Br_2$ (~75)[d]	~48
CH_2I_2 (29.2 ± 1)	$\ddot{C}I_2$ (~92)[d]	~57
CH_3I (3.3)	$\ddot{C}HI$ (~95)[d]	~28

[a]S. W. Benson, *Thermochemical Kinetics*, 2nd ed., Wiley, New York, 1976, unless otherwise indicated.
[b]Relative to a hypothetical noninteracting model, see text.
[c]Reference 12a.
[d]J. J. DeCorpo and J. L. Franklin, *J. Chem. Phys.*, **54,** 1885 (1971).

highly substituted the alkene, the higher the electron density, and the faster the reaction with the electrophile. This data demonstrates that methylene itself is highly indiscriminate, whereas the halocarbenes exhibit large selectivities, comparable to those found in epoxidation reactions.

It is possible to put the selectivities of various carbenes on a more quantitative basis.[172,173] Usually one determines the relative rates of addition of a carbene $\ddot{C}XY$ to various olefins by competition experiments. Moss[173] chose the rate of addition to isobutene as standard. One then determines k_i/k_{ib} for $\ddot{C}XY$, where k_i is the rate constant for addition to an olefin i; k_{ib} is the rate constant for isobutene, arbitrarily chosen as 1. Next we define a "standard carbene," namely $\ddot{C}Cl_2$, and determine again k_i/k_{ib} for $\ddot{C}Cl_2$. The selectivity of any carbene $\ddot{C}XY$ can now be related to that of $\ddot{C}Cl_2$ by means of a diagram:

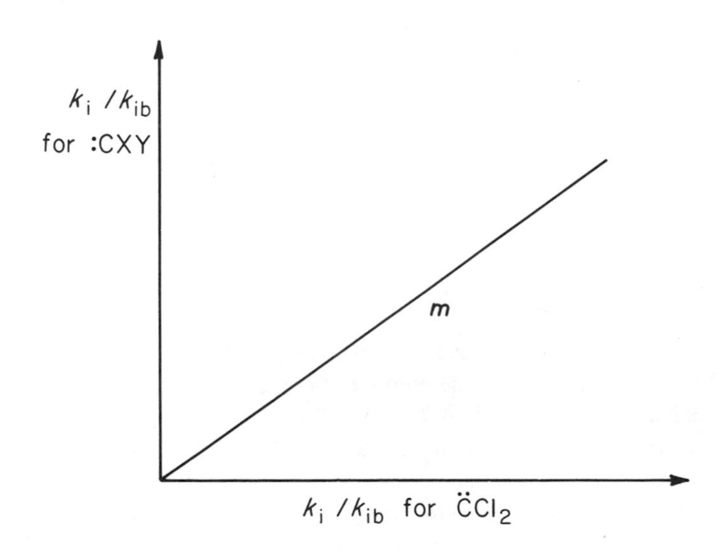

TABLE 4.8. Relative Rates of Epoxidation and Carbene Addition to Olefins[a]

Olefin	Relative Rates of Reaction			
	$R{-}C(=O)OOH$	$\ddot{C}H_2$	$\ddot{C}FCl$	$\ddot{C}Cl_2$
$H_2C{=}CH_2$	1	1	—	—
$H_3C(H)C{=}CH_2$	25	1.15	10[b]	10[b]
$H_3C(H)C{=}C(H)CH_3$	500	0.95	110	136
$(H_3C)_2C{=}C(H)CH_3$	6,500	1.6	7,500	2,800
$(H_3C)_2C{=}C(CH_3)_2$		1.34	35,600	6,740
Precursor		$H_2C{=}C{=}O$ $h\nu$, gas-phase	$(Cl_2FC)_2C{=}O$ + $KOC(CH_3)_3$ −1 °C	$CHCl_3$ + $KOC(CH_3)_3$ − 10 °C
Product	$R_2C{-}CR_2$ (epoxide, O bridge)	$R_2C{-}CR_2$ (cyclopropane, CH_2 bridge)	$R_2C{-}CR_2$ (cyclopropane, CFCl bridge)	$R_2C{-}CR_2$ (cyclopropane, CCl_2 bridge)

[a]For a larger collection of data, see R. A. Moss in *Carbenes,* M. Jones, Jr., and R. A. Moss, Eds., Wiley, New York, 1973, Vol. I, p. 153.
[b]Reaction with 1-butene.

TABLE 4.9. Carbene Selectivity Indices[a]

Carbene	m Value (25°C)
$\ddot{C}H_2$	⩽0.5
CH_3—$\ddot{C}$—Cl	0.50
Br—$\ddot{C}$—Br	0.65
Ph—$\ddot{C}$—Br	0.70
Ph—$\ddot{C}$—Cl	0.83
Ph—$\ddot{C}$—F	0.89
CH_3—S—$\ddot{C}$—Cl	0.91
Cl—$\ddot{C}$—Cl	≡1.00
Cl—$\ddot{C}$—F	1.28
F—$\ddot{C}$—F	1.48

[a]From Ref. 173.

The slope of the correlation line m is an index of carbene selectivity. If $m = 1$, $\ddot{C}Cl_2$ and $\ddot{C}XY$ have equal reactivities and selectivities toward the alkenes i. If $m > 1$, $\ddot{C}XY$ is more selective than $\ddot{C}Cl_2$; if $m < 1$, $\ddot{C}XY$ is less selective. The m values listed in Table 4.9 were obtained.[173]

In like manner, it is possible to compare the selectivities of any other electrophilic species. Some values are given below.[172] Since these reactions have not all been performed at the same temperature, they are only qualitatively comparable:

	Relative Selectivity
$\ddot{C}H_2$ (225°C)	~0.1
ICH_2ZnI (Simons–Smith reagent) (0°C)	0.25
O atoms (25°C)	0.60
$\ddot{C}Br_2$ (−10°C)	0.82
$\ddot{C}Cl_2$ (0°C)	≡1.00
CH_3COOOH (26°C)	1.11
Br_2 (25°C)	1.90

The carbene selectivity indices given in Table 4.9 are in agreement with the normal reactivity–selectivity principle as discussed in Section 1.6. The thermodynamically stabilized difluorocarbene ($\ddot{C}F_2$) shows the highest selectivity. Conjugation with a phenyl group is less effective than lone-pair donation from fluorine or chlorine in stabilizing a carbene. The high-energy methylene ($\ddot{C}H_2$) is virtually indiscriminate. It was also mentioned, however, that *inverse* reactivity–selectivity relationships may be observed (see Section 1.6). Indeed, this is true for the dihalocarbenes also.[174] The carbenes $\ddot{C}F_2$, $\ddot{C}FCl$, $\ddot{C}Cl_2$, $\ddot{C}ClBr$, and $\ddot{C}Br_2$ obey the normal reactivity–selectivity principle at 25°C with $\ddot{C}F_2$ the most selective, $\ddot{C}Br_2$ the least selective in competition for two olefins according to the scheme:

$$:CXY \begin{cases} \xrightarrow{\text{olefin 1}} \text{cyclopropane 1} \\ \xrightarrow[\text{olefin 2}]{} \text{cyclopropane 2} \end{cases}$$

However, with increasing temperature the selectivity of $\ddot{C}F_2$ decreases, and that of $\ddot{C}Br_2$ increases, and at approximately 90°C all the dihalocarbenes mentioned show *the same* selectivity; that is, the isoselective temperature is approximately 90°C. Above this temperature, inverse selectivity is obeyed: $\ddot{C}F_2$ continues to become less selective; $\ddot{C}Br_2$ becomes more selective. Because of the temperature effect on selectivity, the indices reported in Table 4.9 are only valid at 25°C.[174]

The carbenes shown in Tables 4.8 and 4.9 all react preferably with the more highly alkylated (i.e., nucleophilic) alkenes. In other words, the carbenes behave as electrophiles. Accordingly, the halocarbenes have negative Hammett ρ values for the addition to substituted styrenes:[175]

$$\ddot{C}Br_2 \quad \rho = -0.44$$
$$\ddot{C}Cl_2 \quad \rho - -0.62$$
$$\ddot{C}F_2 \quad \rho = -0.57$$

As already mentioned, electrophilic and nucleophilic character are not absolute terms: *Any carbene can show both electrophilic and nucleophilic characteristics in its addition reactions with olefin.*[67] Whether one or the other character will be dominant can be evaluated from a frontier MO consideration.[67,176] Figure 4.6 shows

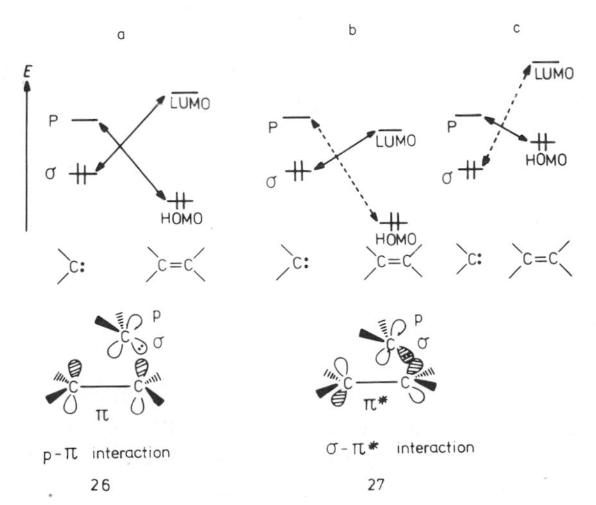

Figure 4.6. Three possible LUMO(carbene)–HOMO(olefin)(p-π) and HOMO(carbene)–LUMO(olefin) (σ-π*) interaction schemes.

three different sets of orbital interactions together with the orbital pictures **26** and **27**. Both the electrophilic p-π interaction **(26)** and the nucleophilic σ-π^* interaction **(27)** can take place simultaneously during the approach of the carbene to the olefin. The relative ordering of the levels determines which effect will be dominant.

In the halocarbenes just mentioned, the energies of the p orbitals are raised due to back donation from the lone pairs, thereby making the carbenes less electrophilic than they would otherwise have been. The σ orbitals are, however, lowered in energy due to electron withdrawal in the σ core (see **162**). Thus a situation such as that shown in Figure 4.6(c) obtains, and the carbenes behave overall as electrophiles, adding preferentially to electron-rich (high-energy HOMO) alkenes. If we raise the p level without unduly lowering the σ level, a nucleophilic carbene should result. This is obtained in dimethoxycarbene, which adds to *electrophilic* but not to nucleophilic double bonds:[177]

A Hammett treatment of the addition to substituted phenyl isocyanates gave a ρ value of $+2.0 \pm 0.5$, confirming the nucleophilic character of the carbene. In MO terms, this corresponds to a situation such as that in Figure 4.6(b). The same obtains for diamino- and amino(alkylthio)carbenes, which also gave positive ρ values for the addition reaction with arylisocyanates:[178]

$\rho = +1.7 \pm 0.5$ $\rho = +1.9 \pm 0.5$

Since the nucleophilic carbenes do not add to "normal" olefins, a Moss correlation cannot be carried out. However, the carbene selectivity indices described in Table 4.9 can be related to the resonance (σ_R^+) and inductive (σ_I) substituent constants for the carbene substituents through the equation:[173]

$$m_{CXY} = -1.10 \sum_{X,Y} \sigma_R^+ + 0.53 \sum_{X,Y} \sigma_I - 0.31 \qquad (12)$$

This equation, then, allows the calculation of the m value for any carbene $\ddot{C}XY$ when the substituent constants are known. For dimethoxycarbene, one obtains $m_{(CH_3O)_2C}$ (calc.) $= 2.22$; that is, if $(CH_3O)_2\ddot{C}$ were to undergo electrophilic addition, it would be highly selective, more so than difluorocarbene.

For methoxychlorocarbene, $CH_3O—\ddot{C}—Cl$, one calculates from Eq. (12) an m value of 1.59. This is highly interesting since the carbene—and the m value—are intermediate between the typically electrophilic halocarbenes and the nucleophilic dimethoxy- and aminocarbenes. In fact, it was found[173] that *methoxychlorocarbene adds both to electron-rich and electron-poor* olefins. For example,

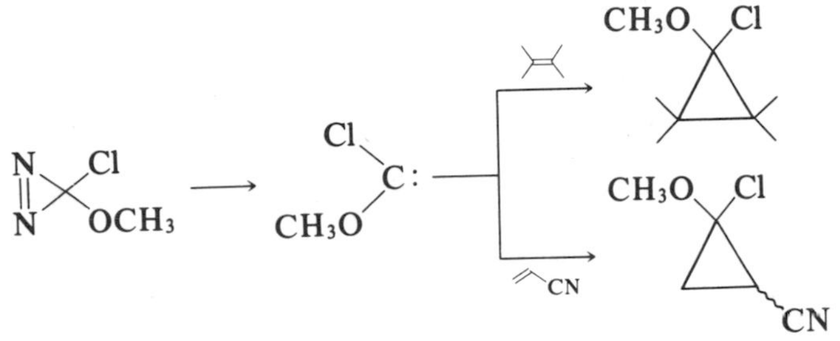

The rate of addition is accelerated *both* by electron-withdrawing and electron-releasing olefin substituents, that is, the carbene is both a nucleophile and an electrophile, aptly termed an ambiphile. In MO terms, Figure 4.6(a) applies, in which the interactions depicted in formulas **26** and **27** are both important in the transition state. Electron-withdrawing substituents on the alkene shift the interaction in the direction of increased σ-LUMO overlap [Figure 4.6(b)]; electron-releasing substituents have the opposite effect [Figure 4.6(c)].

Nitrenes, too, are stabilized by lone-pair substituents, aminonitrenes being prime examples.

$$R_2\ddot{N}—\ddot{N}: \longleftrightarrow R_2\overset{\oplus}{N}=\overset{\ominus}{\ddot{N}}:$$

Phthalimidoylnitrene (**163**) behaves as a singlet nitrene, adding stereospecifically to electrophilic as well as nucleophilic alkenes,[19c,179] a fact that may indicate both

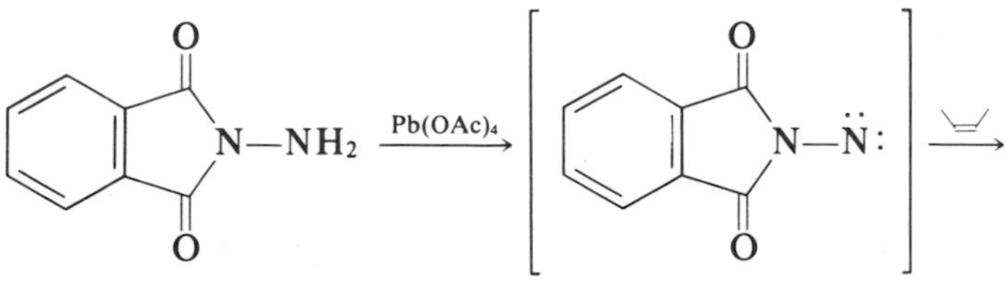

163

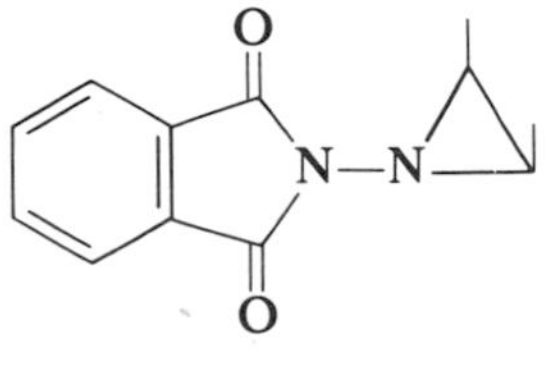

nucleophilic and electrophilic properties. The carbonyl groups in **163** render this nitrene more electrophilic and more reactive than dialkylaminonitrenes. The latter do not normally add to alkenes and should perhaps better be described as 1,1-dialkyldiazenes. Nitrogen extrusion from such compounds was discussed in Chapter 3. The diazene **164,** isolable at low temperatures, dimerizes to the tetrazene **165** already at $-90°C$.[180]

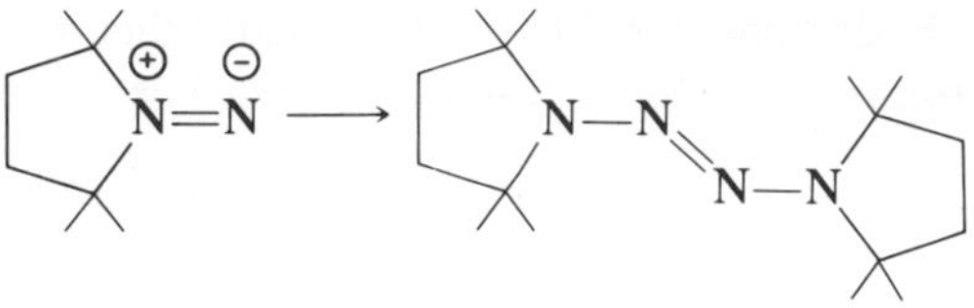

164 **165**

An extraordinary and rare case of addition of a dialkylaminonitrene to a double bond is observed in the following reaction:

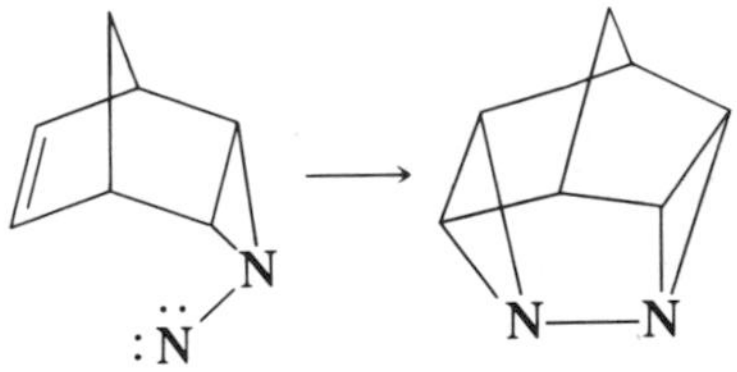

(82%)

This reaction effectively competes with dimerization and is presumably aided by the favorable steric arrangement.[181]

The compounds F—S≡N and Cl—S≡N are known as thiazynes and show little or no evidence of a nitrene character. In a sense, the sulfur atom is completely effective in stabilizing these "nitrenes":[182]

$$R-\ddot{S}-\ddot{N}: \longleftrightarrow R-\ddot{S}\equiv N:$$

Arylsulfenylnitrenes **(166)** can be generated, however, by oxidation of sulfenamides, and are found to add nonstereospecifically to β-methylstyrenes:

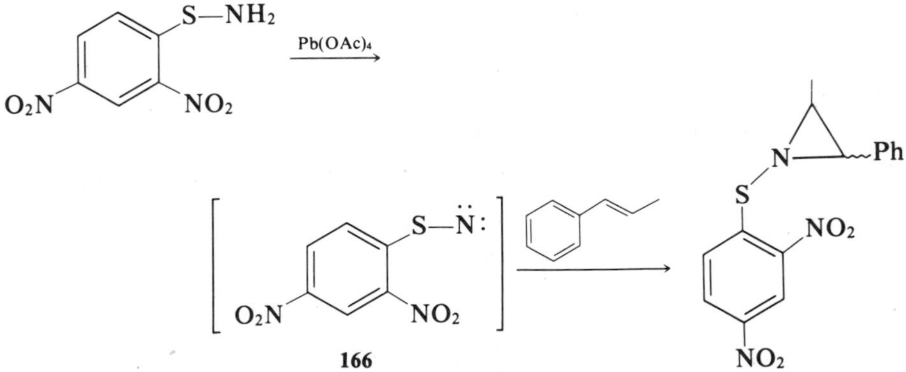

$S-NH_2$; O_2N; NO_2 —$Pb(OAc)_4$→

[$S-\ddot{N}:$; O_2N; NO_2] **166** → N; Ph; S; NO_2; NO_2

Further studies indicated the presence of a singlet nitrene and the absence of the triplet, but another, unidentified intermediate was also invoked.[182]

4.6.3. π-Conjugation

The conjugation of a singlet carbene with a flanking π system will create a delocalized system **(167)** isoelectronic with the allyl cation. The carbene LUMO is raised in energy, and the carbene becomes less electrophilic. If, however, the

167 **168**

flanking π system is itself electrophilic, that is, it has a low-lying LUMO, a system isoelectronic with the allyl anion may obtain **(168).** Here, the carbene HOMO is stabilized, whereas the vacant p orbital (the carbene LUMO) remains unaffected. It was already mentioned that arylcarbenes can behave as either nucleophiles or electrophiles, depending on the nature of the π system. It is extremely interesting that two "simple" vinylcarbenes **169** and **170** have been found to behave as *nucleophiles* in the addition to styrenes, giving positive ρ values:[183]

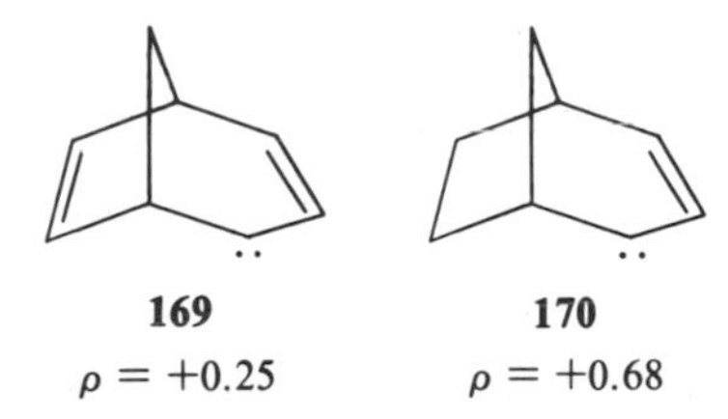

169 $\rho = +0.25$ **170** $\rho = +0.68$

The small magnitude of the ρ value suggests, however, that these carbenes are but a little more "nucleophilic" than they are "electrophilic"; that is, a situation such as that in Figure 4.6(a), shifted a little in the direction of Figure 4.6(b) applies. For geometry reasons, these carbenes can only adopt conformation **167,** and the σ approach to styrene is favored sterically.

The electrophilicity of a carbene can be further reduced by incorporation of the vacant p orbital into an aromatic system. Thus cyclopropenylidene **(171)** and cycloheptatrienylidene **(172)** are Hückel aromatics in the σ^2 configuration, obeying the $4n + 2$ rule:

171 172

This leads to a stabilization of the singlet states and a *destabilization* of the triplet states: The latter would be $4n + 3$ electron systems having one electron in an antibonding orbital. The observed chemistry[184,185] supports the conclusion that these carbenes are *nucleophilic* singlets. Cycloheptatrienylidene gives a positive ρ value (+1.05) for the addition to styrenes:[185]

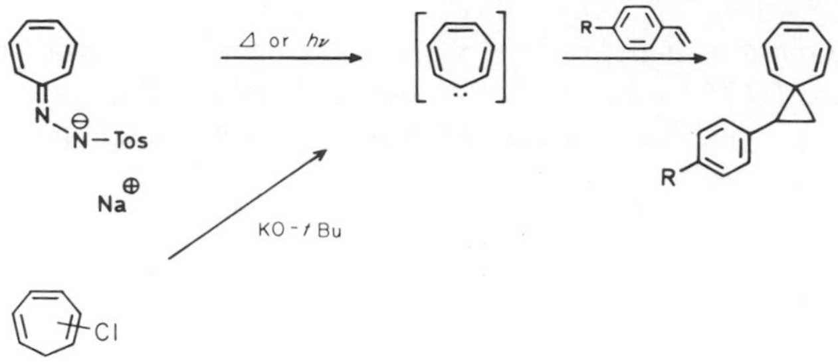

It should be noted that *this nucleophilic property of cycloheptatrienylidene is an intrinsic nucleophilicity of carbenes in general.* It comes to the fore because resonance **(172)** raises the LUMO energy of the carbene, thereby making it *less electrophilic*. Phenomenologically, the effect is indistinguishable from that of lowering the alkene-LUMO. The intereaction corresponds to Figure 4.6(b). It might be possible to turn cycloheptatrienylidene into an electrophile by offering it a substrate with a very high-lying HOMO—shifting the interaction toward that of Figure 4.6(c). Possible candidates are the (formally) antiaromatic molecules, cyclobutadiene and 1,4-dihydropyrazines:[176a]

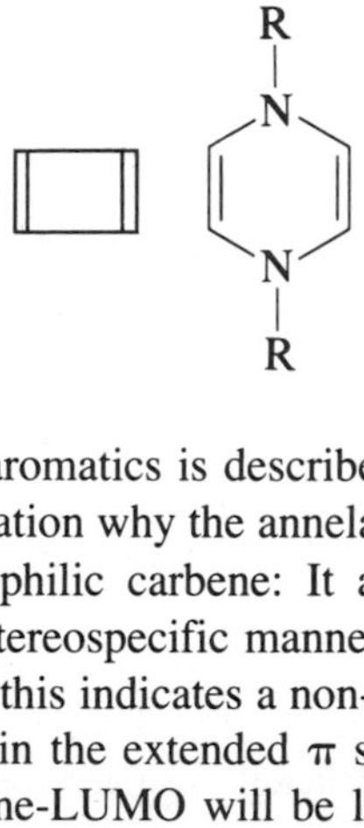

The electronic structure of antiaromatics is described in Chapter 6.

We may now offer an explanation why the annelated cycloheptatrienylidene **173** does not behave like a nucleophilic carbene: It adds to both electron-rich and electron-poor olefins, in a nonstereospecific manner, and also undergoes insertion reactions into C—H bonds. All this indicates a non-nucleophilic triplet carbene.[187a] The explanation can be sought in the extended π system in **173.** This has a low-lying LUMO, so that the carbene-LUMO will be lowered relative to that of **172.**

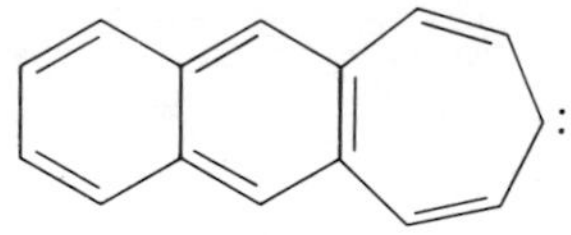

173

The carbene-HOMO, the σ orbital, remains largely unaffected. The carbene **173** will, therefore, be *more electrophilic* than **172,** not less nucleophilic! The lowering of the LUMO has the further effect of reducing the energy gap between the σ and p orbitals, resulting in a triplet ground state.

Having said this, one must be aware that cycloheptatrienylidene **(172)** has never been directly observed, and there is some uncertainty as to its existence. As mentioned in Section 4.4.5, the *stable* form of this species appears to be the allene **(129).** It is likely that much of the chemistry observed is that of the allene, and the importance of a possible equilibrium **129** ⇌ **172** is difficult to assess.[133,186]

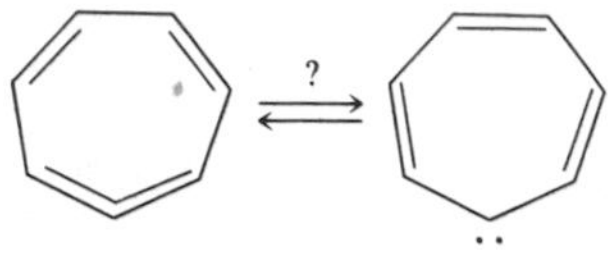

?

129 **172**

The allene (singlet) form of the annelated carbene **173** would be destabilized due to loss of the aromatic resonance energy of naphthalene:

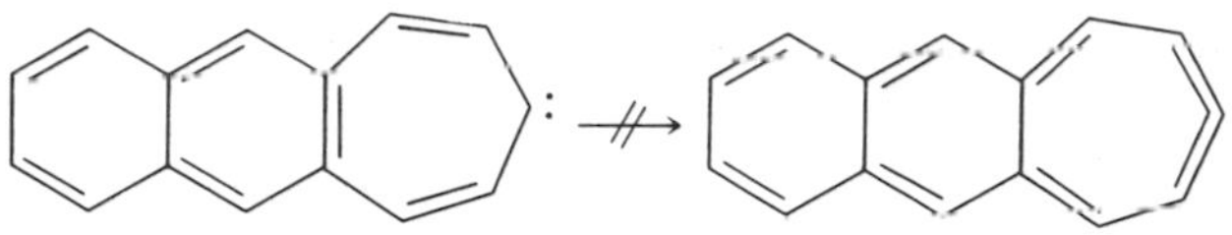

173

The incorporation of a carbene into a $4n$ π system will have an effect opposite to that in cycloheptatrienylidene: Cyclopentadienylidene would be antiaromatic in the σ^2 configuration **(174),** but aromatic in the unusual p^2 configuration **(175).** The latter would react like an electrophilic vinyl cation, and the observed negative Hammett ρ value (-0.46) supports this conclusion.[187b]

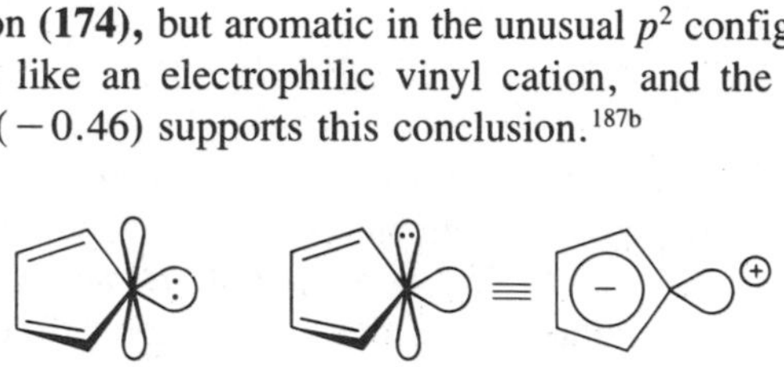

($4n$)
174

($4n + 2$)
175

The promotion of two electrons into the formerly vacant p orbital in **175** corresponds to the creation of a *doubly excited* singlet state (see Figure 4.1 for the S_0 and S_1 states; in a normal carbene, the S_2 state corresponding to **175** will lie even higher). The energy gained through aromatic delocalization in **175** does not suffice to overcome the inherent singlet destabilization, and cyclopentadienylidene has, in fact, a triplet ground state (see Table 4.5 for the ESR spectrum).

In contrast to cyclopentadienylidene, the five-membered heterocyclic carbenes **176** and **177** are again $4n + 2$ π systems in the σ^2 configuration. These carbenes may be considered to be stabilized both by cyclic conjugation and by lone-pair donation. They are highly stabilized and nucleophilic and do not normally add to olefins.[188]

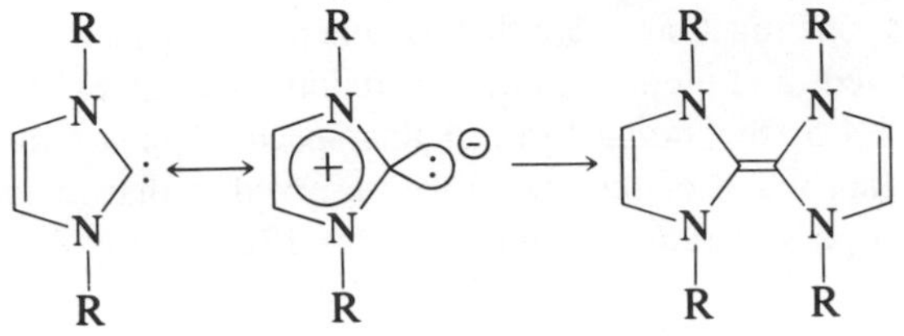

176

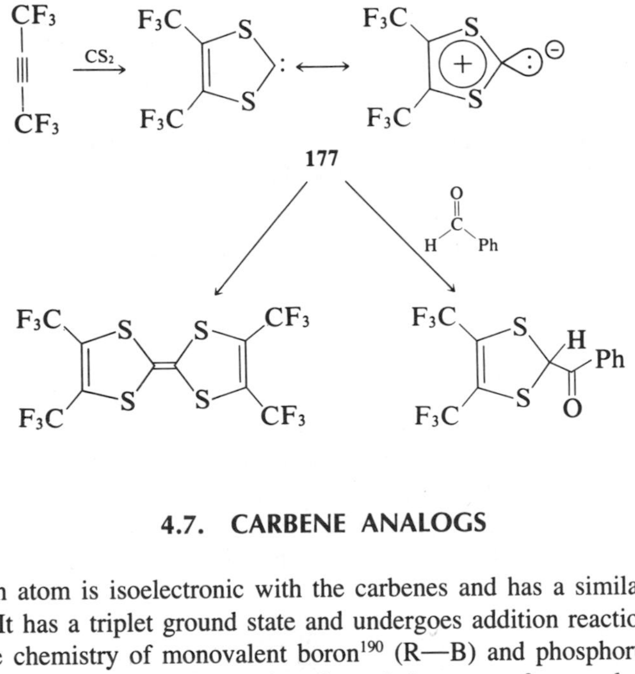

177

4.7. CARBENE ANALOGS

The oxygen atom is isoelectronic with the carbenes and has a similar but higher reactivity. It has a triplet ground state and undergoes addition reactions with olefins.[189] The chemistry of monovalent boron[190] (R—B) and phosphorus[191] (R—P) compounds is still in its infancy, but that of the group four analogs, the silylenes,[192,193] germylenes,[193,194] and stannylenes[194] is flourishing. We shall be concerned primarily with the silylenes.

We have seen that, in agreement with Hund's first rule, diradicals, carbenes, and nitrenes tend to have tirplet ground states. The rule is only a law for atoms and diatomic molecules, however, and silylene, SiH_2, has, in fact, a singlet ground sate.[195] Calculations indicate a singlet–triplet splitting of aproximately 10 kcal/mol,[196] that is, about the same as for methylene, but with opposite sign. Silicon chemistry is often like carbon chemistry in reverse: Singlets are ground states; silaolefins (silenes) are unstable and rearrange to silylenes; silavinylidenes are more stable than silaacetylenes.

Silylenes can be generated, inter alia, by gas-phase pyrolysis of disilanes,

$$H_3CO{-}Si(CH_3)_2{-}Si(CH_3)_2{-}OCH_3 \xrightarrow{\Delta} :Si(CH_3)_2 + (CH_3)_2Si(OCH_3)_2$$

and by photolysis or pyrolysis of polysilanes:[192]

$$(Me_3Si)_2SiR_2 \xrightarrow{\Delta} Me_3Si{-}SiMe_3 + :SiR_2$$

The latter type of reaction has been put to elegant use in the synthesis of a *stable* disilene (disilaolefin):[197,198]

$$3Ar_2SiCl_2 \xrightarrow{[C_{10}H_8]^{\ominus}\,Li^{\oplus}} \text{cyclo-}(Ar_2Si)_3$$

$$\text{cyclo-}(Ar_2Si)_3 \xrightarrow{h\nu} Ar_2Si{=}SiAr_2\ (\mathbf{178}) + Ar_2Si{:}\ (\mathbf{179})$$

$$(Me_3Si)_2SiAr_2 \xrightarrow[-Si_2Me_6]{h\nu} Ar_2Si{:}\ (\mathbf{179})$$

$$\mathbf{179} \xrightarrow{\times 2} \mathbf{178}$$

Ar = 2,6-dimethylphenyl (H_3C groups at both ortho positions)

Whereas normally disilenes are extremely unstable compounds, falling in the realm of reactive intermediates, the sterically shielding substituents in **178** prevent the compound from oligomerizing, thus stabilizing it kinetically.

Dimerization of carbenes and nitrenes is normally observed only with the relatively unreactive nucleophilic carbenes and the arylnitrenes. Silylenes dimerize efficiently (e.g., **179** → **178**), and in the absence of sterical shielding of the disilene, this leads to oligomerization.

Like carbenes, silylenes add to C—C double bonds. Although the silacyclopropanes (**180**) so formed are unstable, the reaction has been shown to be stereospecific, thus mimicking the concerted addition of a singlet carbene:[199]

$$(H_3C)_2Si: + \text{(2-butene)} \longrightarrow \mathbf{180} \xrightarrow{CH_3OH} H_3C{-}Si(CH_3)(OCH_3){-}CH(CH_3)CH_2CH_3$$

180

Air-sensitive but otherwise stable silacyclopropenes (**181**) can be prepared by silylene addition to acetylenes;[200] the reaction is photochemically reversible:

$$CH_3O{-}Si(CH_3)_2{-}Si(CH_3)_2{-}OCH_3 \xrightarrow[\text{vacuum}]{600°C} (H_3C)_2Si: \xrightarrow[\text{gas-phase}]{-\equiv-} \mathbf{181}$$

181

Silacyclopentenes (**182**) are obtained from the addition of silylenes to 1,3-dienes, but a concerted 1,4-addition still remains to be proven. In all cases studied to date, it appears that the reaction proceeds in two steps, forming first the normal 1,2-addition product (**183**).[192]

$$:SiR_2 + \text{(1,3-butadiene)} \longrightarrow \mathbf{182}\ (\text{via } \mathbf{183}:\ R_2Si\text{-vinylsilacyclopropane})$$

182

183

Dimethylsilylene (**185**) can be directly observed spectroscopically when generated in a matrix at 10 K by photolysis of the hexamer (**184**).[201] The silylene **185** undergoes secondary matrix photolysis to 2-silapropene (**186**). The same compound is formed also by flash vacuum pyrolytic cycloreversion of compound **187** and identified by its IR and UV spectra when the pyrolysate is condensed in an argon matrix.[202] The unsubstituted silaethylene has been prepared in the same manner.[203]

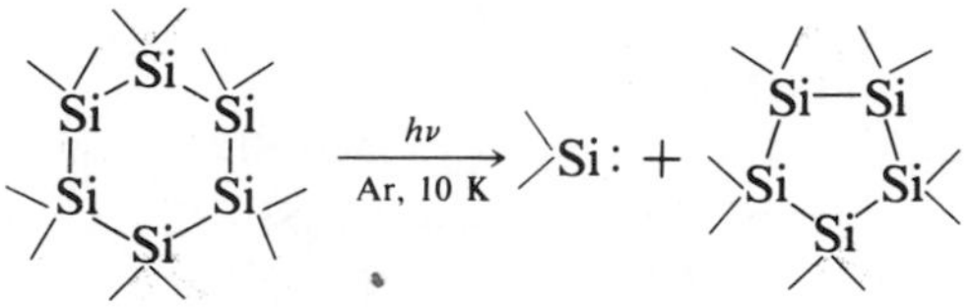

$h\nu$, Ar, 10 K

184 185

The silylene–silene rearrangement **185** → **186** is both thermally[204] and photochemically[202] reversible. The third isomer in the series, the silylcarbene **188**, can be generated by matrix photolysis of the diazo-compound **189**, and again it isomerizes to silapropene (**186**).[205] Silaolefins like **186** are extremely unstable compounds[193,206] that dimerize, in the present case to disilacyclobutane **190**, on warming. Yet the use of sterically blocking groups has allowed the synthesis of a

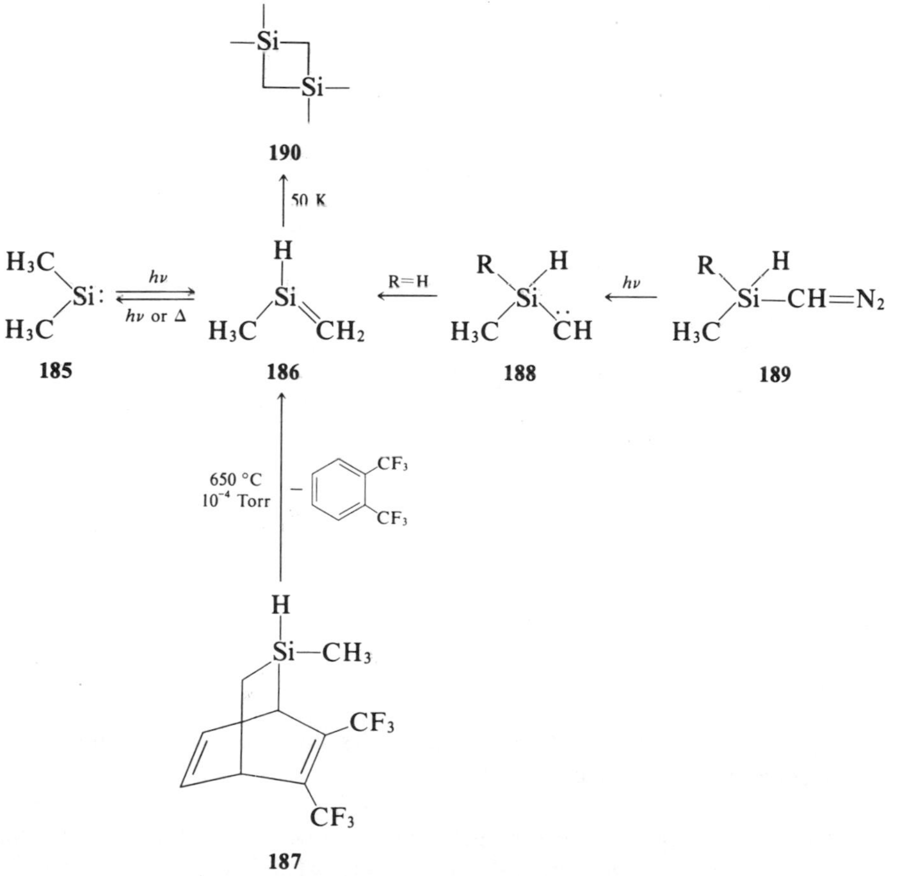

190

50 K

H_3C, H_3C Si: ; $h\nu$; $h\nu$ or Δ ; H, Si, H_3C, CH_2 ; R=H ; R, H, Si, H_3C, CH ; $h\nu$; R, H, Si—CH=N_2, H_3C

185 186 188 189

650 °C, 10^{-4} Torr, − (o-bis(trifluoromethyl)benzene, CF_3, CF_3)

H, Si—CH_3, CF_3, CF_3

187

Scheme 20

thermally stable, isolable but air-sensitive silaolefin (**192**). The driving force in this synthesis (from **191**) is the high affinity of silicon for oxygen:[207]

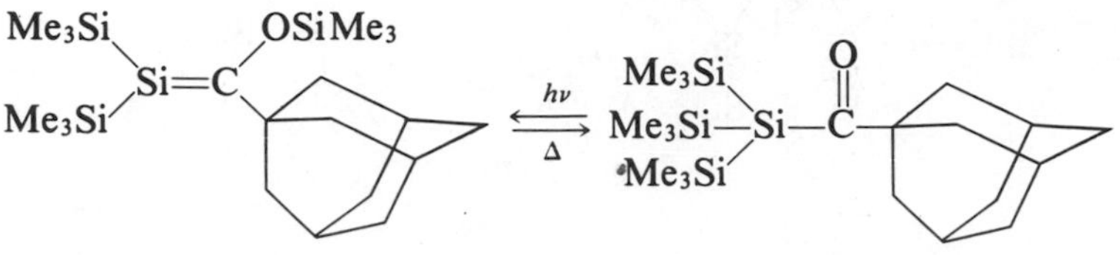

$$(Me_3Si)_2Si{=}C(OSiMe_3)(\text{1-adamantyl}) \underset{\Delta}{\overset{h\nu}{\rightleftarrows}} (Me_3Si)_3Si{-}C({=}O){-}(\text{1-adamantyl})$$

192 **191**

The chemistry illustrated in Scheme 20 demonstrates that, in contrast to carbene chemistry, the silylene and the silaolefin are of comparable energies, possibly with the silylene as the most stable species.[208]

The related disilaolefin–silylsilylene rearrangement has also been established experimentally,[209] and calculations agree that the silylene should be the most stable species.[210]

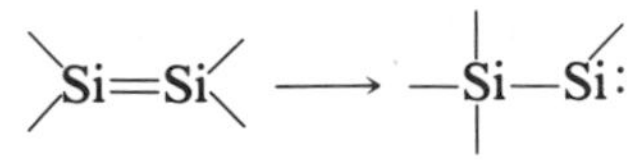

$$>Si{=}Si< \longrightarrow -\overset{|}{\underset{|}{Si}}-\dot{S}i\!:$$

This "reverse" energy ordering continues in the triple-bonded series: Silaacetylene (**193**) is still unknown, and there is experimental[211] and theoretical[212] support for the contention that the silylene (**194**) is thermodynamically more stable:

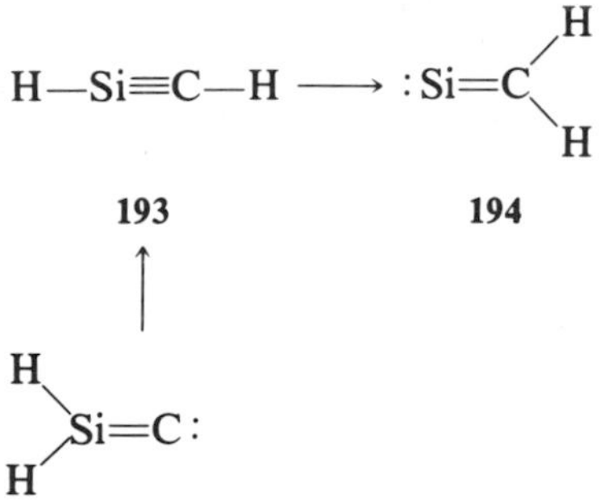

$$H{-}Si{\equiv}C{-}H \longrightarrow :Si{=}CH_2$$

193 **194**

$$H_2Si{=}C: \ (\textbf{195}) \longrightarrow \textbf{193}$$

195

An experimental access to this isomeric series might be achieved through the generation of a silavinylidene (**195**).

Germylenes (e.g., **196**) can be generated by mild thermolysis of 7-germanorbornadienes and undergo stereospecific addition to 1,3-dienes and bisallenes:[213]

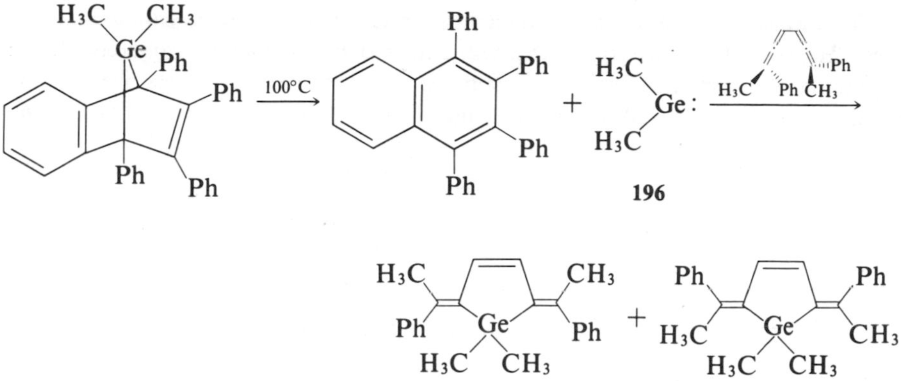

196

This makes a singlet reaction most plausible, in agreement with expectations that germylenes, too, should have singlet ground states,[214] and strongly suggests that concerted 1,4-addition is taking place.

As one descends through group IV of the periodic table, the carbene analogs become not only heavier, but more stable. The range of known stannylenes (R_2Sn)[194,215] spans from the unstable dimethylstannylene **(197)**, which can be trapped in low-

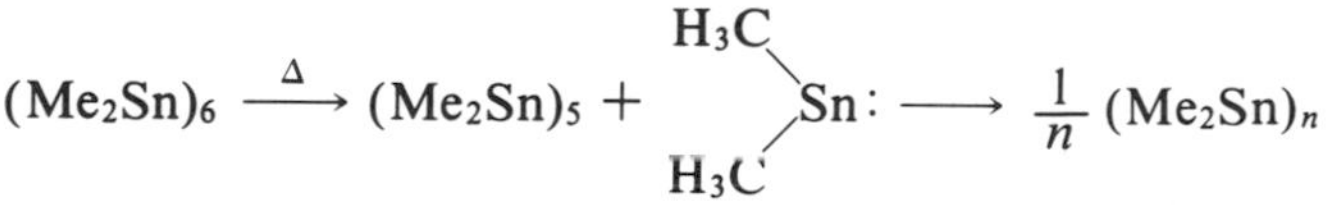

$$(Me_2Sn)_6 \xrightarrow{\Delta} (Me_2Sn)_5 + (H_3C)_2Sn: \longrightarrow \frac{1}{n}(Me_2Sn)_n$$

197

temperature matrices[194] to isolable compounds such as **198** (Ref. 216) and **199** (Ref. 217), the latter being kinetically stabilized by sterical hindrance toward dimerization. Although the "bent" dimer **200** exits in the crystalline state, the monomer **199** is stable in solution at room temperature. The corresponding germanium and lead monomers have also been prepared.[217]

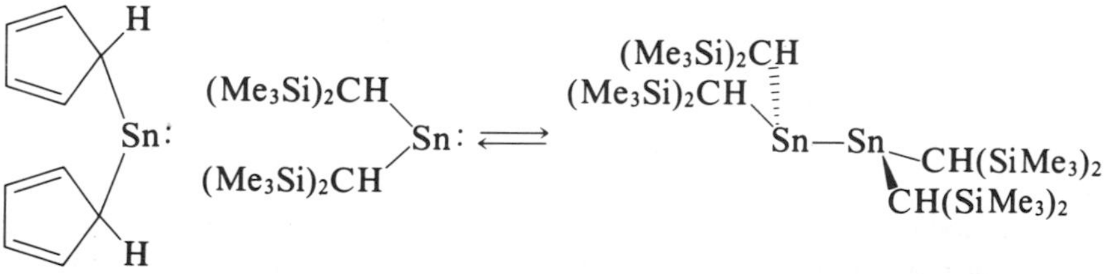

198 199 200

From the preceding discussion, it is apparent that two factors contribute to the unusual chemistry of the heavy carbene analogs: (a) a relatively high stability of the silylenes, germylenes, stannylenes, and plumbylenes compared with carbenes and (b) a weakness of the π bonds to the heavier elements compared with the C=C bond. Heats of formation of the Si, Ge, Sn, and Pb analogs of isobutene have recently been determined through measurements of the proton affinities (PA) of the double-bonded species (**202**) formed by proton abstraction from the cations **201** in

$$(CH_3)_3M{-}Cl \xrightarrow{e^-} \underset{\mathbf{201}}{(CH_3)_3M^{\oplus}} + e^- + Cl^-$$

$$\downarrow -H^{\oplus} \text{ base}$$

$$\text{H-base}^{\oplus} + \underset{\mathbf{202}}{(CH_3)_2M{=}CH_2}$$

$$M = \text{Si, Ge, Sn, Pb}$$

an ion cyclotron resonance spectrometer.[218] From the proton affinities, the heats of formation are determined from the equation:

$$\Delta H_f^0(\mathbf{202}) = \Delta H_f^0(\mathbf{201}) - \Delta H_f^0(H^+) - PA(\mathbf{202}).$$

From these data, π bond energies of 38, 43, 45, and 30 kcal/mol were derived for **202** (M = Si, Ge, Sn, Pb, respectively). Since the π bond energy in isobutene itself is about 58 kcal/mol,[219] it is seen that the Si=C π bond is about 20 kcal/mol weaker.

4.8. PROBLEMS

1. Write a mechanism for the following reaction:

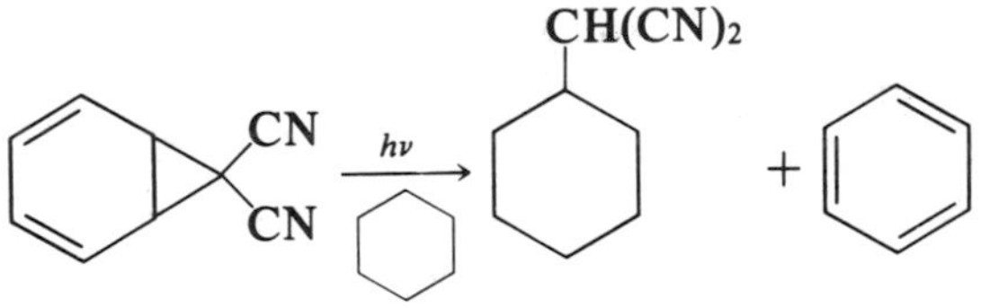

(7,7-dicyanobicyclo[4.1.0]hepta-2,4-diene (CN, CN) $\xrightarrow[\text{cyclohexane}]{h\nu}$ cyclohexyl–CH(CN)₂ + benzene) (Ref. 220)

2. Flash vacuum pyrolysis (FVP) of **203** gave an unstable product (**204**), which consumed one molar equivalent of H_2 on catalytic hydrogenation. **204** was converted to acrolein at 25°C. Give a structure for **204**.

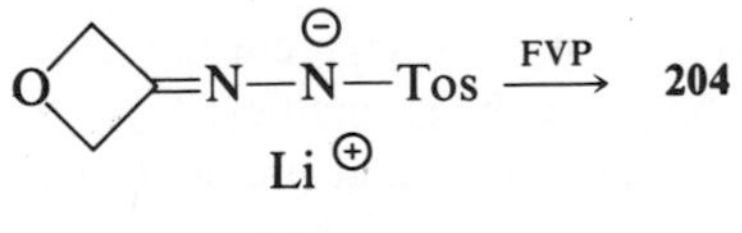

(oxetan-3-ylidene)=N–N$^{\ominus}$–Tos Li$^{\oplus}$ (**203**) $\xrightarrow{\text{FVP}}$ **204** (Ref. 221)

3. Explain the formation of the following reaction products.[222]

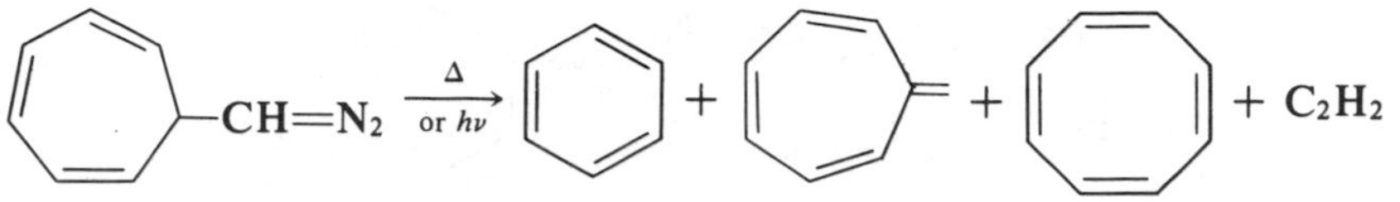

$$\text{C}_7\text{H}_7\text{—CH=N}_2 \xrightarrow[\text{or } h\nu]{\Delta} \text{C}_6\text{H}_6 + \text{C}_7\text{H}_6 + \text{C}_8\text{H}_8 + \text{C}_2\text{H}_2$$

4. Explain the synthesis of isonitriles from primary amines with chloroform and base under the conditions of phase transfer catalysis.[223]

$$R\text{—}NH_2 \xrightarrow[\text{NaOH(aq)}]{\text{CHCl}_3} R\text{—}NC$$

benzyltriethyl-
ammonium chloride

5. The thermal gas-phase decomposition of *n*-butyl diazoacetate **(205)** in a stream of isobutene gives, as the main product, the cyclopropane **206** (15–20%), together with the lactone **207** (2–3%).

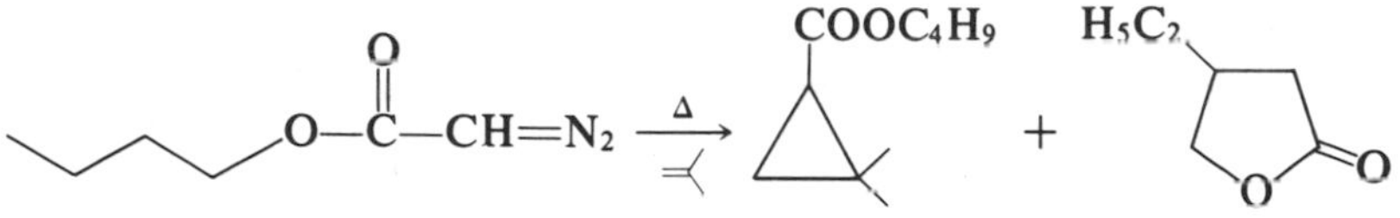

$$\text{C}_4\text{H}_9\text{O—}\overset{\text{O}}{\overset{\|}{\text{C}}}\text{—CH=N}_2 \xrightarrow{\Delta} \text{(cyclopropane–COOC}_4\text{H}_9\text{)} + \text{(lactone, H}_5\text{C}_2\text{)}$$

205 **206** **207**

The yield of **207** increased to 12% in the absence of isobutene.
The diazo-compound **208** gave a 20% yield of **209** in the absence of isobutene.

$$\text{O—}\overset{\text{O}}{\overset{\|}{\text{C}}}\text{—CH=N}_2 \xrightarrow{\Delta}$$

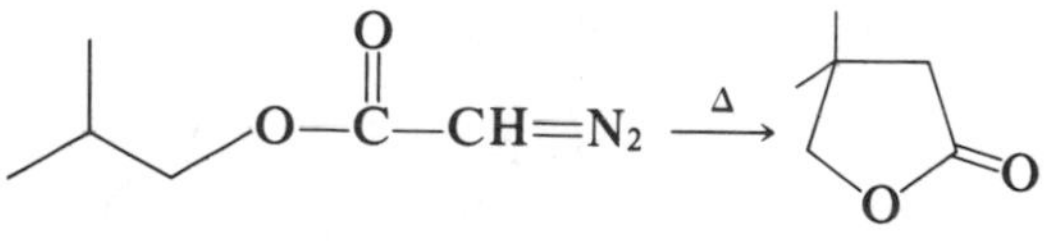

208 **209**

In contrast, ethyl diazoacetate **(210)** and isobutene gave a mixture of the cyclobutanone **211** and the cyclopropane **212**.

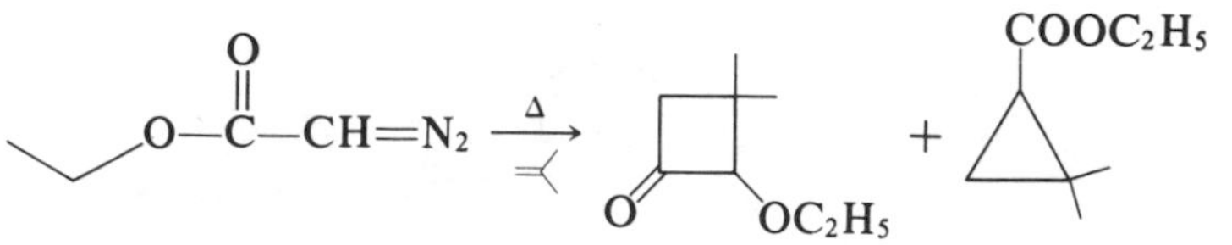

$$\text{O—}\overset{\text{O}}{\overset{\|}{\text{C}}}\text{—CH=N}_2 \xrightarrow{\Delta} \text{(cyclobutanone, OC}_2\text{H}_5\text{)} + \text{(cyclopropane–COOC}_2\text{H}_5\text{)}$$

210 **211** **212**

Explain these observations.[224]

6. Write a mechanism for the following reaction:[225]

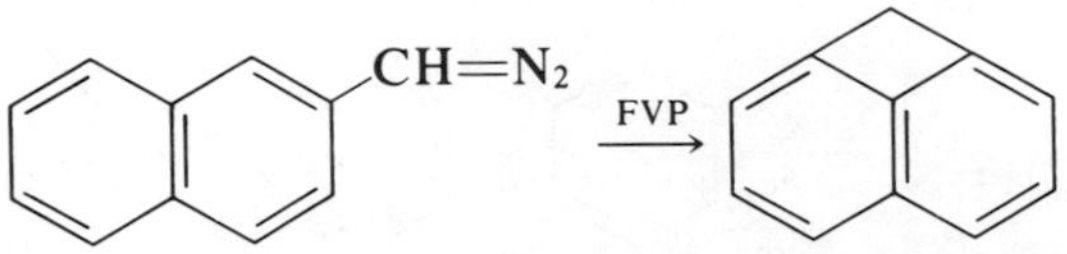

CH=N$_2$ FVP →

7. The heterocyclic diazoketone **213** decomposes at room temperature with evolution of N_2 to give a compound X (^{1}H-NMR; δ 7.5(s,1H), 7.3–7.7(m,5H); IR: 1760 cm^{-1}; mass spectrum: M^+ 186), which reacts with ethanol to give **214** quantitatively. Give a reasonable structure for X.[226]

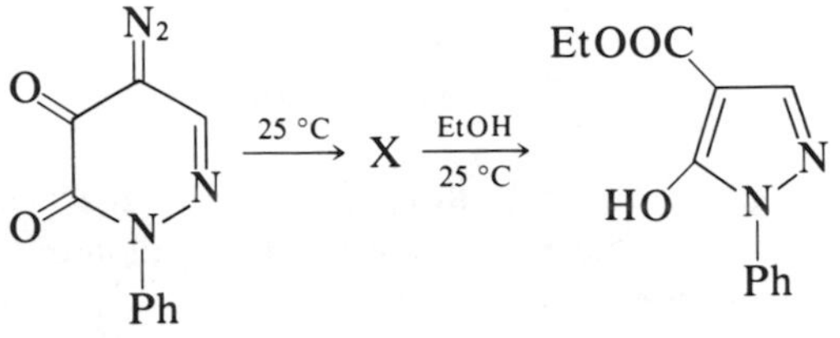

N$_2$ O O N N Ph 25 °C → X EtOH 25 °C → EtOOC HO N N Ph

213 **214**

8. Complete the following sequence.[227]

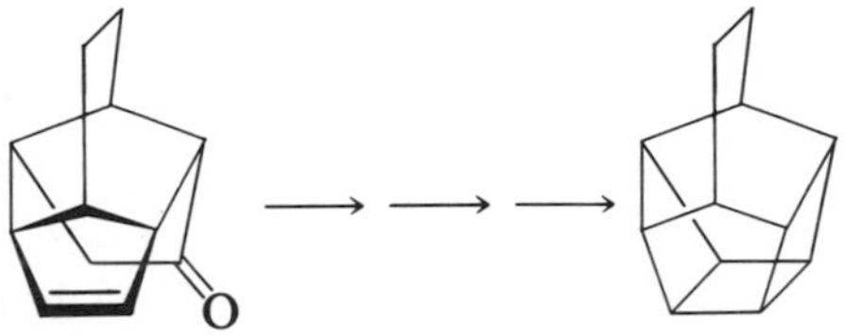

O → → →

9. Explain the following labeling results:[228]

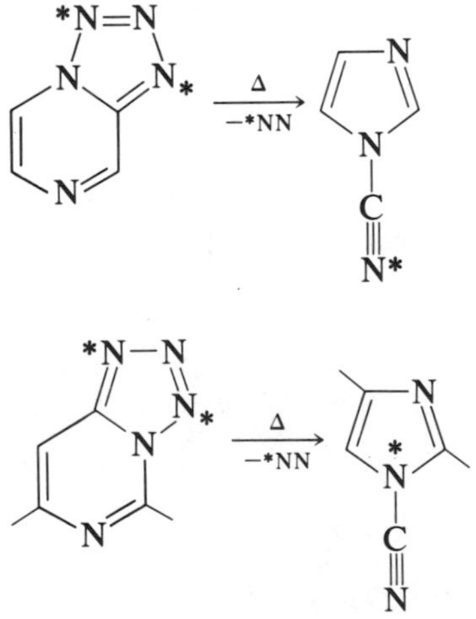

N=N N N Δ −*NN → N N C N*

N−N N N N Δ −*NN → N * N C N

$*N = ^{15}N$

10. The gas-phase pyrolysis of **215** gives a mixture of dichlorocarbene and chlorocarbene that reacts with ethylene to give the cyclopropanes **216** and **217** together with the olefins **218–220**. The cyclopropane/olefin ratios are pressure dependent. As the pressure was increased from 10 to 160 Torr, the ratio **216/(218 + 219)** increased from 1 to 25; at the same time, the ratio **217/220** increased from 1 to 2.5. Explain these observations.[229]

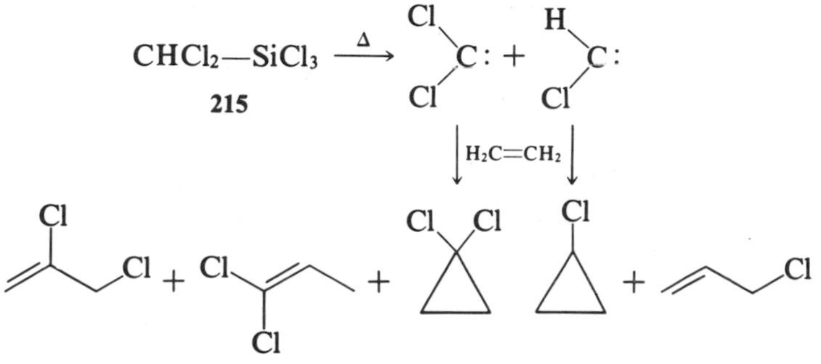

$CHCl_2$—$SiCl_3$ $\xrightarrow{\Delta}$ Cl_2C: + HClC:

215

$H_2C{=}CH_2$

218 **219** **216** **217** **220**

11. Thermolysis and photolysis of the azide **221** give varying amounts of the carbazole **222** and the phenanthridine **223**. Thermolysis at 360°C gives a 100% yield of the carbazole. Sensitized photolysis (with acetophenone) gives 18% of the carbazole and 36% of the phenanthridine. Explain these results in terms of spin multiplicities.[230]

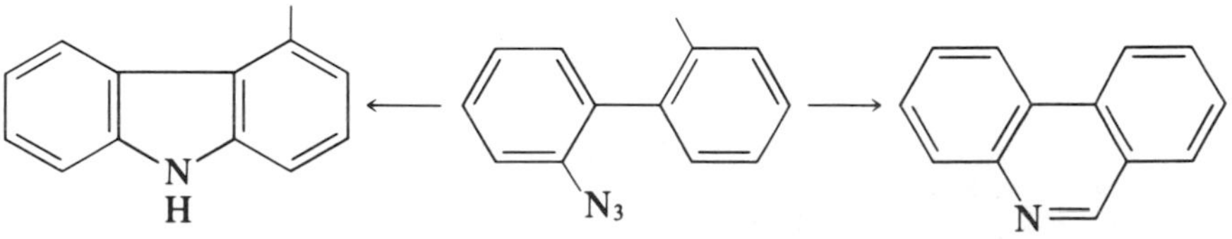

N H N_3 N

222 **221** **223**

12. The two azepines **225** and **226** are obtained *in the same ratio* in the photolysis of the azide **224** and in the thermal deoxygenation of the nitro compound **227** with diethyl methylphosphonite in the presence of diethylamine. Write reaction mechanisms.[231]

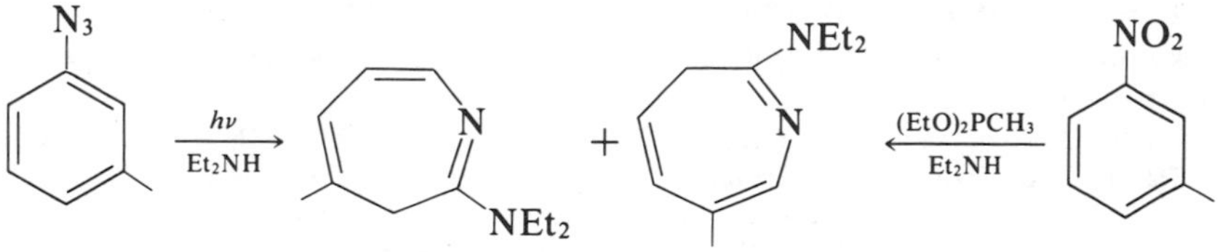

N_3 $\xrightarrow[Et_2NH]{h\nu}$ N NEt_2 + NEt_2 N $\xleftarrow[Et_2NH]{(EtO)_2PCH_3}$ NO_2

224 **225** **226** **227**

13. Treatment of **228** with methyllithium in ether at $-40°C$ gave **229** as the major product. The movements of the carbon atoms were ascertained by ^{12}C labeling (i.e., carbon depleted in ^{13}C). Explain the formation of **229**.[232]

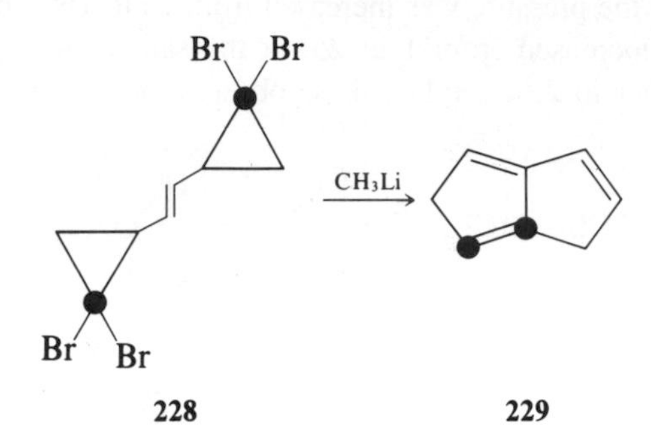

REFERENCES AND NOTES

1. R. A. Bernheim, H. W. Bernard, P. S. Wang, L. S. Wood, and P. S. Skell, *J. Chem. Phys.*, **53,** 1280 (1970); E. Wasserman, W. A. Yager, and V. Kuck, *Chem. Phys. Lett.*, **7,** 409 (1970).
2. G. Herzberg, *Proc. R. Soc. London,* **A262,** 291 (1961); G. Herzberg and J. W. C. Johns, *ibid.*, **295,** 107 (1964).
3. J. F. Harrison and L. C. Allen, *J. Am. Chem. Soc.*, **91,** 807 (1969).
4. C. F. Bender, H. F. Schaefer, D. R. Franceschetti, and L. C. Allen, *J. Am. Chem. Soc.*, **94,** 6888 (1972).
5. N. Bodor, M. J. S. Dewar, and J. S. Wasson, *J. Am. Chem. Soc.*, **94,** 9095 (1972).
6. (a) W. L. Hase, R. J. Phillips, and J. W. Simons, *Chem. Phys. Lett.*, **12,** 161 (1971). (b) H. M. Frey, *Chem. Commun.*, **1972,** 1024. (c) H. M. Frey and G. J. Kennedy, *Chem. Commun.*, **1975,** 233. (d) H. M. Frey and G. J. Kennedy, *J. Chem. Soc., Faraday Trans. 1,* **73,** 164 (1977). (e) W. L. Hase and P. M. Kelly. *J. Chem. Phys.*, **66,** 5093 (1977).
7. M. J. S. Dewar, R. C. Haddon, and P. K. Weiner, *J. Am. Chem. Soc.*, **96,** 253 (1974).
8. P. F. Zittel, G. B. Ellison, S. V. O'Neal, E. Herbst, W. C. Lineberger, and W. P. Reinhardt, *J. Am. Chem. Soc.*, **98,** 3731 (1976); P. C. Engelking, R. R. Corderman, J. J. Wendoloski, G. B. Ellison, S. V. O'Neal, and W. C. Lineberger, *J. Chem. Phys.*, **74,** 5460 (1981). For a theoretical reinterpretation of the photodetachment experiment, bringing the singlet–triplet splitting in line with other experiments and with theory, see D. Feller, L. E. McMurchie, N. T. Borden, and E. R. Davidson, *J. Chem. Phys.*, **77,** 6134 (1982), and references therein.
9. J. A. Meadows and H. F. Schaefer, *J. Am. Chem. Soc.*, **98,** 4383 (1976).
10. C. W. Bauschlicher, H. F. Schaefer, and P. S. Bagus, *J. Am. Chem. Soc.*, **99,** 7106 (1977).
11. (a) B. O. Roos and P. M. Siegbahn, *J. Am. Chem. Soc.*, **99,** 7716 (1977). (b) E. R. Davidson, D. Feller, and P. Phillips, *Chem. Phys. Lett.*, **76,** 416 (1980). (c) P. Saxe, H. F. Schaefer, and N. C. Handy, *J. Phys. Chem.*, **85,** 745 (1981). (d) E. R. Davidson, L. E. McMurchie, and S. J. Day, *J. Chem. Phys.*, **74,** 5491 (1981).

12. (a) R. K. Lengel and R. N. Zare. *J. Am. Chem. Soc.*, **100,** 7495 (1978). (b) C. C. Hayden, D. M. Neumark, K. Shobatake, R. K. Sparks, and Y. T. Lee, *J. Chem. Phys.*, **76,** 3607 (1982).

13. S. W. Benson, *Thermochemical Kinetics*, 2nd ed., Wiley, New York, 1976.

14. G. Herzberg and J. W. C. Johns, *J. Chem. Phys.*, **54,** 2276 (1971); see also G. Herzberg, Nobel Prize Lecture, *Angew. Chem.*, **84,** 1126 (1972). For recent observations of vibrational and rotational transitions of ground-state triplet methylene, see T. J. Sears, P. R. Bunker, and A. R. W. McKellar, *J. Chem. Phys.*, **75,** 4731 (1981); T. J. Sears, P. R. Bunker, A. R. W. McKellar, K. M. Evenson, D. A. Jennings, and J. M. Brown, *ibid.* **77,** 5348 (1982); T. J. Sears, P. R. Bunker, and A. R. W. McKellar, *ibid.*, **77,** 5363 (1982); P. Jensen, P. R. Bunker, and A. R. Hoy, *ibid.*, **77,** 5370 (1982); F. J. Lovas, R. D. Suenram, and K. M. Evenson, *Astrophys. J.*, **267,** L131 (1983).

15. V. Franzen and H.-I. Joschek, *Justus Liebigs Ann. Chem.*, **633,** 7 (1960).

16. H. E. O'Neal and S. W. Benson, *J. Phys. Chem.*, **72,** 1866 (1968); J. M. Birchall, R. N. Haszeldine, and D. W. Roberts, *J. Chem. Soc. Perkin Trans. 1*, **1973,** 1071.

17. C. Wentrup, *Tetrahedron,* **30,** 1301 (1974).

18. For general treatises, see W. Kirmse, *Carbene Chemistry,* 2nd ed., Academic Press, New York, 1971; *Carbenes,* M. Jones, Jr., and R. A. Moss, Eds., Wiley, New York, 1973, Vol. I and 1975, Vol. II; J. Hine, *Divalent Carbon,* Ronald Press, New York, 1964.

19. For general treatises, see (a) *Nitrenes*, W. Lwowski, Ed., Wiley-Interscience, New York, 1970. (b) *Azides and Nitrenes,* E. F. V. Scriven, Ed., Academic Press, New York, 1984. (c) T. L. Gilchrist and C. W. Rees, *Carbenes, Nitrenes, and Arynes,* Nelson, London, 1969.

20. W. D. Crow and C. Wentrup, *Tetrahedron Lett.*, **1967,** 4379.

21. D. A. Bamford and C. H. Bamford, *J. Chem. Soc.*, **1941,** 30; J. S. Swenton, T. J. Ikeler, and G. LeRoy Smyser, *J. Org. Chem.*, **38,** 1157 (1973).

22. C. Wentrup, *Tetrahedron,* **27,** 1027 (1971); S. Saito and C. Wentrup, *Helv. Chim. Acta,* **54,** 273 (1971).

23. O. M. Nefedov and A. I. Dyachenko, *Angew. Chem.*, **84,** 527 (1972); *Angew. Chem. Int. Ed. Engl.*, **11,** 507 (1972).

24. D. Seebach, H. Siegel, K. Müllen, and K. Hiltbrunner, *Angew. Chem.*, **91,** 844 (1979); *Angew. Chem. Int. Ed. Engl.*, **18,** 784 (1979); H. Siegel, K. Hiltbrunner, and D. Seebach, *ibid.*, **91,** 845 (1979) and **18,** 785 (1979).

25. G. Köbrich, *Angew. Chem.*, **84,** 557 (1972); *Angew. Chem. Int. Ed. Engl.*, **11,** 473 (1972).

26. R. A. Moss and F. G. Pilkiewicz, *J. Am. Chem. Soc.*, **96,** 5632 (1974); R. A. Moss, M. A. Joyce, and F. G. Pilkiewicz, *Tetrahedron Lett.*, **1975,** 2425.

27. (a) A. M. Trozzolo, *Acc. Chem. Res.*, **1,** 329 (1968). (b) For absorption and fluorescence spectra of diphenylcarbene at 2–4 K, see R. J. M. Anderson, B. E. Kohler, and J. M. Stevenson, *J. Chem. Phys.*, **71,** 1559 (1979).

28. G. L. Closs and B. E. Rabinov, *J. Am. Chem. Soc.*, **98,** 8190 (1976).

29. N. J. Turro, J. A. Butcher, R. A. Moss, W. Guo, R. C. Munjal, and M. Fedorynski, *J. Am. Chem. Soc.*, **102,** 7576 (1980); N. J. Turro, G. F. Lehr, J. A. Butcher, R. A. Moss, and W. Guo, *ibid.*, **104,** 1754 (1982).

30. D. Griller, C. R. Montgomery, J. C. Scaiano, M. S. Platz, and L. Hadel, *J. Am. Chem. Soc.*, **104,** 6813 (1982).

31. P. C. Wong, D. Griller, and J. C. Scaiano, *J. Am. Chem. Soc.*, **103,** 5934 (1981).

32. J. J. Zupancic and G. B. Schuster, *J. Am. Chem. Soc.*, **102,** 5958 (1980); **103,** 944 (1981); B.-E. Brauer, P. B. Grasse, K. J. Kaufmann, and G. B. Schuster, *ibid.*, **104,** 6814 (1982).

33. R. S. Berry, in *Nitrenes,* W. Lwowski, Ed., Wiley, New York, 1970, p. 13.

34. A. Reiser, G. Bowes, and R. J. Horne, *Trans. Faraday Soc.*, **62,** 3162 (1966); A. Reiser and H. M. Wagner, in *The Chemistry of the Azido Group*, S. Patai, Ed., Interscience, London, 1971, p. 444; V. A. Smirnov and S. B. Brichkin, *Chem. Phys. Lett.*, **87,** 548 (1982).

35. M. Sumitani, S. Nagakura, and K. Yoshihara, *Bull. Chem. Soc. Jpn.*, **49,** 2995 (1976).

36. A. K. Malchev, P. G. Mikaeljan, and O. M. Nefedov, *Dokl. Akad. Nauk. SSSR,* **201,** 901 (1971).

37. D. E. Milligan and M. E. Jacox, *J. Chem. Phys.*, **47,** 703 (1967); L. Andrews, *ibid.*, **48,** 979 (1968); *Tetrahedron Lett.*, **1968,** 1423. From the IR data ∢Cl—C—Cl = 100 ± 9° is derived.

38. P. H. Mueller, N. G. Rondan, K. N. Houk, J. F. Harrison, D. Hooper, B. H. Willen, and J. F. Liebman, *J. Am. Chem. Soc.*, **103,** 5049 (1981); C. W. Bauschlicher, *ibid.*, **102,** 5492 (1980); S. P. So, *J. Chem. Soc. Faraday Trans. 2,* **75,** 820 (1979); N. C. Baird and K. F. Taylor, *J. Am. Chem. Soc.*, **100,** 1333 (1978).

39. M. S. Baird, I. R. Dunkin, N. Hacker, M. Poliakoff, and J. J. Turner, *J. Am. Chem. Soc.*, **103,** 5190 (1981).

40. A. Gilles, J. Masanet, and C. Vermeil, *Chem. Phys. Lett.*, **25,** 346 (1974); see also R. S. Berry in *Nitrenes,* W. Lwowski, Ed., Wiley, New York, 1970, p. 13; J. F. Harrison and G. Shalhoub, *J. Am. Chem. Soc.*, **97,** 4172 (1975).

41. D. C. Doetschman and C. A. Hutchison, *J. Chem. Phys.*, **56,** 3964 (1972).

42. A. M. Trozzolo and E. Wasserman, in *Carbenes,* R. A. Moss and M. Jones, Jr., Eds., Wiley, New York, 1975, Vol. II, p. 185.

43. N. J. Turro, M. Aikawa, J. A. Butler, and G. W. Griffin, *J. Am. Chem. Soc.*, **102,** 5127 (1980).

44. V. P. Senthilnathan and M. S. Platz, *J. Am. Chem. Soc.*, **103,** 5503 (1981); C.-T. Lin and P. P. Gaspar, *Tetrahedron Lett.*, **21,** 3553 (1980); M. S. Platz, V. P. Senthilnathan, B. B. Wright, and C. W. McCurdy, *J. Am. Chem. Soc.*, **104,** 6494 (1982).

45. (a) N. Filipescu and J. R. DeMember, *Tetrahedron,* **24,** 5181 (1968); (b) D. S. Wulfman, G. Linstrumelle, and C. F. Cooper, in *The Chemistry of Diazonium and Diazo Groups,* S. Patai, Ed., Wiley, Chichester, 1978, Vol. II. p. 821. (c) See also H. Tomioka, S. Suzuki, and Y. Izawa, *J. Am. Chem. Soc.*, **104,** 1047 (1982).

46. P. Scheiner, in *Selective Organic Transformations,* B. S. Thyagarajan, Ed., Wiley, New York, 1970, Vol. 1, p. 327.

47. R. L. Smith, A. Manmade, and G. W. Griffin, *Tetrahedron Lett.*, **1970,** 663. The identity of diphenylcarbene derived from tetraphenyloxirane and diphenyldiazomethane has also been established by direct spectroscopic observation (see Ref. 43).

48. E. Wigner and E. Witmer, *Z. Phys.*, **51,** 859 (1928); F. A. Matsen and J. D. Klein, *Adv. Photochem.*, **7,** 1 (1969).

49. M. Jones, Jr., and K. R. Rettig, *J. Am. Chem. Soc.*, **87,** 4013, 4015 (1965); M. Jones, Jr., et al., *ibid.*, **94,** 7469 (1972).

50. W. H. Pirkle and G. F. Koser, *Tetrahedron Lett.*, **1968,** 3959; *J. Am. Chem. Soc.*, **90,** 3598 (1968); G. F. Koser, *J. Org. Chem.*, **42,** 1474 (1977). The ESR spectrum of **21** was observed by G. A. Nikiforov and V. V. Ershov, *Izvest. Akad. Nauk. SSSR Ser. Khim.* (*Engl. Transl.*), **10,** 2241 (1967).

51. M. I. Komendantov, V. Ya. Bespalov, O. A. Bezrukova, and R. R. Bekmukhametov, *J. Org. Chem. USSR,* **11,** 25 (1975); **9,** 431 (1973).

52. R. S. Hutton and H. D. Roth, *J. Am. Chem. Soc.*, **100,** 4324 (1978).

53. K. S. Kim and H. F. Schaefer, *J. Am. Chem. Soc.*, **102,** 5389 (1980).

54. H. D. Roth, *Acc. Chem. Res.*, **10,** 85 (1977).

55. The possibility of a singlet ground state for **23** has also been suggested: H. Tomioka, H. Okuno, and Y. Izawa, *J. Chem. Soc., Perkin Trans. 2,* **1980,** 1636.

56. (a) L. Salem and C. Rowland, *Angew. Chem.,* **84,** 86 (1972); *Angew. Chem. Int. Ed. Engl.,* **11,** 92 (1972). (b) H. E. Hunziker, *J. Chem. Phys.,* **50,** 1288 (1969).

57. Although triplet halocarbenes are unknown, the likelihood of small singlet–triplet splittings with the singlets as the ground states in carbenes such as CHF, CHBr, CHCl, and CBr_2 opens the possibility of observing *triplet* reactions by choosing substrates with low reactivity toward singlet carbenes: M. Jones, Jr., V. J. Tortorelli, P. P. Gaspar, and J. B. Lambert, *Tetrahedron Lett.,* **1978,** 4257, and references therein.

58. P. S. Skell, S. J. Valenty, and P. W. Humer, *J. Am. Chem. Soc.,* **95,** 5041 (1973). This paper also gives an example of an *internal* heavy atom effect, causing rapid intersystem crossing in the mercuricarbene CH_3—Hg—C̈—CN.

59. A. R. Forrester and J. S. Sadd, *Chem. Commun.,* **1976,** 631.

60. Review: C. Wentrup, *Top. Curr. Chem.,* **62,** 173 (1976).

61. W. Braun, A. M. Bass, and M. Pilling, *J. Chem. Phys.,* **52,** 5131 (1970); A. H. Laufer, *Rev. Chem. Intermed.,* **4,** 225 (1981).

62. (a) C. W. Jefford, J. Mareda, J.-C. Gehret, T. Kabengele, W. D. Graham, and U. Burger, *J. Am. Chem. Soc.,* **98,** 2585 (1976); K. N. Houk, N. G. Rondan, M. N. Paddon-Row, C. W. Jefford, P. T. Huy, P. D. Burrow, and K. D. Jordan, *ibid.,* **105,** 5563 (1983). (b) G. W. Klumpp and P. M. Kwantes, *Tetrahedron Lett.,* **22,** 831 (1981). (c) U. Burger and G. Gandillon, *ibid.,* **20,** 4281 (1979).

63. A. Reiser, F. W. Willets, G. C. Terry, V. Williams, and R. Marley, *Trans. Faraday Soc.,* **64,** 3265 (1968).

64. W. D. Crow and C. Wentrup, *Tetrahedron Lett.,* **1968,** 6149.

65. P. S. Skell and R. C. Woodworth, *J. Am. Chem. Soc.,* **78,** 4496 (1956).

66. R. Hoffmann, *J. Am. Chem. Soc.,* **90,** 1475 (1968); R. Hoffmann, D. M. Hayes, and P. S. Skell, *J. Phys. Chem.,* **76,** 664 (1972); N. Bodor, M. J. S. Dewar, and J. S. Wasson, *J. Am. Chem. Soc.,* **94,** 9095 (1972).

67. C. Mayor and C. Wentrup, *J. Am. Chem. Soc.,* **97,** 7467 (1975).

68. A. S. Kende, P. Hebeisen, P. J. Sanfilippo, and B. H. Toder, *J. Am. Chem. Soc.,* **104,** 4244 (1982).

69. (a) W. Lwowski, *Angew. Chem.,* **79,** 922 (1967); *Angew. Chem. Int. Ed. Engl.,* **6,** 897 (1967); P. P. Nicholas, *J. Org. Chem.,* **40,** 3396 (1975). (b) R. A. Abramovitch, S. R. Challand, and E. F. V. Scriven, *J. Org. Chem.,* **37,** 2705 (1972). (c) S. Tamura, H. Imaizumi, Y. Hashida, and K. Matsui, *Bull. Chem. Soc. Jpn.,* **54,** 301 (1981).

70. Y. Yamamoto, S.-I. Murahashi, and I. Moritani, *Tetrahedron Lett.,* **1973,** 589.

71. W. J. Baron, M. E. Hendrick, and M. Jones, Jr., *J. Am. Chem. Soc.,* **95,** 6286 (1973); see also P. P. Gaspar, B. L. Whitsel, M. Jones, Jr., and J. B. Lambert, *J. Am. Chem. Soc.,* **102,** 6108 (1980); G. L. Closs, *Top. Stereochem.,* **3,** 193 (1968).

72. L. E. Helgen, Ph.D. dissertation, Yale University, 1965; W. Kirmse and M. Buschoff, *Chem. Ber.,* **102,** 1098 (1969).

73. G. Smolinsky and B. I. Feuer, *J. Org. Chem.,* **31,** 3882 (1966).

74. C. D. Gutsche, G. L. Bachman, W. Udell, and S. Bäuerlein, *J. Am. Chem. Soc.,* **93,** 5172 (1971).

75. H. F. Schaefer, *Chem. Brit.,* **11,** 227 (1975); C. P. Baskin, C. F. Bender, C. W. Bauschlicher, and H. F. Schaefer, *J. Am. Chem. Soc.,* **96,** 2709 (1974); C. W. Bauschlicher, C. F. Bender, and H. F. Schaefer, *ibid.,* **98,** 3072 (1976); C. W. Bauschlicher, K. Haber, H. F. Schaefer, and C. F. Bender, *ibid.,* **99,** 3610 (1977); H. Kollmar and V. Staemmler, *Theor. Chim. Acta,* **51,** 207 (1979). For the corresponding NH reactions see T. Fueno, V. Bonačić-Koutecký, and J. Koutecký, *J. Am. Chem. Soc.,* **105,** 5547 (1983).

76. (a) G. L. Closs and L. E. Closs, *J. Am. Chem. Soc.*, **91,** 4549 (1969). (b) G. L. Closs and A. D. Trifunac, *ibid.*, **91,** 4554 (1969).

77. A gas-phase study of the reaction between the tritiated carbene CHT and deuterated methyl halides CD_3X also demonstrated halogen abstraction by the singlet and a slower deuterium atom abstraction by the triplet: P. S.-T. Lee and F. S. Rowland, *J. Phys. Chem.*, **84,** 3243 (1980).

78. H. Iwamura, Y. Imahashi, M. Ōki, K. Kushida, and S. Satoh, *Chem. Lett.*, **1974,** 259.

79. T. Sasaki, K. Kanematsu, A. Kakehi, I. Ichikawa, and K. Hayakawa, *J. Org. Chem.*, **35,** 426 (1970); J. Streith and J. M. Cassal, *C. R. Acad. Sci. Ser. C*, **264,** 1307 (1967).

80. H. Iwamura, Y. Imahashi, and K. Kushida, *Tetrahedron Lett.*, **1975,** 1401; *see also* G. K. Agopian, D. W. Brown, and M. Jones, Jr., *ibid.*, **1976,** 2931; H. Iwamura, Y. Imahashi, K. Kushida, K. Aoki, and S. Satoh, *Bull. Chem. Soc. Jpn.*, **49,** 1690 (1976).

81. H. Tomioka, *J. Am. Chem. Soc.*, **101,** 256 (1979).

82. R. A. Moss and M. A. Joyce, *J. Am. Chem. Soc.*, **100,** 4475 (1978).

83. V. P. Senthilnathan and M. S. Platz, *J. Am. Chem. Soc.*, **102,** 7637 (1980).

84. W. Kirmse, *Justus Liebigs Ann. Chem.*, **9,** 666 (1963); W. Kirmse, K. Loosen, and H.-D. Sluma, *J. Am. Chem. Soc.*, **103,** 5935 (1981); D. Bethell, A. R. Newall, and D. Whittaker, *J. Chem. Soc. B*, **1971,** 23; H. Tomioka, S. Suzuki, and Y. Izawa, *J. Am. Chem. Soc.*, **104,** 3156 (1982).

85. M. S. Platz, *J. Am. Chem. Soc.*, **101,** 3398 (1979); **102,** 1192 (1980).

86. M. S. Platz and J. R. Burns, *J. Am. Chem. Soc.*, **101,** 4425 (1979); M. S. Platz, G. Carrol, F. Pierrat, J. Zayas, and S. Auster, *Tetrahedron*, **38,** 777 (1982).

87. The photolysis of benzoyl azide causes Curtius rearrangement to phenyl isocyanate. The latter photolyzes to phenylnitrene and CO: E. Wasserman, *Prog. Phys. Org. Chem.*, **8,** 319 (1971).

88. N. Bodor and M. J. S. Dewar, *J. Am. Chem. Soc.*, **94,** 9103 (1972); V. Menéndez and J. M. Figuera, *Chem. Phys. Lett.*, **18,** 426 (1973); E. P. Kyba, *J. Am. Chem. Soc.*, **99,** 8330 (1977).

89. M. B. Sohn and M. Jones, Jr., *J. Am. Chem. Soc.*, **94,** 8280 (1972); I. Moritani, Y. Yamamoto, and S.-I. Murahashi, *Tetrahedron Lett.*, **1968,** 5697. Intermolecular chemistry with "dimethylcarbene" can be carried out when it is generated as a *carbenoid* in the reaction:

$$Me_2CBr_2 + BuLi \xrightarrow{-70^\circ C} Me_2CBrLi$$

P. Fischer and G. Schaefer, *Angew. Chem.*, **93,** 895 (1981); *Angew. Chem. Int. Ed. Engl.*, **20,** 863 (1981).

90. J. E. Gano, R. H. Wettach, M. S. Platz, and V. P. Senthilnathan, *J. Am. Chem. Soc.*, **104,** 2326 (1982).

91. A. Nickon and W. H. Werstiuk, *J. Am. Chem. Soc.*, **94,** 7081 (1972); A. Nickon, F. Huang, R. Weglein, K. Matsuo, and H. Yagi, *ibid.*, **96,** 5264 (1974); E. P. Kyba and C. W. Hudson, *ibid.*, **98,** 5696 (1976); L. Seghers and H. Shechter, *Tetrahedron Lett.*, **1976,** 1943; L. S. Press and H. Shechter, *J. Am. Chem. Soc.*, **101,** 509 (1979); A. Nickon and J. K. Bronfenbrenner, *ibid.*, **104,** 2022 (1982).

92. Y. Yamamoto and I. Moritani, *Tetrahedron*, **26,** 235 (1970); M. T. H. Liu, O. Banjoko, Y. Yamamoto, and I. Moritani, *ibid.*, **31,** 1645 (1975).

93. C. Wentrup, unpublished work (1970).

94. H. Tomioka, H. Ueda, S. Kondo, and Y. Izawa, *J. Am. Chem. Soc.*, **102,** 7817 (1980).

95. C. Wentrup and W. Reichen, *Helv. Chim. Acta*, **59,** 2615 (1976); C. Wentrup and H.-

W. Winter, *Angew. Chem.*, **90,** 643 (1978); *Angew. Chem. Int. Ed. Engl.*, **17,** 609 (1978); H.-W. Winter and C. Wentrup, *ibid.*, **92,** 743 (1980) and **19,** 720 (1980).

96. R. F. C. Brown, *Pyrolytic Methods in Organic Chemistry,* Academic Press, New York, 1980.

97. H. D. Hartzler, in *Carbenes,* R. A. Moss and M. Jones, Jr., Eds., Wiley, New York, 1975, Vol. II. p. 43.

98. P. J. Stang, *Chem. Rev.*, **78,** 383 (1978).

99. Calculations on the vinylidene–acetylene rearrangement: C. E. Dykstra and H. F. Schaefer, *J. Am. Chem. Soc.*, **100,** 1378 (1978); M. P. Conrad and H. F. Schaefer, *ibid.*, **100,** 7820 (1978); L. Radom, *Aust. J. Chem.*, **31,** 1 (1978); R. Krishnan, M. J. Frisch, J. A. Pople, and P. von R. Schleyer, *Chem. Phys. Lett.*, **79,** 408 (1981) and **81,** 421 (1981).

100. T. R. Fields and P. J. Kropp, *J. Am. Chem. Soc.*, **96,** 7559 (1974); S. S. Hixson, J. C. Tausta, and J. Borovsky, *ibid.*, **97,** 3230 (1975); O. P. Strausz, R. J. Nordstrom, D. Salahub, R. K. Gosavi, H. E. Gunning, and I. G. Csizmadia, *ibid.*, **92,** 6395 (1970); R. Srinivasen and K. H. Brown, *J. Am. Chem. Soc.*, **100,** 4602 (1978); M. G. Steinmetz, R. T. Mayes, and J.-C. Yang, *ibid.*, **104,** 3518 (1982).

101. For a thermally induced retro-1,2-rearrangement proceeding at 110°C in solution, see T. H. Chan and D. Massuda, *J. Am. Chem. Soc.*, **99,** 936 (1977).

102. R. A. Abramovitch and E. P. Kyba, in *The Chemistry of the Azido Group,* S. Patai, Ed., Interscience, London, 1971, p. 221; E. P. Kyba and R. A. Abramovitch, *J. Am. Chem. Soc.*, **102,** 735 (1980).

103. F. D. Lewis and W. H. Saunders, in *Nitrenes,* W. Lwowski, Ed., Wiley, New York, 1970, p. 47; F. C. Montgomery and W. H. Saunders, *J. Org. Chem.*, **41,** 2368 (1976).

104. Calculations on methylnitrene: D. R. Yarkoni, H. F. Schaefer, and S. Rothenberg, *J. Am. Chem. Soc.*, **96,** 5974 (1974); C. Trindle and J. K. George, *Int. J. Quantum. Chem.*, **10,** 21 (1976); J. Demuynck, D. J. Fox, Y. Yamaguchi, and H. F. Schaefer, *J. Am. Chem. Soc.*, **102,** 6204 (1980).

105. D. E. Milligan, *J. Chem. Phys.*, **35,** 1491 (1961); M. E. Jacox and D. E. Milligan, *J. Mol. Spectrosc.*, **56,** 333 (1975); see also I. R. Dunkin and P. C. P. Thomson, *Tetrahedron Lett.*, **21,** 3813 (1980).

106. J. A. Leermakers, *J. Am. Chem. Soc.*, **55,** 3098 (1933); F. O. Rice and C. J. Grelecki, *J. Phys. Chem,* **61,** 830 (1957); W. Pritzkow and D. Timm., *J. Prakt. Chem.*, **32,** 178 (1966).

107. C. Wentrup, S. Fischer, and M. Winnewisser, unpublished work (1979).

108. H. Bock, R. Dammel, and L. Horner, *Chem. Ber.*, **114,** 220 (1981).

109. W. Ried and H. Mengler, *Fortschr. Chem. Forsch.*, **5,** 1 (1965).

110. I. G. Csizmadia, J. Font, and O. P. Strausz, *J. Am. Chem. Soc.*, **90,** 7360 (1968); K.-P. Zeller, H. Meier, H. Kolshorn, and E. Müller, *Chem. Ber.*, **105,** 1875 (1972).

111. Review: H. Meier and K.-P. Zeller, *Angew. Chem.*, **87,** 52 (1975); *Angew. Chem. Int. Ed. Engl.*, **14,** 32 (1975).

112. Review: C. Wentrup, *Adv. Heterocycl. Chem.*, **28,** 231 (1981).

113. K.-P. Zeller, *Chem. Ber.*, **108,** 3566 (1975); Z. Majerski and C. S. Redvanly, *Chem. Commun.*, **1972,** 694; K.-P. Zeller, *Z. Naturforsch.*, **31B,** 586 (1976); see also U. Timm, K.-P. Zeller, and H. Meier, *Tetrahedron,* **33,** 453 (1977).

114. U. Timm, K.-P. Zeller, and H. Meier, *Chem. Ber.*, **111,** 1549 (1978).

115. F. Kaplan and G. K. Meloy, *J. Am. Chem. Soc.*, **88,** 950 (1966).

116. C. Wentrup and H. Dahn, *Helv. Chim. Acta,* **53,** 1637 (1970); G. Paliani, S. Sorriso, and R. Cataliotti, *J. Chem. Soc. Perkin Trans. 2,* **1976,** 707.

117. H. Tomioka, H. Kitagawa, and Y. Izawa, *J. Org. Chem.*, **44,** 3072 (1979).

118. H. Tomioka, H. Okuno, and Y. Izawa, *J. Org. Chem.*, **45,** 5278 (1980).

119. Evidence *against* the convertibility of oxirenes to oxocarbenes has also been adduced: Y. Ogata, Y. Sawaki, and T. Ohno, *J. Am. Chem. Soc.*, **104,** 216 (1982). Evidence for the matrix isolation of perfluoroalkyloxirenes has been reported: M. Torres, J. L. Bourdelande, A. Clement, and O. P. Strausz, *ibid.*, **105,** 1698 (1983); see, however, D. Lemal et al., *ibid.*, in press (1983–1984).

120. Review: M. Torres, E. Lown, and O. P. Strausz, *Heterocycles,* **11,** 697 (1978).

121. (a) A. Krantz and J. Laureni, *J. Org. Chem.*, **44,** 2730 (1979). (b) M. Torres, A. Clement, J. E. Bertie, H. E. Gunning, and O. P. Strausz, *J. Org. Chem.*, **43,** 2490 (1978). (c) J. E. Bertie, H. E. Gunning, and O. P. Strausz, *Nouv. J. Chim.*, **3,** 149 (1979).

122. M. Regitz, B. Arnold, D. Danion, H. Schubert, and G. Fusser, *Bull. Soc. Chim. Belg.*, **90,** 615 (1981).

123. C. Thétaz and C. Wentrup, *J. Am. Chem. Soc.*, **98,** 1258 (1976); C. Wentrup, *Helv. Chim. Acta,* **55,** 1613 (1972).

124. C. Wentrup, E. Wentrup-Byrne, and P. Müller, *Chem. Commun.*, **1977,** 210.

125. R. Schulz, A. Schweig, C. Wentrup, and H.-W. Winter, *Angew. Chem.*, **92,** 846 (1980); *Angew. Chem. Int. Ed. Engl.*, **19,** 821 (1980).

126. W. D. Crow and C. Wentrup, *Tetrahedron Lett.*, **1968,** 6149; C. Wentrup, *Chem. Commun.*, **1969,** 1386.

127. R. C. Joines, A. B. Turner, and W. M. Jones, *J. Am. Chem. Soc.*, **91,** 7754 (1969); C. Wentrup and K. Wilczek, *Helv. Chim. Acta,* **53,** 1459 (1970); W. M. Jones, R. C. Joines, J. A. Myers, T. Mitsuhashi, K. E. Krajca, E. E. Waali, T. L. Davis, and A. B. Turner, *J. Am. Chem. Soc.*, **95,** 826 (1973).

128. See also W. J. Baron, M. Jones, Jr., and P. P. Gaspar, *J. Am. Chem. Soc.*, **92,** 4739 (1970); E. Hedaya and M. E. Kent, *ibid.*, **93,** 3283 (1971).

129. N. M. Lân and C. Wentrup, *Helv. Chim. Acta,* **59,** 2068 (1976).

130. T. T. Coburn and W. M. Jones, *J. Am. Chem. Soc.*, **96,** 5218 (1974).

131. C. Wentrup, C. Mayor, J. Becker, and H. J. Lindner, *Tetrahedron,* **40** (1984).

132. O. L. Chapman, *Pure Appl. Chem.*, **51,** 331 (1979).

133. P. R. West, O. L. Chapman, and J.-P. LeRoux, *J. Am. Chem. Soc.*, **104,** 1779 (1982). Using the chirality test for cyclic allenes described for cyclohexa-1,2-diene in Chapter 5, the existence of cyclohepta-1,2,4,6-tetraene (**129**) in solution has also been demonstrated: J. W. Harris and W. M. Jones, *J. Am. Chem. Soc.*, **104,** 7329 (1982).

134. C. Wentrup and H.-W. Winter, *J. Am. Chem. Soc.*, **102,** 6159 (1980).

135. W. D. Crow and C. Wentrup, *Chem. Commun.*, **1969,** 1387; R. Harder and C. Wentrup, *J. Am. Chem. Soc.*, **98,** 1259 (1976).

136. L. Skattebøl, *Chem. Ind.* (*London*), **1962,** 2146; K. H. Holm and L. Skattebøl, *Tetrahedron Lett.*, **1977,** 2347; *J. Am. Chem. Soc.*, **99,** 5480 (1977). Review: W. M. Jones and U. H. Brinker, in *Pericyclic Reactions,* A. P. Marchand and R. E. Lehr, Eds., Academic Press, New York, 1977, Vol. I, p. 110.

137. W. W. Schoeller and U. H. Brinker, *J. Am. Chem. Soc.*, **100,** 6012 (1978).

138. For further examples and discussion, see R. A. Moss and M. Jones, Jr., in *Reactive Intermediates*, Wiley, New York, 1981, Vol. 2, p. 113. There is evidence that carbenoids rather than free carbenes are involved in the Skattebøl rearrangement: P. M. Warner, Abstracts, 186th Natl. Meeting, American Chemical Society, Washington, D.C., 1983, ORGN 193.

139. S. D. Isaev, A. G. Yurchenko, F. N. Stepanov, G. G. Kolyada, S. S. Novikov, and N. F. Karpenko, *J. Org. Chem. USSR,* **9,** 745 (1973).

140. L. Friedman and H. Shechter, *J. Am. Chem. Soc.*, **82,** 1002 (1960); G. Maier and M. Strasser, *Tetrahedron Lett.*, **1966,** 6453; see also U. Langer and H. Musso, *Justus Liebigs Ann. Chem.*, **1976,** 1180.

141. B. M. Trost and R. M. Cory, *J. Am. Chem. Soc.*, **93,** 5572 (1971).

142. J. E. Baldwin and J. A. Walker, *Chem. Commun.*, **1972,** 354.

143. H. Tsurata, K. Kurabayashi, and T. Mukai, *Tetrahedron Lett.*, **1967,** 3775.

144. J. J. Looker, *J. Org. Chem.*, **36,** 1045, 2681 (1971).

145. C. Wentrup, *Tetrahedron,* **27,** 1281 (1971).

146. A. Hassner and I. W. Fowler, *Tetrahedron Lett.*, **1967,** 1545.

147. R. J. Bailey and H. Shechter, *J. Am. Chem. Soc.*, **96,** 8116 (1974).

148. G. Märkl, H. Hauptmann, and J. Advena, *Angew. Chem.*, **84,** 440 (1972); *Angew. Chem. Int. Ed. Engl.*, **11,** 441 (1972).

149. R. G. Bergman and V. J. Rajadhyaksha, *J. Am. Chem. Soc.*, **92,** 2163 (1970).

150. V. V. Kane, A. D. Wolff, and M. Jones, Jr., *J. Am. Chem. Soc.*, **96,** 2643 (1974).

151. R. F. C. Brown, F. W. Eastwood, and G. L. McMullen, *Chem. Commun.*, **1975,** 328.

152. P. Yates and R. O. Loufty, *Acc. Chem. Res.*, **8,** 209 (1975); N. J. Turro et al., *ibid.*, **5,** 92 (1972); W. D. Stohrer, P. Jacobs, K. H. Kaiser, G. Wiech, and G. Quinkert, *Top. Curr. Chem.*, **46,** 181 (1974).

153. H. Quast and P. Eckert, *Justus Liebigs Ann. Chem.*, **1974,** 1727; J. Michl et al., *Pure Appl. Chem.*, **55,** 315 (1983).

154. C. Wentrup, A. Damerius, and W. Reichen, *J. Org. Chem.*, **43,** 2037 (1978).

155. H. M. Frey, *J. Am. Chem. Soc.*, **79,** 1259 (1957).

156. G. W. Taylor and J. W. Simons, *Int. J. Chem. Kinet.*, **3,** 25 (1971).

157. P. J. Robinson and K. A. Holbrook, *Unimolecular Reactions*, Wiley, New York, 1972.

158. G. B. Kistiakowsky and B. B. Saunders, *J. Phys. Chem.*, **77,** 427 (1973).

159. H. Akimoto, K. Obi, and I. Tanaka, *Bull. Chem. Soc. Jpn.*, **46,** 2267 (1973).

160. H. M. Frey, *Prog. React. Kinet.*, **2,** 132 (1964); J. M. Figuera, E. Fernández, and M. J. Avila, *J. Phys. Chem.*, **78,** 1348 (1974).

161. The formation of *triplet* acetylene by dimerization of triplet methylene with expulsion of H_2 has also been suggested because of direct spectroscopic detection of the triplet acetylene: H. R. Wendt, H. Hippler, and H. E. Hunziker, *J. Phys. Chem.*, **70,** 4044 (1979).

162. D. J. DeFrees and W. J. Hehre, *J. Phys. Chem.*, **82,** 391 (1978).

163. J. D. Rynbrandt and B. S. Rabinovitch, *J. Chem. Phys.*, **54,** 2275 (1971); *J. Phys. Chem.*, **74,** 4175 (1970).

164. B. S. Rabinovitch, R. F. Kubin, and R. E. Harrington, *J. Chem. Phys.*, **38,** 405 (1963); Compare E. A. Hardwidge, C. W. Larson, and B. S. Rabinovitch, *J. Am. Chem. Soc.*, **92,** 3278 (1970).

165. D. C. Montague and F. S. Rowland, *Chem. Commun.*, **1972,** 193; D. C. Montague, *ibid.*, **1972,** 615; F. S. Rowland, P. S.-T. Lee, D. Montague, and R. L. Russell, *Disc. Faraday Soc.*, **52,** 111 (1972).

166. C. Wentrup, in *Reactive Intermediates,* R. A. Abramovitch, Ed., Plenum Press, New York, 1980, Vol. 1, p. 263.

167. J. E. Baldwin and R. A. Smith, *J. Am. Chem. Soc.*, **89,** 1886 (1967).

168. R. A. Abramovitch and R. G. Sutherland, *Fortschr. Chem. Forsch.*, **16,** 1 (1970).

169. A. G. Anastassiou, J. N. Shepelavy, H. E. Simmons, and F. D. Marsh, in *Nitrenes,* W. Lwowski, Ed., Wiley, New York, 1970, p. 305.

170. R. J. Sundberg and R. H. Smith, *Tetrahedron Lett.*, **1971,** 267. The "nitrenoid" formed

from nitrosobenzene and triethylphosphite adds to benzene to give 1-phenylazepine in 34% yield:

$$\text{PhNO} \xrightarrow{(EtO)_3P} \left[:\overset{\ominus}{\ddot{N}}(Ph)\text{—O—}\overset{\oplus}{P}(OEt)_3 \right] \xrightarrow[-PO(OEt)_3]{C_6H_6} \text{1-phenylazepine}$$

171. L. Krbechek and H. Takimoto, *J. Org. Chem.*, **33,** 4286 (1968); G. R. Cliff, E. W. Collington, and G. Jones, *J. Chem. Soc. C,* **1970,** 1490.

172. P. S. Skell and M. S. Cholod, *J. Am. Chem. Soc.*, **91,** 7131 (1969).

173. R. A. Moss, *Acc. Chem. Res.*, **13,** 58 (1980); R. A. Moss, W. Guo, and K. Krogh-Jespersen, *Tetrahedron Lett.*, **23,** 15 (1982).

174. B. Giese, W.-B. Lee, and J. Meister, *Liebigs Ann. Chem.*, **1980,** 725; B. Giese and W.-B. Lee, *Angew. Chem.*, **92,** 864 (1980); *Angew. Chem. Int. Ed. Engl.*, **19,** 835 (1980).

175. R. R. Kostikov, A. P. Molchanov, G. V. Golovanova, and I. G. Zenkevich, *J. Org. Chem. USSR,* **13,** 1712 (1977); R. A. Moss and C. B. Mallon, *J. Am. Chem. Soc.*, **97,** 344 (1975).

176. (a) C. Wentrup, *Reaktive Zwischenstufen,* Georg Thieme Verlag, Stuttgart, 1979, Vol. I, p. 164; Vol. II, p. 407–411. (b) N. G. Rondan, K. N. Houk, and R. A. Moss, *J. Am. Chem. Soc.*, **102,** 1770 (1980).

177. R. W. Hoffmann and M. Reiffen, *Chem. Ber.*, **109,** 2565 (1976).

178. R. W. Hoffmann, B. Hagenbruch, and D. M. Smith, *Chem. Ber.*, **110,** 23 (1977); see also D. Seebach, *ibid.*, **105,** 487 (1972) for di(methylthio)carbene, and M. Nitsche, D. Seebach, and A. K. Beck, *ibid.*, **111,** 3644 (1978) for di(phenylthio)carbene and dimerization reactions.

179. M. Edwards, T. L. Gilchrist, C. J. Harris, and C. W. Rees, *J. Chem. Res. (S),* **1979,** 114; R. S. Atkinson and J. R. Malpass, *J. Chem. Soc. Perkin Trans. 1,* **1977,** 2242; R. S. Atkinson and C. W. Rees, *J. Chem. Soc. C,* **1969,** 772. For addition to azobenzene, see L. Hoesch, *Helv. Chim. Acta,* **64,** 38 (1981). For intramolecular C—H insertion of a related aminonitrene, see R. S. Atkinson, J. R. Malpass, and K. L. Woodthorpe, *Chem. Commun.*, **1981,** 160.

180. W. D. Hinsberg, P. G. Schultz, and P. B. Dervan, *J. Am. Chem. Soc.*, **104,** 766 (1982).

181. L. Hoesch, N. Egger, and A. S. Dreiding, *Helv. Chim. Acta,* **61,** 795 (1978).

182. R. S. Atkinson and B. D. Judkins, *Chem. Commun.*, **1979,** 832 and 833; R. S. Atkinson, B. D. Judkins, and N. Khan, *J. Chem. Soc. Perkin Trans. 1,* **1982,** 2491; **1981,** 2615.

183. S.-I. Murahashi, K. Okumura, T. Naota, and S. Nagase, *J. Am. Chem. Soc.*, **104,** 2466 (1982).

184. W. M. Jones, M. E. Stowe, E. E. Wells, and E. W. Lester, *J. Am. Chem. Soc.*, **90,** 1849 (1968); T. Mitsuhashi and W. M. Jones, *Chem. Commun.*, **1974,** 103.

185. L. W. Christianson, E. E. Waali, and W. M. Jones, *J. Am. Chem. Soc.*, **94,** 2118 (1972); B. L. Duell and W. M. Jones, *J. Org. Chem.*, **43,** 4901 (1978).

186. W. M. Jones, *Acc. Chem. Res.*, **10,** 353 (1977); J. W. Harris and W. M. Jones, *J. Am. Chem. Soc.*, **104,** 7329 (1982).

187. (a) A. Hackenberger and H. Dürr, *Tetrahedron Lett.*, **1979,** 4541. (b) H. Dürr and F. Werndorff, *Angew. Chem.*, **86,** 413 (1974); *Angew. Chem. Int.* Ed. Engl., **13,** 483 (1974).

188. R. Richter and H. Ulrich, *J. Org. Chem.*, **36,** 2005 (1971); J. M. Brown and B. D. Place, *Chem. Commun.*, **1971,** 533; H. Quast and S. Hünig, *Chem. Ber.*, **99,** 2017 (1966); H. D. Hartzler, *J. Am. Chem. Soc.*, **95,** 4379 (1973).

189. R. J. Cvetanovíc, *Can. J. Chem.*, **38,** 1678 (1960); T. H. Varkony, S. Pass, and Y. Mazur, *Chem. Commun.*, **1975,** 457; J. J. Havel and K. H. Chan, *J. Am. Chem. Soc.*, **97,** 5800 (1975),.

190. B. G. Ramsey and D. M. Anjo, *J. Am. Chem. Soc.*, **99,** 3182 (1977).

191. U. Schmidt, *Angew. Chem.*, **87,** 535 (1975); *Angew. Chem. Int. Ed. Engl.*, **14,** 523 (1975); M. Yoshifuji, I. Shima, N. Inamoto, K. Hirotsu, and T. Higuchi, *J. Am. Chem. Soc.*, **103,** 4587 (1981); G. Bertrand, C. Couret, J. Escudie, S. Majid, and J.-P. Majoral, *Tetrahedron Lett.*, **23,** 3567 (1982); J. Escudié, C. Couret, and J. Satgé, *J. Roy. Neth. Chem. Soc.*, **98,** 461 (1979).

192. P. P. Gaspar, in *Reactive Intermediates,* M. Jones, Jr., and R. A. Moss, Eds., Wiley, New York, Vol. 1, 1978; Vol. 2, 1981.

193. L. E. Gusel'nikov and N. S. Nametkin, *Chem. Rev.*, **79,** 529 (1979).

194. W. P. Neumann, *Nach. Chem. Techn. Lab.*, **30,** 190 (1982); P. Bleckmann, H. Maly, R. Minkwitz, W. P. Neumann, B. Watta, and G. Olbrich, *Tetrahedron Lett.*, **23,** 4655 (1982).

195. O. F. Zeck, Y. Y. Su, G. P. Gennaro, and Y.-N. Tang, *J. Am. Chem. Soc.*, **96,** 5967 (1974); P. S. Skell and E. J. Goldstein, *ibid.*, **86,** 1442 (1964).

196. J. H. Meadows and H. F. Schaefer, *J. Am Chem. Soc.*, **98,** 4383 (1976).

197. R. West, M. J. Fink, and J. Michl, *Science*, **214,** 1343 (1981).

198. S. Masamune, Y. Hanzawa, S. Murakami, T. Bally, and J. F. Blount, *J. Am. Chem. Soc.*, **104,** 1150 (1982).

199. V. J. Tortorelli and M. Jones, Jr., *J. Am. Chem. Soc.*, **102,** 1425 (1980).

200. R. T. Conlin and P. P. Gaspar, *J. Am. Chem. Soc.*, **98,** 3715 (1976).

201. T. J. Drahnak, J. Michl, and R. West, *J. Am. Chem. Soc.*, **101,** 5427 (1979).

202. H. P. Reisenauer, G. Mihm, and G. Maier, *Angew. Chem.*, **94,** 864 (1982); *Angew. Chem. Int. Ed. Engl.*, **21,** 854 (1982).

203. G. Maier, G. Mihm, and H. P. Reisenauer, *Angew. Chem.*, **93,** 615 (1981); *Angew. Chem. Int. Ed. Engl.*, **20,** 597 (1981); P. Rosmus, H. Bock, B. Solouki, G. Maier, and G. Mihm, *ibid.*, **93,** 616 (1981) and **20,** 598 (1981).

204. R. T. Conlin and D. L. Wood, *J. Am. Chem. Soc.*, **103,** 1843 (1981); T. J. Drahnak, J. Michl, and R. West, *ibid.*, **103,** 1845 (1981); T. J. Barton, S. A. Burns, and G. T. Burns, *Organometallics,* **1,** 210 (1982).

205. The experiment was actually performed with dimethylsilylcarbene (**188,** R = CH_3), giving 2-methyl-2-silapropene: O. L. Chapman, C.-C. Chang, J. Kolc, M. E. Jung, J. A. Lowe, T. J. Barton, and M. L. Tumey, *J. Am. Chem. Soc.*, **98,** 7844 (1976); M. R. Chedekel, M. Skoglund, R. L. Kreeger, and H. Shechter, *ibid.*, **98,** 7846 (1976).

206. B. Coleman and M. Jones, Jr., *Rev. Chem. Intermed.*, **4,** 297 (1981).

207. A. G. Brook, S. C. Nyburg, F. Abdesaken, B. Gutekunst, G. Gutekunst, R. K. M. R. Kallury, Y. C. Poon, Y.-M. Chang, and W. Wong-Ng, *J. Am. Chem. Soc.*, **104,** 5667 (1982).

208. For theoretical calculations, see Y. Yoshioka and H. F. Schaefer, *J. Am. Chem. Soc.*, **103,** 7366 (1981) and references therein.

209. W. D. Wulff, W. F. Goure, and T. J. Barton, *J. Am. Chem. Soc.*, **100,** 6236 (1978).

210. L. C. Snyder and Z. R. Wasserman, *J. Am. Chem. Soc.*, **101,** 5222 (1979).

211. H. Leclercq and I. Dubois, *J. Mol. Spectrosc.*, **76,** 39 (1979).

212. J. N. Murrel, H. W. Kroto, and M. F. Guest, *Chem. Commun.*, **1977,** 619; M. S. Gordon and R. D. Koop, *J. Am. Chem. Soc.*, **103,** 2939 (1981); M. S. Gordon, and J. A. Pople, *ibid.*, **103,** 2945 (1981).

213. M. Schriewer and W. P. Neumann, *Angew. Chem.*, **93,** 1089 (1981); *Angew. Chem. Int. Ed. Engl.*, **20,** 1019 (1981).

214. J.-C. Barthelat, B. Saint Roch, G. Trinquier, and J. Satgé, *J. Am. Chem. Soc.*, **102,** 4080 (1980).

215. W. P. Neumann, "The Stannylenes R_2Sn," in *Organometallic and Coordination Chemistry of Ge, Sn, and Pb,* M. Gielen and P. G. Harrisan, Eds., Freund Publications, Tel Aviv, 1978.

216. E. O. Fischer and H. Grubert, *Z. Naturforsch.*, **11B**, 423 (1956).

217. P. J. Davidson, D. H. Harris, and M. F. Lappert, *J. Chem. Soc. Dalton Trans.*, **1976,** 2268; T. Fjeldberg, A. Haaland, M. F. Lappert, B. E. R. Schilling, R. Seip, and A. J. Thorne, *Chem. Commun.*, **1982,** 1407.

218. W. J. Pietro and W. J. Hehre, *J. Am. Chem. Soc.*, **104,** 4329 (1982).

219. K. W. Egger and A. T. Cocks, *Helv. Chim. Acta,* **56,** 1516 (1973). The π bond energy is defined as the difference in enthalpy of formation between the double-bonded molecule and the hypothetical 1,2-diradical; see Section 3.2.

220. E. Ciganek, *J. Am. Chem. Soc.*, **89,** 1458 (1967); G. W. Griffin, *Angew. Chem.*, **83,** 604 (1971); *Angew. Chem. Int. Ed. Engl.*, **10,** 537 (1971).

221. P. C. Martino and P. B. Shevlin, J. Am. Chem. Soc., **102,** 5430 (1980).

222. H. E. Zimmerman and L. R. Sousa, *J. Am. Chem. Soc.*, **94,** 834 (1972).

223. W. P. Weber and G. W. Gokel, *Tetrahedron Lett.*, **1972,** 1637; G. W. Gokel, R. P. Widera, and W. P. Weber, *Org. Synth.*, **55,** 96 (1976). For other examples, see M. Makosza, *Pure Appl. Chem.*, **43,** 439 (1975).

224. I. E. Dolgii, E. A. Shapiro, G. V. Lun'kova, and O. M. Nefedov, *Izvest. Akad. Nauk. SSSR* (*Engl. Transl.*), **1979,** 1527; O. M. Nefedov, I. E. Dolgii, and E. A. Shapiro, *ibid.*, **1978,** 1052.

225. J. Becker and C. Wentrup, *Chem. Commun.*, **1980,** 190; see also T. A. Engler and H. Shechter, *Tetrahedron Lett.*, **23,** 2715 (1982).

226. B. Stanovnik, M. Tišler, J. Bradač, B. Budič, B. Koren, and B. Mozetič-Reščič, *Heterocycles,* **12,** 457 (1979).

227. G. Sedelmeier, H. Prinzbach, and H.-D. Martin, *Chimia,* **33,** 329 (1979).

228. C. Wentrup, C. Thétaz, and R. Gleiter, *Helv. Chim. Acta,* **55,** 2633 (1972).

229. H. Heydtmann, K. Eichler, and B. Hildebrandt, *Z. Phys. Chem.* (*Frankfurt*), **113,** 1 (1978).

230. B. Iddon, O. Meth-Cohn, E. F. V. Scriven, H. Suschitzky, and P. T. Gallagher, *Angew, Chem.*, **91,** 965 (1979); *Angew. Chem. Int. Ed. Engl.*, **18,** 900 (1979).

231. T. de Boer, J. I. G. Cadogan, H. M. McWilliam, and A. G. Rowley, *J. Chem. Soc. Perkin Trans. 2,* **1975,** 554.

232. U. H. Brinker and I. Fleischhauer, *Tetrahedron,* **37,** 4495 (1981).

Chapter 5

STRAINED RINGS

forbear the Ring, . . . or claim it, and challenge the Power . . . in its dark hold beyond the valley of shadows.

—J. R. R. Tolkien, *The Lord of the Rings, Part III, The Return of the King,* 2nd ed., Houghton Mifflin Company, Boston, 1965.

5.1. STRAIN IN GENERAL

If sufficient strain is built into a molecule, it can become so unstable that it acquires the character of a reactive intermediate. The strain energy (E_s) is defined as the difference between the free energy of the real molecule and that of a hypothetical, unstrained model. We shall be concerned here primarily with *enthalpies,* ignoring entropy contributions to the free energies of molecules. In conformity with most of the literature, the term "strain energy" will, however, be used in place of the more correct "strain enthalpy."[1]

Strain can arise from enforced deviations from "normal" bond angles and bond lengths. For example, cyclopropane is strained due to severe distortion from the ideal tetrahedral C—C—C angle of 109.5 to one of 60°:

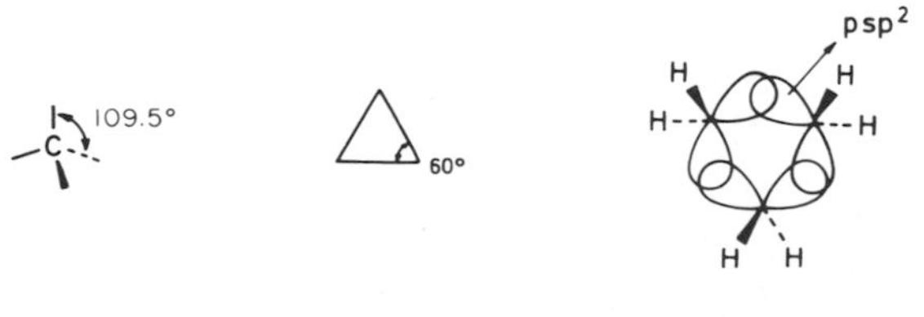

109.5° 60° psp² H H H H H H

Cyclopropane may be said to counteract this distortion by rehybridization, using orbitals of a higher p order than sp^3 in forming carbon–carbon bonds. Thus, "bent" σ bonds are formed, and cyclopropane acquires some double bond character, thereby making it susceptible to electrophilic addition reactions.

The strain energy in cyclopropane can be estimated from its known enthalpy of formation (12.74 kcal/mol, determined by combustion calorimetry). The enthalpy of formation of the hypothetical unstrained model is thrice the thermochemical group increment for the CH_2 group (see Chapter 1), that is $3 \times (-4.95) = -14.85$ kcal/mol. The difference, 27.5 kcal/mol, is the strain energy.

Some further strain energies are collected in Tables 1.3 and 5.1. Intuition predicts that cyclopropene will be more highly strained than cyclopropane. The strain energy of an olefin can be determined as outlined above or by using experimental enthalpies or hydrogenation. If the enthalpy of hydrogenation of a "normal" (i.e., unstrained)

$$\underset{\mathbf{1}}{R-CH=CH-R} + H_2 \longrightarrow \underset{\mathbf{2}}{R-CH_2-CH_2-R}$$

$$\Delta H_1 = \Delta H_f^\circ(2) - \Delta H_f^\circ(1)$$

olefin has a value ΔH_1 and that of a strained olefin has the value ΔH_2, then

$$\mathbf{3} + H_2 \longrightarrow \mathbf{4}$$

$$\Delta H_2 = \Delta H_f^\circ(4) - \Delta H_f^\circ(3)$$

$\Delta H_2 - \Delta H_1$ is the strain energy due to the double bond in **3.** In order to obtain the *total* strain energy of **3,** that of the product **(4)** must be added. This treatment tacitly supposes that the σ strain energies of **3** and **4** are identical, a supposition that will hardly be strictly correct since the geometries are different. The data so obtained will be precise enough for our purposes, however.

The total strain energy of cyclopropene thus found is approximately 54 kcal/mol. Considering that the bond strength of a normal C—C single bond is about 83 kcal/mol, it would appear that only about 30 kcal/mol of additional energy would be required to break such a bond in cyclopropene. Indeed, this reaction takes place with an activation energy of around 35 kcal/mol:[2]

$$\mathbf{3} \overset{\Delta}{\rightleftharpoons} \mathbf{5} \longleftrightarrow \mathbf{6}$$

$$\mathbf{7} \xrightarrow{h\nu,\ 6\text{-}15\ K} \mathbf{5}$$

TABLE 5.1. Strain Energies

Compound	E_s (kcal/mol)[a]
	27.5
	54
	66
	166
	17
	101
	35
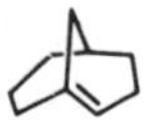	23
	6
	7
	27
	62
	73
	6.5

[a]Total strain energies. References for further data: E. M. Engler, J. D. Andose, and P. v. R. Schleyer, *J. Am. Chem. Soc.*, **95,** 8005 (1973); P. v. R. Schleyer, J. E. Williams, and K. R. Blanchard, *ibid.*, **92,** 2377 (1970); N. L. Allinger and J. T. Sprague, *ibid.*, **94,** 5734 (1972); R. B. Turner, B. J. Mallon, M. Tichy, W. v. E. Doering, W. R. Roth, and G. Schröder, *ibid.*, **95,** 8605 (1973); R. B. Turner, A. D. Jarrett, P. Goebel, and B. J. Mallon, *ibid.*, **95,** 790 (1973); R. B. Turner, P. Goebel, B. J. Mallon, W. v. E. Doering, J. F. Coburn, and M. Pomerantz, *ibid.*, **90,** 4315 (1968); W. F. Maier and P. v. R. Schleyer, *ibid.*, **103,** 1891 (1981).

The intermediate thus formed can be regarded as a resonance hybrid between a carbene **(5)** and a diradical **(6).** The same species was obtained by matrix photolysis of vinyldiazomethane **(7)** and identified by its ESR spectrum.[3]

The strain in cyclopropene is due to angle compression. An olefin may also be strained through torsion around the double bond or by bending the end groups out of the plane. *trans*-Cyclooctene **(8)** is an isolable compound. The strain energy,

16.7 kcal/mol, is about 9 kcal/mol higher than that of *cis*-cyclooctene **(10)**. Obviously, a compound like **8** will adopt a conformation that minimizes the strain. If the olefinic linkage were planar, it could be rather unstrained, but the connecting σ skeleton would experience severe strain due to bond lengthening and angle deformation. The best compromise appears to be the twist-form **9** in which the

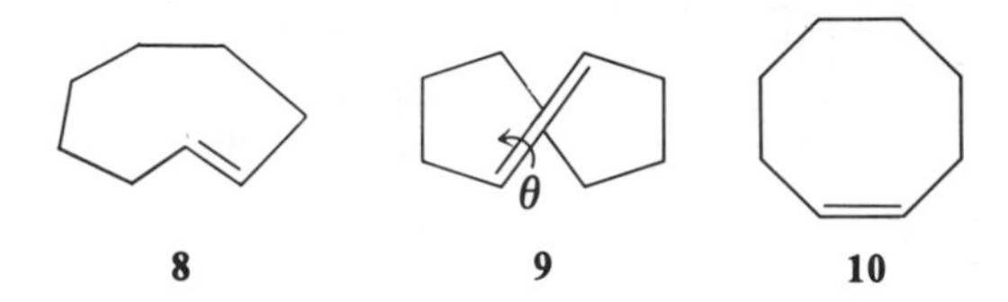

angle of torsion around the double bond, θ, is about 44°.[1] The twisting of the double bond results in a poorer overlap of the p orbitals and hence an increase in the π energy. Accordingly, one should expect the π ionization potential (IP) of *trans*-cyclooctene to be lower than that of *cis*-cyclooctene, an expectation that is borne out in the photoelectron spectra:[4]

IP = 8.82 eV IP = 8.53 eV

Smaller *trans*-cycloalkenes cannot be isolated, and therefore, their strain energies cannot be measured. They can, however, be estimated with considerable confidence with the aid of force field (= molecular mechanics) calculations.[1] The calculated strain energy of *trans*-cycloheptene **(12)** is 27 kcal/mol,[5] that is, about the same as in cyclopropane. The compound can be generated by low-temperature photolysis of *cis*-cycloheptene **(11)** and has a lifetime of about 10 min at 1°C.[6] In spite of the strain, the thermal reversion to the *cis*-compound requires an activation energy of 18.7 kcal/mol.[6]

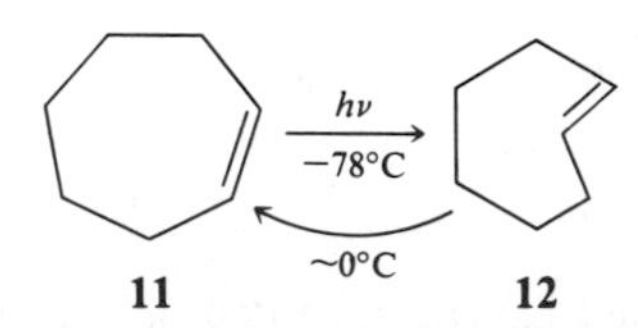

1-Phenyl-*trans*-cycloheptene has been generated in the same way and identified by its UV spectrum.[7] The photochemical formation of twisted *trans*-olefins may be understood to be a consequence of the preferred 90° twisted geometry of the triplet excited states of the olefins.

Even 1-phenyl-*trans*-cyclohexene **(14)** can be generated by photolysis of the *cis*-compound **(13)** and was found to have a lifetime of the order of microseconds at room temperature. Chemical confirmation that the compound observed by UV spectroscopy is indeed **14** is given by the isolation of the dimer **16,** formed in a [2 + 4]-cycloaddition between the two components.[8]

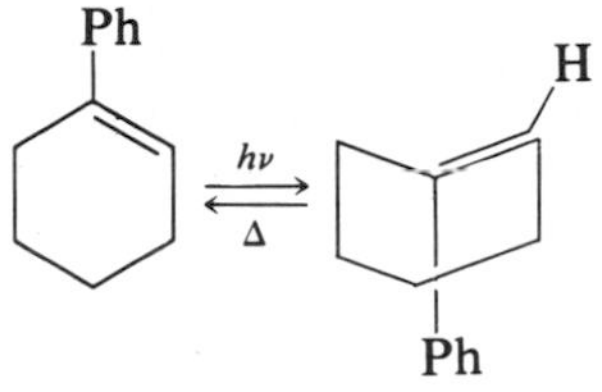

Ph H $h\nu$ Δ Ph

13 **14**

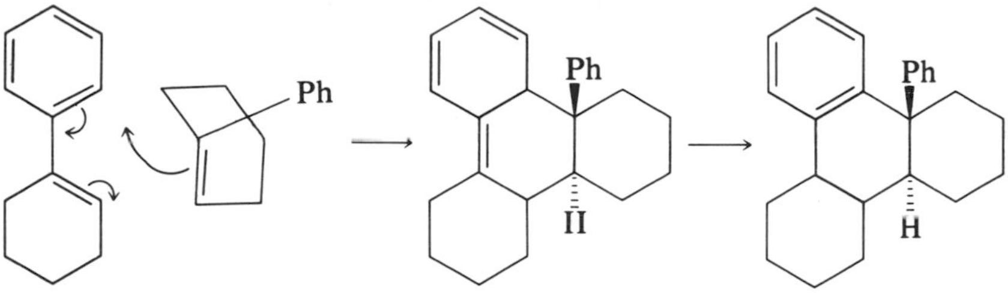

Ph Ph H Ph H

13 **14** **15** **16**

In the compounds considered so far, the strain was a result of the geometry of the ring system. In acyclic compounds, strain may be induced by steric congestion. As shown in Chapter 2, C—C single bonds can be weakened by bulky groups, thus promoting dissociation into free radicals. When sterically demanding groups are attached to an olefin, nonbonded interactions will tend to twist the double bond out of planarity. Tetra-*tert*-butylethylene **(17)** has not been prepared, but force-field calculations indicate a strain energy of a staggering 100 kcal/mol and a torsional angle θ of 45°.[9]

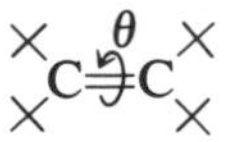

θ C C

17

X-ray crystallography of derivates of 9,9′-bifluorenylidene **(18)** indicate a torsional angle of ca. 40° about the central double bond.[10] The nonplanarity results in reduced double bond character (or increased diradical character) and a very low barrier to rotation (19 kcal/mol for R = CH_3).[11] The dark red color of **18** is a clear indication of an unusual electronic structure. The weak double bond also facilitates the cleavage of **18** to two carbenes **(19),** a process that occurs on gas-phase pyrolysis as indicated by the formation of the hydrogen abstraction products, fluorene **(20)** and 9,9′-bifluorenyl **(21)**.[12] Tetra-1-naphthylethylene can be split into two di-1-naphthylcarbenes for the same reason (see Table 4.2).

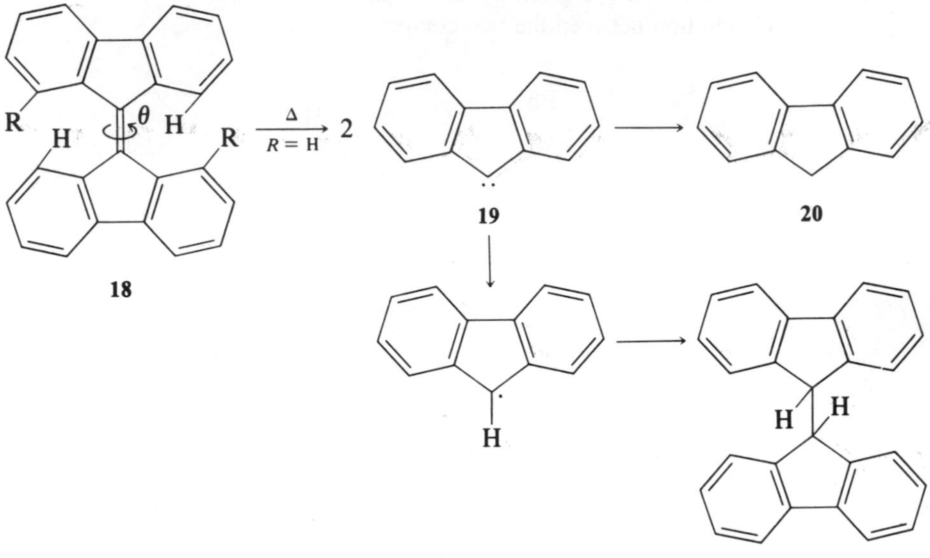

18 **19** **20** **21**

In principle, strain could be relieved from compounds such as **18** by "folding" rather than torsion. Folding could occur in two ways, giving either a *cis*-bent **(22)** or a *trans*-bent structure **(23)**. Although *trans*-bent olefins are unknown, the tin

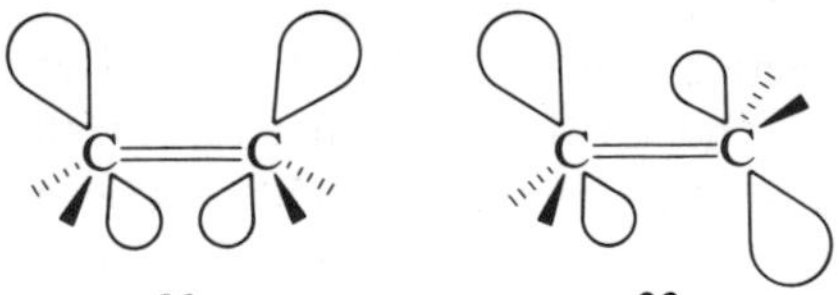

22 **23**

compound **200** in Section 4.7 is an example of such a structure. Many *cis*-bent olefins of the type **22** are known.[1] For example, the stable compound **24** exhibits a dihedral angle of 163°.[13]

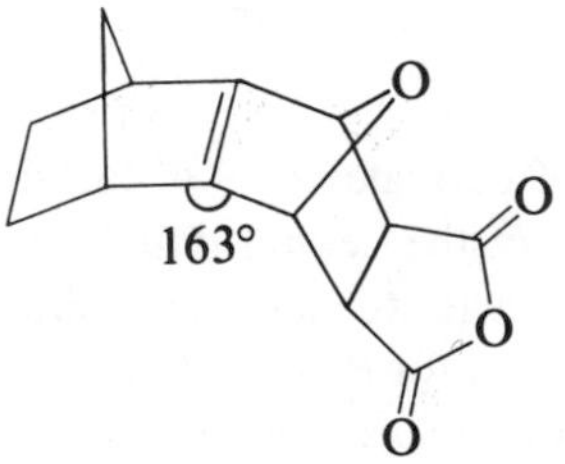

24

In summary, we may identify three major sources of olefinic strain:

1. Angle deformation as in cyclopropene, benzocyclopropene **(25)**,[14] or bicyclo[2.2.0]hex-1(4)-ene **(26)**.[15] Although highly reactive, these are isolable compounds in which the π-bonds are relatively "normal," and the strain is due largely to deformation of the σ framework:

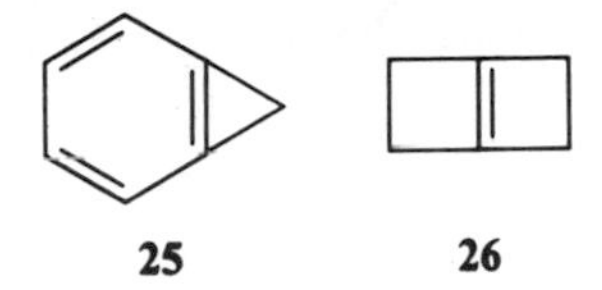

25 **26**

2. Twisting as in tetra-*tert*-butylethylene **(17)** and the bridgehead olefins described in Section 5.2.
3. Bending as in **22**, which will be dealt with in Section 5.3.

5.2. TWISTED BRIDGEHEAD OLEFINS

Bredt's rule[16] asserts that bridgehead double bonds cannot exist, except in large rings, because of deviation from planarity. Nevertheless, many such compounds are now known, some isolable at room temperature, others only as reactive intermediates.[1,17]

Bicyclo[3.3.1]non-1(2)-ene **(29)** is the smallest isolable olefin of this type, and it is seen to contain a *trans*-cyclooctene ring. In fact, the stabilities of the twisted

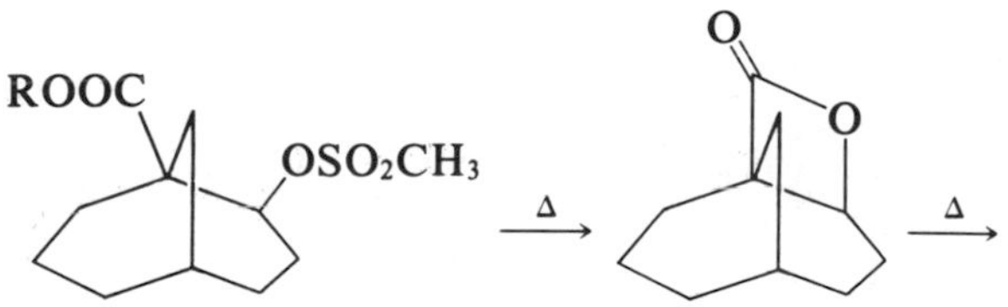

ROOC OSO_2CH_3 $\xrightarrow{\Delta}$ O O $\xrightarrow{\Delta}$

27 **28**

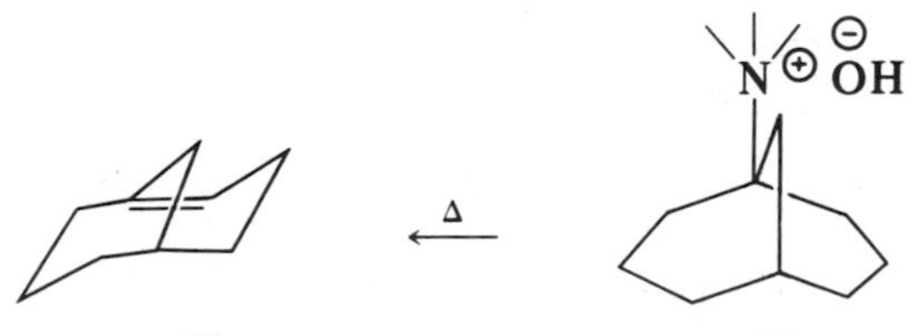

$\overset{\ominus}{OH}$ $N^{\oplus}$

$\xleftarrow{\Delta}$

29 **30**

bridgehead olefins appear to parallel those of the corresponding *trans*-cycloalkenes.[18] This makes it understandable that bicyclo[3.3.1]non-1(2)-ene does *not* exist in form **31**, which would contain a *trans*-cyclohexene ring.[23] As shown above, **29**

is obtained in typical elimination reactions from **27, 28,** or **30.**[19] The measured heat of hydrogenation of **29** allows the determination of the extra strain energy due to the double bond as 12 kcal/mol,[20] which is comparable to that in *trans*-cyclooctene (vide supra). **29** undergoes addition reactions in the normal Markovnikov fashion:

CH_3 ← 1. CH_3Li 2. H_2O — 29 — ROH → OR

If the ring containing the double bond is made smaller, the strain will increase. Two isomeric compounds **33** and **34** were formed by vacuum pyrolysis of the trimethylammonium hydroxide **32** and isolated in a cold trap at −70°C.[21] Compounds **33** and **34** were sufficiently stable to allow spectroscopic identification at

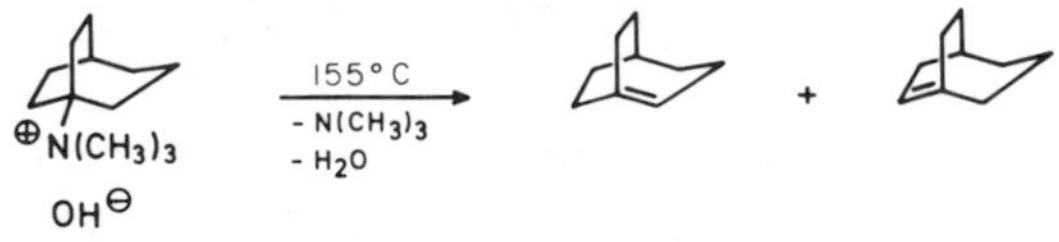

$\oplus N(CH_3)_3$ $OH^{\ominus}$ (32) —155°C, $-N(CH_3)_3$, $-H_2O$→ 33 + 34

−80°C, but on warming to 0°C, they dimerized. The lower stabilities compared to the *isomeric* olefin **29** are apparently due to the presence of *trans*-cycloheptene rings in **33** and **34.**

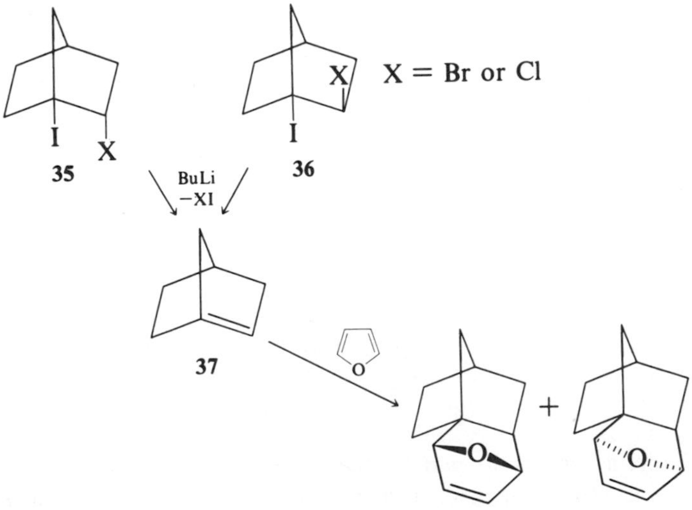

X = Br or Cl

35, **36** —BuLi, −XI→ **37** —(furan)→ **38** + **39**

1-norbornene **(37),** containing a *trans*-cyclohexene ring, can be generated by dehalogenation of either the *endo*- or *exo*-precursors, **35** and **36,** and trapped in a Diels–Alder reaction with furan, giving a mixture of **38** and **39.** The fact that **35** and **36** gave the *same* ratio of **38** and **39** supports the intermediacy of 1-norbornene **(37).**[22]

The double bond strain in **37** is calculated by the force-field method as approximately 35 kcal/mol.[23] The ultimate in double bond twisting obtains in a compound like adamantene **(41)** with a calculated double bond strain energy of 40 kcal/mol.[23] Evidence for the existence of adamantene was first provided by the isolation of the dimers **42** (~90%) from the dehalogenation of **40** with butyl lithium (the *E*-isomer

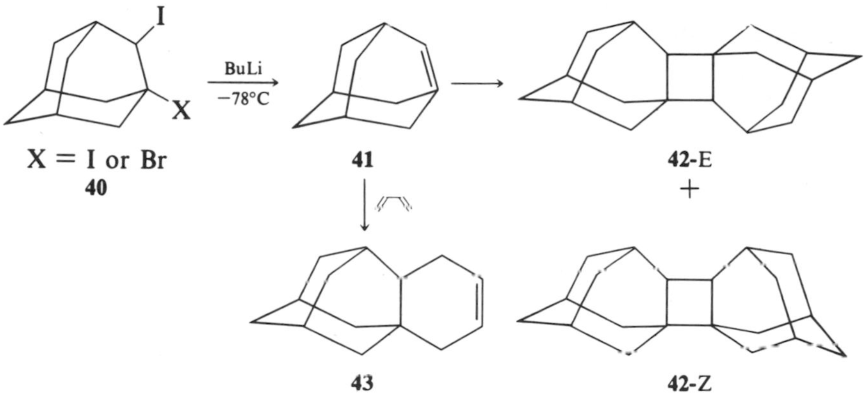

I
X
BuLi
−78°C
X = I or Br
40
41
42-E
+
43
42-Z

is the major constituent).[24] When this and related reactions were carried out in the presence of butadiene, the yield of dimers decreased, and the Diels–Alder adduct **43** was isolated in yields up to 50%.[24,25] Furans were found to be inefficient traps for adamantene produced in this way, but when using a different method, the thermal

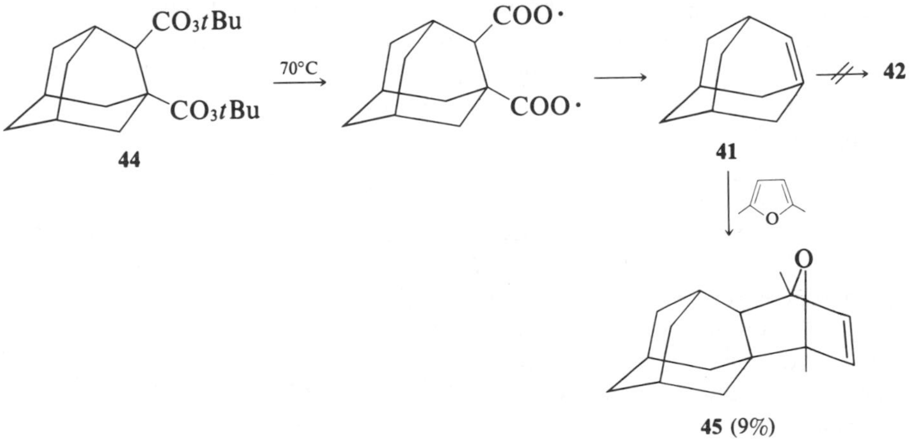

CO_3tBu
CO_3tBu
70°C
COO·
COO·
42
44
41
O
O
45 (9%)

decomposition of the bis-perester **44,** a furan addition product **(45)** was isolated, albeit in low yield. However, no dimers **(42)** were obtained in this reaction.[26]

A third method of producing adamantene involves the dehalogenation of **46** with bis(trimethylsilyl)mercury. This is believed to occur in a concerted four-center manner (see **47**), and when the reaction is performed in the presence of furans, adducts (e.g., **45**) are formed in yields of 16–35%. In the absence of furans, the dimers **(42)** were isolated in low yield.[27,28]

$$\mathbf{46} \xrightarrow[80°C]{(Me_3Si)_2Hg} \left[\mathbf{47}\right] \xrightarrow[-Hg]{-2Me_3SiBr} \mathbf{41}$$

46 (Br, Br) **47** (Br···SiMe₃, Hg, Br···SiMe₃)

41 → (furan) **45**; **41** → (×2) **42**

The preceding somewhat conflicting results may be ascribed, at least in part, to different reaction conditions. When generated at low temperature, adamantene is long-lived enough to dimerize, and its reactivity toward furans is low. In other words, the Diels–Alder reaction with furans appears to have a significant activation energy that is surmounted only at higher temperatures.

Adamantene can be generated in the gas-phase by means of a 1,2-alkyl shift in the carbene **48.**[29] Both dimer formation **(42)** and trapping with butadiene (giving **43**) was observed in this reaction, but, in addition, the diadamantyls **52–54** were isolated. Although the origin of the diadamantyls is not clear, hydrogen abstraction by diadamantyl diradicals **(49–51)** is a possible explanation. The diradicals could be formed either as the initial products of dimerization of **41** or by subsequent thermal cleavage of the dimers **42.** Another route to the radical dimers could be hydrogen abstraction by adamantene itself, to give **55** and **56.** This would imply that adamantene possesses partial radical character, a trait not unexpected of a perpendicular olefin.

Evidence for the discrete existence of adamantene and its radical character has been provided.[30] Diiodoadamantane **(57)** in a stream of argon was allowed to react with sodium (or potassium) vapor, and the products condensed in an argon matrix at 10 K. This allowed the recording of an infrared spectrum, interpreted as being due to adamantene **(41).** Adamantene was stable till at least 70 K; at higher temperatures, it dimerized to the well-known dimers **42,** which were isolated and identified.

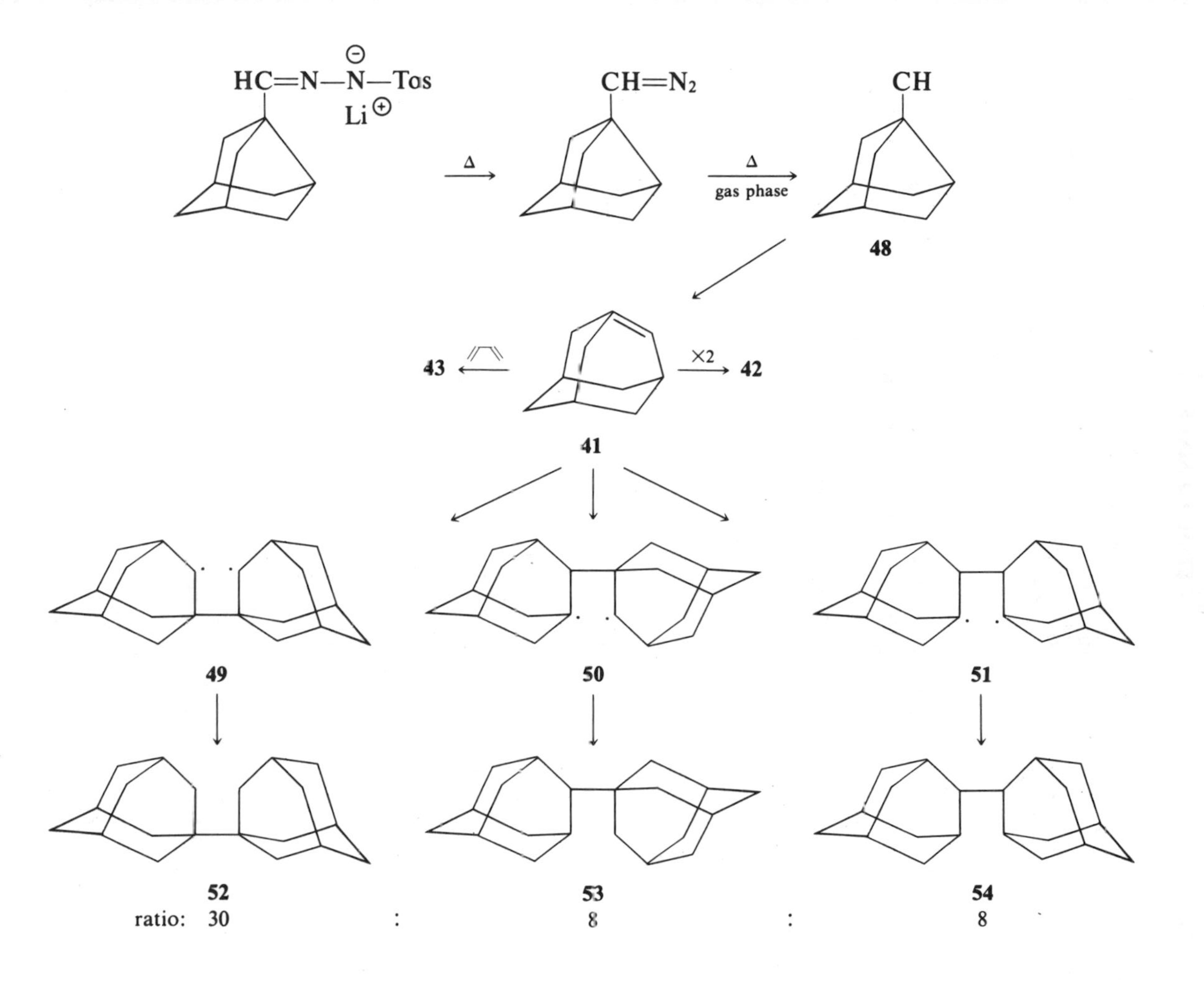

HC=N—N—Tos
⊖
Li⊕
Δ
CH=N$_2$
Δ
gas phase
CH
48
43
×2
42
41
49
50
51
52
53
54
ratio: 30 : 8 : 8

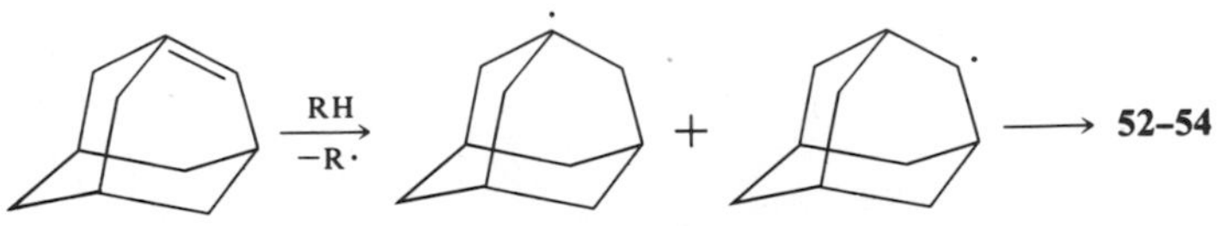

41 $\xrightarrow[-R\cdot]{RH}$ 55 + 56 ⟶ 52–54

The matrix-isolated product also contained a radical, identified by ESR spectroscopy as the 1-adamantyl radical (**55**). A small amount of the radical dimer (**52**) was detected after warming to room temperature. These results indicate that adamantene does, in fact, undergo radical abstraction reactions in the gas phase, giving **55** as the major primary product.

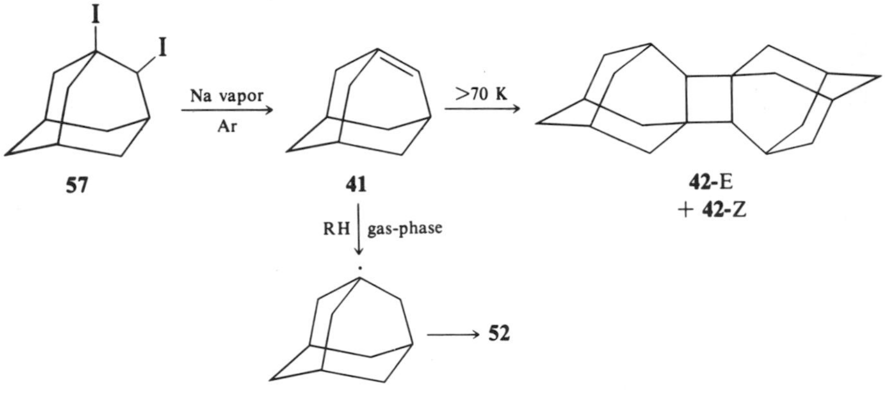

57 $\xrightarrow[Ar]{\text{Na vapor}}$ 41 $\xrightarrow{>70\ K}$ 42-E + 42-Z

41 $\xrightarrow{\text{RH, gas-phase}}$ 55 ⟶ 52

The carbene ring expansion route employed in the generation of adamantene (**41**) from **48** can also be used in the preparation of the more stable homoadamant-3-ene (**59**) from adamantylcarbene (**58**):[23]

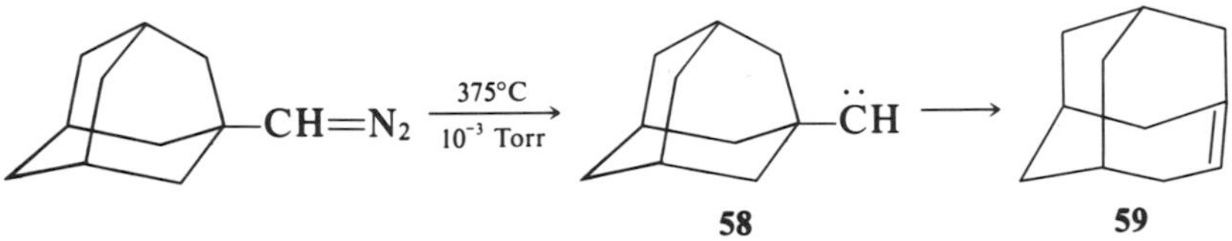

Ad–CH=N_2 $\xrightarrow[10^{-3}\ \text{Torr}]{375°C}$ Ad–$\ddot{C}H$ (58) ⟶ 59

59 was isolated on a NaCl window at −196°C, and the IR spectrum recorded at this temperature showed bands at 3000 (=C—H) and 1610 (C=C) cm^{-1}. For comparison, the C=C stretching vibration of 1-methylcycloheptene is at 1673 cm^{-1}. The shift of 63 cm^{-1} toward lower frequency is ascribed to the reduced C=C bond strength of the *trans*-cycloheptene ring in **59**.

Just as free radicals can be kinetically stabilized by bulky groups (see Chapter 2), adamantyl–homoadamantene (**61**) is a remarkably stable compound, isolable from the gas-phase pyrolysis of diadamantyldiazomethane (**60**):[31]

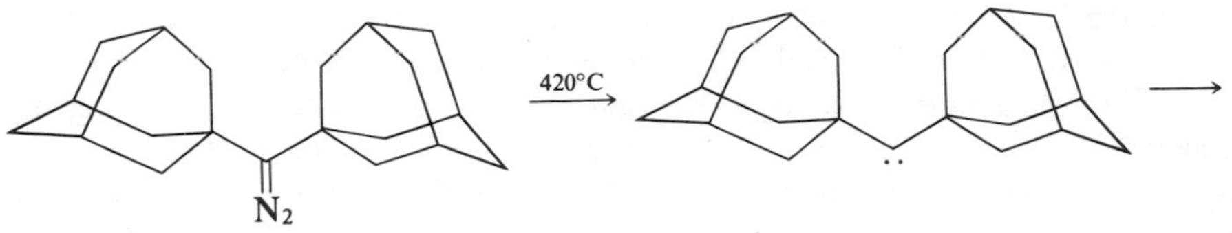

420°C

N_2

60

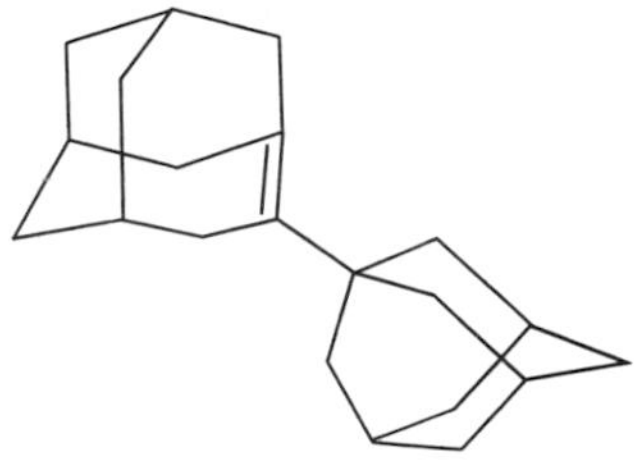

61

Once isolated, compound **61** is stable till +185°C.

5.3. *CIS*-BENT OLEFINS

The severely folded or *cis*-bent olefin **63** has been generated from the dioxolane derivative **62** in boiling tetraglyme. In the presence of diphenylisobenzofuran, **63** is trapped in a Diels–Alder reaction, giving **65.** In the absence of the trapping

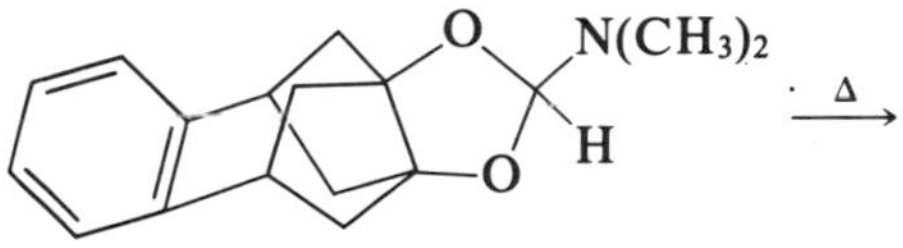

$N(CH_3)_2$ H O O Δ

62

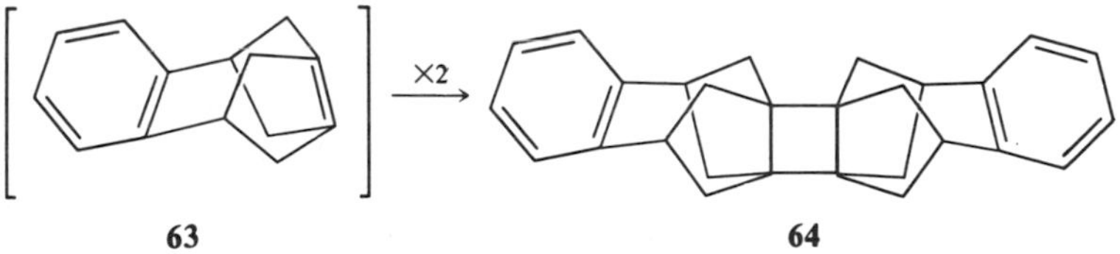

×2

63 **64**

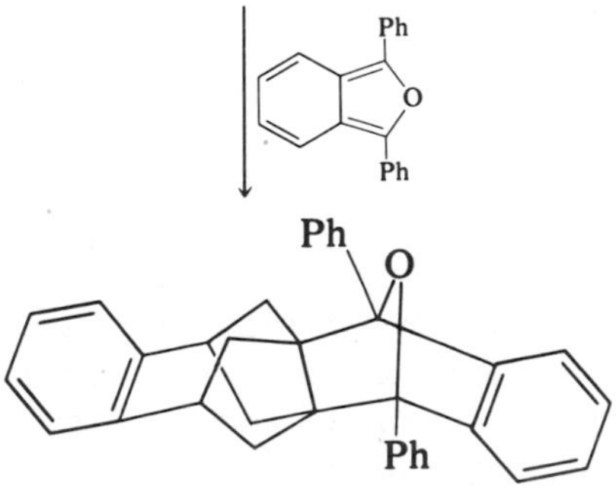

Ph O Ph

Ph O Ph

65

agent, the [2 + 2] dimer **64** is formed.[32] The total strain energy of **63** has been calculated as 45 kcal/mol, and the strain due to the double bond alone as 25 kcal/mol.[23] An even more remarkable *cis*-bent olefin, the quadricyclene **67**, can be generated by HCl elimination from the tetracyclic chloride **66.** Proof of the existence of **67** is given by the isolation of [2 + 4] cycloaddition products with anthracene and 2,5-dimethylfuran:[33]

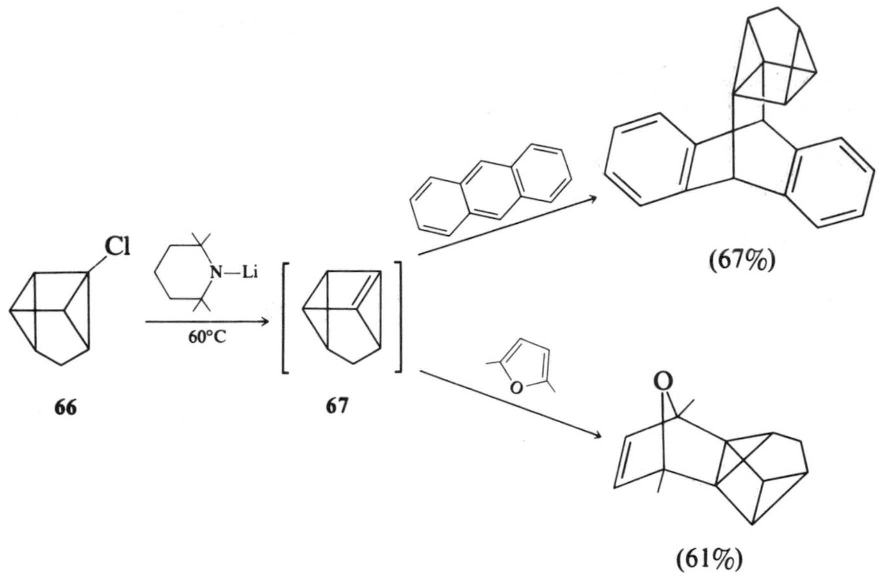

Cl
N—Li
60°C
66
67
(67%)
(61%)

The severely strained olefin **68**, an isomer of benzvalene, can be generated and trapped in a similar manner:[34]

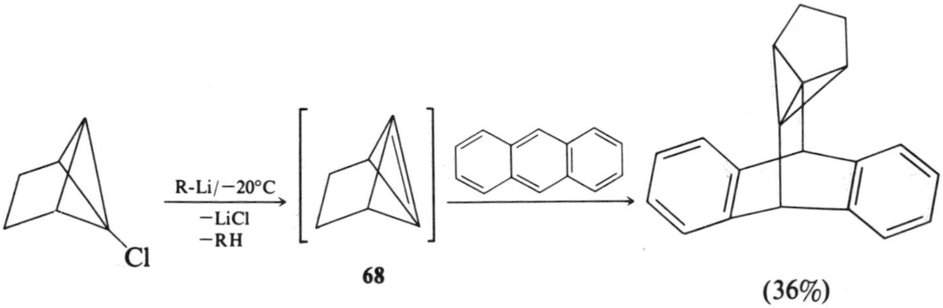

Cl
R-Li/−20°C
−LiCl
−RH
68
(36%)

68 is also an isomer of benzene, but it does not isomerize to benzene. The introduction of an extra double bond as in **69** makes this compound isomeric with benzyne **(70)**, to which it does, in fact, isomerize on attempted generation. The benzyne formed can be trapped with 2,5-dimethylfuran:[35]

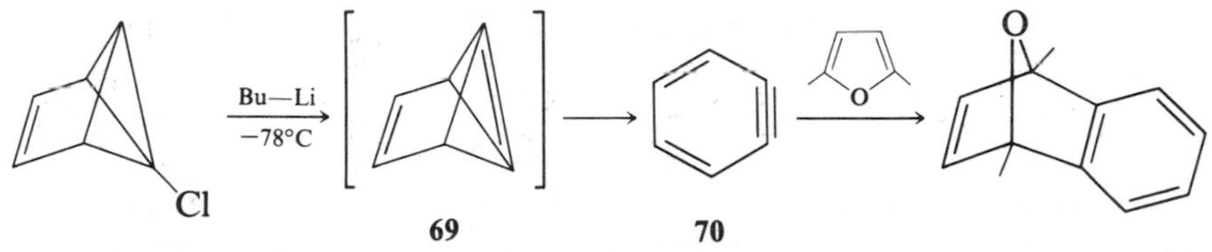

The latter rearrangement indicates that the folded double bond in **69** is more highly strained than the triple bond in benzyne. A similar rearrangement leading to a bent 1,2,3-triene **(75)** occurs in the olefin **73.** When **73** is generated at low temperature (e.g., by HCl elimation from **71** using lithium diisopropylamide), the trapping product **(74)** with diphenylisobenzofuran can be isolated in high yield.

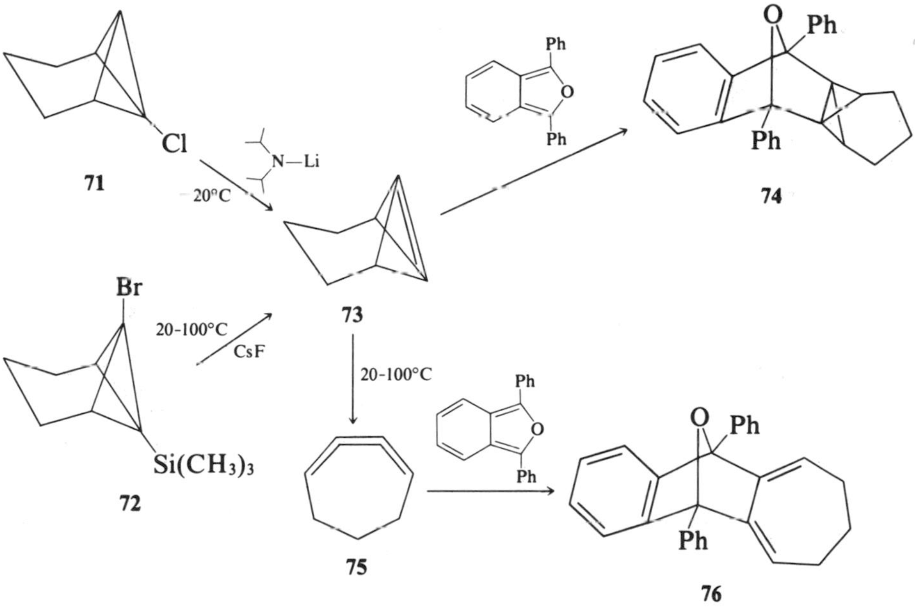

Another method of producing **73** involves the fluoride-ion induced desilylation of compound **72** [giving $(CH_3)_3SiF$ and CsBr], which allows the use of higher temperatures. Under these conditions, **73** is found to rearrange to cyclohepta-1,2,3-triene **(75)**. Since normal allenes and cumulenes are linear—or very nearly so—the central double bond in **75** is highly strained and again undergoes a [2 + 4] cycloaddition with diphenylisobenzofuran, giving **76.** Although neither **73** nor **75** were directly observed, the nature of the adducts (**74** and **76**) leaves little doubt that the intermediates were actually produced. The extent of the rearrangement **73** → **75** increased with increasing temperature, and the adduct **74** did *not* rearrange to **76** under these conditions.[36]

5.4. CYCLOALKYNES AND CYCLIC ALLENES

Cyclic alkynes and allenes with 10 or more ring members may assume relatively unstrained conformations with normal bond lengths and bond angles, but already cyclononyne **(77)**, a stable compound, shows a 20° deviation from linearity at the terminal acetylenic bonds as determined by electron diffraction measurements.[37] Also cyclooctyne[38] **(78)** and even cycloocta-1,5-diyne[39] **(79)** are isolable compounds. The bond angles indicated in the formulas were determined by X-ray crystallography or electron diffraction.[40]

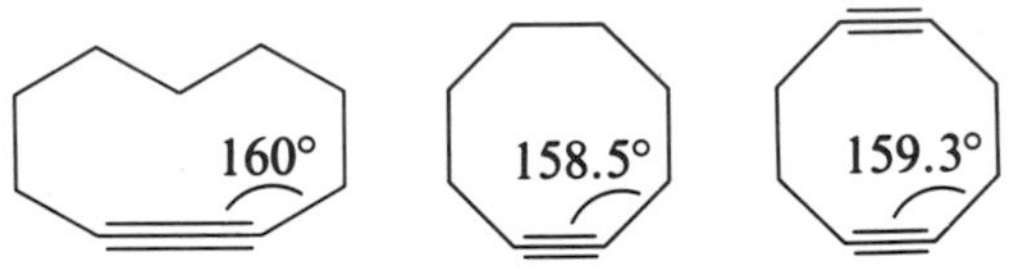

160° 158.5° 159.3°

77 **78** **79**

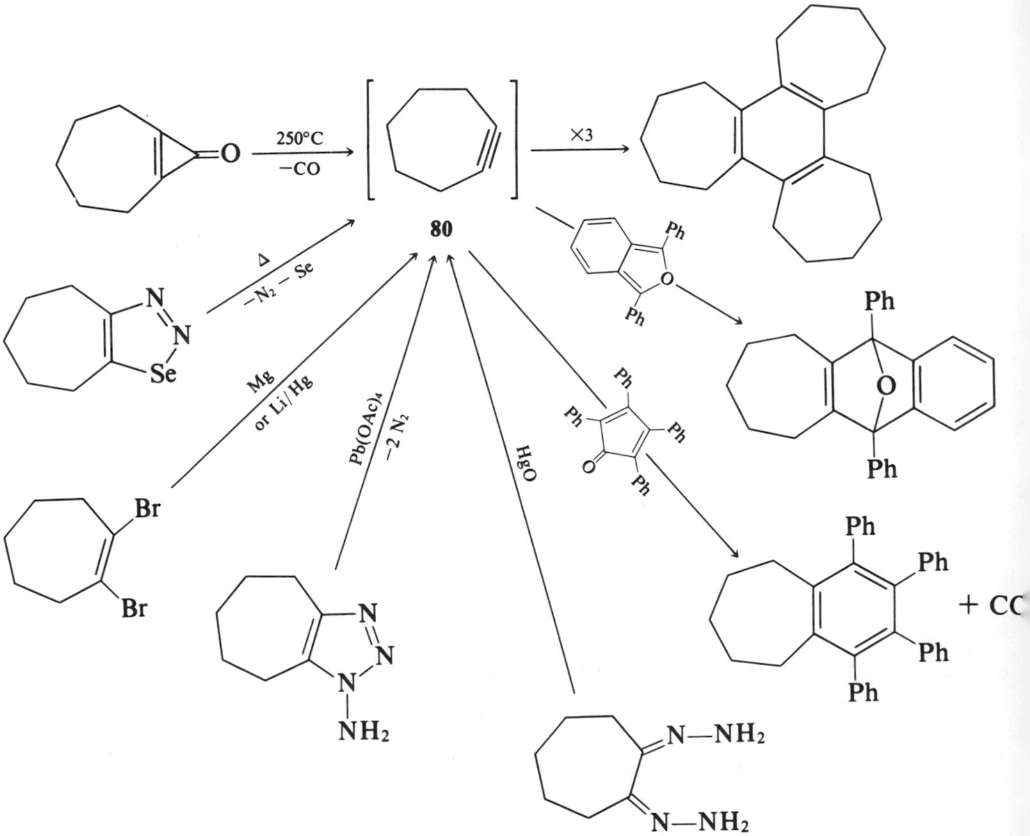

Scheme 1

Cycloheptyne **(80)** is known only as a reactive intermediate that can be generated in several ways and trapped with reactive dienes. In the absence of a trapping agent, trimerization occurs (Scheme 1).[41,42] The strain energy calculated[43] for **80** is 31 kcal/mol (for **78**, 21 kcal/mol; for **77**, 16 kcal/mol). In spite of this significant amount of strain, relatively small modifications suffice to stabilize a cycloheptyne. Thus the sterically shielding methyl groups make **81** and **82** isolable compounds, and the silacycloheptyne **83** is stable in solution at 4°C.[44] The increased stability

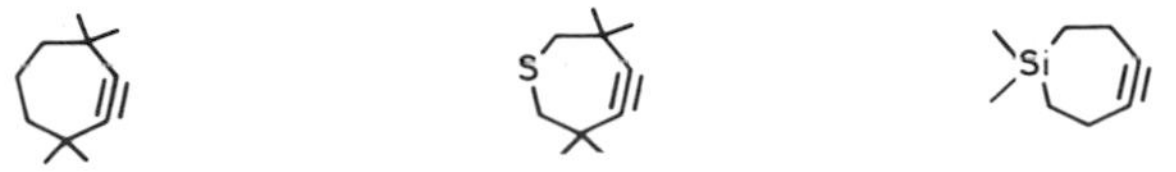

81 **82** **83**

of **83** vis-à-vis **80** may be ascribed to the longer Si—C bonds (~1.81 Å; C—C ~ 1.54 Å), which will result in smaller angle deformations in **83.**

Cyclohexyne **(84)** can be generated in reactions analogous to those shown in Scheme 1.[41,42] Two examples are given in Scheme 2. **84** is trapped by diphenylisobenzofuran **(85)** in yields up to 50%, and by tetracyclone **(86)** in up to 88% yield. It is also trapped by tris(triphenylphosphine)platinum(0) to give a stable, square planar platinum(0) complex **(87).**[45] In the absence of trapping agents, cyclohexyne trimerizes to dodecahydrotriphenylene **(88)**; some tetramer is also formed.[42,46]

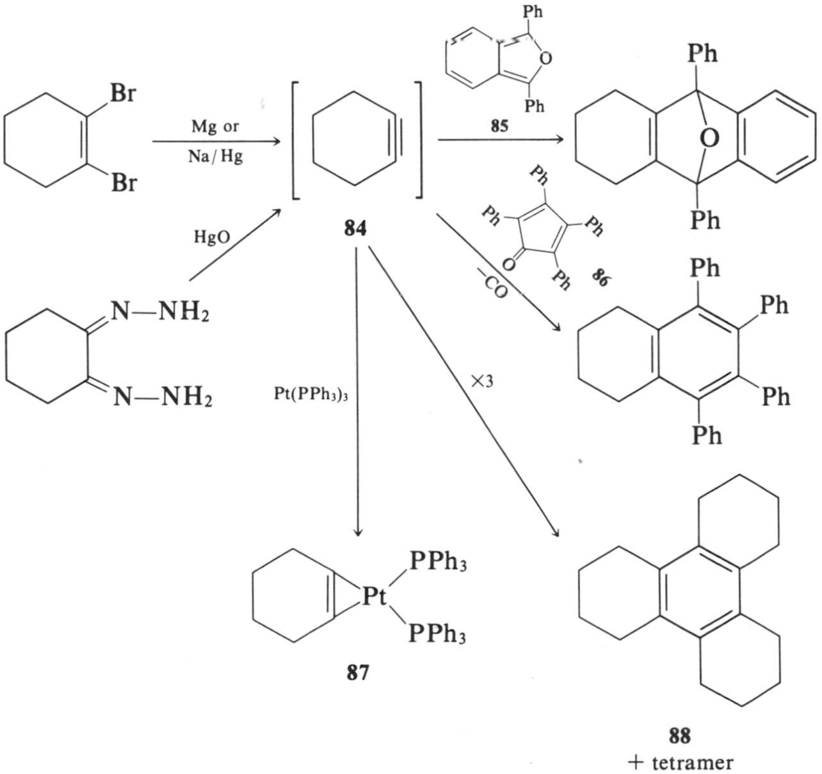

Br Br Mg or Na/Hg HgO N—NH₂ N—NH₂ **84** Ph O Ph **85** Ph Ph Ph Ph O **86** −CO Pt(PPh₃)₃ ×3 PPh₃ Pt PPh₃ **87** **88** + tetramer

Scheme 2

The vinylidene–acetylene rearrangement (see Section 4.4.5) may also be used to generate cycloalkynes. Thus **89** reacts with potassium *tert*-butoxide in the presence of diphenylisobenzofuran to give the adduct **(91)** in 35% yield, presumably via the vinylidene **90** and cyclohexyne:[47]

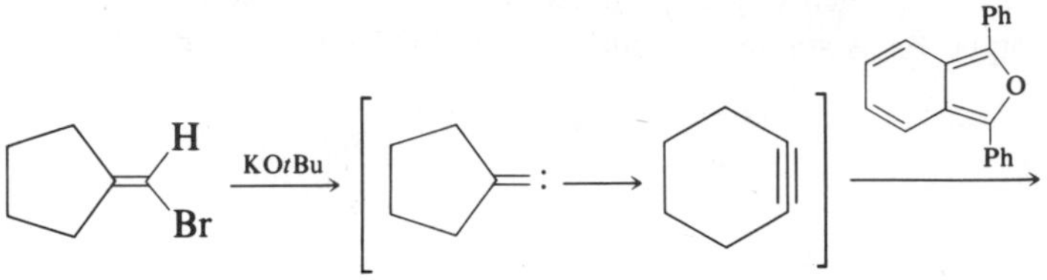

H Br KO*t*Bu Ph O Ph

89 **90** **84**

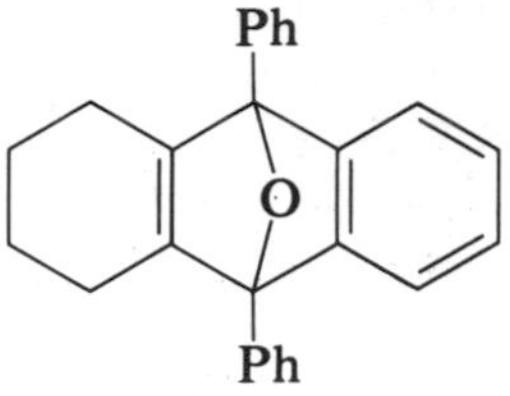

Ph O Ph

91

Vinylidenes are also generated by gas-phase pyrolysis of derivatives of Meldrum's acid (see Section 4.4.5). Hence the observation that cyclohexa-1,3-diene and benzene were formed in a combined yield of 92% in the pyrolysis of **92** appeared to be good evidence for the intermediacy of cyclohexyne:[48]

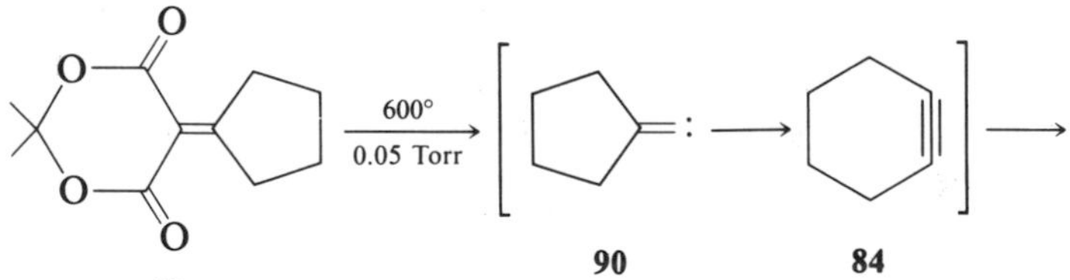

O O O O 600° 0.05 Torr

92 **90** **84**

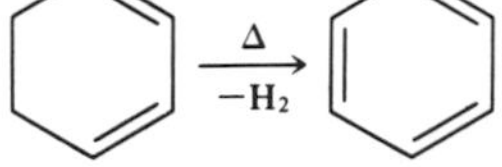

Δ $-H_2$

Cyclohexa-1,3-diene was shown to dehydrogenate to benzene under the conditions of the pyrolysis reaction.[48]

However, the chemistry is more complicated. The initial product of pyrolysis of **92** is the methyleneketene **93**, which can be observed by IR spectroscopy at temperatures below −100°C. Above this temperature, dimerization to **94** takes place. On further pyrolysis, **93** undergoes a 1,3 hydrogen shift giving the vinylketene **95. 95** can be observed by IR spectroscopy and trapped with methanol to give **96**:[49]

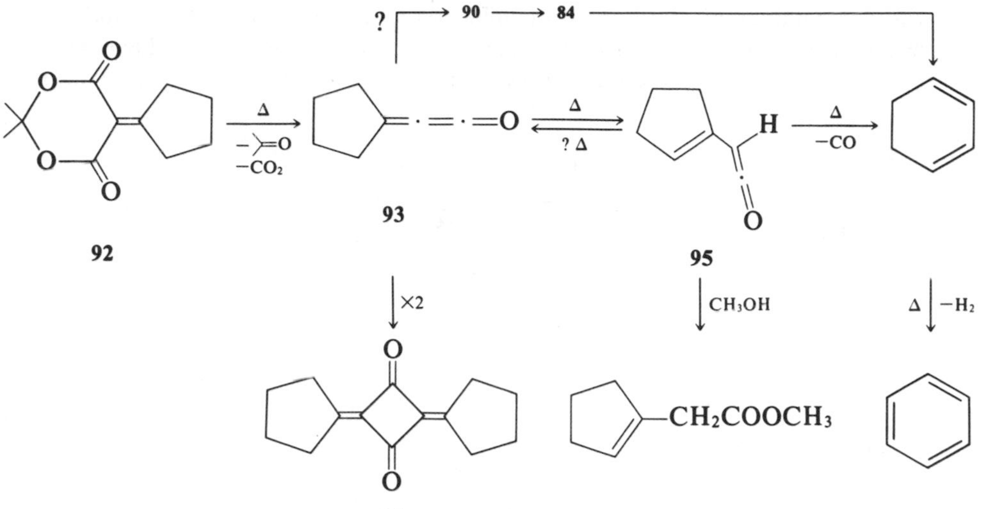

Scheme 3

Therefore, the final products, cyclohexa-1,3-diene and benzene, are logically derived from the decarbonylation of the ketene **95.** This experiment does not tell whether **95** decarbonylates directly to cyclopentenylcarbene or if it isomerizes back to **93,** which then decarbonylates to **90** and hence to cyclohexyne (Scheme 3).

On the other hand, the cycloheptylidene analog (**97**) does give cyclooctyne (**78**) in an isolated yield of 52%,[48] and here, too, the ketene **98** is detectable by IR spectroscopy.[49] Regardless of the mechanism, it thus appears certain that cycloalkynes are formed in these reactions.

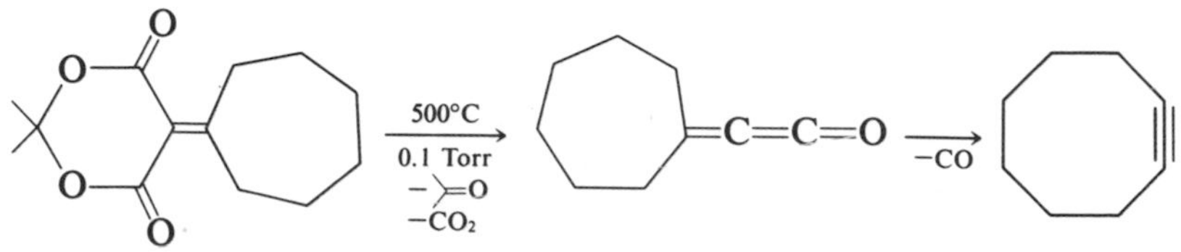

It should be noted that, although the C≡C stretching vibration in symmetrical disubstituted linear acetylenes is inactive in the infrared, the lower symmetry of the cyclic acetylenes permits the observation of weak IR bands for this vibration. For example, cyclooctyne (**78**) shows two weak absorptions at 2260 and 2206 cm^{-1}.[50]

Cyclohexyne may further be generated in solution in base-induced elimination reactions of halogenocyclohexenes. Proof of its formation was obtained in a historically important ^{14}C-labeling study of the reaction of 1,1-dichlorocyclohexane with phenyllithium.[51] The first HCl elimination gives a mixture of two chlorides, **99** and **100**. The second HCl elimination gives two cyclohexynes which, after renewed addition of C_6H_5Li and hydrolysis lead to four distinct phenylcyclohexenes

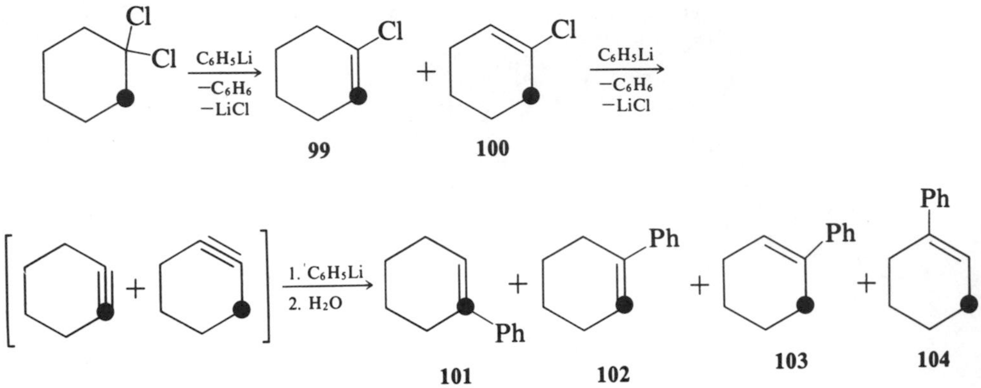

Cl Cl C_6H_5Li $-C_6H_6$ $-LiCl$ Cl + Cl C_6H_5Li $-C_6H_6$ $-LiCl$

99 100

1. C_6H_5Li 2. H_2O Ph Ph Ph Ph

101 102 103 104

● = ^{14}C

(**101–104**). If the cyclohexynes are in fact intermediates, the four phenylcyclohexenes must be formed in 25% (relative) yield each. The experimental result, 23% **101**, is in excellent agreement with this prediction.

In elimination reactions of this kind, one must be aware that also 1,2-dienes may be formed. Whereas **99** and **100** appear to give cyclohexyne almost exclusively with phenyllithium, the analogous reaction with potassium *tert*-butoxide gives a mixture of the acetylene (**84**) and the allene (**105**) (Scheme 4). Although both

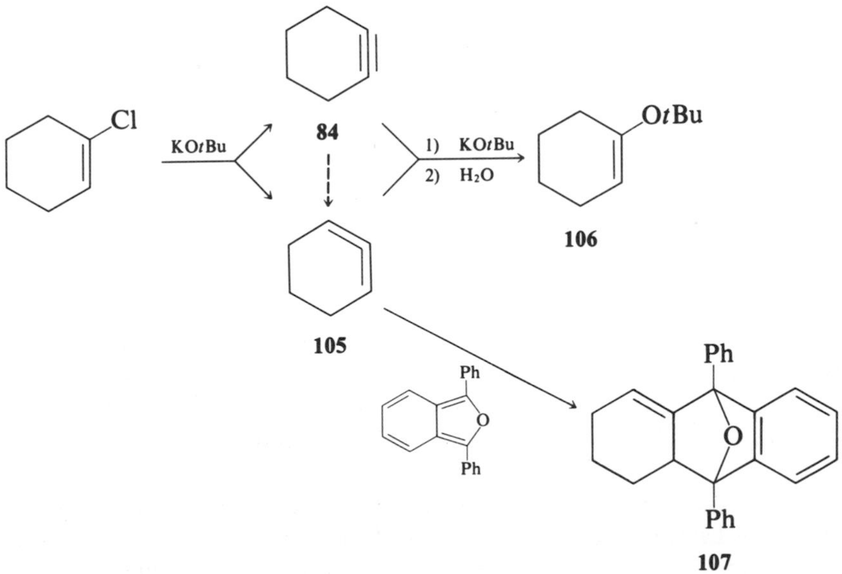

Cl KO*t*Bu 84 1) KO*t*Bu 2) H_2O O*t*Bu 106

105 Ph O Ph Ph O Ph 107

Scheme 4

intermediates react with KO-*t*-Bu to give 1-*tert*-butoxycyclohexene **(106)**, the allene **(105)** is trapped by diphenylisobenzofuran to give a distinct adduct **(107)**.[52] Similarly, 1-chlorocycloheptene reacts with KO-*t*-Bu or sodium pyrolidide to give a mixture (ca. 2:1) of cycloheptyne and cyclohepta-1,2-diene.[53] It is likely that at least in some of these reactions the cycloalkyne is formed initially and undergoes a subsequent base-induced isomerization to the allene (e.g., **84** → **105**).[41]

Since allenes are also formed by ring opening of cyclopropylidenes, the bicyclic carbene **109** is expected to isomerize to cyclohexa-1,2-diene **(110)**. Indeed, the

CH_3Li, $-CH_3Br$, $-LiBr$

108 109 110 111 112

generation of **109** (or a corresponding carbenoid) from the dibromo compound **108** led to the isolation of dimers and tetramers of cyclohexa-1,2-diene, the major product being the dimer **111.**[54] Cyclohexa-1,2-diene **(110)** has been directly observed by its IR spectrum: the flash pyrolysis of the acid chloride **113** with condensation of the products at $-196°C$ gives a sharp absorption at 1886 cm^{-1}, ascribed to **110.**[55] Again, this is about 100 cm^{-1} lower than normal for allenes. A similar

COCl Δ −CO

113 **114** **109** **110**

1886 cm^{-1}

< −100 °C

111

situation was described for cycloheptatetraene and its aza analogs (see Section 4.4.5). Support for the reaction sequence **113** → **114** → **109** → **110** is found in the analogous formation of allene on pyrolysis of cyclopropanecarboxylic acid chloride.[55,56] The IR absorption ascribed to **110** disappeared below $-100°C$ and, at the same time, absorptions due to the dimers appeared. **111** was isolated after warming to room temperature.[55]

The observation of an allene band for **110** in the IR demonstrates that the stable form of the compound is the 1,2-diene shown, and not a diradical or zwitterion **(115)**.

∗ = · or +, −

115

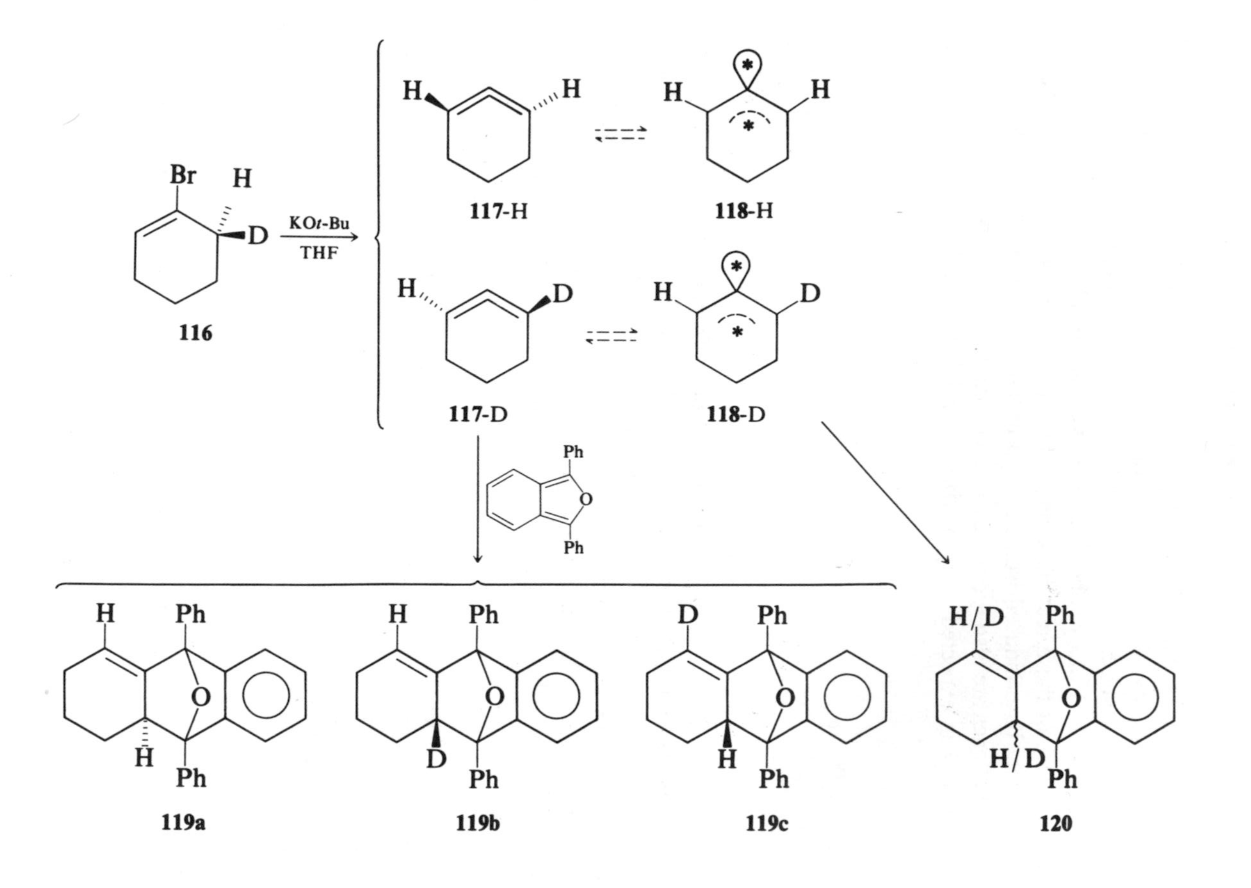
Br
H
D
116
KOt-Bu
THF
H
H
117-H
H
H
118-H
H
D
117-D
H
D
118-D
Ph
O
Ph
H
Ph
H
Ph
119a
H
Ph
D
Ph
119b
D
Ph
H
Ph
119c
H/D
Ph
H/D
Ph
120

This has also been demonstrated in an elegant study of the base-induced elimination of HBr from optically active 1-bromo-6-deuteriocyclohexene (**116**), the idea being that HBr and DBr elimination will give different compounds, belonging to different enantiomeric series (**117**-*H* and **117**-*D*) if the chiral allene is the intermediate. The diradical or zwitterion **118** would be achiral. Since a kinetic isotope effect is expected,[41] the two isotopomers **117**-*H* and **117**-*D* will also form with different rates. Accordingly, the optically active mixture of **117**-*H* and **117**-*D* will give an optically active product mixture, **119 a–c**, with diphenylisobenzofuran, whereas **118** will give inactive **120.** The experiment demonstrated that the optically active allene **117** is, in fact, formed and trapped, but it loses its optical activity when generated at 80°C, indicating that at this temperature an interconversion with the diradical or zwitterion **118** takes place.[57]

The chemistry of cyclopentyne (**124**) will not be treated in detail here, but the evidence for the existence of the compound is similar to that described for cyclohexyne earlier.[41,42,48,58] Cyclopentyne has not been directly observed, but the matrix photolysis of the bis-diazoketone **121** gave the allene **125**, presumably via intermediate cyclopentyne formation:[59]

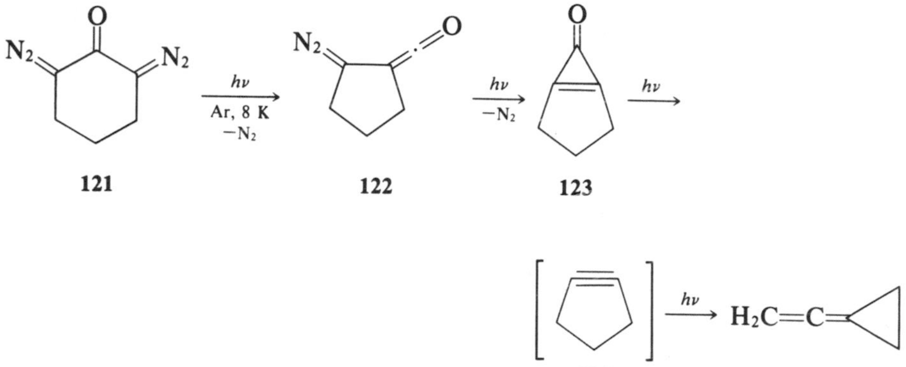

The intermediates **122** and **123** were monitored by IR spectroscopy.

An annelated cyclopentyne, acenaphthyne (**127**) was, however, directly observed by both IR and UV spectroscopy in the analogous matrix photolysis of **126**. **127** reacted with oxygen in the matrix to give **128**, and on warming in the absence of oxygen, the trimer **129** was formed.[59] The formation of a diketone (**128**) in the reaction between (triplet) oxygen and a strained alkyne is not unexpected: the isolable thiacycloheptyne **82** reacts with molecular oxygen to give the dione **132** in a chemiluminescent reaction.[60] Chemiluminescence is typical for dioxetes such as **130**, which undergoes ring opening to an electronically excited state of the dione (**131**). The latter returns to the ground state (**132**) with emission of light.

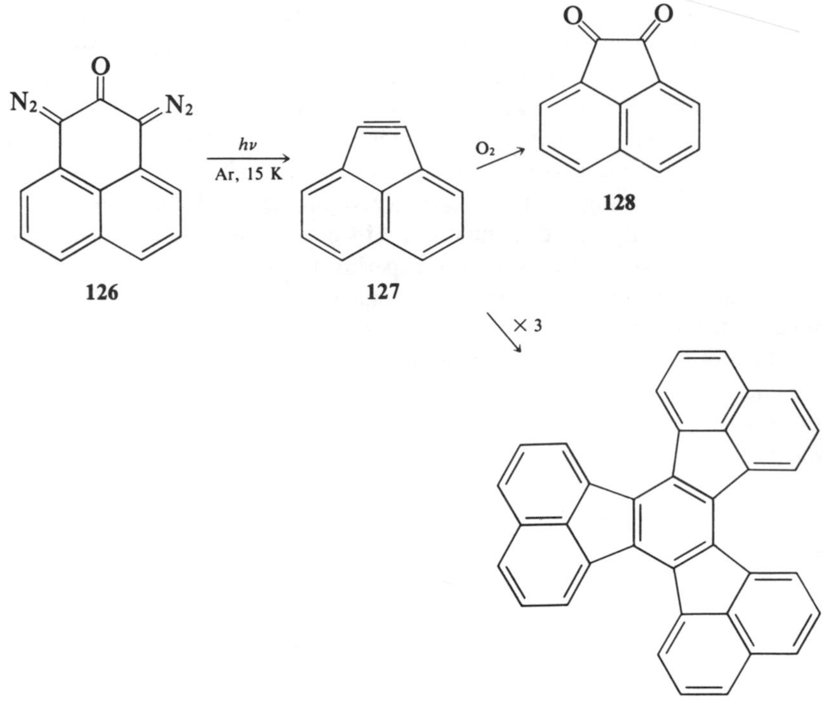

126 $\xrightarrow[\text{Ar, 15 K}]{h\nu}$ 127 $\xrightarrow{O_2}$ 128

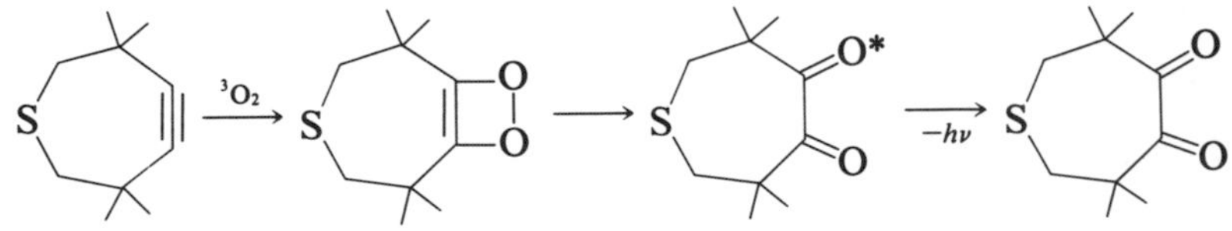

127 $\xrightarrow{\times 3}$ 129

82 $\xrightarrow{^3O_2}$ 130 $\longrightarrow$ 131 $\xrightarrow{-h\nu}$ 132

5.5 BENZYNES

There are two useful books[41,61] dealing with the chemistry of *o*-benzyne (or just benzyne) **(133)** and its derivates. Some of the preparative routes to this intermediate are indicated in Scheme 5. The first rigorous proof for the existence of benzyne was given by the ^{14}C labeling experiment indicated below:[62]

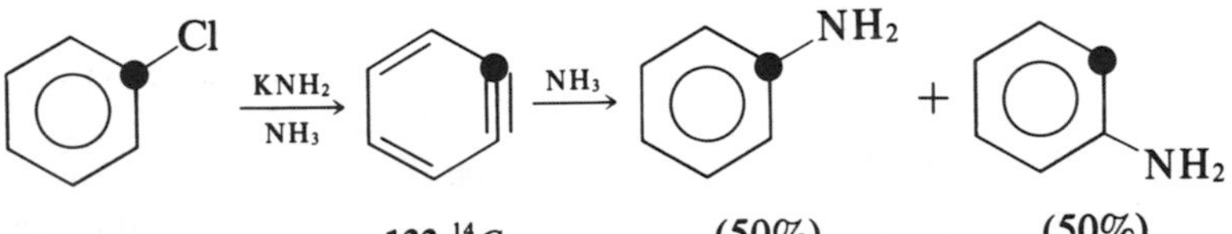

$\xrightarrow[NH_3]{KNH_2}$ **133-**^{14}C $\xrightarrow{NH_3}$ (50%) + (50%)

The observation that equal amounts of the two labeled anilines were formed demonstrated that a symmetrical intermediate, that is, benzyne, had been present. More

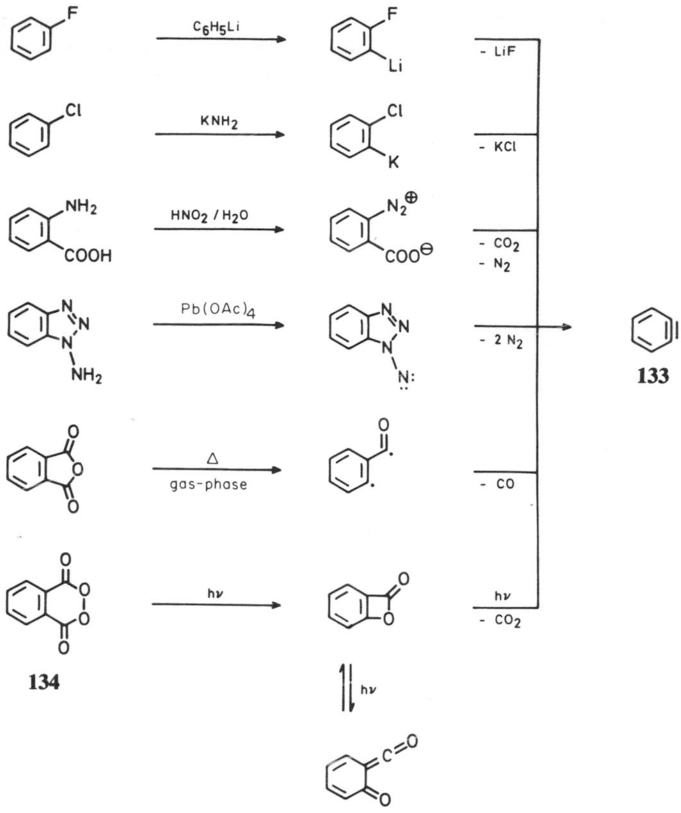

Scheme 5

recently, benzyne has been directly observed in the matrix photolysis of phthaloyl peroxide **(134)** at 8 K (see Scheme 5).[63] A related matrix photolysis (Scheme 6) allowed the preparation of benzyne in a higher concentration so that the weak C≡C stretching vibration at 2085 cm^{-1} could be observed in the IR spectrum.[64]

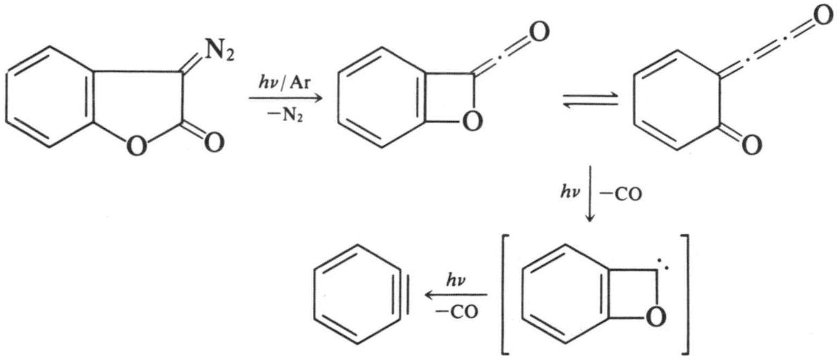

Scheme 6

One of the most typical and synthetically useful reactions of benzyne is the concerted Diels–Alder reaction with dienes, for example, furan. When benzyne is generated in an argon matrix in the presence of furan, the cycloaddition reaction takes place upon warming to 50 K:[63]

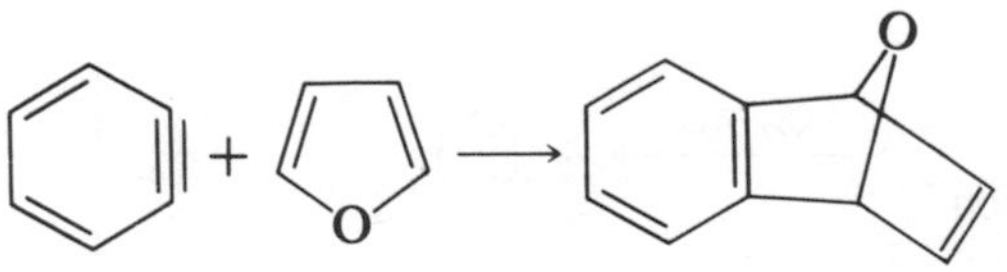

When benzyne is generated in the gas phase in the absence of other reaction partners, it dimerizes to diphenylene; smaller amounts of triphenylene may also be formed:[41]

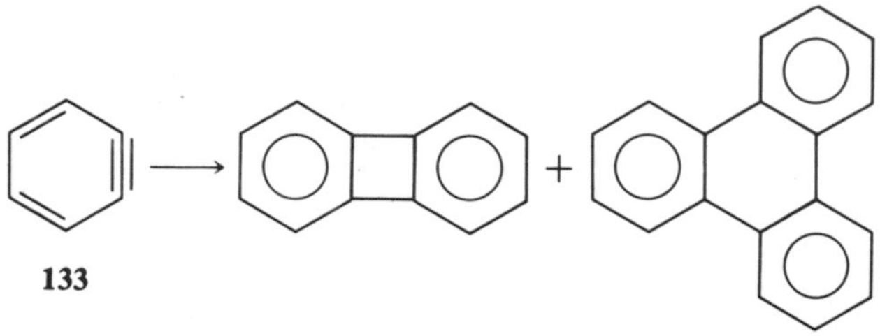

133

As presented above, the benzyne molecule is a singlet, acetylenic compound. The formula **133** implies that the symmetric combination (S) of the two in-plane orbitals forming the strained π bond be of lower energy than the corresponding antisymmetric (A) combination. That is, S is HOMO; A is LUMO. The opposite arrangement (i.e., A as HOMO, S as LUMO) would lead to a diradical ground-state molecule that could exist as either a singlet **(135)** or a triplet **(136)**.

S **133**

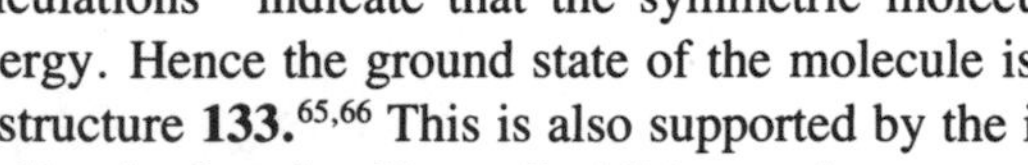

or

A **135** **136**

Theoretical calculations[65] indicate that the symmetric molecular orbital is, in fact, of lowest energy. Hence the ground state of the molecule is best represented by the acetylenic structure **133.**[65,66] This is also supported by the infrared spectrum quoted earlier[64] and by the fact that [2 + 4] additions to benzyne are stereospecific, whereas [2 + 2] additions are not. The latter are thermally "forbidden" and presumably take place via 1,4-diradicals **(137)**:[67]

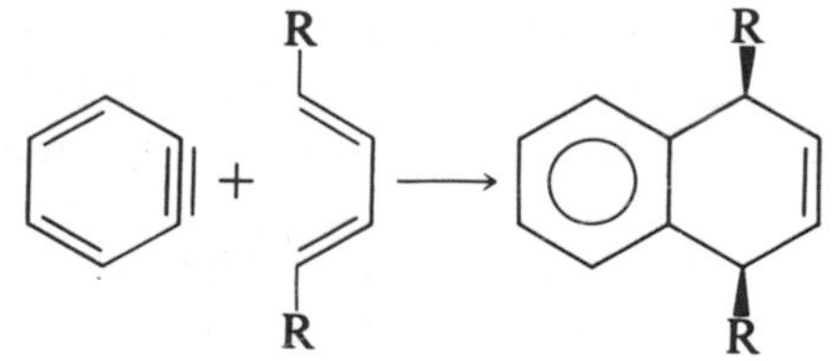

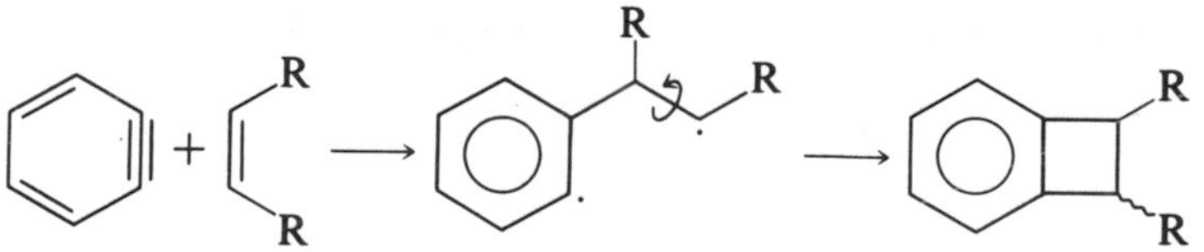

137

The enthalpy of formation of benzyne has been determined experimentally as approximately 120 kcal/mol.[68] This allows an approximate evaluation of the strain energy. The difference between the enthalpies of formation of an acetylene and an alkene is normally about +37 kcal/mol. With an enthalpy of formation of 20 kcal/mol for benzene, a strain-free benzyne would thus have $\Delta H_f^0 \sim 57$ kcal/mol. Accordingly, the strain energy is $\sim 120 - 57 \cong 63$ kcal/mol. The high enthalpy of formation may be held responsible for the occurrence of formally forbidden [2 + 2] cycloaddition reactions.

The removal of two hydrogen atoms from the *meta* and *para* positions of benzene generates *m*- and *p*-benzyne (**138** and **140**), respectively. Although no triple bond is present in these compounds, the benzyne name is retained in analogy with

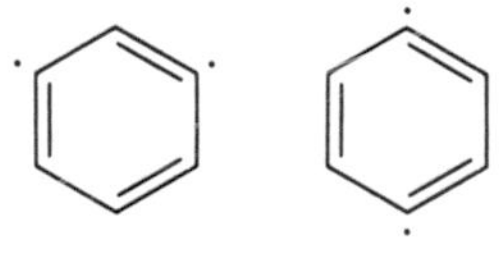

138 **140**

139 **141**

o-benzyne. However, two isomeric forms must also be considered: bicyclo[3.1.0]hexa-1,3,5-triene (**139**) and bicyclo[2.2.0]hexa-1,3,5-triene (butalene) (**141**).

Theoretical calculations[66] indicate the energy ordering *ortho*- < *meta*- < *para*-benzyne. There is no clear decision as to the relative energies of **138** and **139**, but the "open" diradical form of *p*-benzyne (**140**) appears to be considerably lower in energy than butalene (**141**).[66,69] Butalene is destabilized not only by ring strain, but also by antiaromaticity, being a cyclobutadiene derivative (cf. Chapter 6). The "closed" *m*-benzyne **139** is also strained, but it retains a 6 π electron system and should, therefore, benefit from an aromatic resonance stabilization.

There is experimental evidence for the existence of each of these four species, **138–141**, as distinct intermediates. Flash photolysis of the diazonium salt **142** formed from *m*-aminobenzoic acid gave a transient intermediate that was monitored by UV and mass spectrometry.[70] In particular, the mass spectrum demonstrated the presence of a m/e 76 species (i.e., C_6H_4). This experiment does not give any definitive information as to the structure of the C_6H_4 intermediate.

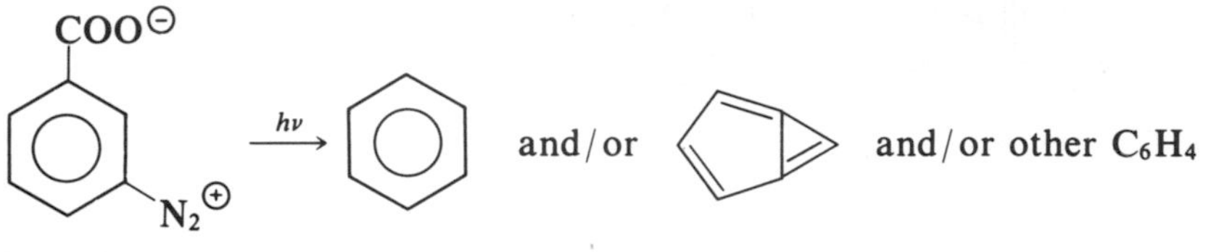

$COO^{\ominus}$ $N_2^{\oplus}$ $\xrightarrow{h\nu}$ and/or and/or other C_6H_4

142

The discrete bicyclohexatriene structure **(139)** has been generated by base-catalyzed HBr elimination from the dibromo compound **143**.[71] With the aid of deuterium, ^{13}C, and alkyl group labeling, it was shown that only **139** (and not **138**)

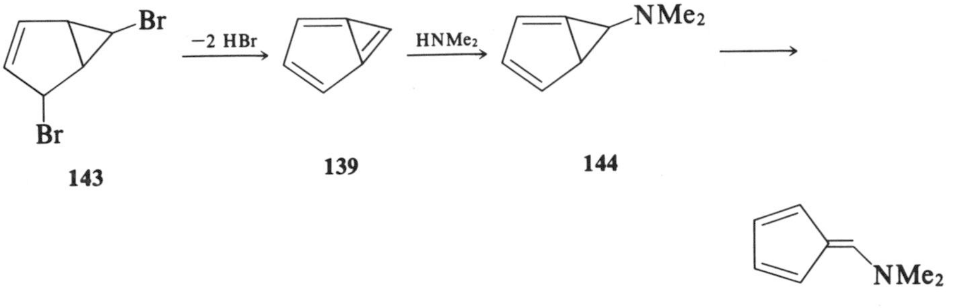

Br Br $\xrightarrow{-2\ HBr}$ $\xrightarrow{HNMe_2}$ NMe_2 $\longrightarrow$

143 **139** **144**

NMe_2

145

can be an intermediate in this reaction. **139** is trapped by dimethylamine to give **144**, which then undergoes ring opening to give the fulvene **145** as the final product.

In contrast, the benzologue **146** undergoes base-induced elimination to give substituted naphthalenes **(148–150)** as products, thereby implicating the *m*-benzyne **147** as an intermediate.[72] Although more work in this area is needed, it would appear that derivatives of **138** and **139** are capable of independent existence.

Evidence for the existence of *p*-benzyne **(140)** is provided by the observation of specific thermal deuterium scrambling (**151** $\rightleftharpoons$ **153**; E_a = 32 kcal/mol) in the diyne **151**.[73] The dideuterio-*p*-benzyne **(152)** behaves as a diradical, undergoing hydrogen and chlorine atom abstractions (Scheme 7). Further evidence for the diradical character is found in the reactions of the dipropyl analog **155**.[74]

Formed by cyclization of the diyne **154**, **155** reopens to **156** in competition with

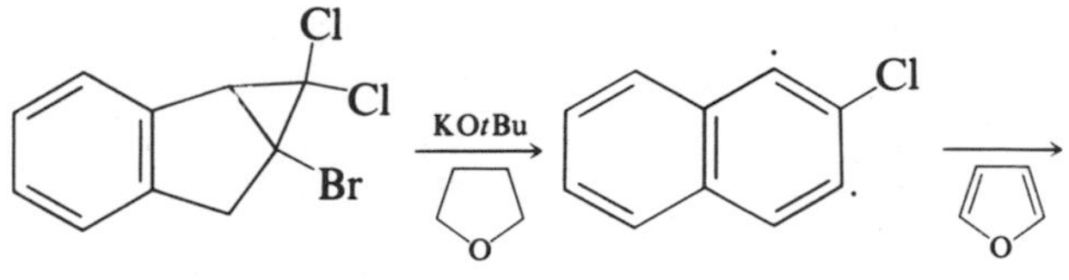

Cl Cl Br KO*t*Bu Cl

146 147

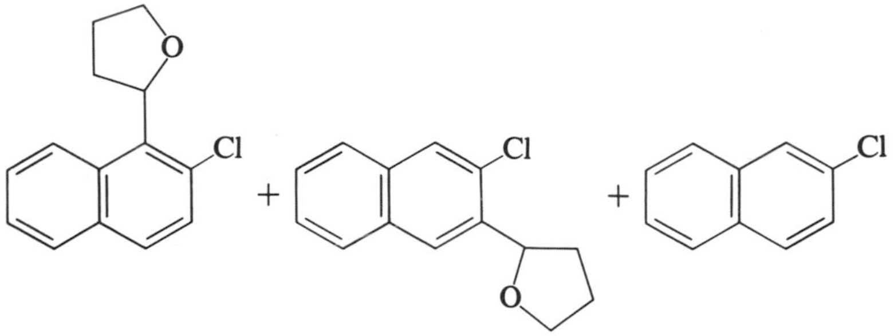

O Cl Cl O Cl

148 149 150

a [1,5]-hydrogen shift giving the new diradical **157** (Scheme 8). Repetition of the latter process gives **158**. **157** and **158** abstract hydrogen from an added hydrogen donor, cyclohexa-1,4-diene, to give dipropylbenzene (**159**). In the absence of the hydrogen donor, **158** cyclizes to **160** or disproportionates intramolecularly to **161**. The kinetics of the reaction implicated the *p*-benzyne **155** as an intermediate, and CIDNP studies demonstrated that the products are formed from *singlet* precursors.[75]

It will be recalled from Chapter 2 that 1,5-shifts in free radicals occur more

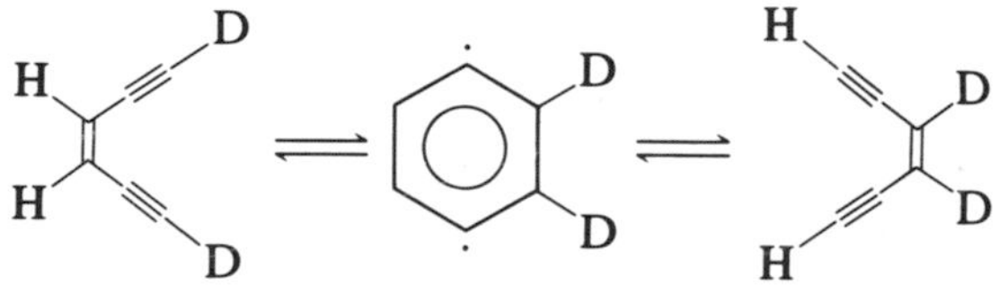

H D H D D D H D H D D D

151 152 153

($\Delta H_f^0 \cong 126$ kcal/mol)

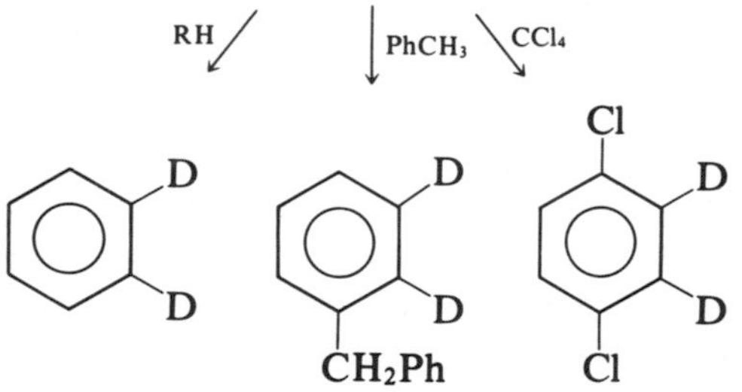

RH PhCH$_3$ CCl$_4$

D D D D Cl D D

CH$_2$Ph Cl

Scheme 7

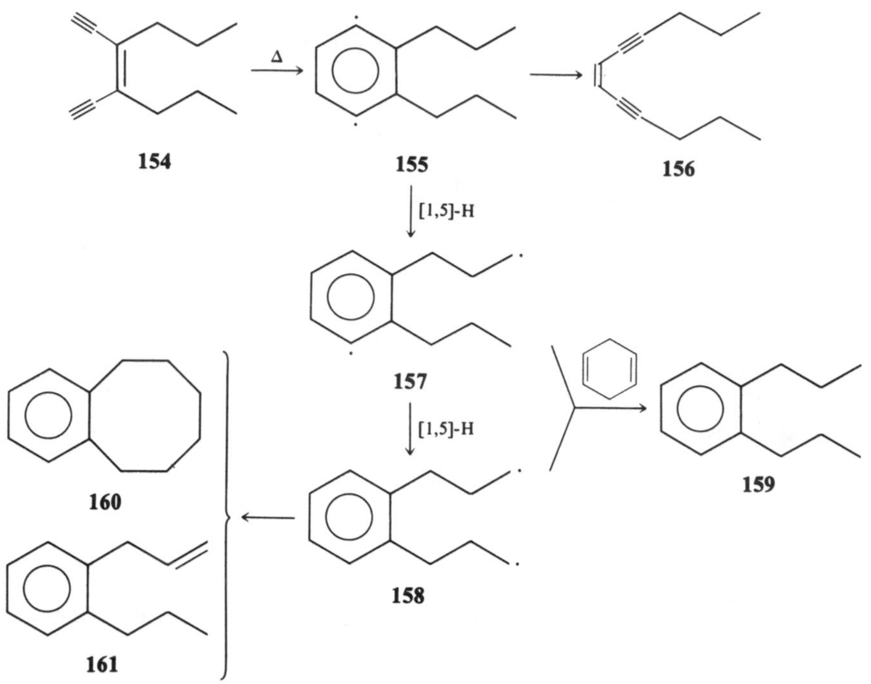

Δ

154 155 156

[1,5]-H

157

[1,5]-H

158

159

160

161

Scheme 8

readily than do 1,4-shifts. In agreement with this, the similarly generated diethyl-*p*-benzyne **162** shows little tendency to undergo hydrogen shifts:[74]

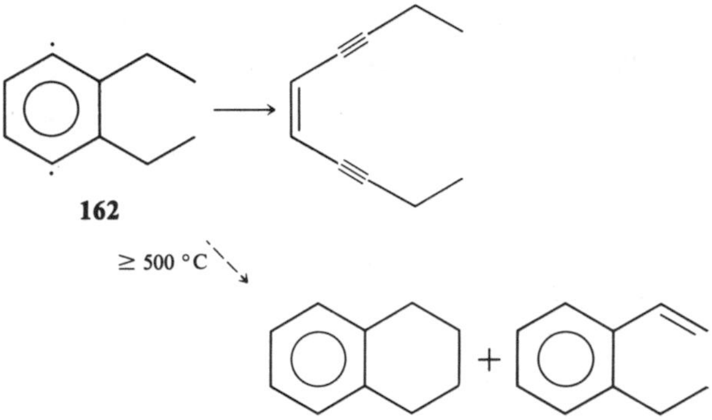

162

≥ 500 °C

+

The chemistry just summarized clearly demonstrates the existence and diradical nature of *p*-benzyne. This species was also generated by flash photolysis of *p*-diazoniobenzenecarboxylate (**163**) and monitored by UV and mass spectrometry.[76] A surprisingly long-lived species, persisting for up to 2 min in the gas-phase, was observed. Again, this does not give any direct information as to the structure, but perhaps the longevity was due to interconversion with the diyne **164**.

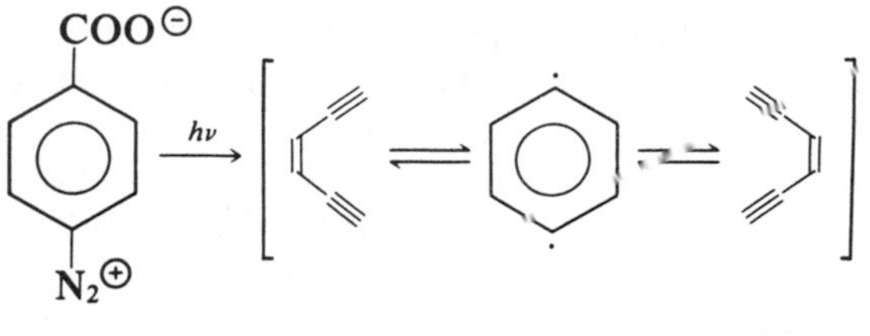

163 164 140 164

The chemical approach to butalene (**141**) involves the elimination of HCl from the preformed bicyclohexadiene **165**.[77] Butalene is trapped in the presence of dimethylamine, but the Dewar benzene **166** so formed is unstable and isomerizes to

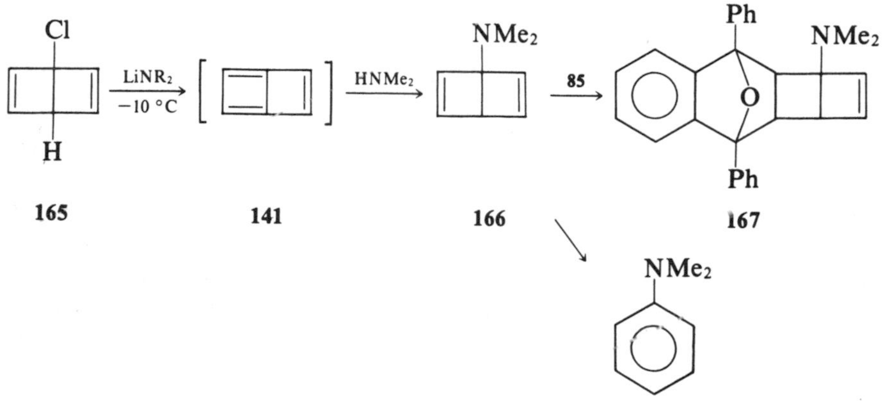

165 141 166 167

168

dimethylaniline (**168**). However, the adduct **167** can be isolated in modest yield when the reaction is run in the presence of diphenylisobenzofuran (**85**).

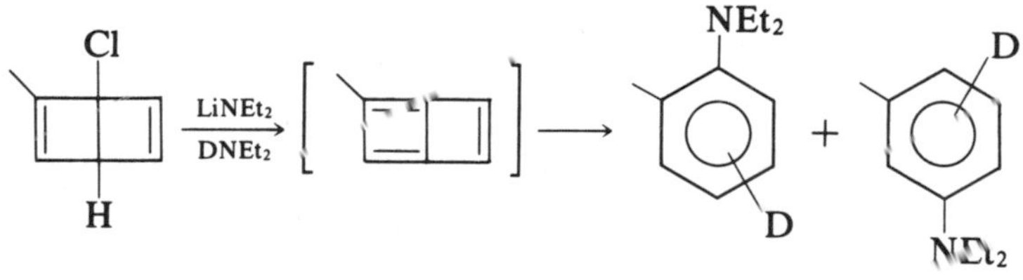

169 170

	171	172
d_0	53%	21%
d_1	31%	42%
d_2	14%	27%
d_3	2%	9%

The substituted butalene **169** can give rise to two isomeric anilines **171** and **172** by addition of diethylamine accross the central bond. The results obtained using *N*-deuteriodiethylamine show, however, that the reaction is more complicated. Some hydrogen/deuterium exchange takes place, and the large amount of undeuterated product indicates the occurrence of a direct substitution of the starting material **169** that bypasses the butalene.[78]

Although many other arynes and heterocyclic analogs[79] are known, we shall limit the presentation to one further case, 1,8-dehydronaphthalene or 1,8-naphthyne **(174)**, which can be generated by lead tetraacetate oxidation of the aminotriazine **173**.[80] In this case, extended Hückel calculations[65] predict the antisymmetric MO

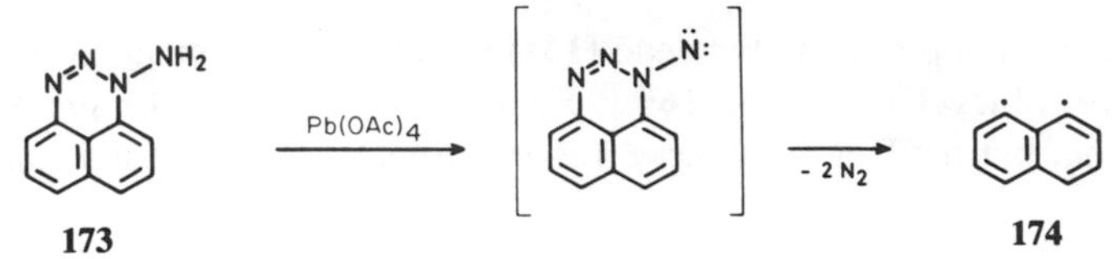

$Pb(OAc)_4$; $-2\,N_2$

173 **174**

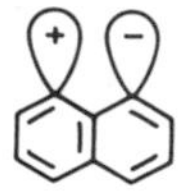

A

(A) to be the HOMO. Accordingly, the molecule should behave as a diradical. Most importantly, the thermal [2 + 2] cycloaddition with an olefin will be an allowed, concerted process. This is borne out by experiment: **174** adds stereospecifically to olefins and shows a marked preference for [2 + 2] addition even with dienes. The [2 + 4] cycloaddition observed for *o*-benzyne has now become forbidden.[61,80]

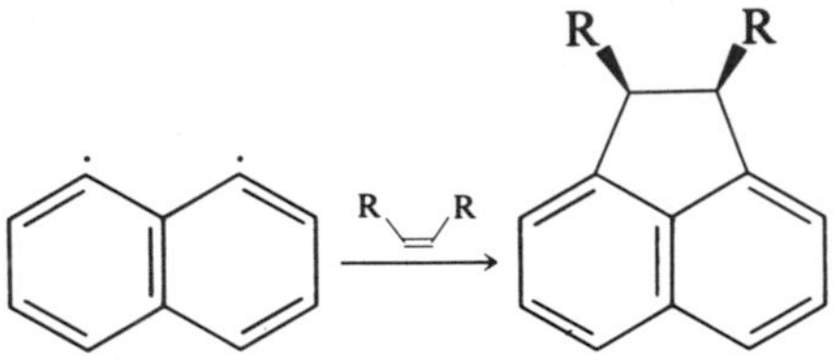

174 R = Cl; CH_3

5.6. STRAINED σ-BONDS

5.6.1 Propellanes

Propellanes[81] are tricyclic hydrocarbons in which the three rings are joined through an interior σ-bond. As an example, consider [2.2.2]propellane **(175)**. In this com-

175 **176**

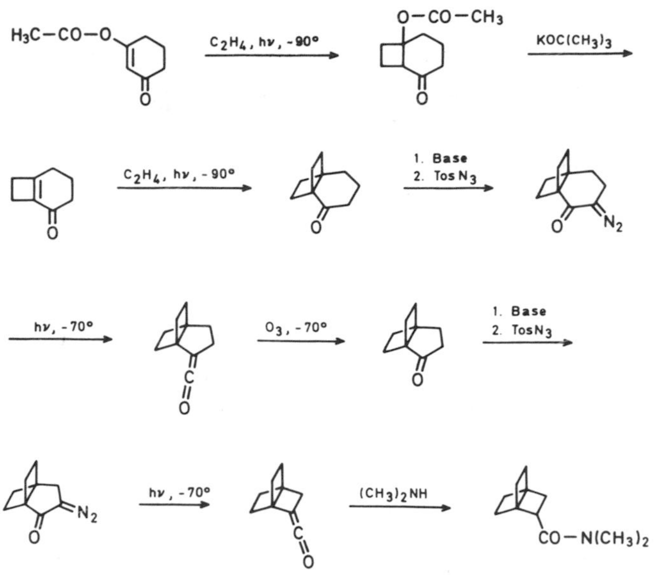

Scheme 9

pound, the interior C—C bond must be nearly perpendicular to the other bridgehead bonds, and the bridgehead carbon atoms cannot, therefore, be sp^3 hybridized. Rather, they will tend toward sp^2 hybridization, in which case the central bond will be a *p-p* σ-bond (see **176**), a highly unusual situation. The formula **176** suggests that such molecules might be highly susceptible to electrophilic attack at the bridgehead positions, and this, indeed, is observed.

A derivative of **175** has been synthesized in a series of photochemical [2 + 2] cycloadditions and Wolff rearrangements (Scheme 9),[82] and it is a remarkably stable compound that undergoes ring opening **(175 → 178)** with an activation energy of approximately 22 kcal/mol. **175** and its derivatives might well owe their existence to the fact that the direct ring opening to **178** is symmetry forbidden (a $[2_s + 2_s]$ cycloreversion). Hence this reaction probably occurs *via* the 1,4-diradical **177**.[83]

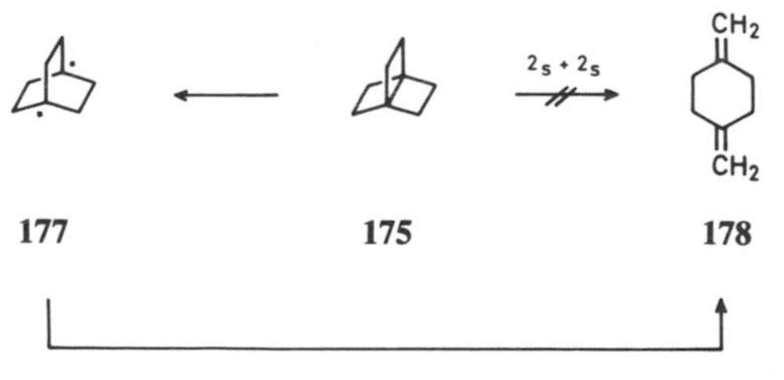

The [3.2.1]propellane **180** may be obtained by carbene addition to the olefin **179** or by reduction of the dibromide **181** with sodium in ethanol (a Wurtz reaction).[84]

CH_2N_2 / Cu_2Cl_2 ; Br_2 / Na/EtOH

Br Br

179 **180** **181**

$BrCCl_3$; O_2

Br ; O—O···

(polymer)

CCl_3 ; ···O—O

182 **183**

In agreement with the expectation that the inner bridgehead bond in **180** should have a high degree of *p* character, the chemistry is dominated by electrophilic addition reactions: **180** reacts with bromine to **181**, with bromotrichloromethane to **182**, and with oxygen to a polymeric peroxide **183**. In the absence of electrophiles, the half-life of **180** is about 20 h at 195°C. The [2.2.2]propellane **175** is much less stable thermally; the activation energy toward ring opening cited above (22 kcal/mol) corresponds to a half-life of approximately $2 \cdot 10^{-3}$ s at 195°C. Obviously, the stability of **180** is not due to a lack of strain (a stain energy of 67 kcal/mol has been measured[85]) but to a lack of energetically feasible reaction channels.[83,86]

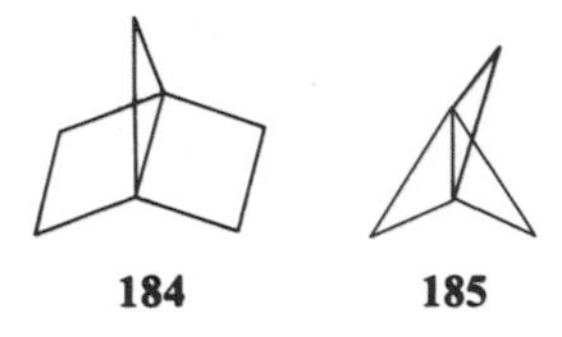

184 **185**

180 is an example of a compound showing ''inverted tetrahedral'' symmetry at the bridgehead carbons; that is, all four bridgehead bonds lie in one hemisphere. The distortion becomes worse still in the smaller propellanes, for example, [2.2.1]propellane **(184)** and [1.1.1]propellane **(185)**. **184** is a highly reactive compound which can only be observed in an argon matrix at low temperature, and it has been conservatively estimated that the strength of the inner bridgehead bond can be no more than 20 kcal/mol.[87] In contrast, **185** is an amazingly stable compound, having a half-life of about 5 min at 114°C.[88] This becomes even more remarkable when it is considered that the calculated strain energies of **184** and **185** are very similar: 108 and 103 kcal/mol, respectively.[89] An explanation[89] may be found in the fact that the ''opening'' of **184** to the relatively unstrained norbornane **186** entails a calculated strain relief of about 94 kcal/mol, whereas the corresponding

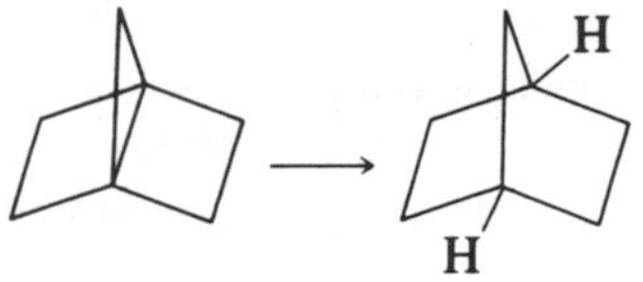

H

H

184 186

strain relief ~ 94 kcal/mol

H

H

185 187

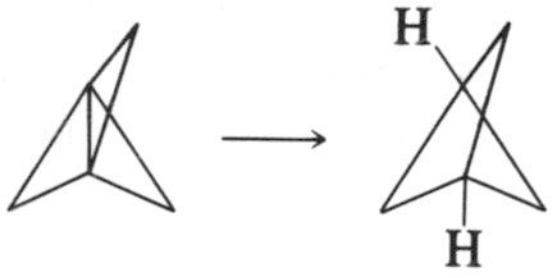

strain relief ~ 36 kcal/mol

reaction of **185** gives the still highly strained bicyclo[1.1.1]pentane **187** with a calculated strain relief of only 36 kcal/mol.

Further ab initio calculations[90] indicate that the central bond in **185** is longer, and the electron density lower, than for a C—C bond in cyclopropane. There is experimental evidence for a lack of electron density in the bridgehead bond in another propellane system: the [3.1.1]propellane derivative **188**. This compound, obtained in a Diels–Alder reaction between the strained olefin **73** and 1,2,3-trimethylisoindole, has been examined by X-ray crystallography and found to possess quite a normal bridgehead–bridgehead bond length of 1.574 Å. However, the

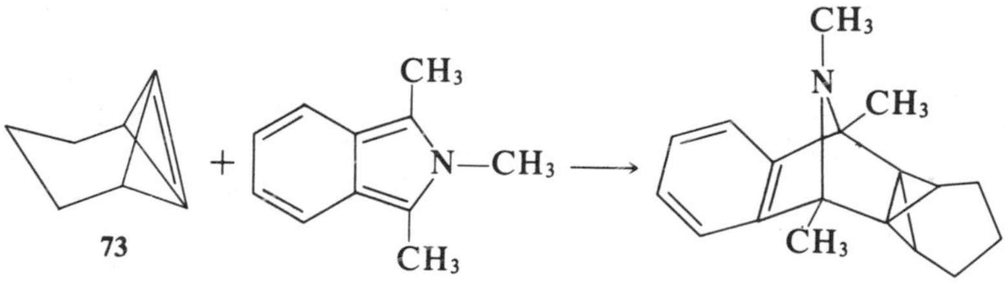

73 + (CH_3, N—CH_3, CH_3) ⟶ (CH_3, N, CH_3, CH_3)

188

electron density maps also obtained in this study failed to demonstrate any significant electron density between the bridgehead carbons.[91] Although the normal rules of valence require a bond, this bond apparently has no electron density! With stable compounds of this type now available, one may expect further studies to clarify the nature of this apparently novel type of "bonding" in carbon compounds.

5.6.2. Tetrahedranes

The beauty of the platonic solids is in itself sufficient reason for organic chemists to attempt the preparation of the corresponding hydrocarbons.[92] The smallest such hydrocarbon, tetrahedrane **(189)** presents a formidable synthetic obstacle. Theoretical calculations[93] predict a strain energy of about 134 kcal/mol, which is far in excess of the strength of an ordinary C—C bond (~83 kcal/mol). However, this strain energy is distributed over six C—C bonds, leaving "only" about 22 kcal/mol of strain in each. Thus it might be possible to observe tetrahedrane at very low temperatures, provided that no low-energy pathways for rearrangement are available.

Calculations indicate that the rearrangement of tetrahedrane to cyclobutadiene (**191**) is exothermic. The reaction is, however, "forbidden" by the rules of orbital symmetry and may, therefore, proceed via the diradical **190**.[93] The reaction is predicted to have an activation energy of 15–30 kcal/mol and to be the lowest energy pathway for rearrangement of tetrahedrane.

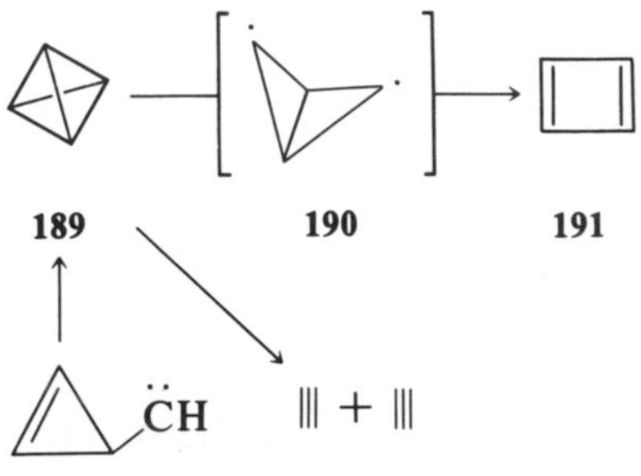

189 **190** **191**

ĊH

192

The cleavage to two molecules of acetylene is also exothermic (by ca. 26 kcal/mol), but this reaction is calculated to require a higher energy of activation. Ring opening of **189** to cyclopropenylcarbene (**192**) is, however, endothermic by as much as 35 kcal/mol according to the calculations,[93] thereby making the carbene a possible precursor of tetrahedrane.

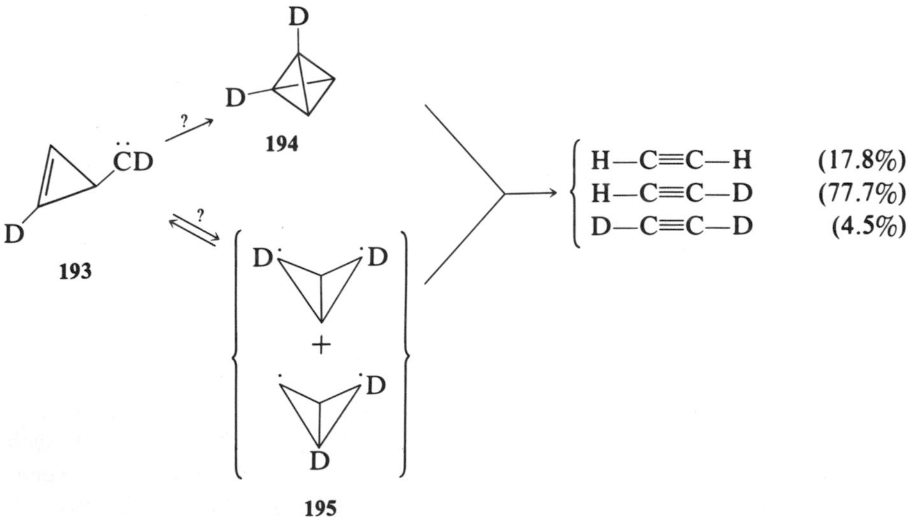

D D

194

ĊD

D

193

D D

\+

D

D

195

H—C≡C—H (17.8%)
H—C≡C—D (77.7%)
D—C≡C—D (4.5%)

In fact, early attempts to demonstrate the existence of tetrahedrane took advantage of the cleavage of carbene **192** into two molecules of acetylene. The generation of the deuterium-labeled carbene **193** led to the formation of singly and doubly deuterated and nondeuterated acetylene, thus implying that tetrahedrane **194** and/or the diradicals **195** are intermediates.[94]

These experiments do not really prove that tetrahedranes are formed. In related work with the ketene **196**, a precursor of the carbene **197**, no direct spectroscopic evidence for the formation of dimethyltetrahedrane (**198**) could be adduced, even at a temperature as low as 10 K. Thus it is entirely uncertain whether the tetrahedrane is a precursor of the end products **199** and **200**, although the corresponding cyclobutadiene **201** was excluded.[95]

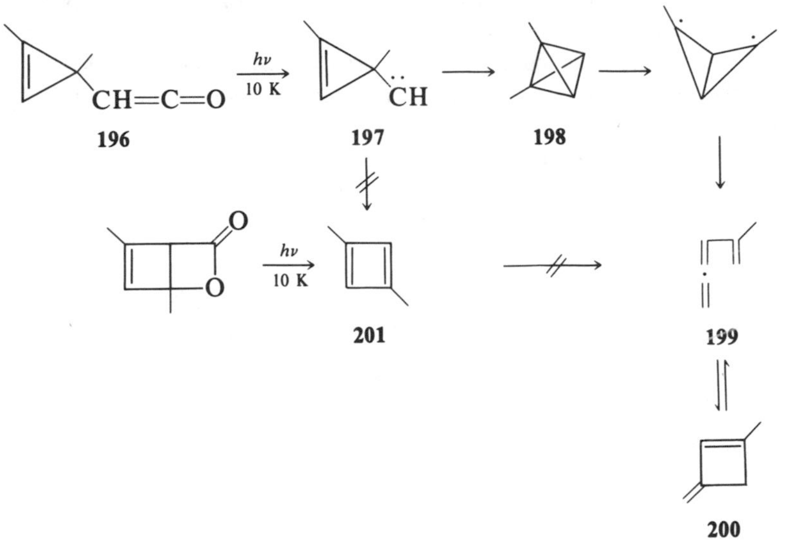

CH=C=O $\xrightarrow[10\ K]{h\nu}$ CH

196 **197** **198**

$\xrightarrow[10\ K]{h\nu}$

201 **199**

200

One might well conclude from the preceding discussion that there is little prospect of isolating a tetrahedrane at room temperature. Nevertheless, that is exactly what has been achieved.[96] Tetra-*tert*-butylcyclopentadienone (**202**), the synthesis of which is no easy task, is photochemically converted to the tricyclopentanone **203**. Further

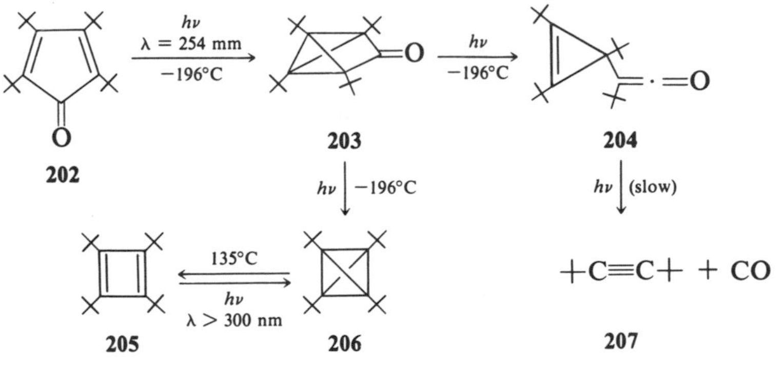

$h\nu$, $\lambda = 254$ mm, $-196°C$; $h\nu$, $-196°C$

202 **203** **204**

$h\nu$ $-196°C$; $h\nu$ (slow)

135°C; $h\nu$, $\lambda > 300$ nm

+C≡C+ + CO

205 **206** **207**

photolysis of **203** at temperatures between -100 and -196°C gave a mixture of the ketene **204** and tetra-*tert*-butyltetrahedrane **206**. The ketene **204** underwent slow photochemical decarbonylation to the acetylene **207**. The tetrahedrane **206** could be isolated in 40% yield and had a melting point of 135°C. At this temperature, it was converted to tetra-*tert*-butylcyclobutadiene (**205**), a reaction that is photochemically reversible. The spectroscopic properties of the compound melting at 135°C are in full agreement with the tetrahedrane structure.[97] The question is, Why is this tetrahedrane so stable? The answer lies in the steric requirements of the four *tert*-butyl groups:[96] we have already seen that a compound like tetra-*tert*-butylethylene would be severely twisted due to steric interactions between the *tert*-butyl groups. The tetrahedron is the optimal framework for the arrangement of four sterically demanding groups. Any other arrangement would result in a closer proximity of these groups, thus raising the strain energy. Consequently, the cyclobutadiene **205** is destabilized relative to **206**, and the energy difference between the two is less than for the unsubstituted molecules. Any attempt at ring opening of **206** to **205** will increase the steric repulsion between the groups, thus raising the activation energy for the isomerization to the cyclobutadiene. One may say that the strain energy of tetrahedrane (about 134 kcal/mol) is played against the strain energy of four *tert*-butyl groups.

5.7. PROBLEMS

1. Would you expect the following base–induced E2 elimination reaction[98] to give *cis*- or *trans*-cyclooctene in a kinetically controlled reaction?

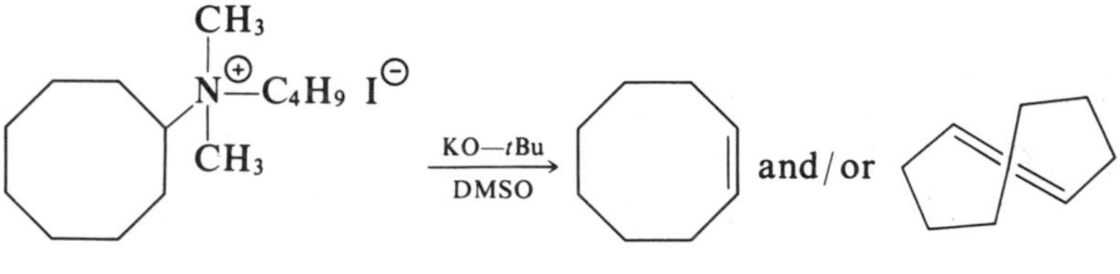

CH_3 / $N^{\oplus}$—C_4H_9 $I^{\ominus}$ / CH_3 (cyclooctyl) $\xrightarrow[\text{DMSO}]{\text{KO—}t\text{Bu}}$ and/or

(Consider both *syn* and *anti* elimination as mechanistic possibilities.)

2. Write the structure for the product of the following Diels–Alder cycloaddition reaction:[99]

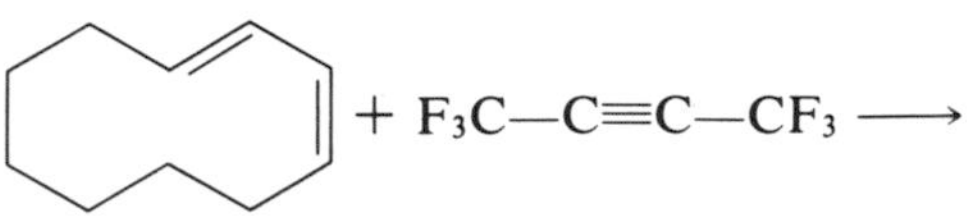

$+ F_3C—C{\equiv}C—CF_3 \longrightarrow$

3. The benzene isomer 3,3′-dicyclopropenyl (**208**) has an estimated enthalpy of formation $\Delta H_f^0 \sim 139$ kcal/mol.[1] Calculate with the aid of group increments the strain energy of **208**.[100]

208

4. Naphthalene undergoes high-temperature 1,2-carbon scrambling:

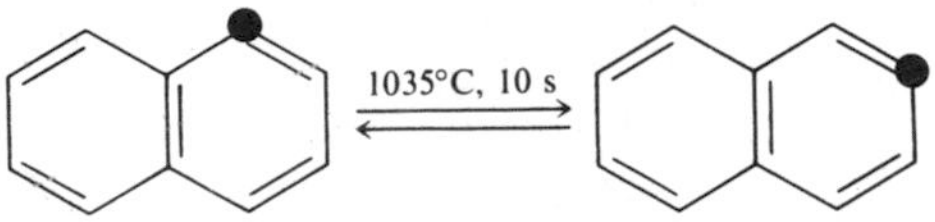

1035°C, 10 s

The enthalpy of formation of naphthalene is 36 kcal/mol, and the free energy of activation for the automerization is estimated as $\leqslant 86$ kcal/mol. The mechanism of the reaction is unknown.[101]

One possible mechanism is:

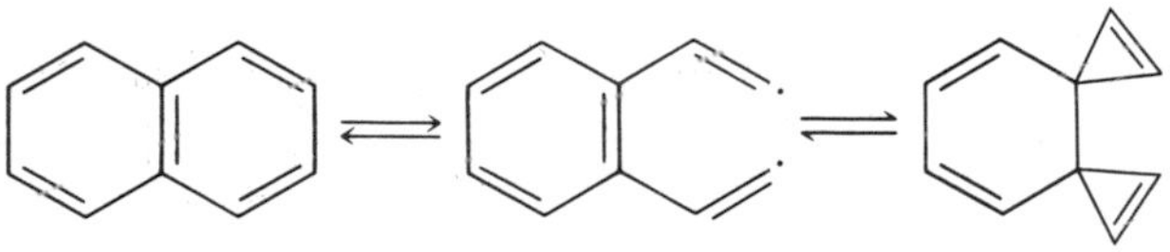

209

Assuming that the strain energy of **209** equals that of **208**, calculate ΔH_f^0**(209)** and determine whether **209** is a viable intermediate in the automerization of naphthalene.[102]

5. Write a mechanism for the following reaction:[103]

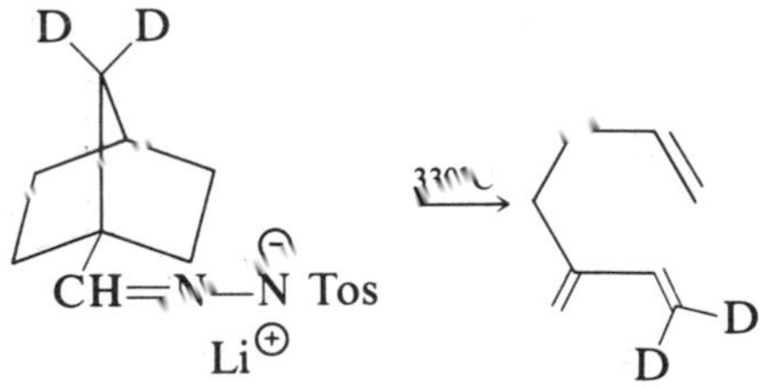

D D

CH=N—N Tos

Li⊕

330°C

D

D

6. Write a detailed mechanism for the reaction

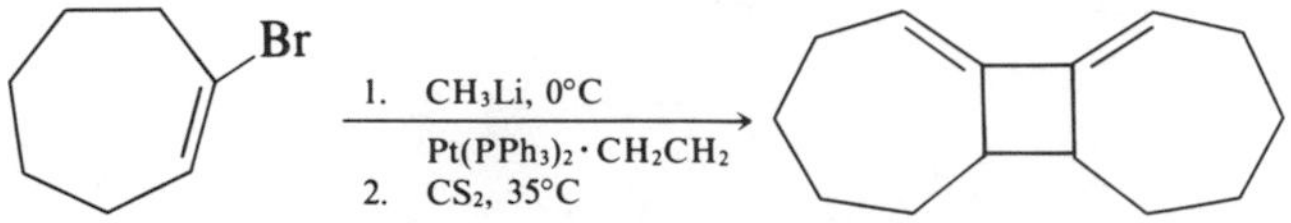

Br

1. CH_3Li, 0°C
$Pt(PPh_3)_2 \cdot CH_2CH_2$
2. CS_2, 35°C

in which an intermediate product can be isolated.[104]

7. Predict the product of the following transformations[105]

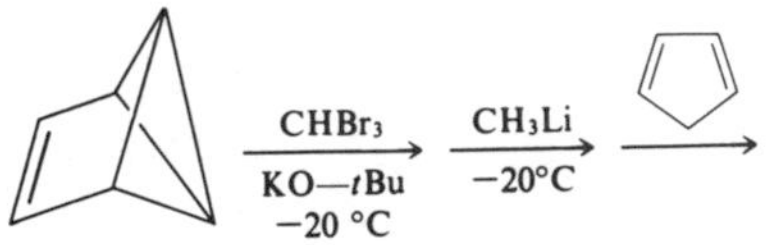

$CHBr_3$
KO—tBu
−20 °C

CH_3Li
−20°C

8. Predict the product of the reaction:[106]

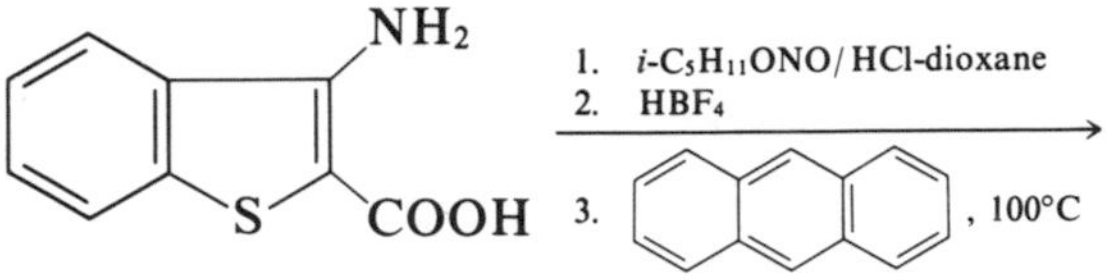

NH_2
S
COOH

1. i-$C_5H_{11}ONO$/ HCl-dioxane
2. HBF_4
3. , 100°C

9. Suggest a mechanism for the following reaction:[107]

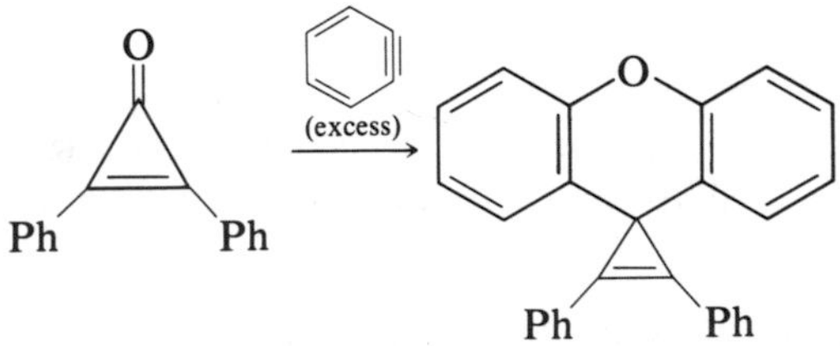

O
Ph Ph
(excess)
O
Ph Ph

10. Give structures for the products A and B in the following sequence and explain why C is thermodynamically more stable than B.[108]

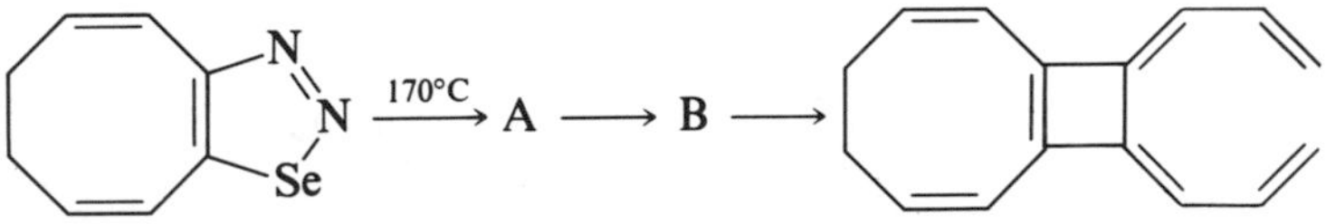

N
N
Se
170°C → A → B →

C

11. Predict the product of the following reaction:[109]

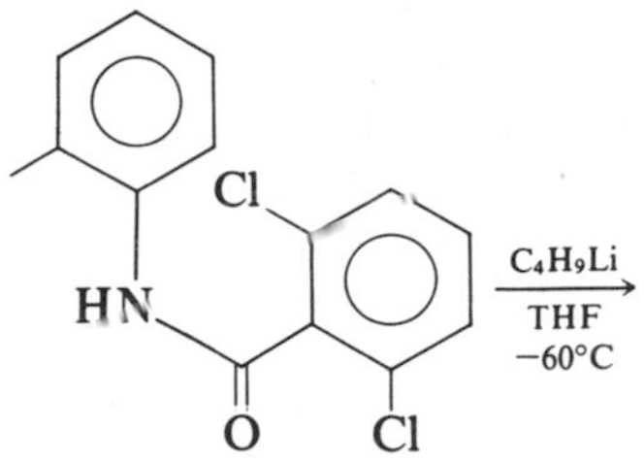

$$\xrightarrow[\text{THF}, -60^\circ\text{C}]{C_4H_9Li}$$

REFERENCES AND NOTES

1. For a detailed treatise, see A. Greenberg and J. F. Liebman, *Strained Organic Molecules*, Academic Press, New York, 1978.
2. Cf. C. Wentrup, *Top. Curr. Chem.*, **62,** 173 (1976), Table 3.
3. R. S. Hutton, M. L. Marion, H. D. Roth, and E. Wasserman, *J. Am. Chem. Soc.*, **96,** 4680 (1974). For substituted vinylcarbenes, see D. R. Arnold, R. W. Humphreys, W. J. Leigh, and G. E. Palmer, *ibid.*, **98,** 6225 (1976).
4. C. Batich, O. Ermer, E. Heilbronner, and J. R. Wiseman, *Angew. Chem.*, **85,** 302 (1973); *Angew. Chem. Int. Ed. Engl.*, **12,** 312 (1973).
5. N. L. Allinger and J. T. Sprague, *J. Am. Chem. Soc.*, **94,** 5734 (1972).
6. Y. Inoue, T. Ueoka, T. Kuroda, and T. Hakushi, *Chem. Commun.*, **1981,** 1031.
7. R. Bonneau, J. Joussot-Dubien, J. Yarwood, and J. Pereyre, *Tetrahedron Lett.*, **1977,** 235.
8. R. Bonneau, J. Joussot-Dubien, L. Salem, and A. J. Yarwood, *J. Am. Chem. Soc.*, **98,** 4329 (1976); W. G. Dauben, H. C. H. A. Van Riel, C. Hauw, F. Leroy, J. Joussot-Dubien, and R. Bonneau, *ibid.*, **101,** 1901 (1979).
9. U. Burkert, *Tetrahedron,* **37,** 333 (1981); D. Lenoir, H. Dauner, and R. M. Frank, *Chem. Ber.*, **113,** 2636 (1980).
10. N. A. Bailey and S. E. Hull, Chem. Commun., **1971,** 960; H. L. Ammon and G. L. Wheeler, *J. Am. Chem. Soc.*, **97,** 2326 (1975).
11. I. Agranat, M. Rabinovitz, and A. Weitzen-Dagan, *Chem. Commun.*, **1972,** 732; see also *Tetrahedron Lett.*, **1974,** 1241.
12. C. Wentrup, unpublished work, Lausanne, 1970.
13. J.-P. Hagenbuch, P. Vogel, A. A. Pinkerton, and D. Schwarzenbach, *Helv. Chim. Acta,* **64,** 1818 (1981).
14. E. Vogel, W. Grimme, and S. Korte, *Tetrahedron Lett.*, **1965,** 3625; W. E. Billups, A. J. Blakeney, and W. Y. Chow, *Org. Synth.*, **55,** 12 (1976).
15. J. Casanova, J. Bragin, and F. D. Cottrell, *J. Am. Chem. Soc.*, **100,** 2264 (1978).
16. J. Bredt, *Justus Liebigs Ann. Chem.*, **437,** 1 (1924).
17. Reviews: G. Köbrich, *Angew. Chem.*, **85,** 494 (1973); *Angew. Chem. Int. Ed. Engl.*, **12,** 464 (1973); R. Keese, ibid., **87,** 568 (1975)/**14,** 528 (1975); G. Szeimies, *React. Intermed. (Plenum)*, **3,** 299 (1983); K. J. Shea, *Tetrahedron,* **36,** 1683 (1980).
18. J. R. Wiseman, *J. Am. Chem. Soc.*, **89,** 5966 (1967).
19. J. A. Marshall and H. Faubl, *J. Am. Chem. Soc.*, **92,** 948 (1970); J. R. Wiseman and W. A. Pletcher, *ibid.*, **92,** 956 (1970).
20. P. M. Lesco and R. B. Turner, *J. Am. Chem. Soc.*, **90,** 6888 (1968).

21. J. R. Wiseman and J. A. Chang, *J. Am. Chem. Soc.*, **91,** 7775 (1969); J. A. Chang and J. R. Wiseman, *ibid.*, **94,** 8627 (1972).

22. R. Keese and E. P. Krebs, *Angew. Chem.*, **83,** 254 (1971); **84,** 540 (1972); *Angew. Chem. Int. Ed. Engl.*, **10,** 262 (1971); **11,** 518 (1972).

23. D. J. Martella, M. Jones, Jr., P. v. R. Schleyer, and W. F. Maier, *J. Am. Chem. Soc.*, **101,** 7634 (1979); W. F. Maier and P. v. R. Schleyer, *ibid.*, **103,** 1891 (1981).

24. W. Burns, D. Grant, M. A. McKervey, and G. Step, *J. Chem. Soc. Perkin Trans. 1*, **1976,** 234; D. Lenoir and J. Firl, *Justus Liebigs Ann. Chem.*, **1974**, 1467.

25 D. Lenoir, W. Kornrumpf, and H. P. Fritz, *Chem. Ber.*, **116,** 2390 (1983).

26. A. H. Alberts, J. Strating, and H. Wynberg, *Tetrahedron Lett.*, **1973,** 3047.

27. J. I. G. Cadogan and R. Leardini, *Chem. Commun.*, **1979,** 783.

28. For further work on adamantene, see J. E. Gano and L. Eizenberg, *J. Am. Chem. Soc.*, **95,** 972 (1973); G. W. Wood, B. T. Kiremire, J. E. Gano, N. Marron, A. Falick, and P. Tecon, *J. Org. Chem.*, **45,** 4990 (1980); H. Schwarz, M. T. Reetz, W. F. Maier, C. Wesdemiotis, I. Chatziiosifidis, and M. Schilling, *Angew. Chem.*, **91,** 1019 (1979); *Angew. Chem. Int. Ed. Engl.*, **18,** 952 (1979).

29. D. J. Martella, M. Jones, Jr., and P. v. R. Schleyer, *J. Am. Chem. Soc.*, **100,** 2896 (1978).

30. R. T. Conlin, R. D. Miller, and J. Michl, *J. Am. Chem. Soc.*, **101,** 7637 (1979).

31. S. F. Sellers, T. C. Klebach, F. Hollowood, M. Jones, Jr., and P. v. R. Schleyer, *J. Am. Chem. Soc.*, **104,** 5492 (1982).

32. R. Greenhouse, W. T. Borden, K. Hirotsu, and J. Clardy, *J. Am. Chem. Soc.*, **99,** 1664 (1977).

33. O. Baumgärtel, J. Harnisch, G. Szeimies, M. Van Meerssche, G. Germain, and J.-P. Declercq, *J. Am. Chem. Soc.*, **101,** 3370 (1979); *Chem. Ber.*, **116,** 2205 (1983).

34. U. Szeimies-Seebach, J. Harnisch, G. Szeimies, M. Van Meerssche, G. Germain, and J.-P. Declercq, *Angew. Chem.*, **90,** 904 (1978); *Angew. Chem. Int. Ed. Engl.*, **17,** 848 (1978).

35. G. Szeimies, personal communication (1982).

36. H.-G. Zoch, G. Szeimies, R. Römer, and R. Schmitt, *Angew. Chem.*, **93,** 894 (1981); *Angew. Chem. Int. Ed. Engl.*, **20,** 877 (1981); H.-G. Zoch, G. Szeimies, R. Römer, G. Germain, and J.-P. Declercq, *Chem. Ber.*, **116,** 2285 (1983).

37. V. Typke, J. Haase, and A. Krebs, *J. Mol. Struct.*, **56,** 77 (1979).

38. A. T. Blomquist and L. H. Liu, *J. Am. Chem. Soc.*, **75,** 2153 (1953); H. Meier, *Synthesis*, **1972,** 235.

39. E. Kloster-Jensen and J. Wirz, *Helv. Chim. Acta*, **58,** 162 (1975).

40. R. A. G. De Graff, S. Gorter, C. Romers, H. N. C. Wong, and F. Sondheimer, *J. Chem. Soc. Perkin Trans. 2*, **1981,** 478 and references therein.

41. Review: R. W. Hoffmann, *Dehydrobenzene and Cycloalkynes*, Academic Press, New York, 1967.

42. Reviews: A. Krebs, in *Chemistry of Acetylenes*, H. G. Viehe, Ed., Dekker, New York, 1969, p. 987; M. Nakagawa, in *The Chemistry of the Carbon–Carbon Triple Bond*, S. Patai, Ed., Wiley, Chichester, 1978, Part 2, p. 635.

43. N. L. Allinger and A. Y. Meyer, *Tetrahedron*, **31,** 1807 (1975).

44. S. F. Karaev and A. Krebs, *Tetrahedron Lett.*, **1973,** 2853.

45. M. A. Bennett, G. B. Robertson, P. O. Whimp, and T. Yoshida, *J. Am. Chem. Soc.*, **93,** 3797 (1971); G. B. Robertson and P. O. Whimp, *ibid.*, **97,** 1051 (1975).

46. G. Wittig and U. Mayer, *Chem. Ber.*, **96,** 342 (1963).

47. K. L. Erickson and J. Wolinsky, *J. Am. Chem. Soc.*, **87,** 1142 (1965).

48. G. J. Baxter and R. F. C. Brown, *Aust. J. Chem.*, **31,** 327 (1978).

49. C. Wentrup, H.-M. Berstermann, and G. Gross, unpublished results (1981).

50. H. Meier, H. Petersen, and H. Kolshorn, *Chem. Ber.*, **113,** 2398 (1980).

51. F. Scardiglia and J. D. Roberts, *Tetrahedron,* **1,** 343 (1957); L. K. Montgomery, F. Scardiglia, and J. D. Roberts, *J. Am. Chem. Soc.*, **87,** 1917 (1965).

52. A. T. Bottini, F. P. Corson, R. Fitzgerald, and K. A. Frost, *Tetrahedron,* **28,** 4883 (1972); G. Wittig and P. Fritze, *Justus Liebigs Ann. Chem.*, **711,** 82 (1968).

53. A. T. Bottini, K. A. Frost, B. R. Anderson, and V. Dev, *Tetrahedron,* **29,** 1975 (1973).

54. W. R. Moore and W. R. Moser, *J. Am. Chem. Soc.*, **92,** 5469 (1970).

55. C. Wentrup, G. Gross, A. Maquestiau, and R. Flammang, *Angew. Chem.*, **95,** 551 (1983); *Angew. Chem. Int. Ed. Engl.*, **22,** 542 (1983).

56. H. Bock, T. Hirabayashi, and S. Mohmand, *Chem. Ber.*, **114,** 2595 (1981).

57. M. Balci and W. M. Jones, *J. Am. Chem. Soc.*, **102,** 7607 (1980).

58. L. Fitjer, U. Kliebisch, D. Wehle, and S. Modaressi, *Tetrahedron Lett.*, **23,** 1661 (1982); J. C. Gilbert and M. E. Baze, *J. Am. Chem. Soc.*, **105,** 664 (1983).

59. O. L. Chapman, J. Gano, P. R. West, M. Regitz, and G. Maas, *J. Am. Chem. Soc.*, **103,** 7033 (1981).

60. N. J. Turro, V. Ramamurthy, K.-C. Liu, A. Krebs, and R. Kemper, *J. Am. Chem. Soc.*, **98,** 6758 (1976).

61. T. L. Gilchrist and C. W. Rees, *Carbenes, Nitrenes, and Arynes,* Nelson, London, 1969.

62. J. D. Roberts, H. E. Simmons, L. A. Carlsmith, and C. W. Vaughan, *J. Am. Chem. Soc.*, **75,** 3290 (1953).

63. O. L. Chapman, K. Mattes, C. L. McIntosh, J. Pacansky, G. V. Calder, and G. Orr, *J. Am. Chem. Soc.*, **95,** 6134 (1973).

64. O. L. Chapman, C.-C. Chang, J. Kolc, N. R. Rosenquist, and H. Tomioka, *J. Am. Chem. Soc.*, **97,** 6586 (1975); J. W. Laing and R. S. Berry, *ibid.*, **98,** 660 (1976); I. R. Dunkin and J. E. MacDonald, *Chem. Commun.*, **1979,** 772.

65. R. Hoffmann, A. Imamura, and W. J. Hehre, *J. Am. Chem. Soc.*, **90,** 1499 (1968).

66. J. O. Noell and M. D. Newton, *J. Am. Chem. Soc.*, **101,** 51 (1979).

67. M. Jones, Jr., and R. H. Levin, *J. Am. Chem. Soc.*, **91,** 6411 (1969).

68. H.-F. Grützmacher and J. Lohmann, *Justus Liebigs Ann. Chem.*, **705,** 81 (1967); S. K. Pollack and W. J. Hehre, *Tetrahedron Lett.*, **1980,** 2483; see also H. M. Rosenstock, J. T. Larkins, and J. A. Walker, *Int. J. Mass. Spectrom. Ion Phys.*, **11,** 309 (1973).

69. M. J. S. Dewar and W.-K. Li, *J. Am. Chem. Soc.*, **96,** 5569 (1974).

70. R. S. Berry, J. Clardy, and M. E. Schafer, *Tetrahedron Lett.*, **1965,** 1003.

71. W. N. Washburn, R. Zahler, and I. Chen, *J. Am. Chem. Soc.*, **100,** 5863 (1978).

72. W. E. Billups, J. D. Buynak, and D. Butler, *J. Org. Chem.*, **45,** 4636 (1980).

73. R. G. Bergman, *Acc. Chem. Res.*, **6,** 25 (1973).

74. T. P. Lockhart, P. B. Comita, and R. G. Bergman, *J. Am. Chem. Soc.*, **103,** 4082 (1981).

75. T. P. Lockhart and R. G. Bergman, *J. Am. Chem. Soc.*, **103,** 4091 (1981).

76. R. S. Berry, J. Clardy, and M. E. Schafer, *Tetrahedron Lett.*, **1965,** 1011.

77. R. Breslow, J. Napierski, and T. C. Clarke, *J. Am. Chem. Soc.*, **97,** 6275 (1975).

78. R. Breslow and P. L. Khanna, *Tetrahedron Lett.*, **1977,** 3429.

79. T. Kauffmann, *Angew. Chem.*, **77,** 557 (1965); *Angew. Chem. Int. Ed. Engl.*, **4,** 543 (1965); M. G. Reinecke, *Tetrahedron,* **38,** 427 (1982).

80. C. W. Rees and R. C. Storr, *J. Chem. Soc. C,* **1969,** 760, 765; R. W. Hoffmann, G. Guhn, M. Preiss, and B. Dittrich, *ibid.*, **1969,** 769.

81. D. Ginsburg, *Acc. Chem. Res.*, **2,** 121 (1969); D. Ginsburg, *Propellanes: Structure and Reactions,* Verlag Chemie, Weinheim, 1975.

82. P. E. Eaton and G. H. Temme, *J. Am. Chem. Soc.*, **95,** 7508 (1973).

83. W.-D. Stohrer and R. Hoffmann, *J. Am. Chem. Soc.*, **94,** 779 (1972).

84. K. B. Wiberg and G. J. Burgmaier, *J. Am. Chem. Soc.*, **94,** 7396 (1972) and references therein.

85. K. B. Wiberg, H. A. Connon, and W. E. Pratt, *J. Am. Chem. Soc.*, **101,** 6970 (1979).

86. D. H. Aue and R. N. Reynolds, *J. Org. Chem.*, **39,** 2315 (1974).

87. F. W. Walker, K. B. Wiberg, and J. Michl, *J. Am. Chem. Soc.*, **104,** 2056 (1982).

88. K. B. Wiberg and F. H. Walker, *J. Am. Chem. Soc.*, **104,** 5239 (1982).

89. J. Michl, G. J. Radziszewski, J. W. Downing, K. B. Wiberg, F. H. Walker, R. D. Miller, P. Kovacic, M. Jawdosiuk, and V. Bonačić-Koutecký, *Pure Appl. Chem.*, **55,** 315 (1983).

90. R. F. W. Bader, T.-H. Tang, Y. Tal, and F. W. Biegler-König, *J. Am. Chem. Soc.*, **104,** 940 (1982); see also M. D. Newton and J. M. Schulman, *ibid.*, **94,** 773 (1972); M. D. Newton, in *Electronic Structure Theory,* H. F. Schaefer, Ed., Plenum, New York, 1977, Vol. III, p. 223.

91. P. Chakrabarti, P. Seiler, J. D. Dunitz, A.-D. Schlüter, and G. Szeimies, *J. Am. Chem. Soc.*, **103,** 7378 (1981).

92. W. Grahn, *Chem. in Unserer Zeit,* **15,** 52 (1981); see also *Chem. Eng. News.*, August 16, 1982, p. 25.

93. (a) H. Kollmar, *J. Am. Chem. Soc.*, **102,** 2617 (1980) and references therein. (b) A. Schweig and W. Thiel, *J. Am. Chem. Soc.*, **101,** 4742 (1979).

94. L. B. Rodewald and H.-K. Lee, *J. Am. Chem. Soc.*, **95,** 623, 3084 (1973); P. B. Shevlin and A. P. Wolf, *ibid.*, **92,** 406, 5291 (1970).

95. G. Maier and H. P. Reisenauer, *Chem. Ber.*, **114,** 3916 (1981).

96. G. Maier, S. Pfriem, U. Schäfer, K.-D. Malsch, and R. Matusch, *Chem. Ber.*, **114,** 3965 (1981).

97. G. Maier, S. Pfriem, K.-D. Malsch, H.-O. Kalinowski, and K. Dehnicke, *Chem. Ber.*, **114,** 3988 (1981).,

98. R. D. Bach and J. W. Knight, *Tetrahedron Lett.*, **1979,** 3815.

99. P. G. Gassman, S. R. Korn, T. F. Bailey, T. H. Johnson, J. Finar, and J. Clardy, *Tetrahedron Lett.*, **1979,** 3401.

100. Using group increments from Table 1.2, one calculates ΔH_f^0(**208**) (strain-free) = $4C_d(C)(H) + 2C(C_d)_2(C)(H) \cong 4C_d(C)(H) + 2C(C_d)(C)_2(H)$ = 31.4 kcal/mol. With ΔH_f^0(**208**) = 139 kcal/mol, this gives $E_s \cong 108$ kcal/mol.

101. L. T. Scott, *Acc. Chem. Res.*, **15,** 52 (1982).

102. One calculates ΔH_f^0(**209**) (strain-free) $\cong$ 68 kcal/mol. With $E_s = 108$ kcal/mol, this gives ΔH_f^0(**209**) $\cong$ 176 kcal/mol, which is about 54 kcal/mol above the transition state for naphthalene automerization. Hence **209** is not a viable intermediate.

103. A. D. Wolf and M. Jones, Jr., *J. Am. Chem. Soc.*, **95,** 8209 (1973).

104. J. P. Visser and J. E. Ramakers, *Chem. Commun.*, **1972,** 178.

105. M. Christl, *Angew. Chem.*, **93,** 515 (1981); *Angew. Chem. Int. Ed. Engl.*, **20,** 529 (1981).

106. M. G. Reinecke and H. H. Ballard, *Tetrahedron Lett.*, **1979,** 4981.

107. See R. W. Hoffmann, *Dehydrobenzene and Cycloalkynes,* Academic Press, New York, 1967, p. 205.

108. H. Meier, T. Echter, and O. Zimmer, *Angew. Chem.*, **93,** 901 (1981); *Angew. Chem. Int. Ed. Engl.*, **20,** 865 (1981).

109. W. L. Houlihan, Y. Uike, and V. A. Parrino, *J. Org. Chem.*, **46,** 4515 (1981); see also W. M. Best and D. Wege, *Tetrahedron Lett.*, **22,** 4877 (1981).

Chapter 6

ANTIAROMATICS

Es ist gewiss eine auffallende Erscheinung, daβ man in zahlreichen Verbindungen einen durch Aneinanderlagerung von 3 Mol. Acetylen gebildeten Kern, den Benzolkern, annimmt, während von einem auf gleiche Weise durch Aneinanderlagerung von 2 Mol. Acetylen gebildeten Kern bisher nicht die Rede gewesen ist. In dieser hypothetischen Verbindung könnte man die Kohlenstoffatome auf folgende Art [*i*]

HC=CH
| |
HC=CH

[*i*]   [*ii*]

gebunden . . . oder man könnte auch eine tetraedrische Gruppierung [*ii*] des Kohlenstoffs annehmen . . . .

—H. Limpricht, *Ber. Dtsch. Chem. Ges.*, **2**, 211 (1869).

Limpricht appears to have been the first to postulate the existence of C_4H_4—"tetrol"—as a lower homologue of benzene. He tentatively suggested that pyrrole was "aminotetrol." It took 100 years before cyclobutadiene was established as a free molecule and longer still before the first tetrahedrane had been synthesized (see Chapter 5).

6.1. ELECTRONIC STRUCTURE

Cyclobutadiene (**1**) (Refs. 1–3) and the isoelectronic three-membered heterocycles, oxirene (**2**), 1*H*-azirine (**3**), and thiirene (**4**), are highly unstable molecules, not so much because of ring strain, but because of a quantum mechanical effect: antiaromaticity (for the three-membered heterocycles, see Chapter 4). Although the

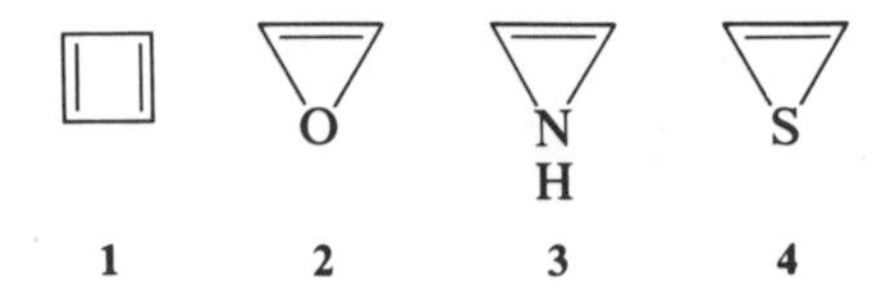

enthalpy of formation of cyclobutadiene is not known, the best calculations indicate a value of approximately 100 kcal/mol.[3,4] The hypothetical strain-free molecule would have an enthalpy of formation equal to four $C_d(C_d)(H)$ group increments (see Table 1.2), that is, about 27 kcal/mol. From a comparison with cyclobutene and cyclobutane, one would expect that the strain energy of **1** due to angle deformation would not be much higher than 33 kcal/mol. The major part of the enthalpy of formation of cyclobutadiene must, therefore, be due to antiaromaticity, accounting for approximately 40 kcal/mol.

ΔH_f^0	~100	37.5	6.3 kcal/mol
E_S		30	27 kcal/mol

The Hückel molecular orbital energies[5] of cyclobutadiene are depicted in Figure 6.1. If the molecule is square planar **(1a)**, two electrons will occupy two degenerate nonbonding orbitals, and according to Hund's first rule, we would expect a triplet ground state. If the molecule is rectangular **(1b)**, however, the degeneracy of the nonbonding orbitals will be lifted, and, if the splitting is large enough, a singlet ground state will obtain. The lifting of the degeneracy is often ascribed to the pseudo-Jahn–Teller effect, that is, if in a molecule two orbitals are accidentally degenerate, the molecule will distort itself in such a way as to lift the degeneracy. Applied to cyclobutadiene, this means that the molecule should exist as a nonsquare singlet **(1b)**.[6]

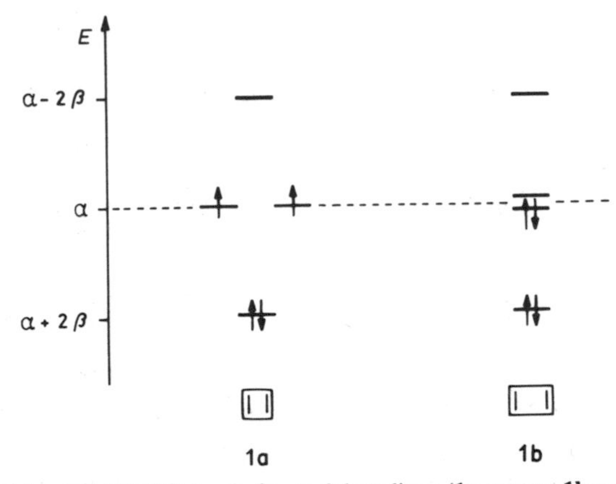

Figure 6.1. Hückel MO levels for cyclobutadiene (**1a** square; **1b** rectangular).

Electronically, cyclobutadiene has much in common with the carbenes. In both cases, the instability and high reactivity is due to the presence of high-lying HOMOs and/or low-lying LUMOs. The well-known Hückel rules[5] predict that all fully conjugated cyclic molecules possessing $4n$ π electrons will be antiaromatic and have electronic structures similar to that of cyclobutadiene, whereas the $4n + 2$ π electron systems are aromatic. By this token, cyclopropenyl anion (**5**) and cyclopentadienyl cation (**6**) are expected to be antiaromatic, and, in fact, both are highly

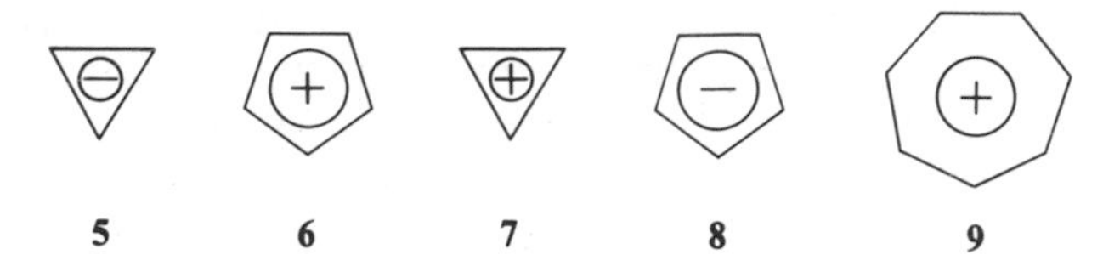

reactive and difficult to prepare.[7,8] The instability of **5** is reflected in the high pK_a value for cyclopropene,[7] and **6** exists as a ground state triplet. **6** can be generated in solution by treatment of 5-bromocyclopentadiene with SbF_5 at 78 K, and from its ESR spectrum a D/hc value of 0.1868 cm^{-1} was derived.[8] Also pentachloro-

SbF_5 / $-SbF_5Br^{\ominus}$ (Br) → **6**

cyclopentadienyl cation is a triplet, but the pentaphenyl analog is a ground state singlet with the triplet state lying only 1 kcal/mol higher. Accordingly, the triplet can be thermally populated, and hence an ESR spectrum can be recorded.[8]

In contrast, cyclopropenyl cation (**7**), cyclopentadienyl anion (**8**), and cycloheptatrienyl (tropylium) cation (**9**) are aromatic $4n + 2$ π electron systems, the latter two isoelectronic with benzene.[5,9]

Cyclooctatetraene (**10**) is a $4n$ π system, but it avoids some potential antiaromaticity by being nonplanar (**11**),[10] and the molecule is best described as a polyene. Cyclooctatetraene undergoes ring inversion, possibly via the planar form **10**, with

OH

10 **11** **12**

a free energy of activation $\Delta G^{\ddagger} \sim$ 13–15 kcal/mol. The observation of a lower barrier ($\Delta G^{\ddagger} \sim$ 5.7 kcal/mol) in the tetrabenzo-derivative **12** supports the contention that planar cyclooctatetraene is antiaromatic: delocalization in **12** reduces the antiaromatic character of the transition state.[11]

In terms of energy, the magnitudes of aromaticity and antiaromaticity rapidly decrease as the rings become larger. Quantum-chemical calculations indicate that the resonance energies of the [$4n + 2$]-annulenes (benzene, [10]-annulene, etc.) are positive, as expected, but for the [$4n$]-annulenes (cyclobutadiene, cyclooctatetraene, [12]-annulene etc.), they are negative.[12] Already [14]- and [16]-annulene have only minute positive and negative resonance energies, respectively, and the higher annulenes are better described as nonaromatic polyenes.

The annulenes can be distinguished by their NMR spectra.[13] All [$4n$]-annulenes up to $n = 6$ show the effects of a paramagnetic ring current (high-field absorption of outer protons in the NMR spectrum); all planar [$4n + 2$]-annulenes show a diamagnetic ring current (low-field absorption of outer protons).

When low-lying molecular orbitals are available, antiaromatic compounds may be transformed into aromatic ones by removal or addition of two electrons. Thus both cyclooctatetraene dication and dianion and [12]-annulene dianion exhibit aromatic properties, for example, in the form of low-field proton NMR absorptions.[14]

Referring to Figure 6.1, it can be seen that the dication of cyclobutadiene should be relatively stable since the two electrons occupying the nonbonding MOs are now removed. In fact, the dication **13** is readily obtained and observable by NMR spectroscopy:[15]

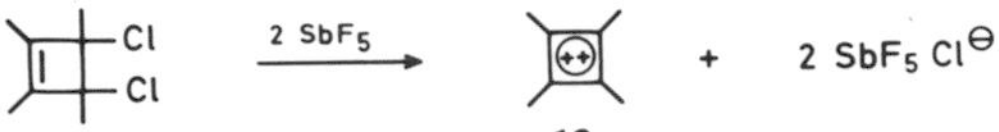

Cl, Cl —($2\ SbF_5$)→ **13** + $2\ SbF_5Cl^{\ominus}$

13 is a $4n + 2$ π electron system ($n = 0$). The dianion of cyclobutadiene **(14)** is also *formally* a $4n + 2$ system (isoelectronic with benzene), but reference to Figure 6.1 shows that *there is no low-lying bonding MO that can accept the two additional electrons*. Thus **14** would have four electrons in anti- or nonbonding orbitals, and it is expected, therefore, to be at least as unstable as cyclobutadiene itself. It is not the *number* of the electrons, but the *energies* of the MOs that count. In agreement

14

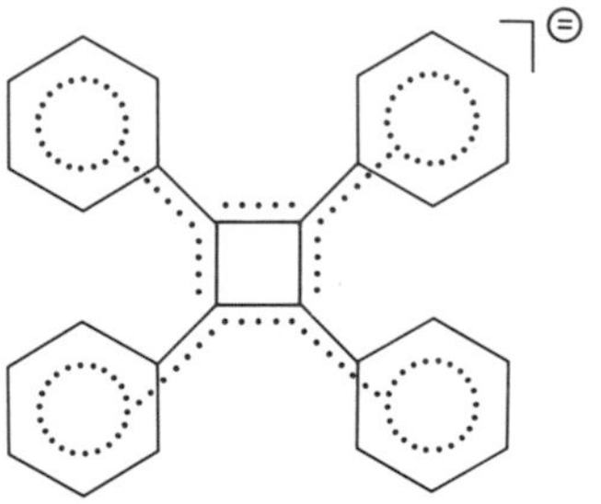

15

with this, cyclobutadiene dianions are extremely difficult to generate,[16] and only when the negative charges can be delocalized over four phenyl rings (**15**) or similar substituents is it possible to prepare and observe such ions.[17] An NMR study of **15** in conjunction with MNDO calculations indicated that only 36% of the charge resides in the four-membered ring; the rest is delocalized over the phenyl groups.[17a]

6.2. SYNTHESIS AND STRUCTURE OF CYCLOBUTADIENES

From the many early attempts at synthesis of cyclobutadienes,[1] it is apparent that these molecules tenaciously resist formation. Methylenecyclobutene (**16**) cannot be converted to methylcyclobutadiene (**17**),[18]

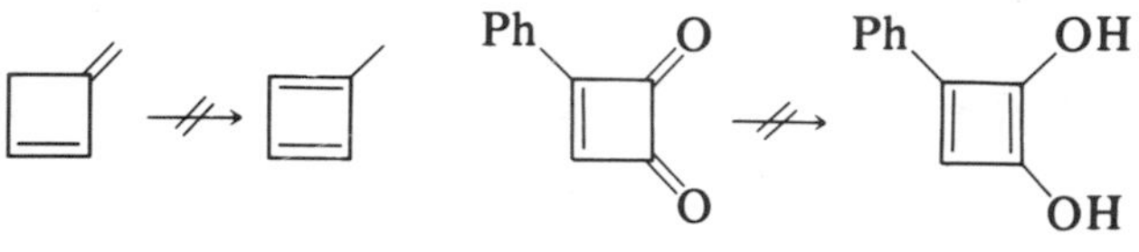

Ph O Ph OH

O OH

16 **17** **18** **19**

and the dione **18** cannot be reduced to a "hydroquinone" such as **19**.[19] Similarly, the electrochemical oxidation of the hydroquinone dianion **20** to the quinone **21** was found to be much more difficult than the corresponding transformation **22** → **23**.[7c]

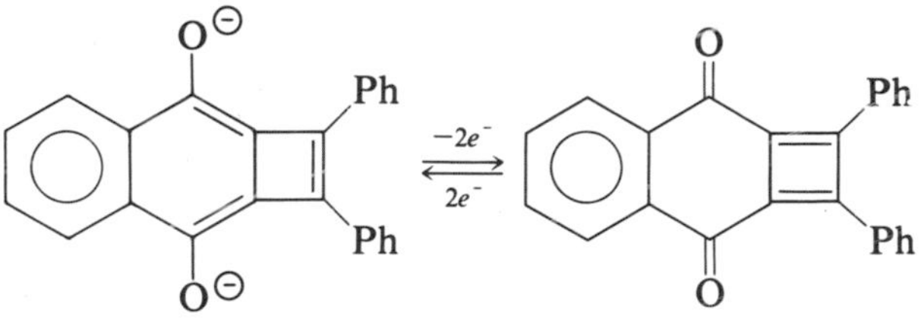

$O^{\ominus}$ O

Ph Ph

$-2e^-$

$2e^-$

Ph Ph

$O^{\ominus}$ O

20 **21**

$O^{\ominus}$ O

$-2e^-$

$2e^-$

$O^{\ominus}$ O

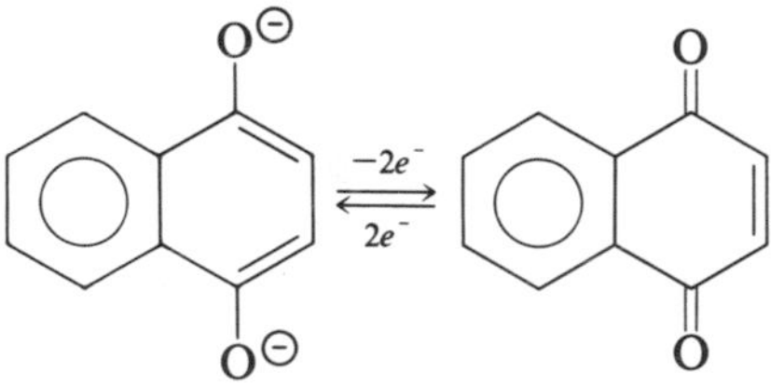

22 **23**

Although the stable ion **20** can be written in another resonance form incorporating a cyclobutadiene ring, the structure shown is expected to be the main contributor.[20a] The X-ray structure of the stable benzocyclobutadiene derivative **24** also shows a pronounced dimethylenecyclobutene character (**24a**); the alternative resonance structures **24b** and **24c** are disfavored due to the presence of a full cyclobutadiene ring.[20b] Thus **20** avoids antiaromaticity in the structure shown, but the oxidation product **21** necessarily contains a cyclobutadiene ring. From the difference in the oxidation potentials for the two reactions **20** → **21** and **22** → **23**, it was deduced that the cyclobutadiene **21** is destabilized by roughly 15–20 kcal/mol.[7c]

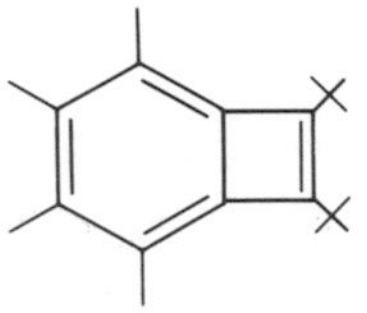

24a 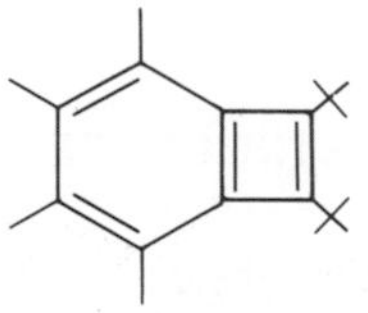 **24b** 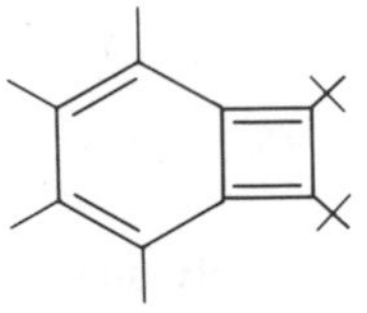 **24c**

Progress toward the free, unsubstituted cyclobutadiene was made with the observation that a dimer **(26)** was obtained from the treatment of **25** with lithium amalgam.[21] With nickel tetracarbonyl, the square planar cyclobutadiene complex

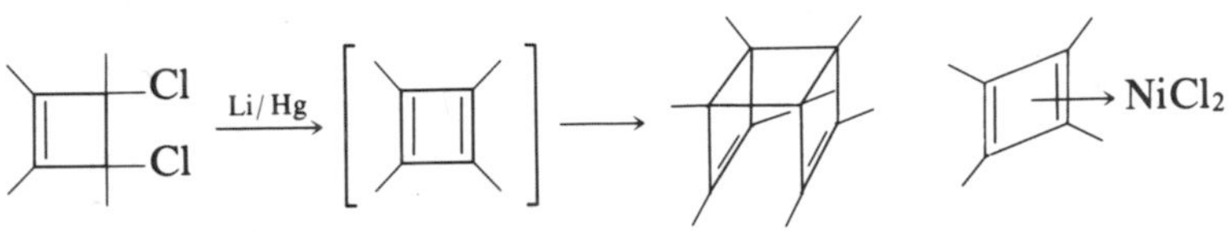

Cl Cl Li/Hg → NiCl$_2$

25 **26** **27**

27 was formed.[22] The analogous iron tricarbonyl complex of cyclobutadiene **(28)** was produced from dichlorocyclobutene with $Fe_2(CO)_9$:[23]

Cl Cl $Fe_2(CO)_9$ → $Fe(CO)_3$

28

Cyclobutadiene is liberated from the latter complex by oxidation with Ce(IV) and trapped in *sterospecific* [2 + 2] cycloadditions, for example, with diethyl maleate ($E = COOCH_3$):[23a,24]

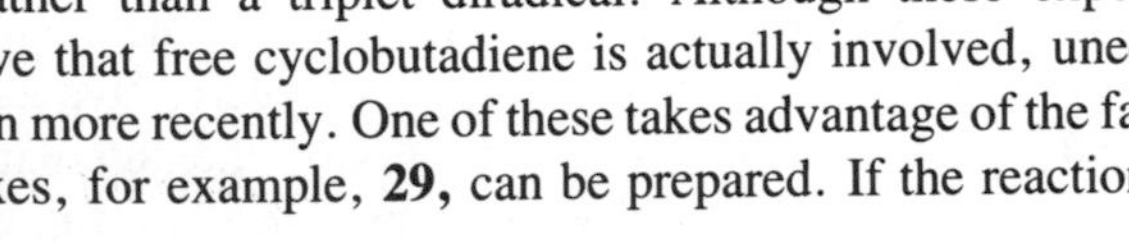

$Fe(CO)_3$ $\xrightarrow{Ce^{4+}}$ E E → E E

28

The stereospecificity of this and related reactions indicates that cyclobutadiene reacts as a singlet rather than a triplet diradical. Although these experiments do not rigorously prove that free cyclobutadiene is actually involved, unequivocal proofs have been given more recently. One of these takes advantage of the fact that optically active complexes, for example, **29,** can be prepared. If the reaction leading to the

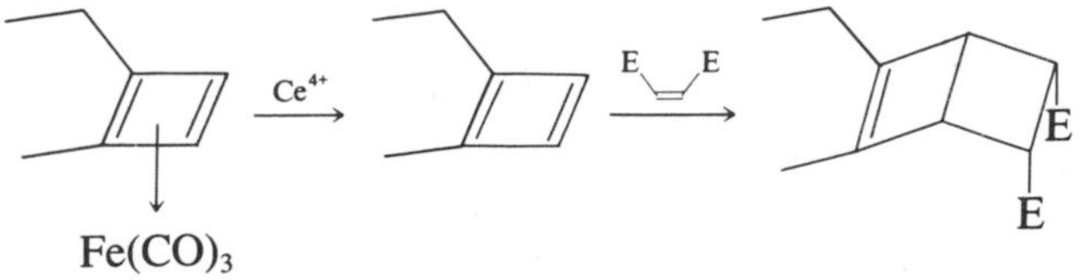

Ce^{4+}

E E

E

E

$Fe(CO)_3$

29 Optically active

30 Inactive

31 Inactive

cycloadduct **31** takes place on complexed cyclobutadiene, the adduct will be optically active. If free cyclobutadiene (**30**) is involved, the adduct must be racemic. In the event, racemic **31** was obtained.[25]

Another approach employed the "three-phase-test" in which the two reaction partners **32** and **33** are anchored on two different, solid polymers P. In order to react, the free cyclobutadiene has to traverse a third, liquid, phase. The addition product **34** was isolated in 96% yield.[26]

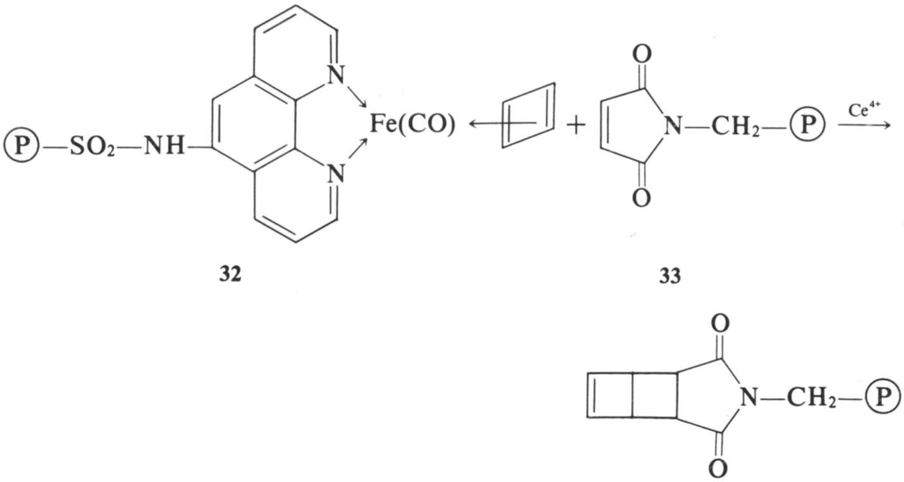

Ⓟ—SO_2—NH— N N Fe(CO) + O O N—CH_2—Ⓟ Ce^{4+}

32

33

O O N—CH_2—Ⓟ

34

In 1972–1973, the free parent cyclobutadiene was generated by four different groups using low-temperature photochemical matrix isolation (Scheme 1), that is, under conditions permitting the IR and UV spectroscopic examination of the compound.[27–30] Cyclobutadiene (**1**) dimerized to the *syn*-dimer **35** on warming. On prolonged photolysis at low temperature, it was split into two molecules of acetylene. The IR spectrum of **1** immediately sparked off extensive experimental and theoretical investigation.[3] The original spectrum[27,29] showed four bands, in agreement with group theoretical predictions for square planar cyclobutadiene with D_{4h} symmetry (**1a**). For the rectangular form **1b** with D_{2h} symmetry, a total of seven bands is predicted.[31] The four bands were interpreted as follows: ~ 3000 (CH

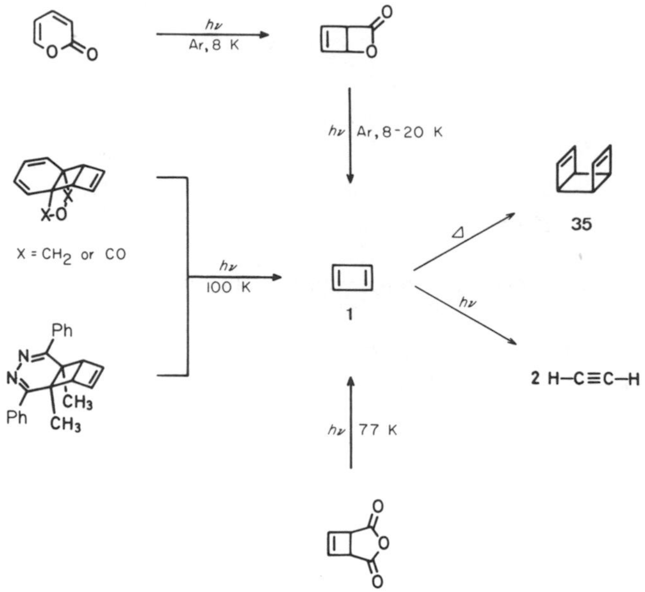

Scheme 1

stretch), 1236 (in-plane C—C def.), 653 (in-plane CH def.), and 573 (out-of-plane CH def.) cm^{-1}. Furthermore, it was reported that the two differently dideuterated precursors **36** and **37** gave the *same* dideuteriocyclobutadiene **(38)**, as evidenced by the observation of identical IR spectra.[29,32] This again indicated a D_{4h} symmetry **(38a)**. If the D_{2h} species **(38b)** were present, the two rectangular forms would have to interconvert rapidly, even at 8 K, in order to produce identical IR spectra.

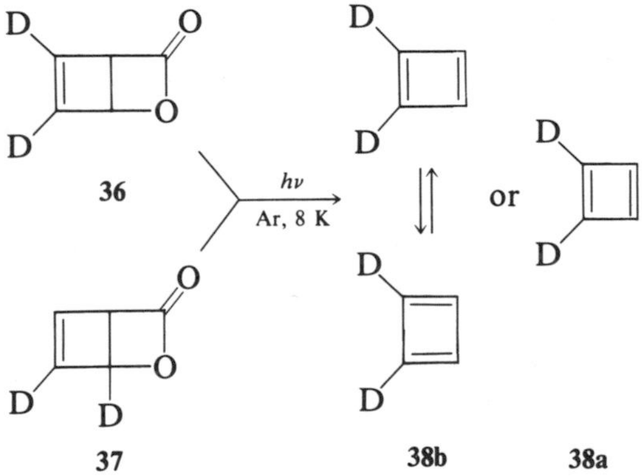

The D_{4h} symmetry is, however, at variance with the stereospecific addition reactions mentioned earlier. Square planar **1a** is expected to be a triplet diradical (see Figure 6.1), to react nonstereospecifically, and to be observable by ESR spectroscopy. Yet all attempts to observe an ESR spectrum of **1** failed.[3] The idea of a square planar ground state had to be abandoned, however, when it was discovered that only *three* and not four IR bands had been observed: the 653-cm^{-1} band was shown not to belong to cyclobutadiene since it was absent when cyclobutadiene was generated from precursors where CO_2 is not formed as a by-product.[3,33] Free, matrix-isolated CO_2 shows a band at 660 cm^{-1}. Accordingly, the 653-cm^{-1} band previously ascribed to cyclobutadiene was in reality due to a CO_2 vibration arising from the proximity of cyclobutadiene in the matrix. This may be visualized in terms of the formation of a weakly bound complex of the type **39.**[33] The proximate cyclobutadiene causes a lifting of the degeneracy of the O—C—O

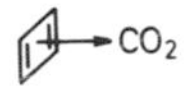

$\rightarrow CO_2$

39

bending mode with the consequence that two bands rather than one are observed for CO_2.

With only three bands observed, it was clear that not all the allowed vibrations of cyclobutadiene—square or rectangular—had been recorded. In further work using Fourier transform IR spectroscopy, two additional weak bands at 1523 and 723 cm^{-1} were detected, together with a further, very weak band at approximately 1000 cm^{-1}, the assignment of the latter to cyclobutadiene being tentative. Consequently, the IR spectrum was now in accord with the *rectangular structure* **1b**, and this was supported by theoretical calculations of both equilibrium structure and infrared spectra.[3]

Since the ground state of cyclobutadiene is rectangular, the results obtained with the dideuterated cyclobutadienes **38** must be reinterpreted to mean that either the starting materials **36** and **37** interconvert photochemically,[32] or the rectangular cyclobutadienes **38b** interconvert rapidly, even at 8 K.

An elegant proof that dideuteriocyclobutadiene is in fact nonsquare has been given by generation from the two precursors **40** and **41** (Scheme 2).[34a] Since the complex **40** is square planar, it can only give square cyclobutadiene *or*, if the molecule in nonsquare, a mixture of the two rectangles, **42** and **43**. On the other hand, **41** is expected to give initially only one rectangular form, **43** (for decomposition of azo-compounds, see Chapter 2). The cyclobutadienes were trapped with either methyl acrylate or *Z*-3-cyanoacrylate to give the four adducts **44–47**. In order to facilitate the NMR examination of the deuterium distribution, the adducts were converted to the four iodolactones **48–51** by consecutive treatment with KOH, CO_2, and I_2/KI.

The evaluation of the NMR spectra of the product mixtures revealed that different ratios of **48** and **50** were obtained in the two reactions. Accordingly, square cyclobutadiene cannot be a common intermediate in the two reactions. The corollary is that cyclobutadiene is rectangular. A further conclusion is that, although the two

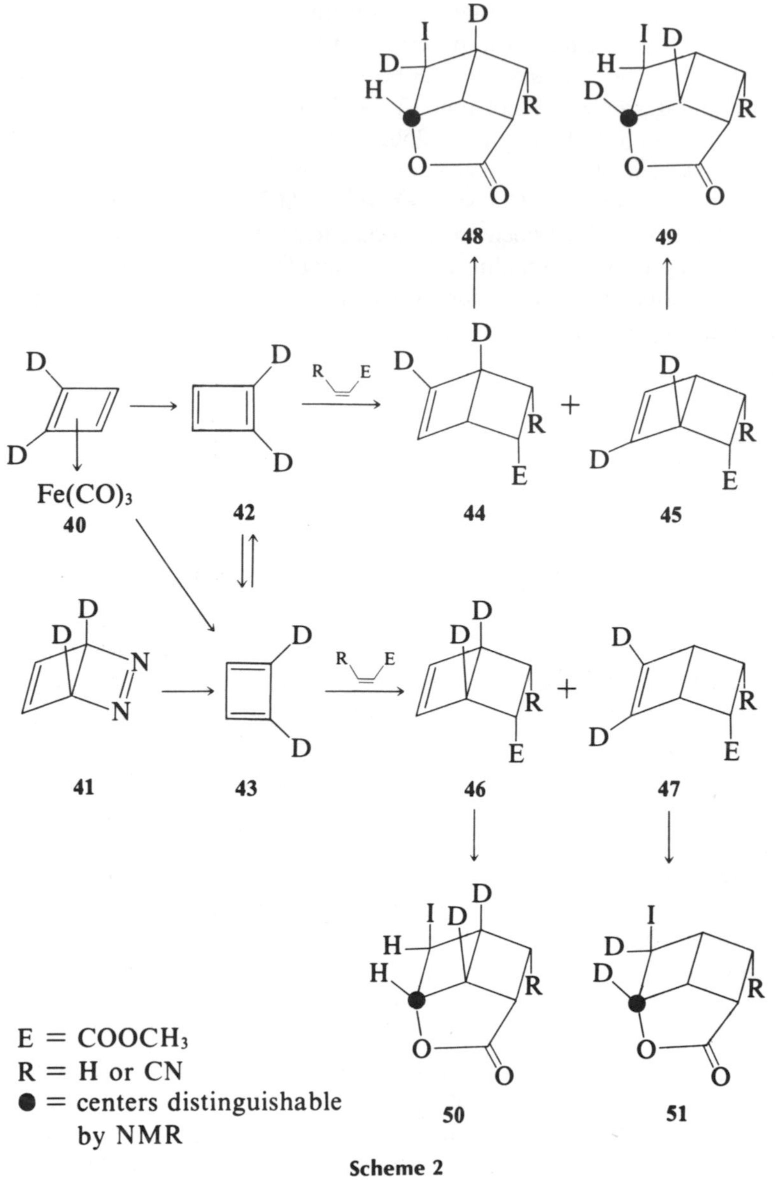

E = $COOCH_3$
R = H or CN
● = centers distinguishable by NMR

Scheme 2

rectangular forms **42** and **43** interconvert, the cycloaddition reaction with methyl *Z*-3-cyanoacrylate is faster than the interconversion.

Ab initio calculations[35] indicate that cyclobutadiene is indeed a ground-state rectangular singlet, and that the rectangular forms interconvert via the square with an activation energy between 4 and 12 kcal/mol. Experimentally, it appears that the activation energy may be considerably lower. The barrier height has been directly

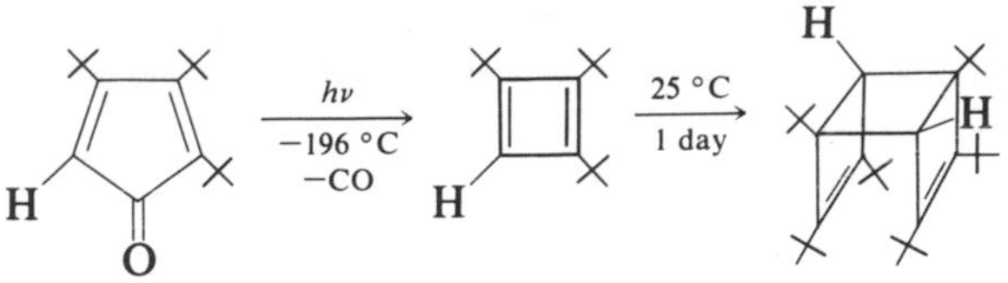

$h\nu$, −196 °C, −CO; 25 °C, 1 day

52 **53**

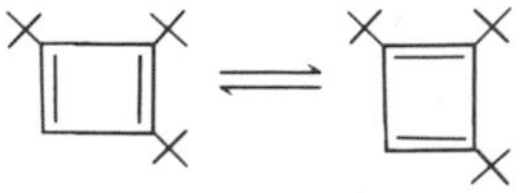

53a **53b**

probed by NMR spectroscopy of a relatively stable cyclobutadiene, the tri-*tert*-butyl compound **53**, obtained by photolysis of **52**. **53** is sufficiently stable to be observed by proton NMR at room temperature. The spectrum shows three signals, the chemical shift of the single ring proton (δ = 5.38) being in agreement with a paramagnetic ring current as observed in other [$4n$]-annulenes.[36] The fact that only three ^{1}H-NMR signals are observed implies that the two rectangular forms of **53** interconvert rapidly at room temperature. It was not possible to "freeze out" the interconversion, even at temperatures as low as −190°C. This requires that the activation energy for the process be no more than 2.5 kcal/mol.[37]

There is no doubt that **53** is in fact rectangular in the ground state.[38] As for the parent cyclobutadiene itself, it has been suggested on the basis of activation parameters for the processes shown in Scheme 2 that the interconversion of rectangles, **42** $\rightleftharpoons$ **43,** takes place mainly by tunneling.[34b] This conclusion has been shown to be incorrect, however, since it is possible to generate oriented cyclobutadiene rectangles in an argon matrix using linearly polarized light.[34c] This experiment not only rules out tunneling but also unequivocally establishes a rectangular ground state.

Other stable cyclobutadiene derivatives are known for which X-ray structure analyses have demonstrated rectangular geometries. The isolable thiacycloheptyne **54** (see also Chapter 5) dimerizes to the cyclobutadiene complex **55** in the presence of palladium chloride. Treatment of **55** with ethylenebis(diphenylphosphane) liberates the sterically protected cyclobutadiene **56.** This compound forms yellow

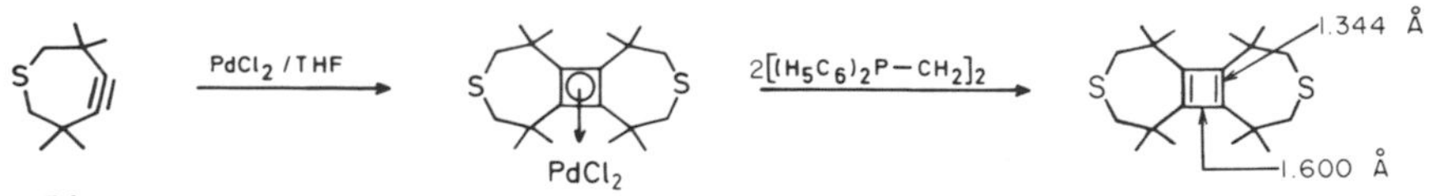

$PdCl_2$ /THF; $2[(H_5C_6)_2P-CH_2]_2$

$PdCl_2$

1.344 Å

1.600 Å

54 55 **56**

crystals, decomposing at 240°C.[39a] The X-ray structure (see formula **56**) clearly shows alternating bond lengths.[39b]

The X-ray analysis of methyl tri-*tert*-butylcyclobutadienecarboxylate (**57**) likewise demonstrates bond alternation and a slightly distorted rectangular geometry.[40]

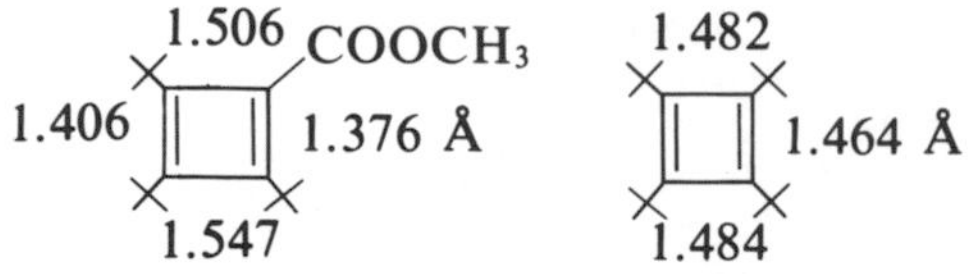

1.506 COOCH$_3$ 1.482

1.406 1.376 Å 1.464 Å

1.547 1.484

57 **58**

In contrast, tetra-*tert*-butylcyclobutadiene (**58**), a stable compound described in connection with tetra-*tert*-butyltetrahedrane (p. 301), shows no remarkable degree of bond alternation at room temperature. However, a renewed X-ray analysis carried out at −160°C revealed an approximately 0.08-Å bond alternation. Furthermore, this molecule was found not to be planar, both at room temperature and at −160°C: The torsional angle between two neighboring *tert*-butyl groups is 28°, and that between two opposite ring bonds is 7°.[41] It would appear that the peculiar geometry and small degree of bond alternation is due to steric compression resulting from the four *tert*-butyl groups: Apart from the tetrahedron (to which **58** is isomerized photochemically; see Chapter 5), the next best arrangement of four sterically demanding groups is the square.[35c] The square is distorted for the same reason that tetra-*tert*-butylethylene is predicted to be highly twisted (see Section 5.1).

In conclusion, cyclobutadiene and most of its substituted derivatives exist in a singlet rectangular ground state.

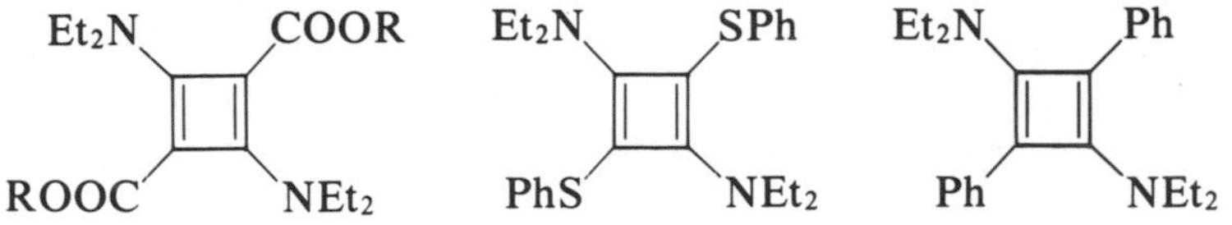

Et$_2$N COOR Et$_2$N SPh Et$_2$N Ph

ROOC NEt$_2$ PhS NEt$_2$ Ph NEt$_2$

59 **60** **61**

Et$_2$N—C≡C—COOR

↓ BF$_3$ · Et$_2$O

H NEt$_2$

ROOC ⊕ BF$_4^{\ominus}$

Et$_2$N COOR

↓ NaH

ROOC NEt$_2$ ⟷ ROOC NEt$_2$ ⟷ RO— O$^{\ominus}$ $^{\oplus}$NEt$_2$ ⟷ etc.

Et$_2$N COOR Et$_2$N COOR Et$_2$N COOR

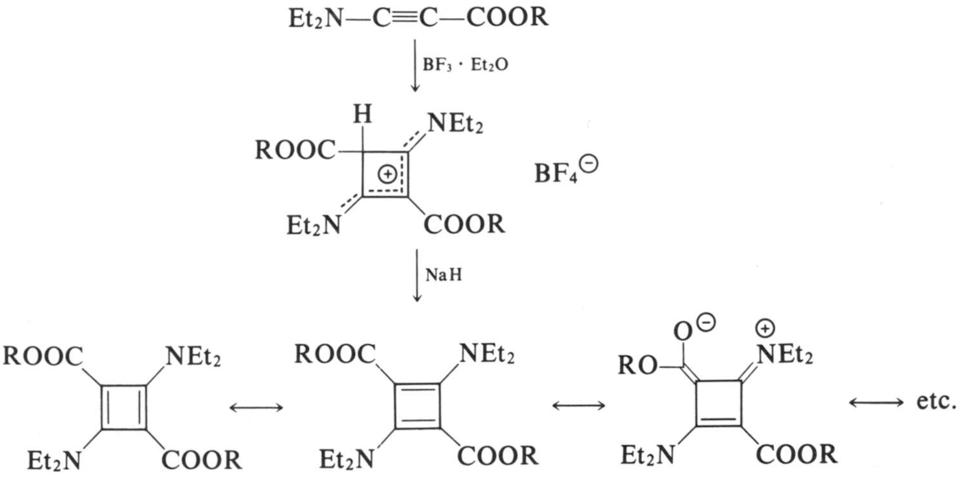

59a **59b** **59c**

The compounds **59–61** constitute another series of remarkably stable cyclobutadiene derivatives.[42] In all likelihood, the stability is due to a combination of steric protection and resonance delocalization. **59**, a "push-pull" substituted cyclobutadiene, may be described in terms of the resonance structures **59a–c**.

So far azacyclobutadiene (azete) **(62)** has not been directly observed, and on attempted preparation, even in an argon matrix at 10 K, it fragments into acetylene and HCN.[43] Analogous results were obtained with alkyl-substituted derivatives.[43]

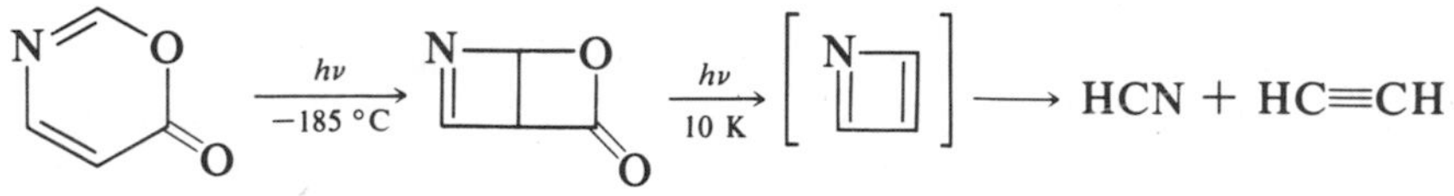

$h\nu$, −185 °C; $h\nu$, 10 K; → HCN + HC≡CH

62

A "push-pull" substituted azete **(64)** has, however, been prepared in 30% yield by flash vacuum pyrolysis of **63.** The red **64** was stable in solution at −80°C and disappeared on warming to room temperature. In this compound the amino groups may be regarded as donors, and the ring nitrogen atom as the acceptor group.[44]

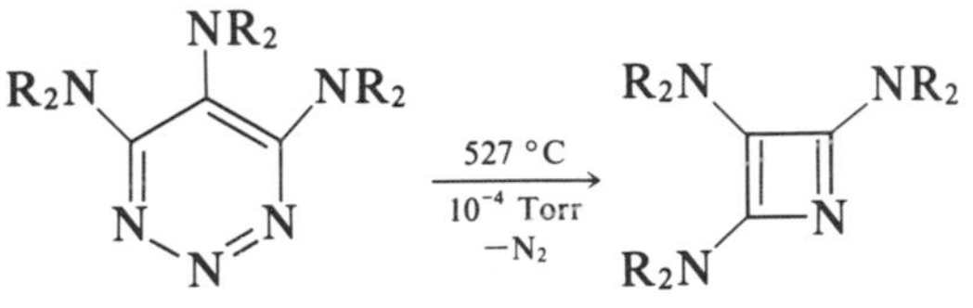

NR_2, R_2N, NR_2; 527 °C, 10^{-4} Torr, $-N_2$

63 $R = CH_3$ **64**

Also substituted benzazetes **(66)** are reasonably stable compounds, isolable as red crystals at −80°C following flash vacuum pyrolysis or photolysis of 1,2,3-benzotriazines **(65)**.[45] At temperatures above 500°C in the gas-phase, decomposition to benzyne **(67)** and a nitrile occurs. Benzyne dimerizes to the completely stable cyclobutadiene derivative, biphenylene **(68)** (cf. Section 5.5).

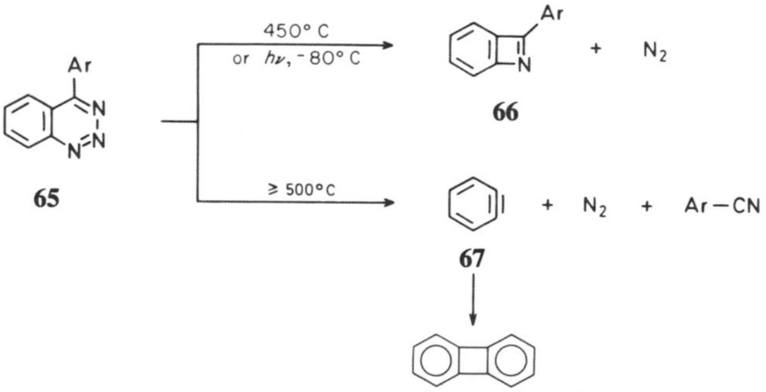

450° C or $h\nu$, −80° C → **66** + N_2

65 ≥ 500°C → **67** + N_2 + Ar—CN

68

The stability of **66** is probably due in part to delocalization over the aryl substituent, which also offers some steric protection. Unsubstituted benzazete is expected to be quite unstable in analogy with the unsubstituted benzocyclobutadiene (**70**). The latter can be prepared by heterogeneous gas-phase deiodination of **69** and

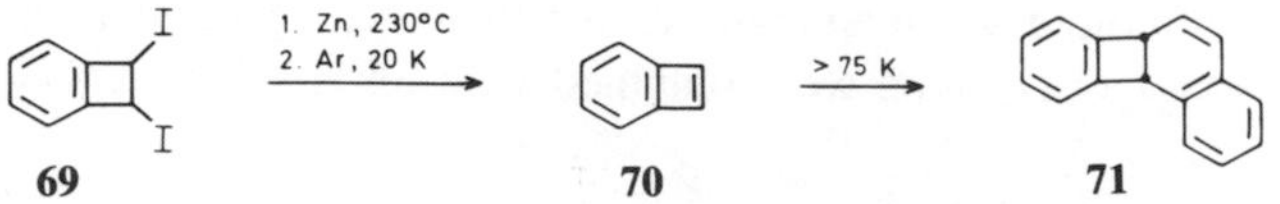

1. Zn, 230°C
2. Ar, 20 K
> 75 K

69 **70** **71**

trapped in an argon matrix at 20 K. Warming of the isolated material above 75 K caused dimerization to the well-known[1] dimer (**71**).[46] As already mentioned, benzocyclobutadiene, too, is stabilized by sterically demanding substituents (see **24**).

6.3. PROBLEMS

1. The oxidation of tetra-*tert*-butyltetrahedrane with aluminum chloride in CH_2Cl_2 gives the radical cation, not of the tetrahedrane, but of the corresponding cyclobutadiene:

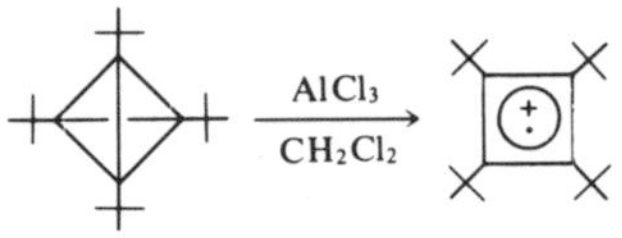

$AlCl_3$ / CH_2Cl_2

Explain why, apparently, in the radical cationic series, the cyclobutadiene is more stable relative to the tetrahedrane than in the neutral series.[47]

2. 1,2-Dimethylenecyclobutane (**72**) reacts with maleic anhydride to give the Diels–Alder adduct **73**. The diphenyldimethylenecyclobutene **74** does not react under these conditions, however.[48] Explain.

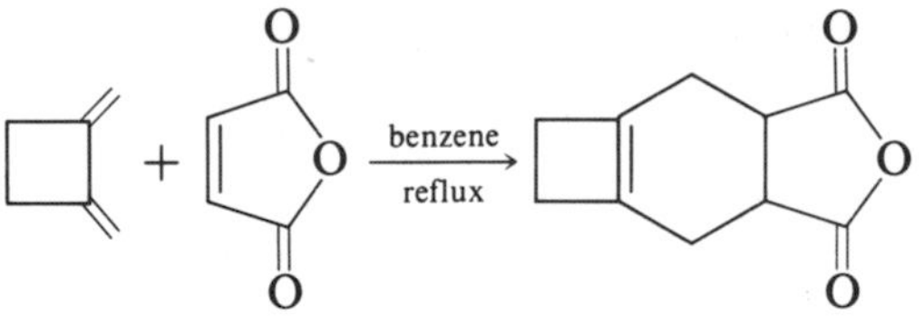

benzene / reflux

72 **73** (94%)

Ph Ph → no reaction

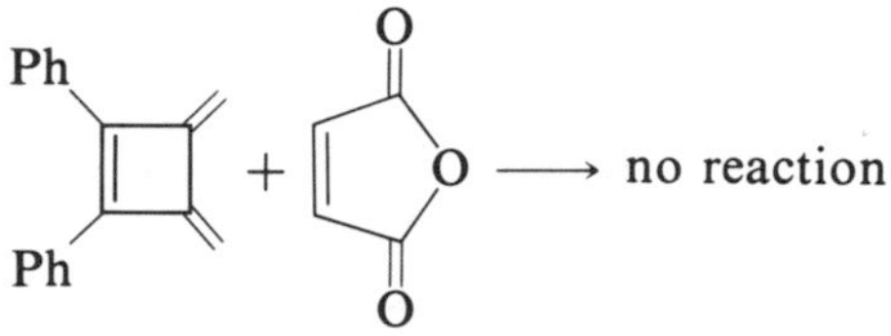

74

3. The ^{13}C-labeled α-pyrone **75,** a precursor of cyclobutadiene, is photochemically equilibrated with the isotopomer **76.** When the photolysis was carried out in an Ar matrix at 8 K, IR absorptions due to transient intermediates were initially recorded at 2068, 2074, 2081, and 2089 cm^{-1}. During photolysis, an *additional* four absorptions developed at 2121, 2129, 2133, and 2136 cm^{-1}.[32] Give the mechanism for the reaction **75** ⇄ **76**.

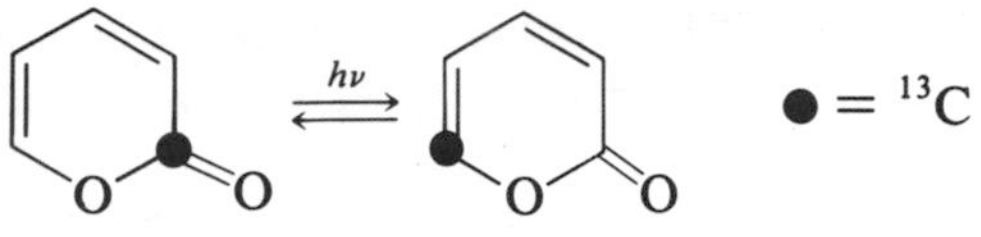

$h\nu$

● = ^{13}C

75 **76**

4. The relative rates of the silver ion–assisted solvolyses of cyclopentyl and cyclopentadienyl iodides are indicated below. Calculate a lower limit for the destabilization of the cyclopentadienyl cation in kcal/mol.[49]

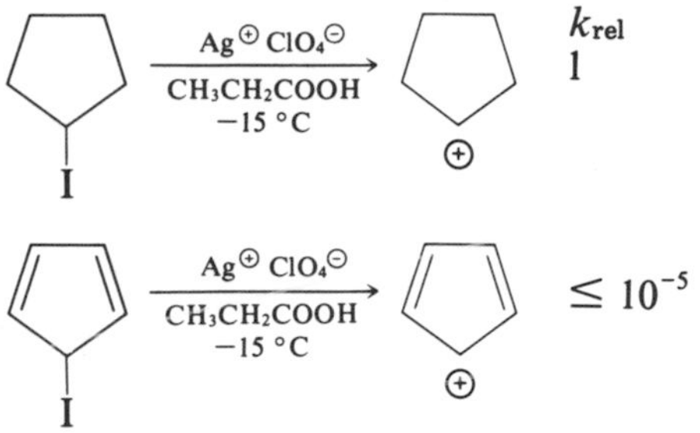

$Ag^{\oplus}\ ClO_4^{\ominus}$, CH_3CH_2COOH, −15 °C — k_{rel} 1

$Ag^{\oplus}\ ClO_4^{\ominus}$, CH_3CH_2COOH, −15 °C — $\leq 10^{-5}$

5. The ionization potentials (IP) of the cyclopentyl, cyclopentenyl, and cyclopentadienyl radicals have been measured by mass spectrometry. They are given below together with the enthalpies of formation of the radicals and the corresponding hydrocarbons (RH).

	$\Delta H_f^0(R\cdot)$ (kcal/mol)	IP (eV)	$\Delta H_f^0(RH)$ (kcal/mol)
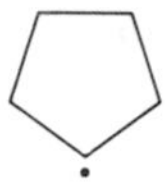 cyclopentyl	24	7.47	−18.46
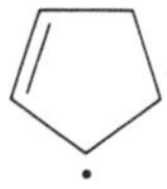 cyclopentenyl	38	7.00	7.87
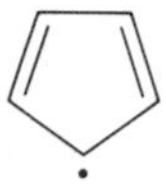 cyclopentadienyl	61	8.41	32.00

From these data, calculate (in kcal/mol) the enthalpies of formation of the cations, the stabilization of the cyclopentenyl cation, and the destabilization of the cyclopentadienyl cation.[50]

Compare with the value derived in Problem 4.

6. The treatment of the quaternary ammonium salt **77** with base results in the formation of the hydrocarbon **78**. In the presence of cyclopentadiene, the product **79** is obtained. Flash pyrolysis of **79** at 200°C gives **78**. The photolysis of **78** at −196°C (λ = 254 nm) gives a transient absorbing at approximately 335 nm in the UV. This absorption disappears on warming, with concomitant regeneration of **78.**[51]

Give the structure and explain the instability of the transient.

$I^{\ominus}$ $^{\oplus}N-CH_3$ Ag_2O 20° C 200° C 1.5 s

77 **79** **78**

REFERENCES

1. M. P. Cava and M. J. Mitchell, *Cyclobutadiene and Related Compounds,* Academic Press, New York, 1967.
2. Review: G. Maier, *Angew. Chem.,* **86,** 491 (1974); *Angew. Chem. Int. Ed. Engl.,* **13,** 425 (1974).
3. Review: T. Bally and S. Masamune, *Tetrahedron,* **36,** 343 (1980).
4. M. J. S. Dewar and M. L. McKee, *Pure Appl. Chem.,* **52,** 1431 (1980); M. J. S. Dewar and H. W. Kollmar, *J. Am. Chem. Soc.,* **97,** 2933 (1975).
5. A. Streitwieser, Jr., *Molecular Orbital Theory for Organic Chemists,* Wiley, New York, 1961.
6. For the difficulties engendered in the application of Jahn–Teller distortion to cyclobutadiene, see T. Bally and S. Masamune, *Tetrahedron,* **36,** 343 (1980).
7. (a) R. Breslow, J. Brown, and J. Gajewski, *J. Am. Chem. Soc.,* **89,** 4383 (1967). (b) M. R. Wasietewski and R. Breslow, *J. Am. Chem. Soc.,* **98,** 4222 (1976). (c) R. Breslow, *Acc. Chem. Res.,* **6,** 393 (1973).

8. M. Saunders, R. Berger, A. Jaffe, J. M. McBride, J. O'Neill, R. Breslow, J. M. Hoffman, C. Perchonock, E. Wasserman, R. S. Hutton, and V. J. Kuck, *J. Am. Chem. Soc.*, **95,** 3017 (1973).

9. R. Breslow, J. T. Groves, and G. Ryan, *J. Am. Chem. Soc.*, **89,** 5048 (1967); D. G. Farnum, G. Metha, and R. G. Silberman, *ibid.*, **89,** 5048 (1967).

10. M. Traetteberg, *Acta Chem. Scand.*, **20,** 1724 (1966).

11. H. P. Figeys and A. Dralants, *Tetrahedron Lett.*, **1971,** 3901.

12. B. A. Hess and L. J. Schaad, *Tetrahedron Lett.*, **1972,** 5113.

13. F. Sondheimer, *Acc. Chem. Res.*, **5,** 81 (1972).

14. T. J. Katz, *J. Am. Chem. Soc.*, **82,** 3784 (1960); J. F. M. Oth and G. Schröder, *J. Chem. Soc. B,* **1971,** 904; G. A. Olah, J. S. Staral, and L. A. Paquette, *J. Am. Chem. Soc.*, **98,** 1267 (1976).

15. G. A. Olah, J. M. Bollinger, and A. M. White, *J. Am. Chem. Soc.*, **91,** 3667 (1969); G. A. Olah and J. S. Staral, *ibid.*, **98,** 6290 (1976); see also K. Krogh-Jespersen, P. v. R. Schleyer, J. A. Pople, and D. Cremer, *ibid.*, **100,** 4301 (1978).

16. J. S. McKennis, L. Brener, J. R. Schweiger, and R. Pettit, *Chem. Commun.*, **1972,** 365.

17. (a) G. Boche, H. Etzrodt, M. Marsch, and W. Thiel, *Angew. Chem.*, **94,** 141 (1982); *Angew. Chem. Int. Ed. Engl.*, **21,** 133 (1982). (b) See also P. J. Garratt and R. Zahler, *J. Am. Chem. Soc.*, **100,** 7753 (1978).

18. D. R. Howton and E. R. Buchman, *J. Am. Chem. Soc.*, **78,** 4011 (1956); D. E. Applequist and J. D. Roberts, *ibid.*, **78,** 4012 (1956).

19. E. J. Smutny, M. C. Caserio, and J. D. Roberts, *J. Am. Chem. Soc.*, **82,** 1793 (1960).

20. (a) M. L. Herr, *Tetrahedron,* **32,** 2835 (1976); (b) W. Winter and H. Straub, *Angew. Chem.*, **90,** 142 (1978); *Angew. Chem. Int. Ed. Engl.*, **17,** 127 (1978).

21. R. Criegee and G. Louis, *Chem. Ber.*, **90,** 417 (1957).

22. R. Criegee and G. Schröder, *Justus Liebigs Ann. Chem.*, **623,** 1 (1959); J. D. Dunitz, H. C. Mez, O. S. Mills, and H. M. M. Shearer, *Helv. Chim. Acta.*, **45,** 647 (1962).

23. (a) G. F. Emerson, L. Watts, and R. Pettit, *J. Am. Chem. Soc.*, **87,** 131 (1965). (b) R. P. Dodge and V. Schomacher, *Nature,* **186,** 798 (1960); *Acta Crystallogr.*, **18,** 614 (1965).

24. L. Watts, J. D. Fitzpatrick, and R. Pettit, *J. Am. Chem. Soc.*, **88,** 623 (1966); J. C. Barborak, L. Watts, and R. Pettit, *ibid.*, **88,** 1328 (1966).

25. R. H. Grubbs and R. A. Grey, *J. Am. Chem. Soc.*, **95,** 5765 (1973); R. H. Grubbs and T. A. Pancoast, *ibid.*, **99,** 2382 (1977); E. K. G. Smith, *Angew. Chem.*, **85,** 820 (1973); *Angew. Chem. Int. Ed. Engl.*, **12,** 777 (1973).

26. J. Rebek and F. Gaviña, *J. Am. Chem. Soc.*, **97,** 3453 (1975).

27. C. Y. Lin and A. Krantz, *Chem. Commun.*, **1972,** 1111; A. Krantz, C. Y. Lin, and M. D. Newton, *J. Am. Chem. Soc.*, **95,** 2744 (1973).

28. S. Masamune, M. Suda, H. Ona, and L. M. Leichter, *Chem. Commun.*, **1972,** 1268.

29. O. L. Chapman, D. De La Cruz, R. Roth, and J. Pacansky, *J. Am. Chem. Soc.*, **95,** 1337 (1973).

30. G. Maier and B. Hoppe, *Tetrahedron Lett.*, **1973,** 861.

31. G. Herzberg, *Molecular Spectra and Molecular Structure,* Van Nostrand, New York, 1945, p. 92.

32. See also B.-S. Huang, R. G. S. Pong, J. Laureni, and A. Krantz, *J. Am. Chem. Soc.*, **99,** 4154 (1977).

33. G. Maier, H.-G. Hartan, and T. Sayrac, *Angew. Chem.*, **88,** 252 (1976); *Angew. Chem. Int. Ed. Engl.*, **15,** 226 (1976); R. G. S. Pong, B.-S. Huang, J. Laureni, and A. Krantz, *J. Am. Chem. Soc.*, **99,** 4153 (1977).

34. (a) D. W. Whitman and B. K. Carpenter, *J. Am. Chem. Soc.*, **102,** 4272 (1980). (b) B. K. Carpenter, *ibid.*, **105,** 1700 (1983). (c) J. Michl, J. G. Radziszewski, and C. Wentrup, *ibid.*, **106** (1984).

35. (a) W. T. Borden, E. R. Davidson, and P. Hart, *J. Am. Chem. Soc.*, **100,** 388 (1978). (b) J. A. Jafri and M. D. Newton, *ibid.*, **100,** 5012 (1978). (c) W. T. Borden and E. R. Davidson, *ibid.*, **102,** 7958 (1980).

36. S. Masamune, N. Nakamura, M. Suda, and H. Ono, *J. Am. Chem. Soc.*, **95,** 8481 (1973); G. Maier and A. Alzérreca, *Angew. Chem.*, **85,** 1056 (1973); *Angew. Chem. Int. Ed. Engl.*, **12,** 1015 (1973).

37. G. Maier, U. Schäfer, W. Sauer, H. Hartan, R. Matusch, and J. F. M. Oth, *Tetrahedron Lett.*, **1978,** 1837.

38. G. Maier, H.-O. Kalinowski, and K. Euler, *Angew. Chem.*, **94,** 706 (1982); *Angew. Chem. Int. Ed. Engl.*, **21,** 693 (1982); see also G. Lauer, C. Müller, K.-W. Schulte, A. Schweig, G. Maier, and A. Alzérreca, *ibid.*, **87,** 194 (1975) and **14,** 172 (1975).

39. (a) A. Krebs, H. Kimling, and R. Kemper, *Justus Liebigs Ann. Chem.*, **1978,** 431. (b) H. Irngartinger and H. Rodewald, *Angew. Chem.*, **86,** 783 (1974); *Angew. Chem. Int. Ed. Engl.*, **13,** 740 (1974).

40. L. T. J. Delbaere, M. N. G. James, N. Nakamura, and S. Masamune, *J. Am. Chem. Soc.*, **97,** 1973 (1974).

41. H. Irngartinger, N. Riegler, K.-D. Malsch, K.-A. Schneider, and G. Maier, *Angew. Chem.*, **92,** 214 (1980); *Angew. Chem. Int. Ed. Engl.*, **19,** 211 (1980); H. Irngartinger and M. Nixdorf, *Angew. Chem.*, **95,** 415 (1983); *Angew. Chem. Int. Ed. Engl.*, **22,** 403 (1983).

42. R. Gompper, J. Kroner, G. Seybold, and H.-U. Wagner, *Tetrahedron*, **32,** 629 (1976); R. Gompper, S. Mensch, and G. Seybold, *Angew. Chem.*, **87,** 711 (1975); *Angew. Chem. Int. Ed. Engl.*, **14,** 704 (1975); R. Gompper and G. Seybold, *Angew. Chem.*, **80,** 804 (1968); *Angew. Chem. Int. Ed. Engl.*, **7,** 824 (1968).

43. G. Maier and U. Schäfer, *Liebigs Ann. Chem.*, **1980,** 798.

44. G. Seybold, U. Jersak, and R. Gompper, *Angew. Chem.*, **85,** 918 (1973); *Angew. Chem. Int. Ed. Engl.*, **12** 847 (1973); H.-U. Wagner, *ibid.*, **85,** 920 (1973) and **12,** 848 (1973).

45. B. M. Adger, C. W. Rees, and R. C. Storr, *J. Chem. Soc. Perkin Trans. 1*, **1975,** 45; C. W. Rees, R. C. Storr, and P. J. Whittle, *Chem. Commun.*, **1976,** 411.

46. O. L. Chapman, C. C. Chang, and N. R. Rosenquist, *J. Am. Chem. Soc.*, **98,** 261 (1976).

47. (a) H. Bock, B. Roth, and G. Maier, *Angew. Chem.*, **92,** 213 (1980); *Angew. Chem. Int. Ed. Engl.*, **19,** 209 (1980). (b) See also Q. B. Broxterman, H. Hogeveen, and D. M. Kok, *Tetrahedron Lett.*, **22,** 173 (1981).

48. A. T. Blomquist and J. A. Verdol, *J. Am. Chem. Soc.*, **77,** 1806 (1955); A. T. Blomquist and Y. C. Meinwald, *ibid.*, **81,** 667 (1959).

49. R. Breslow and J. M. Hoffman, *J. Am. Chem. Soc.*, **94,** 2110 (1972). The rate ratio $\geqslant 10^5$ corresponds to a $\geqslant 7$ kcal/mol difference in the energies of activation.

50. F. P. Lossing and J. C. Traeger, *J. Am. Chem. Soc.*, **97,** 1579 (1975). A destabilization of the cyclopentadienyl cation by about 8 kcal/mol (due to antiaromaticity) and a stabilization of the cyclopentenyl cation by about 24 kcal/mol (due to allylic resonance) are calculated.

51. K. Hafner, R. Dönges, E. Goedecke, and R. Kaiser, *Angew. Chem.*, **85,** 362 (1973); *Angew. Chem. Int. Ed. Engl.*, **12,** 337 (1973); R. Dönges, K. Hafner, and H. J. Lindner, *Tetrahedron Lett.*, **1976,** 1345; K. Hafner, *Nachr. Chem. Tech. Lab.*, **28,** 222 (1980); see also R. Bloch, R. A. Marty, and P. de Mayo, *J. Am. Chem. Soc.*, **93,** 3071 (1971); N. C. Baird and R. M. West, *ibid.*, **93,** 3072 (1971).

INDEX